SERIES ON SEMICONDUCTOR
SCIENCE AND TECHNOLOGY

Series Editors

R. J. Nicholas University of Oxford
H. Kamimura University of Tokyo

AF740183

SERIES ON SEMICONDUCTOR SCIENCE AND TECHNOLOGY

---

1. M. Jaros: *Physics and applications of semiconductor microstructures*
2. V. N. Dobrovolsky and V. G. Litovchenko: *Surface electronic transport phenomena in semiconductors*
3. M. J. Kelly: *Low-dimensional semiconductors*
4. P. K. Basu: *Theory of optical processes in semiconductors*
5. N. Balkan: *Hot electrons in semiconductors*
6. B. Gil: *Group III nitride semiconductor compounds: physics and applications*
7. M. Sugawara: *Plasma etching*
8. M. Balkanski and R. F. Wallis: *Semiconductor physics and applications*
9. B. Gil: *Low-dimensional nitride semiconductors*

24 0560530 6

Low-Dimensional Nitride Semiconductors

Edited by

Bernard Gil

Directeur de Recherche au CNRS
Université de Montpellier II

OXFORD
UNIVERSITY PRESS

537.311.33 LOW

22 AUG 2002

OXFORD
UNIVERSITY PRESS

Great Clarendon Street, Oxford OX2 6DP

Oxford University Press is a department of the University of Oxford.
It furthers the University's objective of excellence in research, scholarship,
and education by publishing worldwide in

Oxford New York

Auckland Bangkok Buenos Aires Cape Town Chennai
Dar es Salaam Delhi Hong Kong Istanbul Karachi Kolkata
Kuala Lumpur Madrid Melbourne Mexico City Mumbai Nairobi
São Paulo Shanghai Taipei Tokyo Toronto

with an associated company in Berlin

Oxford is a registered trade mark of Oxford University Press
in the UK and in certain other countries

Published in the United States
by Oxford University Press Inc., New York

© Oxford University Press, 2002

The moral rights of the author have been asserted

Database right Oxford University Press (maker)

First published 2002

All rights reserved. No part of this publication may be reproduced,
stored in a retrieval system, or transmitted, in any form or by any means,
without the prior permission in writing of Oxford University Press,
or as expressly permitted by law, or under terms agreed with the appropriate
reprographics rights organization. Enquiries concerning reproduction
outside the scope of the above should be sent to the Rights Department,
Oxford University Press, at the address above.

You must not circulate this book in any other binding or cover
and you must impose this same condition on any acquirer.

A Catalogue record for this title is available from the British Library

Library of Congress Cataloging in Publication Data

(Data available)

ISBN 0 19 850974 X

10 9 8 7 6 5 4 3 2 1

Typeset by Cepha Imaging Pvt. Ltd.

Printed in Great Britain
on acid-free paper by
Biddles Ltd, Guildford & King's Lynn

Preface

This is the second book of a series dedicated to nitride semiconductors. The preceding volume, *Group III nitride semiconductor compounds*, was completed in the first half of 1997. More than four years later, Oxford University Press asked me to assemble a second volume. A lot has happened in four years: a 10000 hour lifetime was attained for the 'blue' laser in 1998 and it 'went commercial' in 1999; blue, green and white light-emitting diodes can now be found here and there in our cities or homes; considerable progress has been made in the area of solar blind ultraviolet light detectors; transistor designers are actively developing advanced devices. Most of this recent progress is due to an extensive utilization of low-dimensional micro- and nano-structures in nitride devices. For this reason, the title of the present book is *Low-dimensional nitride semiconductors*. The book is designed to be accessible to the widest community, including students at the beginning of their PhD studies who are not yet familiar with the whole world of semiconductor physics.

Jean-Yves Duboz and Jean-Charles Garcia agreed to write a short introductory chapter entitled 'GaN of the history of a scientific explosion driven by applications and markets' as a very welcome practical introduction for newcomers. It is followed by a tutorial chapter written by one of the leading theorists working in the field, Friedhelm Bechstedt. Reviews of the most modern growth techniques are provided by Izabella Grzegory, Sylvester Porowsky, Tanya Paskova, Bo Monemar, Hiroshi Amano, Satoshi Kamiyama, Isamu Akasaki, Nicolas Grandjean, Benjamin Damilano and Jean Massies. I managed to entice Pierre Ruterana out of his laboratory to write a lovely chapter on electron microscopy that I hope you will enjoy. Although I wrote the chapter devoted to GaN–AlGaN nanostructures, I asked Kevin O'Donnell and Yoichi Kawakami to review what has become known as the 'InGaN enigma'. Xue Bing Zhang, my smart post-doctoral associate, shares with me a chapter on the electron–phonon interaction in low-dimensional nitrides. These materials are promising for ultrafast spectroscopy as shown by Alexey Kavokin and Guillaume Malpuech. The book continues with chapters that review the current status of optoelectronic devices and transistors. Andreas Hangleiter, Hadis Morkoç, Aldo di Carlo and Roberto Cingolani did a really excellent job. Nitrides are not restricted to wide bandgap semiconductors; their mixed solutions with other III–Vs have enormous potential, both in areas of electronics and infra-red optoelectronics. Therefore, I have asked my friend Eric Tournié to join me in writing a chapter dedicated to these interesting and challenging compounds. The photographs on the cover have been kindly provided by Kevin O'Donnell, Bruno Daudin and Jean Luc Rouvière.

I really want to thank, deeply and warmly, all of the colleagues named above for offering you this book. Lastly, I wish to thank Soenke Adlung from the Oxford University Press for believing in nitrides at the very beginning of their odyssey and for producing this second volume.

I hope you will like it.

Belle Ile en Mer B. G.
August 2001

Contents

Contributors

Isamu Akasaki
High-Tech Research Center,
Meijo University,
1-50 Shioganaguchi,
Tempaku-Ku, Nagoya 468-8502
Japan
e-mail: akasaki@meijo-u.ac-jp

Hiroshi Amano
Department of Materials Science and Engineering,
High-Tech Research Center,
Meijo University,
1-50 Shioganaguchi,
Tempaku-Ku, Nagoya 468-8502
Japan
e-mail: amano@meijo-u.ac-jp

Friedhelm Bechstedt
Institut für Festkörpertheorie und Theoretische Optik,
Friedrich-Schiller-Universität,
Max-Wien-Platz 1,
07743 Jena,
Germany
e-mail: bechsted@ifto.physik.uni-jena.de

Roberto Cingolani
Virginia Commonwealth University,
Department of Electrical Engineering and
Physics Department,
601 W. Main Street,
Po Box 843072,
Richmond, VA 23284-3072,
USA.
e-mail: roberto.cingolani@unile.it

Benjamin Damilano
CRHEA-CNRS, Rue Bernard Grégory,
Sophia-Antipolis,
06560 Valbonne,
France
e-mail: bd@crhea.cnrs.fr

Aldo Di Carlo
INFM and Dept. Ingegneria Electronica,
Università di Roma,
'Tor Vergata'
Rome, Italy
e-mail: dicarlo@ing.uniroma2.it

Jean-Yves Duboz
Laboratoire Central de Recherches,
Thales Research Technology-Domaine ole Corbeville
91404 Orsay,
France
e-mail: jean-yves.duboz@thalesgroup.com

Jean-Charles Garcia
INTEXYS-Photonics,
42 Avenue Général ole Croute
31000 Toulouse,
France
e-mail: jean_charles@club-internet.fr

Bernard Gil
Centre National de la Recherche Scientifique,
Gooup d'Étude des Semiconducteurs,
Université de Montpellier II,
Cas Courrier 074
F-34095 Montpellier Cedex 5,
France
e-mail: gil@ges.univ-montp2.fr

Nicolas Grandjean
CRHEA-CNRS, Rue Bernard Gregory,
Sophia-Antipolis,
06560 Valbonne,
France
e-mail: ng@crhea.cnrs.fr

Izabella Grzegory
High-Pressure Research Center,
Polish Academy of Sciences,
Sokolowska 29/37,
01-142 Warsaw,
Poland
e-mail: izabella@unipress.waw.pc

Andreas Hangleiter
Institute of Technical Physics,
Technical University of Braunschweig,
Mendelssohnstrasse 2
D-38 106-Braunschweig,
Germany
e-mail: a.hangleiter@tu-bs.ole

Yoichi Kawakami
Department of Electronic Science and Engineering,
Kyoto University,
Yoshida-Honmachi,
Sakyo-ku,
Kyoto 606-01,
Japan
e-mail: kawakami@kuee.kyoto-u.ac.jp

Satoshi Kamiyama
Department of Materials Science and Engineering,
High-Tech Research Center,
Meijo University,
1-50 Shioganaguchi,
Tempaku-Ku,
Nagoya 468-8502
Japan
e-mail: kamiyama@meijo-u-ac.jp

Alexey Kavokin
LASMEA-CNRS,
Université Blaise Pascal,
24 av. des Landais,
63177 Aubière,
France
e-mail: kavokin@lasmea.univ-bpclermont.fr

Guillaume Malpuech
LASMEA-CNRS,
Université Blaise Pascal,
24 av. des Landais,
63177 Aubière,
France
e-mail: malpuech@lasmea.univ-bpclermont.fr

Jean Massies
CRHEA-CNRS, Rue Bernard Gregory,
Sophia-Antipolis,
06560 Valbonne,
France
e-mail: jm@crhea-cnrs.fr

Hadis Morkoç
Virginia Commonwealth University,
Department of Electrical Engineering and
Physics Department,
601 W. Main Street,
Po Box 843072,
Richmond, VA 23284-3072,
USA.
e-mail: hmorkoc@vcu.edu

Bo Monemar
Department of Physics and Measurement Technology,
Materials Science Division,
Linkoping University,
S-581 83 Linkoping,
Sweden
e-mail: bom@ifm.liu.se

Kevin P. O'Donnell
University of Strathclyde
Glasgow G4 0NG
Scotland, UK
e-mail: k.p.odonnell@strath.ac.uk

Tanya Paskova

Department of Physics and Measurement Technology,
Materials Science Division,
Linkoping University,
S-581 83 Linkoping,
Sweden
e-mail: tanya@ifm.liu.se

Sylwester Porowski

High Pressure Research Center
Polish Academy of Sciences,
Sokolowska 29/37,
01-142 Warsaw,
Poland
e-mail: sylvek@unipress.waw.pc

Pierre Ruterana

ESCTM-CRISMAT,
Institute des Sciences de la Matière et du Rayonnement,
6Bd Maréchal Juin,
14050 Caen cedex,
France
e-mail: ruterana@ismra.fr

Eric Tournié

Centre National de la Recherche Scientifique,
Centre de Recherche Sur l'Hetero-Epitaxie et ses Applications,
(CNRS-CRHEA)
Rue Bernard Gregory
Parc Sophia-Antipolis,
F-06560 Valbonne,
France
e-mail: et@crhea-cnrs.fr

Plates

Plates fall between pages 206 and 207

Plate I (Fig. 7.28, p. 186) Strain and related composition profiles inside the area of Fig. 7.28, the high indium content areas extend to 2 nm height and 3 nm width. The peak In concentration is 45%.

Plate II (Fig. 8.11, p. 202) In-plane extension of the densities of probability for electron and hole in the case of a 3 nm wide GaN-$Ga_{0.83}Al_{0.17}N$ QW. Aluminium atoms in GaAlN layers are represented using large yellow spots to illustrate the alloy disorder. Note that the in-plane extension of the hole wave function is smaller than the average Al-Al distancy. The alloy disorder is responsible of the interface roughness experienced by the exciton.

Plate III (Fig. 9.2, p. 215) A colourized micrograph of an InGaN epilayer, obtained using the confocal microscope SYCLOPS at the Daresbury Laboratories. The area represented is 500 micron square. Two images are superposed: the green image was obtained using reflected laser light, while the red image shows photoluminescence intensity variations. Note the 'large' hexagonal crystallites and the speckled background.

Plate IV (Fig. 13.11, p. 324) Photograph of an ELOG-based blue LED under current injection. There is only a slight modulation of intensity between almost defect-free and highly defective stripes [73].

Plate V (Fig. 15.6, p. 422) Isosurfaces of the conduction wave functions for (a) $x << x_c$ GaAsN showing nitrogen-localized cluster states within the gap (red), (b) $x \sim x_c$ showing the 'amalgamated state' of localized (red) and the delocalized (Bloch-like) states (green), and (c) $x >> x_c$ extended states (green). (Figure reproduced from ref. [53].)

1 GaN or the history of a scientific explosion driven by applications and markets

Jean-Yves Duboz and Jean-Charles Garcia

In the early 90s, GaN was an exotic semiconductor studied by a few exotic people. The rest of the planet was considering this domain of research as 'scientifically interesting', which in certain minds translates as unrealistic, an endless search for a mythical grail, a black hole for time and money. There was even scientific evidence that GaN would remain forever a material as good as wood for optoelectronics.

In 2001, the GaN activity represents about \$500m and hundreds of people involved in research only. Most of us already use GaN devices (LEDs) in our immediate environment, such as our cars. And most of us have been or are still searching in the GaN area.

What a tremendous change in less than ten years! Such a big bang could not have been driven by science. Applications such as blue LEDs and the prospect of large revenue associated with emerging markets have been the driving force. The blue El Dorado! Furthermore, science was left behind, and devices were produced and used while the physical mechanisms responsible for their good performance remain somewhat mysterious.

Hence, we thought it was now a good time to write a book on the physics of this fascinating compound semiconductor. For sure this book will not be the last one on this subject, as all physical aspects in nitrides have not yet been unveiled. We simply think that our community needs some reference works from time to time that can serve as a basis for further exploration.

So, physics will have the place of honor in this book. However, in this introductory chapter, we want to give credit to the applications that have pulled the nitrides out of the darkness, which goes beyond metaphor when talking about LEDs.

We therefore summarize hereafter the applications envisioned for and allowed by GaN materials, and discuss some economical and technological issues.

1.1 Light-emitting diodes (LEDs)

The total LED market in 2000 is estimated to be in excess of \$3 billion, in rapid and continuous growth, especially for high brightness LEDs (annual growth rate of 22%). A large part (around two-thirds) of this market corresponds to infrared and red LEDs fabricated with GaAs/AlGaAs or GaInP materials. The remaining one-third corresponds to amber, green, and blue LEDs. In the blue and green, there is currently almost no competition and this market is entirely devoted to InGaN materials. In terms of performance, the output power is about the same for green and blue commercial LEDs, and goes from 2 to 7 mW at 20 mA, depending on products, with a typical area of 0.06 mm^2. In the laboratories, much higher performance has been measured: Cree Lighting reported 21 mW for 20 mA at 390 nm wavelength, which corresponds to an impressive 32% external quantum efficiency. The diode is not a simple planar structure that is then simply cut, but a rather complicated one, with a mushroom shape. Indeed, one has to remember that the optical escape cone corresponding to the GaN/air interface in a planar structure is 24° and that 8.7% of emitted photons can actually escape the active layer. Thanks to the metallic reflector below and around the diode and the epoxy dome lens, this escape probability is increased to a much larger value. However, it remains much smaller than unity and limits the external quantum efficiency. This is the reason why various tricks—including the mushroom shape, or surface texturing, or micro-pillar—have to be played. The Cree structure is, in fact, similar to the inverted truncated pyramid developed by Agilent/LumiLeds on AlInGaP, where 55% quantum efficiency was reached. This shows that further improvements can still be expected in the GaN LEDs, all the more since the index of refraction is lower in these materials (nitrides, sapphire, or SiC) than in conventional III–Vs. For yellow LEDs, AlGaInP still leads to better performances than InGaN, and still dominates the market. Indeed, the current status is that the quantum efficiency tends to decrease when the emission wavelength increases, and this becomes significant in the yellow range: at these wavelengths, the quantum efficiency equals 5% for InGaN and 11% for AlGaInP. The reason for this trend is believed to be due to the larger piezoelectric field (both spontaneous and strain induced) in quantum wells with larger In concentration. The field separates electron and hole wave functions and reduces the radiative matrix element. A second and additional reason could be the degradation of the crystallographic quality associated with phase separation that occurs in In rich layers. A good point, however, for yellow InGaN LEDs is the temperature dependence of the quantum efficiency which is much smaller than that of AlGaInP, possibly due to a larger confinement potential.

The nitride LED market is about $500 million in 2000. Major LED companies include Nichia (Japan), Hewlett-Packard (USA), Cree (USA), Agilent/LumiLeds (USA), Emcore/Uniroyal (USA), Toyoda-Gosei (Japan), Siemens-Infineon (Germany), Samsung (Korea), GellCore (USA) and many others, with a total production over 100 million units per month.

Last but not least, white LEDs can be obtained by covering blue LEDs with a phosphorescent dye the composition of which can be adjusted so that the dye luminescence is in the blue, green, or red. In fact, it is adjusted to give a white light by addition of blue and yellow, with an equivalent black body temperature of 6000 °C, close to the sun's temperature. The following table summarizes the performance of various technological approaches.

	Efficacy (lumen/W)	Market share
White GaN LED	20	
Incandescent bulb	16	30%
Halogen lamp	25	20%
Fluorescent lamp	70	40%
Sodium lamp, and others		10%

This comparison indicates that white LEDs are able to penetrate the lighting market, which is a huge market worth about US$10–12 billion. Advantages of LEDs for lighting include size, weight, reliability, low voltage, high colour temperature and new design possibilities. However, the lighting market penetration is still hindered by important issues: first, the total power of an LED is small compared to that of conventional sources, of the order of 1 lumen, while general illumination requires almost 100 lumen. A strong increase of the amount of lumen per die is thus desired and is very likely to occur in the coming years, both by increasing the die area and by increasing the current density. LumiLeds started to work on these high power LEDs with amber AlGaInP inverted pyramid LEDs where 60 lumen were extracted from a 0.25 mm^2 area at a current density of about 200 A/cm^2, which is five times larger than usual current densities. Recently, LumiLeds also reported impressive results on GaN/InGaN LEDs, with 200 mW for 250 mA at a wavelength of 425 nm (22% quantum efficiency), and 525 mW at 1 A.

They finally reported a single white LED with 17 lumen light output. In hard-driven LEDs, problems of reliability, heat dissipation, and packages will then require special attention. Second, the colour rendering of white LEDs has to be improved, better controlled, and stable in time. Technological improvement on phosphors composition, deposition methods, or color mixing in the case of blue–green–red mixing are needed. Finally, the cost of LEDs is still prohibitive. It is of about $0.5 per lumen, while it ranges from $0.05 down to $0.005 per lumen for conventional sources. As the power per die increases, the cost per lumen will come down as part of the costs (packaging, commercializing) do not scale with the die area. A decrease by a factor of 10 should be possible within a few years (low brightness LEDs are already sold in the region of 11 cents by Cree). Then, if the efficacy is comparable with that of other sources, general illumination should become an accessible market. This is forecast to occur in ten years from now. At present, white LEDs are used in specialty lighting such as in cars. These niche applications represent a non-negligible market, which is, however, far smaller than the general lighting market.

1.2 Lasers

The total laser diode market is about $2 billion, with a very fast annual growth rate. About one-half of the market is related to the communications with an exploding growth. Lasers based on GaAs and InP perfectly fit the requirements for 0.98, 1.3, 1.48, or 1.55 μm wavelengths. In large part due to the emergence of nitrides and the associated technology (N sources, for instance), recent and fast improvements in InGaAsN materials have been achieved. In these materials, a limited amount (<3%) of N is incorporated in InGaAs, lattice matched to GaAs or close to. The coupling between the N levels and the InGaAs conduction band leads to an anomalous and large reduction of the bandgap with increasing N content, thus allowing long wavelength emission be reached. A ridge laser based on InGaAsN has first been demonstrated by Infineon at 1.3 μm, and then by the University of Würtzburg at 1.52 μm. A VCSEL at 1.3 μm was demonstrated by Sandia.

On the other extremity of the spectrum, ultraviolet and blue lasers have been demonstrated, based on InGaN/GaN/AlGaN structures. A dozen groups have demonstrated laser emission in InGaN by electrical pumping. Some of them have reached CW operation at room temperature, from a few seconds to hours (Cree, Xerox, Fujitsu, Infineon etc.). Nichia is the only company that has started to ship lasers over the world, with an initial price of about $2000 per unit. The lifetime at room temperature and for an output power of 5 mW is larger than 10 000 hours. Best results measured on separate devices include a maximum power of 420 mW, a threshold current density as low as 1.2 kA/cm^2, a turn-on voltage of 4.3 V, or a 39% quantum efficiency in the linear regime. Improvements are obviously to be expected for the lifetime at higher powers.

Optically pumped nitride VCSELs were demonstrated first by the University of Tokyo in 1998 at 77 K. The cavity was made of a bottom epitaxial mirror (GaN/AlGaN 34%, 35 pairs, R_{max}=97%) and a top dielectric mirror (SiO_2/TiO_2, 6 pairs, R_{max}=98%). Room temperature operation was achieved in 1999 by the Ioffe Institute. In 2000, the Brown University demonstrated an optically pumped VCSEL almost at room temperature based on top and bottom dielectric mirrors deposited after substrate laser lift-off. The same team demonstrated a vertical cavity LED based on the same technology, which seems to be one of the most promising approaches towards an electrically pumped VCSEL. Achieving electrical pumping in GaN VCSEL is a really challenging task, as carrier injection, particularly for holes, is difficult. Lateral injection can be envisioned but the access resistance is difficult to keep small because of the large resistivity of p type GaN (1 Ω cm). Very small VCSELs will probably be needed in order to minimize the access resistance and the injection non-uniformity. In terms of applications, advantages of VCSELs compared to ridge lasers include the integration into arrays for massive parallel operation and an increased modulation bandwidth.

The market for GaN lasers is in the area of storage and printing. In both cases, there is a strong demand for cheap and compact lasers, with moderate powers and a short wavelength. The reason for this is the diffraction, that limits a beam spot to an area equal to the square of the wavelength λ. The smaller the wavelength, the smaller the spot. If one uses this spot to store or print any information on a surface, the density of information goes as λ^{-2}.

The most well known application is optical storage, due to the very popular CD. More than 300 million lasers were produced for CD or DVD players in 2000. The current trend is for higher resolution and higher power, especially for rewritable DVDs. The lasers used in CD players are based on AlGaAs (780 nm). For DVD players, AlGaInP lasers (635 or 650 nm) are now used, and InGaN lasers (410 nm) are expected to be used very soon for High Definition DVDs. The minimum feature size is 0.82 μm for the CD, leading to a storage capacity of 0.65 GBytes. DVDs are based on two layers on both sides of the disk. The minimum feature size is 0.4 μm and 0.24 μm for the standard DVD and the HD-DVD respectively, enabling a capacity of 9.4 GBytes and 30 GBytes. The power requirements are 5 mW CW for reading. CD-write once on a dye layer, DVD or CD-rewritable on phase change materials as AgInSbTe, are currently gaining much popularity. The writing function in this case requires a larger power, in the 35–50 mW range, not necessarily CW. Although InGaN laser is well suited for optical storage, it is noted that competitive technologies exist, with advantages and drawbacks. The achievement of powerful infrared DBR laser diodes at 808 nm and efficient non-linear materials such as $LiNbO_3$ with phase matching realized by period poling makes it possible now to generate 425 nm light by second harmonic generation, with power levels in the 15 mW range. Compared to InGaN laser diodes, it provides a larger power, lower noise, and better beam quality. On the other hand, it requires a precise mechanical alignment, is more expensive, and less compact. As a result, it may be used in more expensive and high-quality players, leaving the higher volume of cheaper players to InGaN lasers that, as a matter of fact, will have to be cheap (about 2% of the player price). This may have some consequences on the technology chosen for the laser fabrication (substrate etc.). At a recent exhibition in Japan, Sony unveiled a prototype DVD recorder based on a 405 nm laser that could write 22.5 Gbytes on a 12 cm diameter disk. This shows how fast high technology can be incorporated in commercial products as soon as large volumes are expected.

Printing and xerography are other applications for InGaN lasers, corresponding to a huge market of $100 billion (the part for the laser only being a small fraction of this). The current standard is 600 dpi (dots per inch), but the trend is for increasing the definition to 1200 dpi. This requires to decrease the spot size from 35 μm to 17 μm. This could be done by improving the optics, with, however, a direct effect on the system price. An alternative is to keep the same optics and reduce the wavelength. This appears to be a cheaper solution. On the low wavelength side, the sensitivity of the organic photoreceptor on the printer drum and the transparency of glass impose some limitations, so that the final compromise is around 400 nm. A power of 5 mW allows for a printer output of 60 pages per minute. A switching speed in the nanosecond range and a good beam quality are additional requirements.

The remaining issue for GaN lasers is the choice of the substrate. While best results were obtained on sapphire substrate with an ELO process, other substrates can be used. SiC provides a better thermal management and a back side electrode, thus simplifying the process. The crystalline quality and surface preparation are, however, not as good as those of sapphire. In addition, the availability and price of SiC substrates are still limitations. GaN substrates offer first of all a high quality homoepitaxy, but also a reasonable thermal conductivity (recent measurements have shown that the thermal conductivity in high-quality GaN was indeed much larger than previously thought, of the order of 2–2.2 W/cm.K, while the previously reported value of 0.5 W/cm.K was due to the presence of dislocations) and an excellent back side contact. The high pressure high temperature growth of bulk GaN shows some limitations on size, production output, etc. Free-standing films fabricated by HVPE growth on a host substrate that is subsequently removed appears to be a more realistic way. Experiments have successfully been carried out on SiC removed by RIE or on sapphire removed by laser lift-off. Films as large as 50 mm in diameter and 150 μm thick have been obtained. The question of how to produce these substrates in a cost effective way remains open. While the production of such substrates remains difficult, it may be easier to remove the substrate on smaller areas (typically a laser area) and wafer bound a few devices only. Such an approach was successfully employed by Xerox to transfer InGaN MQW lasers grown on sapphire onto copper and diamond substrates, with a clear improvement in thermal management.

1.3 Ultraviolet detectors

Due to their direct bandgap, GaN and AlGaN are good candidates for optical detection. When going from GaN to AlN, the bandgap energy increases from 3.43 eV to 6.2 eV, thus covering the 200–365 nm range. The spectral region between 250 and 300 nm is of particular interest. The ozone present at 15–20 km altitude absorbs the incident solar radiation for wavelengths smaller than 300 nm. As a consequence, any radiation emitted on earth or at low altitude between 250 and 300 nm can easily be observed due to the absence of background radiation under the condition that the detector does not see the background radiations above 300 nm. Such detectors are called solar blind. The most straightforward military applications is the missile launching detection. Due to the long distance of detection required, the detectivity has to be extremely large, and the spectral selectivity (how well it rejects solar light) also needs to be as high as possible. In order to produce an image, two-dimensional arrays of detectors are desired. In the civilian area, flame detectors and sensors are needed for monitoring the combustion in large industrial burners and boilers. In order to observe the flame in a daylight environment, solar blind detectors are needed as well. Single detectors with a moderately high detectivity will be suitable, under the condition that they can be produced at a moderate price. This kind of detector can also be used for fire detection, with a better detectivity, however. In the deeper ultraviolet (225–275 nm), pilot landing aid systems have been proposed based on UV lights put on the edges of runways and taxiways, and a UV camera aboard. When its wavelength is much smaller than rain or fog particles, the light can propagate over reasonable distances thanks to the effect of forward scattering. First attempts have shown that the visibility can gain a factor of three when going from the visible to the ultraviolet. A last example of application for UV detectors is the UV A (230–400 nm) and UV B (280–320 nm) dosimeter, for personal use, in UV-rich environment (beach, mountains). Price and compactness are, of course, the main requirements here.

The total possible market for UV detectors remains quite small (less than \$100 millions) and the actual market is even smaller than that.

Many technologies are candidates for UV detection. Indeed, most visible or even infrared sensors can be used in the ultraviolet, with a reduced sensitivity, however. Optical filters can be added to limit the system detection to the desired range. However, rejecting the visible light requires the use of interference filters. These filters are expensive, difficult to fabricate, and degrade the performance in the UV. If UV selectivity is not an important issue, then Silicon is probably the best semiconductor candidate. Some UV enhanced Silicon detectors have a decent response down to 250 nm. For extremely high detectivities, single elements, and moderate visible rejection, photoemissive devices (gas or vaccum cell) are appropriate. Diamond can be a good choice for deep UV, below 225 nm. Indirect bandgap semiconductors (especially those with a wide bandgap as SiC or diamond) do not provide a sharp cutoff, which is a severe drawback when a good spectral selectivity is required. For the precise case of solar blind detectors, AlGaN is the best candidate.

Photoconductors based on AlGaN were obtained in many places, such as Madrid University. Solar blind photoconductors were obtained at Thales, with a moderate responsivity (10–100 A/W for an electric field of 50 kV/cm), due to a small electron mobility in highly resistive AlGaN and a long response time (1 ms). The visible and near UV rejection is limited to a 2–4 decades. The dark currents are relatively small but still high enough to give rise to a strong $1/f$ noise: $i^2_{\mathrm{noise}} = H(I^2/f)$, where the parameter H is in the range of 10^{-5}–10^{-4} for interdigitated finger on a $100\,\mu\mathrm{m} \times 100\,\mu\mathrm{m}$ area. At 1 kHz, the noise equivalent power for such a device is in the range of 100 pW.

Photovoltaic detectors, MSM (interdigitated fingers of metal–semiconductor–metal) detectors, Schottky or pn junctions, have been realized. The response time is short, in the nanosecond range. The response is in the range of 0.01–0.1 A/W. Small (32×32) and medium size (128×128) arrays have been realized by a US collaboration including North Carolina State University at 365 nm (visible blind). A phototransistor with a large response (5000 A/W at 365 nm) and a rejection of the visible reaching the impressive value of 8 decades was demonstrated by Honeywell. Flame detection at 300 nm based on an AlGaN pn junction was demonstrated by Osaka gas. At Thales, solar blind MSM detectors were demonstrated with a noise equivalent power as low as 300 fW at 50 Hz, and a visible and near UV rejection ratio of

5 decades. Interestingly, a similar performance was obtained in GaN/Si and GaN/sapphire.

1.4 Electronics for microwave applications

The RF (radio frequency) semiconductor market is estimated to be around \$5 billion in the year 2000. It is increasing rapidly due to the general increase in communications on earth. In particular, the exploding development of cellular phones (already 400 million units in use, annual growth rate of 30%) and that of satellite communications (the market for RF electronics for satellites is about \$200 million, with an annual growth rate of 25%, and with about 100 new satellites in orbit every year) will keep the demand for new RF products at the highest level. About 70% of this market is occupied by Si, while the remaining part is for GaAs. The part devoted to other semiconductors is almost negligible. This may change in the near future with the emergence of two newcomers: SiC and GaN. The electronics based on GaN is forecast to reach \$500 million within the next 10 years. Compared with conventional semiconductors such as Si and GaAs or InP, GaN shows some advantages including a higher saturation velocity, higher breakdown field, and higher temperature operation range. As a consequence, higher power densities are expected in a GaN transistor (10 W/mm is an accessible value). Larger electron masses, reduced mobility, and contact resistances limit the operation frequency range to about 50 GHz. Let us compare GaN and other materials in the frequency-power area. On the high frequency side, InP-like materials take advantage of small effective masses and record mobilities to achieved record frequencies. On the low frequency side, Si prevails for moderate powers. Thanks to SiGe, the Si industry is now extending towards higher frequencies, up to and above 5 GHz, at the expense, however, of a reduced power. GaAs is well suited for moderate power and frequency ranges. Pseudomorphic and metamorphic structures allow GaAs to reach higher and higher frequencies. SiC is doing well for high power and low frequency. However, it will be limited to frequencies of a few GHz only. In conclusion, GaN is in competition with SiC on its low frequency side for high power and with GaAs on its low power–high frequency side. Thanks to breakthroughs in technology that improves the thermal management (wafer fusion, smart cut process, flip-chip, etc.), all technologies are moving towards higher power. Finally, it is important to recall that the main competitor for GaN in the high frequency–high power domain is the traveling wave tube. For high power RF, TWTs, and semiconductor devices almost equally share the market. For very high frequencies, TWT will maintain its supremacy.

The domain of RF application is wide and in rapid growth. As new applications emerge, there is a continuous shift towards higher frequencies. On the low frequency side, Si and GaAs benefit from the exponential development of cellular phones (0.8 GHz and 1.9 GHz). At higher frequencies, satellite and terrestrial communications (1.6, 2.5, 5.2, 23, 28 GHz) and military applications

(X Band, 8–12 GHz) represent a strong potential of development for GaN. Applications at higher frequencies such as anticolliding radars (76 GHz) are reserved for InP or GaAs.

The cost of an RF system depends, of course, on the power and frequency. As the frequency increases, it becomes more and more difficult to coherently add the individual powers coming from different transistors. Dephasing has to be accounted for by a proper design and the addition of passive elements, mainly selfs and capacitances, that can take up to three-quarters of the area on the wafer. As a result, high frequency–high power modules are expensive. In addition, a cooling system is usually needed. The advantage of wide bandgap materials is manifold: they can provide a larger RF power per unit area which simplifies the design, and also can work at higher temperature, which allows reduction of the cooling. Finally, the power is generated in each transistor under a large bias, which makes the impedance matching between different stages easier. Thus, it is expected that using GaN rather than conventional III–V compounds will lead to the same performance at a lower price, or to a higher performance (with a high price). For many applications, electronics for satellites being the best example, reliability and power added efficiency are major concerns.

Due to the competition with existing technologies, the development of GaN electronics was not as fast as that of GaN optoelectronics. Still, recent progress is impressive, as can be judged from the following state of the art data collected on different devices: The saturation drain–source current is about 1 A/mm in virtue of the very large 2D electron density (10^{13} cm^{-2}) that is made possible by the large conduction band offset at the GaN/AlGaN interface. The transconductance reaches 220 ms/mm (0.25 μm gate). Highest f_T and f_{max} are 67 and 140 GHz respectively. At 10 GHz, Cree reported power densities of 9.1 W/mm in small area devices, and a total power of 40 W (pulsed) in a 12 mm long device, with a PAE (Power Added Efficiency) of 20%. On a smaller device, 62% PAE was reported by Cree together with a power density of 7 W/mm. GaN presents the largest power density and total power ever measured in a semiconductor transistor at 10 GHz. The actual power limitation is given by the thermal management. Up to now, the best solution for thermal management has been to use SiC as a substrate. Finally, we have to consider the availability of large substrates in order to process the GaN transistors in the existing or future production lines. As the actual trend is to go from 4 inches to 6 inches for GaAs (12 inches for Si), having GaN on a large substrate (for instance 6 inches by year 2005) would be a clear bonus. In this regard, Si is a good candidate as a substrate. The effort that is currently under way at Nitronex to grow HEMT structures on 4-inch silicon wafer best illustrates this very promising possibility. After processing, the substrate can be removed if necessary for thermal, conductivity, etc., reasons. Report on a highly conductive substrate (polycrystalline diamond or SiC) would then solve the thermal management.

Finally, even if the way towards large volume commercialization is still long and paved with difficulties, the future looks bright for GaN electronics for high-power high-frequency applications.

1.5 High-temperature electronics

The total available market is gigantic (about $12 billion in 1999). However, due to the lack of devices and to 'work-around' techniques (electronics is kept further away from the hot source or is cooled), the actual market is limited to $400 million. As products begin to emerge, this market will partly become reality, and will shift to higher temperatures. Currently, 87% of this market is for temperatures below 200 °C, which is simply due to the fact that SOI can work up to this temperature. Domains of application include well logging, aerospace, automotive, etc. Emerging markets such as the automotive one require cost-effective electronics. In general, the substrate thermal conductivity is not an issue and Si or sapphire are well adapted. There have been many demonstrations of GaN transistors working at high temperature. Heterojunction bipolar transistors (HBT) were demonstrated to work at high temperature (400 °C). In some cases they even work better at higher temperatures as the base doping increases, with an activation energy equal to the Mg acceptor energy (around 200 meV, which is a clear limitation for HBT at room temperature). MESFETs were also shown to behave well at high temperatures (400 °C–600 °C).

Issues such as reliability, material and contact stability still have to be carefully addressed.

1.6 Conclusion

The field of GaN has been and will remain in the near future driven by optical sources in the violet, blue, and green. The market for such products already exists and corresponds to large production volumes. Other applications are still under development and will join the LEDs and laser as commercial products in a few years: the best example is likely to be the high power–high frequency electronics. However, other applications, that were not dealt with here, such as cold cathodes, piezoelectric devices, SAW (surface acoustic wave) devices, could also take advantage of progress made in the area of GaN growth and technology to become commercial products. This wide potential of applications for a given material is, in fact, one of the most decisive advantages of GaN compared to other semiconductors. From the scientific point of view, this justifies the efforts to fully understand its basics properties. Much work is still needed, science is still moving ahead. Our purpose with this book, is to establish a solid base camp, that can be used by explorers on their way towards further investigations, and we hope that it will be a useful, if not restful, refuge for the GaN community.

2 Nitrides as seen by a Theorist

Friedhelm Bechstedt

2.1 Introduction

The modern solid state physics and materials science try to understand the properties of solids starting from basic notions and to predict the properties of novel materials. For instance, the combination of two or more compounds in heterostructures makes it possible to tailor electronic and optical properties via the quantum confinement effect. In the past experiments have mainly been used to determine the properties of the new solids. Theories accompanied the measurements to tell us why the solids have the properties they have. However, in the last 15 years it became possible to compute lattice constants, bulk moduli, cohesive energies, phonon spectra, and other static and elastic properties of solids starting from no other ingredients than the atomic numbers and masses of the atoms composing the materials.

For that reason the new theoretical approaches are called *first-principles* or *ab initio* methods. They really allow predictions without taking any parameter from the experiment. Most of these techniques are based on the density functional theory (DFT) [1] in the local density approximation (LDA) [2]. The full interaction of the electrons including exchange and correlation is taken into account. The form of the interaction functional is taken from exact results obtained for the homogeneous electron gas [3, 4]. Spin polarization can immediately be included [5]. However, being based on a ground-state theory the *ab initio* methods cannot describe excited states correctly, in particular not electronic excitation energies. The band structures derived from the single-particle Kohn–Sham eigenvalues of the DFT-LDA [2] cannot directly be compared with measured transition energies [6]. Quasiparticle corrections calculated at least in the so-called GW approximation [7, 8] of the exchange-correlation self-energy have to be included to account for the excitation aspect. On the other hand, theories based on DFT and LDA allow accurate calculations of changes of the total energy with atomic displacements. The density-functional perturbation theory (DFPT) [9] provides an opportunity to compute the dynamical matrix and, hence, the lattice vibrations of a crystal, without the knowledge of experimental input parameters apart from atomic numbers and masses.

The group-III nitrides AlN, GaN, and InN, their alloys, and their low-dimensional structures represent outstanding examples of the applicability of the *ab initio* theories mentioned above. One reason is related to the extreme properties, such as high ionicity, very short bond lengths, large energy gap, low compressibility, high thermal conductivity, and high melting temperature of the nitride subgroup of the III–V compounds. These properties also make the growth of high-quality single crystals and the epitaxy of perfect layers complicated. In particular, in the beginning of the nitride research the samples possessed a high density of defects and the studied films were strongly n-type with carrier densities in the range of 10^{18}–$10^{20}\,cm^{-3}$, even when not deliberately doped. Consequently, only insufficient data basis sets existed or exist for several properties. For instance, authors are frequently using elastic coefficients that have been calculated for zinc-blende and wurtzite AlN, GaN, and InN [10] but not measured. Another example may be the determination of electronic properties of wurtzite InN. The fundamental energy gap has been extracted from room-temperature optical absorption data of thin films prepared by reactive radio-frequency sputtering [11] (and references therein). The polycrystalline films consist of oriented needle-like wurtzite crystallites with the *c*-axis parallel to the growth direction and diameters in the range of 10–30 nm. Meanwhile, epitaxial layers of cubic InN have been grown by means of molecular beam epitaxy (MBE) on GaAs(001) substrates [12]. However, the large number of free carriers made photoluminescence measurements of the fundamental gap impossible. Calculations are thus helpful tools to clarify the properties of intrinsic InN crystals.

The theoretical description of the properties of crystals, heterostructures, and mixed crystals consisting of group-III nitrides is a challenge for several reasons. First of all, these compounds contain a first-row element, nitrogen. Due to the lack of *p* electrons the N core is extremely small. Consequently, an expansion of the one-electron eigenfunctions requires many plane waves in order to represent correctly the wave functions in the near-core region. Moreover, semicore *d* shells occur in the compounds GaN and InN. It has been argued [13, 14] that in GaN the semicore *3d* shell of Ga is not inert as in the most Ga-based materials and affects both the electronic and structural properties. The situation is similar for the In *4d* levels in InN [15]. As indicated by Fig. 2.1, the shallow Ga 3*d* and In *4d* levels are resonant with the N 2*s* states. This makes a hybridization of these states and contributions of the shallow *d* electrons to the chemical bonding very likely in GaN and InN. The Ga *3d* and In *4d* states are highly localized in real space. Consequently, methods based on normconserving pseudopotentials cannot readily be applied, because they are difficult to converge in a plane-wave expansion of the eigenfunctions. Wright and Nelson [14] used an energy cutoff of 240 Ry which corresponds to about 9400 plane waves per GaN unit.

Another problem of the theoretical treatment of the group-III nitrides concerns the extreme differences between the cations and the nitrogen anion, in particular for GaN and InN. These differences are clearly demonstrated in Table 2.1 summarizing the atomic weights, the covalent radii, the electronegativities [18] and the effective spin-orbit splittings of the *p* and *d* states [19] of free atoms. The extremely different atomic sizes and the electronegativities should cause considerable charge transfers between cations and anions and, hence, strongly ionic bonds. Altogether, these properties of the free atoms already indicate the zinc-blende-wurtzite polytypism of the group-III nitrides [20].

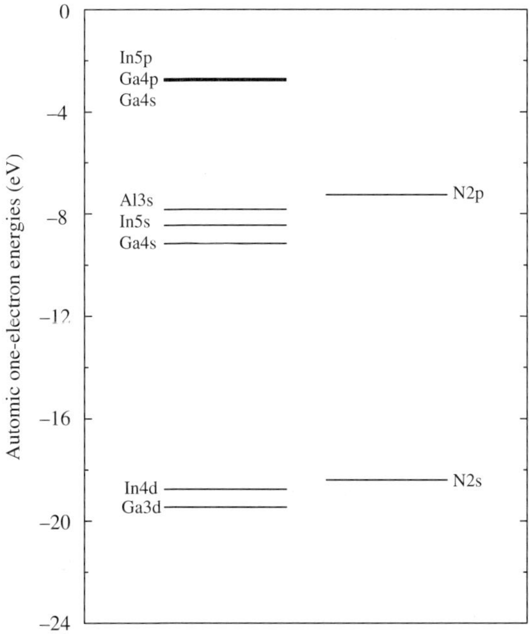

Fig. 2.1 Energy levels of atoms forming the group-III nitrides. The levels are taken from a DFT-LDA calculation of the free atoms using a scalar-relativistic approach [16, 17].

Table 2.1 Properties of atoms from ref. [18]. The effective spin-orbit splittings of *p* and *d* states are taken from ref. [19].

Atom	Atomic weight	Covalent radius (Å)	Electro-negativity	Spin-orbit splitting (meV)
N	14.0067	0.75	3.04	40 (*p*)
Al	26.98154	1.18	1.61	24 (*p*)
Ga	69.72	1.26	1.81	22 (*p*), −220 (*d*)
In	114.82	1.44	1.78	500 (*p*), −530 (*d*)

The spin-orbit splittings of N and In atoms differ by one order of magnitude [19] and, hence, raise the question concerning the spin-orbit influence on the electronic structure of InN crystals close to the valence-band maximum (VBM).

As a consequence of the extreme atomic properties of the constituents, the III–V nitride semiconductors AlN, GaN, and InN have electronic and optical properties that make them attractive candidates for fabricating novel electronic and in particular optoelectronic devices [21]. These devices usually consist of layered structures. Heterostructures involving wurtzite layers are influenced by the phenomenon of spontaneous polarization fields [22], which are not present in layered (unstrained) systems of cubic crystals. The accompanying built-in electric fields, stemming from polarization charges at the heterointerfaces, have a major impact on the electronic and vibronic properties of single and multiple quantum wells [23] or superlattices [24]. The misfit between the lattice constants of the nitrides can induce considerable biaxial strains in the layers. The strain affects the layer properties directly or via the induced piezoelectric polarization fields in addition to the spontaneous ones. In order to tailor further the electrical and optical properties of the layered systems, ternary or quaternary alloys of the group-III nitrides are applied. Alloys such as $In_xGa_{1-x}N$ and $Al_xGa_{1-x}N$ allow a variation of the fundamental energy gaps and effective masses in a wide parameter range. However, the alloying also drastically influences the spontaneous and piezoelectric polarization fields.

Consequently, in this chapter important properties of group-III nitride crystals, their layered structures, and their mixed crystals are studied on the basis of results of theoretical considerations and calculations within the framework of the *ab initio* DFT approach. Three aspects of the nitride systems are discussed in detail: strain influence, alloys, and interplay of different wurtzite layers resulting in spontaneous polarization fields. After a brief description of the computational methods, the results for basic properties of electrons and phonons in bulk nitride crystals are summarized. Then, from the view point of a theorist the aspects of strain, alloying, and layering are discussed.

2.2 Basic properties

2.2.1 Computational approaches

The majority of results presented in this chapter is based on two versions of parameter-free calculations within the framework of DFT-LDA [1, 2]. In the first version the semicore Ga 3*d* and In 4*d* electrons are explicitly treated as valence electrons [25, 26]. Their interaction with the atomic cores is treated by non-normconserving *ab initio* Vanderbilt pseudopotentials. They allow a substantial softening also for first-row elements with the lack of core *p* electrons as well as for the attraction of shallow *d* electrons [27]. As a consequence of the optimization the plane-wave expansion of the one-electron eigenfunctions can be restricted by a kinetic-energy cutoff of 14.4 (AlN), 16.2 (GaN), or 15.5 Ry (InN). The cutoffs have been carefully tested. In the case of alloys the cutoff is substantially increased to 22 Ry to account for possible shorter bond lengths. The many-body electron-electron interaction is described by the Ceperley–Alder functional [3] as parametrized by Perdew and Zunger [4]. The **k**-space integrals are approximated by sums over special points. For instance, 10 or 12 mesh points of the Chadi–Cohen type [28] are used within the irreducible part of the Brillouin zone (BZ) of the zinc-blende or wurtzite structure. In the case of superlattices and alloys more dense meshes of special points of the Monkhorst–Pack type [29] are used. The calculations employ the conjugate-gradient method to minimize the total energy with respect to the atomic coordinates. Explicitly, the Vienna Ab-initio Simulation Package [16, 17] is used.

In the second version [30, 31] the valence electron–ion interaction is treated by soft *ab initio* normconserving pseudopotentials for Ga, Al, and N which are generated according to the scheme of Troullier and Martins [32]. The shallow Ga 3*d* electrons are frozen into the core. Their effect on the other electrons is taken into account [33] via the nonlinear core corrections [34]. Nevertheless, the small N core requires relatively high cutoff energies for the plane-wave expansion of the eigenfunctions up to 80 Ry for the various polytypes and structures based on GaN and AlN. The second version can easily be combined with the self-consistent DFPT [9]. It allows the calculation of dielectric and lattice-dynamical properties. The high-frequency dielectric tensor and the Born effective charge tensor can be directly calculated for each inequivalent atom in the unit cell. Their combination gives the non-analytic part of the dynamical matrix of polar crystals. The force constants follow from the total energies of the systems with displaced atoms.

2.2.2 Structure of polytypes

The strongly ionic III–N compounds crystallize in two polytypes, the wurtzite one with the space group C_{6v}^4 ($P6_3mc$) and the zinc-blende structure with T_d^2 ($F\bar{4}3m$) symmetry. Their bond stackings AB and ABC are shown in Fig. 2.2 in cubic (C) [111] or hexagonal (H) [0001] direction. According to the number of bilayers needed to guarantee the translational symmetry in this direction, the two polytypes are labeled with 3C or 2H. The ground-state properties are obtained by minimizing the total energy with respect to the cell volume. Whereas in the zinc-blende case the volume is directly related to the cubic lattice constant a_0, the wurtzite case is governed by the two hexagonal lattice constants a and c. There is also a third structural degree of freedom. The internal-cell parameter u characterizes the relative length of the III–N bonds parallel to the c-axis [0001]. For both polytypes the volume dependence of the total energy per III–N pair is plotted in Fig. 2.3. It clearly indicates the stability of the wurtzite structure independent of the group-III atom. The zinc-blende polytype is higher in energy by 48 (AlN) or 20 (GaN) meV per cation-N pair [30, 31]. Other values per cation-N pair are 60 (18, 44) for AlN, 24 (11, 8) for GaN, and 142 (9, 18) meV for InN from refs. [26, 20, 35]. Figure 2.3 represents the well-known fact that AlN, GaN, and InN crystallizes in wurtzite structure under ambient conditions. Nevertheless, many papers reported the growth of the metastable zinc-blende polytype meanwhile for all three nitrides under consideration.

The total energy as a function of volume fits the Murnaghan equation of state near its minimum. It also allows the determination of the isothermal bulk modulus B at zero pressure. Results of the structural optimizations are listed in Table 2.2. They are compared with results of converged plane-wave calculations using normconserving Bachelet–Hamann–Schlüter pseudopotentials [15]

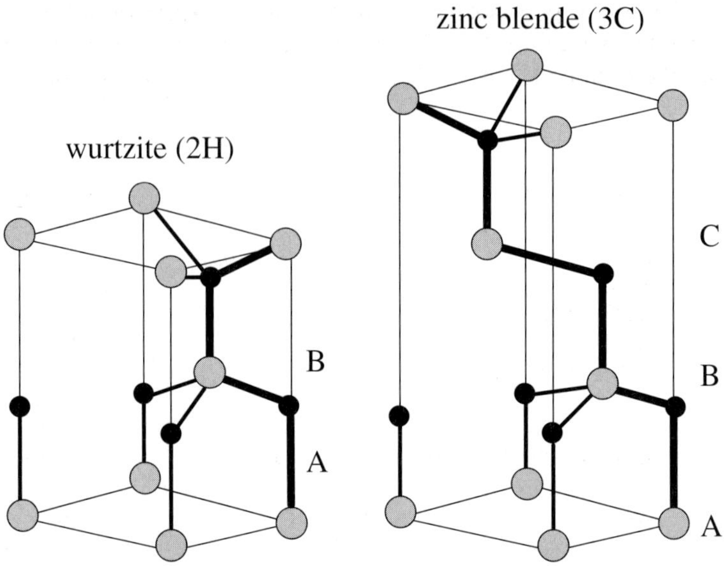

Fig. 2.2 Perspective view of hexagonal unit cells of 2H (wurtzite) and 3C (zinc-blende) polytypes. The bonds in a $(11\bar{2}0)$ plane are indicated by thick solid lines to show the stacking sequence.

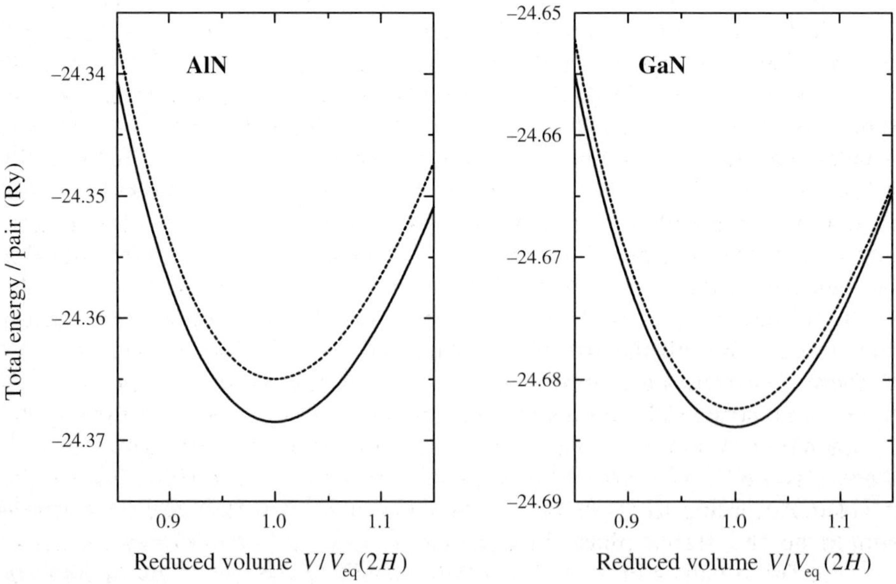

Fig. 2.3 Total energy versus normalized volume. V_{eq} denotes the equilibrium volume. Solid line: wurtzite, dotted line: zinc-blende. (From refs. [30, 31].)

Table 2.2 Structural parameters of zinc-blende and wurtzite III-nitrides [26]. Other calculated values (refs. [15, 35]) and experimental data (third value), taken from the data collection in ref. [15], are given in parenthesis.

Parameter	AlN	GaN	InN
a_0 (Å)	4.34 (4.34, 4.31; 4.38)	4.46 (4.46, 4.52; 4.5)	4.97 (4.93, 5.00; 4.98)
B (GPa)	209 (207, 206; 202)	183 (187, 191; 190)	133 (140, 140; 137)
a (Å)	3.08 (3.08, 3.06; 3.110)	3.15 (3.16, 3.19; 3.190)	3.53 (3.50, 3.54; 3.544)
c/a	1.607 (1.604, 1.617; 1.601)	1.631 (1.626, 1.634; 1.627)	1.632 (1.619, 1.626; 1.613)
u	0.3824 (0.3814, 0.3802; 0.3821)	0.3815 (0.3770, 0.3760; 0.3770)	0.3780 (0.3784, 0.3770; 0.3750)
B (GPa)	203 (205, 209; 202)	195 (202, –; 195)	125 (139, 140; 126)

or Troullier–Martins pseudopotentials [35] as well as experimental data taken from the corresponding collection in ref. [15]. There is an excellent agreement with previous theoretical values, in particular for AlN and GaN. InN is a more complicated material for DFT-LDA calculations as will be discussed below. Nevertheless, good agreement can be also stated for this III-nitride comparing computed structural data with experimental values. As a consequence of the typical overbinding in LDA calculations, the lattice constants are about 1% smaller than the measured values. The discrepancies are even smaller in the zinc-blende case. Somewhat surprising is the excellent reproduction of the bulk moduli by the already mentioned first version of the theory using the ultrasoft non-normconserving pseudopotentials, in particular in the wurtzite case. This fact underlines the good quality of elastic constants derived from the calculations [10] described above. The lattice constants and bulk moduli follow clear chemical trends with the size of the group-III atoms (cf. Table 2.1).

These clear trends are somewhat in contrast to the group-III dependence of the internal structural parameter u and the ratio c/a of lattice constants in the hexagonal case. For GaN and InN the calculated ratios approach the ideal value $c/a=1.633$, whereas the AlN value is smaller. In any case $c/a<1.633$ is fulfilled. According to a general rule [36], for such ratios the wurtzite structure should be the stable phase in agreement with the total-energy results in Fig. 2.3. The u values in Table 2.2 follow seemingly the trend with the atomic sizes. However, if BN is also considered [26] the influence of the bond ionicity becomes obvious. The bond ionicities may be related to the differences 1.43, 1.23, and 1.26 of the electronegativities in Table 2.1 or to the charge-asymmetry coefficients $g=0.794$ (AlN), 0.780 (GaN), and 0.853 (InN) [37].

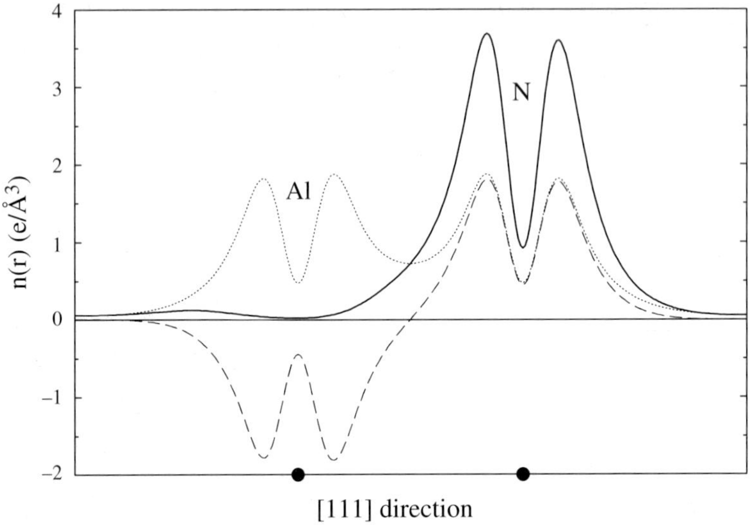

Fig. 2.4 Valence-electron density of cubic AlN along a [111] bond direction (solid line). It is decomposed into a symmetric (dotted) and antisymmetric (dashed) part. (After K. Karch, private communication.)

The latter ones have been computed as square roots of the ratio of the squares of the antisymmetric and symmetric parts of the electron density integrated over volume. These contributions are plotted in Fig. 2.4 for cubic AlN.

2.2.3 Electronic states

Band structures of AlN, GaN, and InN are presented for the two polytypes, 3C and 2H, in Fig. 2.5. They are calculated using the structural parameters in Table 2.2 obtained within the DFT-LDA. Quasiparticle corrections [6, 7, 8] are not taken into account. The smallest energy gaps between the valence-band maximum and the lowest conduction-band minima are listed in Table 2.3. On average the fundamental gaps decrease along the series AlN, GaN, and InN as a consequence of the decreasing covalent bonding and total bond energies. In the InN case the direct gap at Γ is even negative. The sequence of the Γ_{15v}^{3C} and Γ_{1c}^{3C} (Γ_{6v}^{2H} and Γ_{1c}^{2H}) levels is inverted in the zinc-blende (wurtzite) case. This is mainly due to the overestimation of the *pd* repulsion. The negative energy gaps are also observed in several other DFT-LDA calculations [15, 35, 39, 40]. The band structures for GaN and InN also show the flat bands related to the shallow Ga 3*d* and In 4*d* core states. Within the DFT-LDA approach they lie in the same energetical region as the lowest valence bands governed by N 2*s* states. The *d*-bands are below the N 2*s* bands in excitation spectra [41]. The energetical overlap of *s*- and *d*-states gives rise to a hybridization and an unphysical splitting of the lowest valence bands. In wurtzite crystals the conduction-band minima are located at the Γ point in the BZ. In the zinc-blende case the

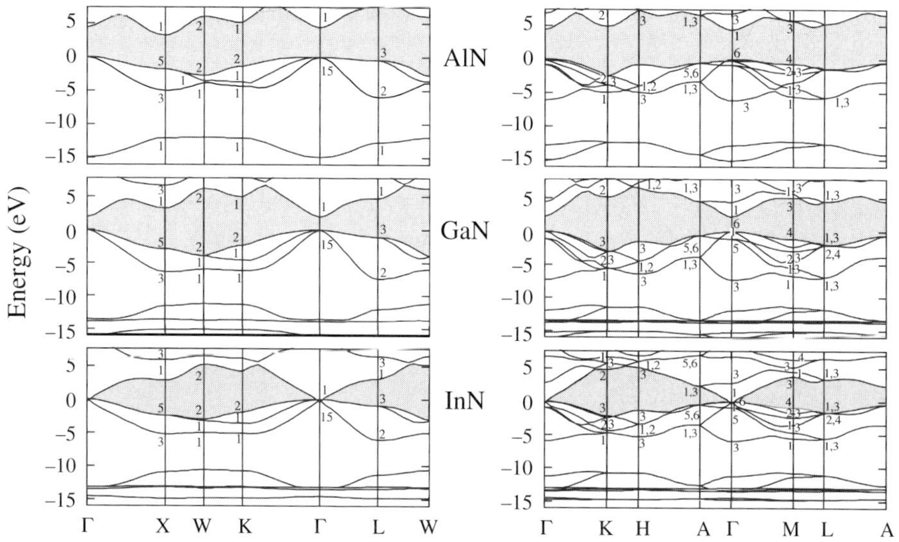

Fig. 2.5 Energy band structures of group-III nitrides crystallized in zinc-blende (left panels) or wurtzite (right panels) structure. The energies are calculated within DFT-LDA. The numbers indicate the irreducible representations at the respective symmetry points. The Γ_{15} (zinc-blende) and Γ_6 or Γ_1 (wurtzite) valence-band maxima are set to zero. The fundamental gap is indicated by the shaded region. (After ref. [38] and U. Grossner, private communication.)

Table 2.3 Fundamental energy gaps derived from Kohn–Sham eigenvalues (in eV). From ref. [38].

Gap	AlN	GaN	InN
3C			
$\Gamma \rightarrow \Gamma$	4.46	1.92	−0.36
$\Gamma \rightarrow X$	3.29	3.24	2.81
$\Gamma \rightarrow K$	5.15	4.97	4.24
$\Gamma \rightarrow L$	7.52	4.76	3.06
2H			
$\Gamma \rightarrow \Gamma$	4.45	2.25	−0.22
$\Gamma \rightarrow K$	4.96	5.18	4.78
$\Gamma \rightarrow \frac{3}{4}\overline{ML}$	4.96	4.21	3.05

band minima of GaN and InN are still at Γ. All these crystals are therefore direct semiconductors. However, cubic AlN becomes an indirect semiconductor with the conduction-band minimum situated at the X point in the fcc BZ. Experimentally, there is a controversy concerning the direct or indirect character of cubic AlN. The energetical position and the decrease of emission intensity is believed to imply the indirect band structure of cubic $Al_xGa_{1-x}N$

in the mid-range of composition x [42] in agreement with the theoretical findings. However, recent cathodoluminescence studies of cubic AlN and $Al_xGa_{1-x}N$ epilayers do not show a significant decrease of the luminescence signal [43]. Hence, a direct character is assumed contrary to the theoretical predictions with 1.17 eV difference in the energies of the conduction-band minima at X and Γ.

Quasiparticle band structures have been calculated for AlN and GaN [44, 45]. They show the characteristic widening of all energy gaps between empty and occupied bands. The quasiparticle corrections vary between 1.2–1.9 eV for AlN and 1.2–1.5 eV for GaN [44]. They give rise to fundamental quasiparticle gaps of 4.7 eV (3C-AlN), 5.8 eV (2H-AlN), 3.1 eV (3C-GaN), and 3.5 eV (2H-GaN) [44]. These values are close to experimental gaps, 6.3 eV (2H-AlN [46]), 3.4 eV (3C-GaN [47]), and 3.5 eV (2H-GaN [48]). The model of Bechstedt and Del Sole [49] also gives reliable values for the quasiparticle corrections of the fundamental gaps, 1.85 eV (3C- and 2H-AlN), 1.61 eV (3C-GaN), 1.63 eV (2H-GaN), 0.88 eV (3C-InN), and 1.03 eV (2H-InN) [50, 51]. Their combination with the DFT-LDA gaps in Table 2.3 results in fundamental gaps of 6.30 (5.14), 3.88 (3.53), and 0.81 (0.52) eV for 2H-(3C-)AlN, -GaN, and -InN. Whereas the 2H-AlN gap is correctly described, an overestimation of the experimental values of about 0.2 eV is observed for both 3C-GaN and 2H-GaN.

The electronic-structure theory seems to fail completely in the case of InN. The experimental value for the optical gap of 2H-InN accepted until now amounts to about 1.9 eV [11]. This is in sharp contrast to the estimated quasi-particle gap of about 0.8 eV. One may speculate that the DFT-LDA calculations underestimate the gap due to the overestimation of the *pd*-repulsion of about 0.4 eV as a consequence of the too shallow In 4*d* bands within the DFT-LDA. Then, the estimated theoretical gap of 2H-InN would be 1.2 eV. Still a discrepancy of about 0.7 eV remains. The experience with quasiparticle calculations of energy gaps of many semiconductors gives no indication for such a huge inaccuracy of the theory [6, 8]. On the other hand, there are also indications for smaller gaps of InN from measurements. O'Donnell [52] measures peak positions of the photoluminescence at about 1.7 eV (i.e. below the 1.9 eV gap) for $In_xGa_{1-x}N$ alloys with an In molar fraction of $x=0.4$ far away from $x=1$. Optical absorption measurements of high-quality samples grown by means of MBE seem to tend to InN energy gaps close to 1 eV, whereas studies of low-quality samples with broad Raman and X-ray diffraction peaks confirm the accepted value close to 2 eV (H. Harima, private communication).

2.2.4 Dielectric constants and lattice vibrations

Mainly due to the lack of sufficiently large high-quality single crystals no studies of the complete phonon branches have been published until the year 2000. Very recently, however, the measurement of the phonon dispersion in GaN became possible by means of inelastic X-ray scattering [53]. Only a variety of

Raman data has been published for the zone-center frequencies of AlN. Until the year 1999 practically no trustworthy phonon frequencies were available for InN. Consequently, in the past there was a real necessity to determine the dielectric and lattice-dynamical properties by the above-mentioned combination of the DFT [1, 2] and a generalized response theory, the DFPT [9]. Dielectric constants, effective ion charges, and phonon dispersion relations have been calculated by Karch *et al.* [30, 31] for AlN and GaN. However, for unknown reasons these authors were unable to describe reasonably the lattice-dynamical properties of InN using normconserving Troullier–Martins pseudopotentials [32]. Only recently this could be done by treating the semi-core In4*d* electrons as valence electrons [54]. Bechstedt, Grossner, and Furthmüller [26, 38] followed another way to obtain the vibrational spectrum of InN. The first version of the total-energy calculations described in Section 2.2.1 has been used to calculate the force constants in an *ab initio* manner as second derivatives of the total energy with respect to atomic displacements within large supercells [26]. Such a total-energy method, as all frozen-phonon methods, does not include the effect of the long-range electric fields accompanying polar vibrations. Nevertheless, it was possible to determine such electric fields, more strictly speaking, the screened Born effective ion charges that generate these macroscopic electric fields, from the asymptotic values of planar force constants far away from the displaced atom.

Several computed values of the high-frequency dielectric constant and of the unscreened as well as the screened Born effective charge are listed in Table 2.4 and compared with experimental values. Whereas the variation of the Born effective charges with the group-III nitride is weak, the electronic dielectric

Table 2.4 High-frequency dielectric constants ε_∞, Born effective charges Z_B, and screened ion charges $Z_B/\sqrt{\varepsilon_\infty}$. In the wurtzite case the two independent tensor components perpendicular ($\perp$) and parallel ($\parallel$) to the *c*-axis are given. The first data set is from refs. [30, 31], the second one for wurtzite crystals from ref. [54]. The experimental data in parenthesis (mostly unspecified with respect to the direction) are taken from the collections in refs. [30, 31, 54]. The screened effective charges of cubic crystals are taken from ref. [26].

Nitride	ε_∞	$\varepsilon_\infty^\perp$	$\varepsilon_\infty^\parallel$	Z_B	$Z_B^\perp$	$Z_B^\parallel$	$Z_B/\sqrt{\varepsilon_\infty}$
AlN	4.46	4.38	4.61	2.56	2.53	2.69	1.16
		5.17	5.36		2.70	2.85	
		(4.84, 4.68)			(2.57)		
GaN	5.41	5.21	5.41	2.65	2.60	2.74	1.11
		6.11	6.38		2.69	2.86	
	(5.2, 5.3, 5.7, 5.35, 5.8)			(2.65, 2.82, 2.74, 3.2)			
InN	–	–	–	–	–	–	1.03
		7.61	7.96		2.78	2.96	
		(8.4)					

constants increase remarkably along the row AlN, GaN, and InN. They follow the trend with the decreasing fundamental energy gaps. As a net effect, also the screened effective charges decrease along AlN, GaN, and InN. However, their variation is more or less restricted to the small interval $Z_B/\sqrt{\varepsilon_\infty} = 1.2 - 1.0$ The transverse Born effective charges of the nitrides are slightly smaller than the corresponding quantities of conventional III–V compounds. However, in the case of the screened dynamic charges this discrepancy increases to about a factor of 2.

The phonon frequencies resulting for the high-symmetry points Γ, *X*, and *L* in the fcc BZ are listed in Table 2.5. Complete dispersion relations are presented for all phonon branches in Fig. 2.6. In Table 2.5 we compare data from the two types of *ab initio* calculations in refs [26] and [30, 31] with results from first-order Raman scattering for cubic epitaxial layers [12, 55, 56, 57]. The agreement of the two *first-principles* calculations is excellent in the case of the acoustic modes. The maximum deviation amounts to 4 cm^{-1}. The deviations in the frequency region of the optical branches are slightly larger. The main reason is related to the screened effective ion charges calculated by completely different approaches within the two methods. In comparison with the full DFPT results [30, 31], the improved frozen phonon calculation [26] underestimates the LO-TO splittings by 20 cm^{-1} (AlN) and 11 cm^{-1} (GaN). The agreement of the frequencies calculated for Γ phonons and the first-order Raman data [12, 55, 56, 57] is also satisfactory. The discrepancy in the case of the TO phonons is a result of the overestimation of the force constants due to

Table 2.5 Phonon frequencies calculated at the high-symmetry points Γ, *X*, and *L* for the zinc-blende polytype of AlN, GaN, and InN from ref. [26]. Frequencies from DFPT calculations [30, 31] (first value) and experimental data (following values) are given in parenthesis. All values are listed in units of cm^{-1}.

Mode	AlN	GaN	InN
Γ_{TO}	665 (662; 655[a])	567 (560; 552[a], 552[b], 555[c])	467 (–; 457[d])
Γ_{LO}	890 (970; 902[a])	746 (750; 743[a], 737[b], 741[c])	596 (–; 588[d])
X_{TA}	341 (340)	197 (195)	116
X_{LA}	588 (590)	351 (353)	231
X_{TO}	668 (674)	623 (628)	518
X_{LO}	716 (734)	695 (709)	567
L_{TA}	226 (230)	138 (139)	78
L_{LA}	585 (582)	349 (345)	227
L_{TO}	655 (656)	587 (585)	488
L_{LO}	735 (750)	708 (720)	573

First-order Raman data:
[a] Harima *et al.* [55]
[b] Siegle *et al.* [56]
[c] Tabata *et al.* [57]
[d] Tabata *et al.* [12]

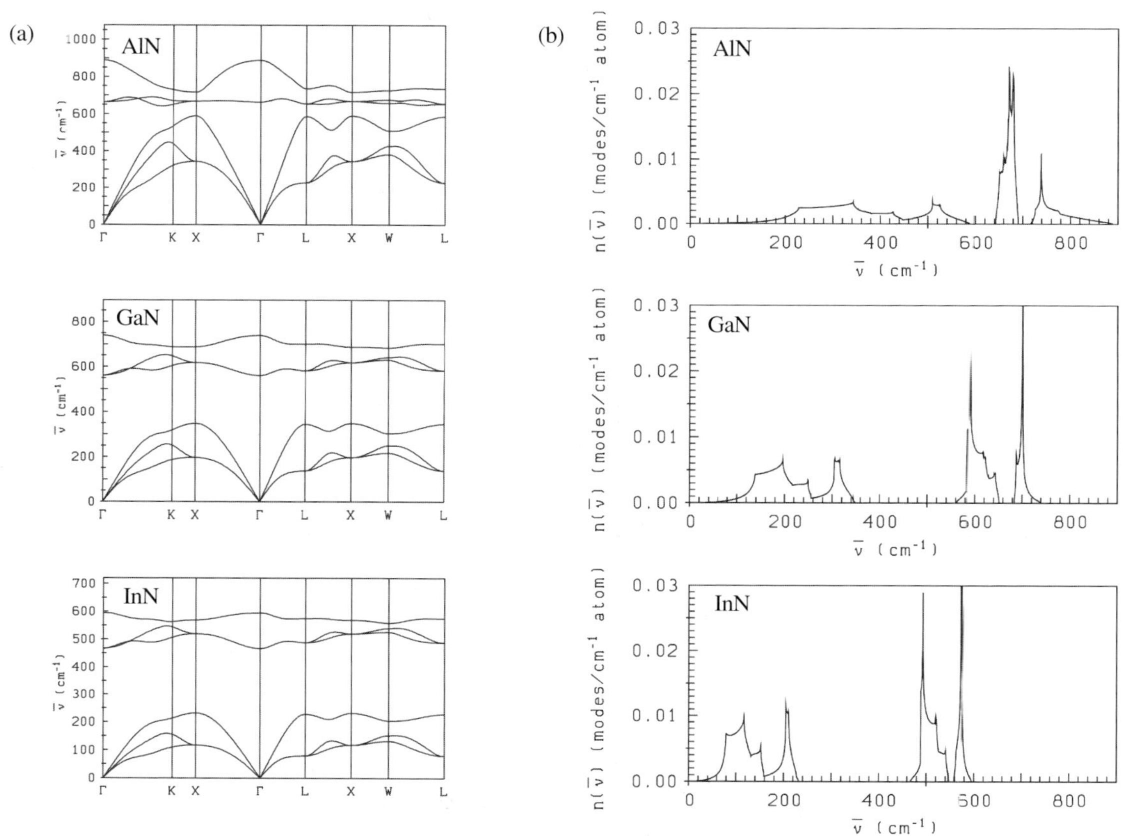

Fig. 2.6 Calculated dispersion of phonon branches (a) and corresponding density of states (b) of the zinc-blende nitrides. (After ref. [26].)

the LDA underestimation of the lattice constants. However, the predictive power of the *ab initio* treatments suggests trusting the results for InN despite the relatively poor material quality of the sample used [12]. The optical phonon frequencies as well as the frequencies of the LA phonons are larger than the corresponding values of conventional III–V semiconductors. This is mainly a consequence of the smaller anion mass, but is also supported by the larger force constants as a consequence of the smaller bond lengths.

The situation is similar for the phonons of the hexagonal polytypes. Table 2.6 summarizes important results for the zone-center phonons obtained by different *ab initio* calculations [26, 30, 31, 54, 58, 59] and compares them with first-order Raman observations [59, 60, 61]. Nine optical branches are expected for an atomic basis of four atoms. For the Γ point (point group symmetry C_{6v}), group theory predicts the modes $A_1+2B_1+E_1$(twofold)$+2E_2$ (twofold), of which the A_1 and E_1 are both Raman-active and infrared-active. The two twofold E_2 modes are only Raman-active, and the B_1 modes are silent. Due to the macroscopic electric field associated with relative atomic displacements, the A_1 and E_1 optical modes are split into LO and TO components. In particular, as a result of the anisotropic nature of the wurtzite structure, these modes exhibit an angular dependence. The mode frequency depends on the angle Θ between the propagation direction **q** and the *c*-axis. Figure 2.7 shows this angular dispersion of the optical modes at the Γ point for GaN. The character of the LO mode varies from A_1 to E_1 with the propagation angle. The twofold degenerate E_1(TO) mode at $\Theta=0$ splits into E_1(TO)

Table 2.6 Zone-center phonon frequencies of wurtzite AlN, GaN, and InN (in units of cm^{-1}). Comparison of results from *ab initio* theories [26, 30, 31, 54, 58, 59] with those of first-order Raman scattering [60, 59, 61]. The data of ref. [60] are derived from low-temperature spectra.

Nitride	Reference	E_2^{low}	B_1^{low}	A_1(TO)	E_1(TO)	E_2^{high}	B_1^{high}	A_1(LO)	E_1(LO)
AlN									
	Theory [30]	238	551	619	677	667	737	893	919
	Theory [54]	233	547	614	666	655	719	875	898
	Theory [58]	247	636	612	679	672	645		
	Exper. [60]	248		614	673	660		894	917
GaN									
	Theory [31]	143	337	541	568	579	720	748	757
	Theory [54]	138	334	550	572	574	690	733	737
	Theory [58]	185	526	544	566	557	584		
	Exper. [60]	144		534	560	569		737	744
InN									
	Theory [26]	78	227	449	475	488	573	594	603
	Theory [54]	83	225	443	467	483	576	586	595
	Theory [59]	104	270	440	472	483	530		
	Exper. [61]	87		447	476	488		586	593
	Exper. [59]			445	472	488		588	

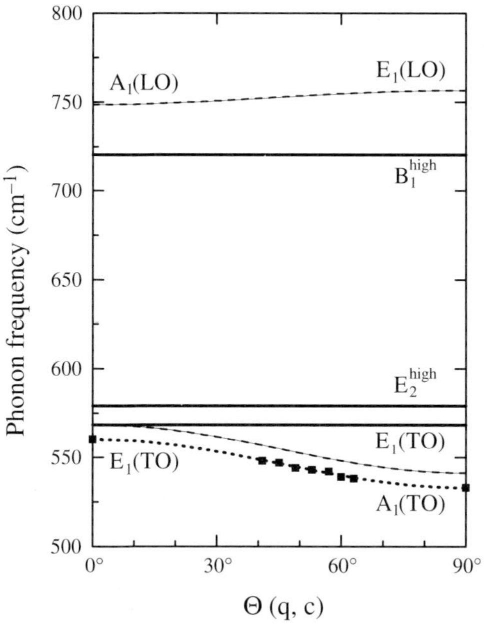

Fig. 2.7 Angular dependence of the calculated optical modes of wurtzite GaN. (After ref. [31].) Experimental results (squares, ref. [62]) are shown for comparison.

and A_1(TO) for propagation perpendicular to the *c*-axis. Thereby, one TO mode frequency remains constant, whereas the other one changes from E_1(TO) to A_1(TO). Such behavior has also been observed experimentally [62].

The overall agreement of the frequencies of the six Raman-active phonons in Table 2.6 obtained from *ab initio* theories [26, 30, 31, 54] and the experimental data [60, 59, 61] is excellent. Apart from the lower E_2 mode the deviations in the measured and calculated frequencies are not bigger than 1%. Most surprising is the good agreement in the InN case. It seems that the puzzle of the Raman frequencies of InN has now finally been solved [26, 54, 59, 61]. The improvement of the sample quality and the application of the accurate parameter-free methods have driven this progress.

2.3 Strain effects

2.3.1 Macroscopic description

An important problem in growing group-III nitride layers on crystalline substrates as sapphire, SiC, ZnO, Si, or GaAs is related to the large lattice mismatch and the difference in the thermal expansion coefficients between epitaxial layers and substrates. Lattice and thermal misfits can cause large biaxial stresses in the epitaxial layers. For instance, a compressive stress results for hexagonal GaN grown on sapphire whereas a tensile one is observed for GaN on 6H-SiC [63]. The situation is more complex in the case of heterostructures or superlattices, e.g. based on the AlN/GaN material combination, for which a mutual influence of the different material layers may occur [64, 65].

In hexagonal crystals biaxial strains (or stresses) in the *xy*-plane perpendicular to the *c*-axis belong to the same class of perturbations as uniaxial ones along the *c*-axis and hydrostatic pressure. All these perturbations conserve the C_{6v}^4 space-group symmetry. The influence of a biaxial strain in the plane perpendicular to the *c*-axis (i.e. the *z*-axis) of the wurtzite structure can be modeled by fixing the in-plane lattice constant *a* at a value different from the equilibrium one, a_{eq}. Then, the diagonal components of the strain tensor (labeled with respect to the *c*-axis) are

$$\epsilon_\perp = \epsilon_{xx} = \epsilon_{yy} = (a - a_{eq})/a_{eq},$$

$$\epsilon_\| = \epsilon_{zz} = (c - c_{eq})/c_{eq}. \tag{2.1}$$

The actual lattice constant *c* follows from the boundary conditions for the stress tensor. Within the framework of *ab initio* calculations it is determined self-consistently by minimizing again the total energy with respect to *c* and *u* but for a fixed value *a*. This procedure simultaneously gives the internal strain, defined as $(u - u_{eq})/u_{eq}$.

The diagonal elements of the strain tensor in eqn (2.1) are related to the diagonal elements $\sigma_\perp = \sigma_{xx} = \sigma_{yy}$ and $\sigma_\| = \sigma_{zz}$ of the stress tensor by the relations [66]

$$\begin{aligned} \sigma_{xx} &= (C_{11} + C_{12})\epsilon_{xx} + C_{13}\epsilon_{zz} - e_{31}E_z, \\ \sigma_{zz} &= 2C_{13}\epsilon_{xx} + C_{33}\epsilon_{zz} - e_{33}E_z, \end{aligned} \tag{2.2}$$

with C_{ij} as the elastic stiffness constants of the group-III nitrides crystallized in wurtzite structure [10]. A collection of the piezoelectric constants e_{ij} can be found in ref. [67]. The built-in electric field is assumed to have only a component E_z parallel to the *c*-axis. This corresponds to the assumption of a biaxial strain in the plane perpendicular to the *c*-axis and the consideration of a uniaxial crystal. The first two terms in expression (2.2) represent the usual relationships of elasticity theory according to Hooke's law. The third term is included to account for the piezoelectric and pyroelectric effects.

The electric field inside a wurtzite layer is governed by the polarization fields in the same and the neighboring layers. The total polarization field in one layer is given by [68]

$$P_z = P_{Tx} + 2e_{31}\epsilon_{xx} + e_{33}\epsilon_{zz} - \varepsilon_0(\varepsilon_\infty^\| - 1)/E_z. \tag{2.3}$$

The first term P_{Tz} on the right-hand side describes the spontaneous polarization field existing for vanishing strain and electric fields. The sum of the second and third terms gives the piezoelectric polarization field. The last contribution represents the electronic screening of the polarization fields. It is therefore governed by the high-frequency dielectric constant in z-direction. The dielectric displacement field is related to the polarization and electric fields by

$$D_z = \varepsilon_0 E_z + P_z. \tag{2.4}$$

The displacement field fulfils the continuity condition at the interfaces. A similar condition holds for the stress field parallel to the c-axis. For finite systems without external charges, the boundary conditions read as

$$\begin{aligned} D_z &= 0, \\ \sigma_{zz} &= 0. \end{aligned} \tag{2.5}$$

For an individual layer only a screened polarization field

$$P_z = \frac{1}{\varepsilon_\infty^{\|}}(P_{Tz} + 2e_{31}\epsilon_{xx} + e_{33}\epsilon_{zz}) \tag{2.6}$$

occurs. The components of the strain tensor are related by

$$\epsilon_{zz} = -2\frac{C_{13} + (e_{33}e_{31}/\varepsilon_0\varepsilon_\infty^{\|})}{C_{33} + (e_{33}^2/\varepsilon_0\varepsilon_\infty^{\|})}\epsilon_{xx} - \frac{1}{C_{33} + (e_{33}^2/\varepsilon_0\varepsilon_\infty^{\|})}\frac{e_{33}}{\varepsilon_0\varepsilon_\infty^{\|}}P_{Tz}. \tag{2.7}$$

The modification of the relation $\epsilon_{zz} = -2(C_{13}/C_{33})\epsilon_{xx}$ by the piezoelectric effect is small. Taking into account typical piezoelectric constants and dielectric constants [68] one finds $e_{33}^2/\varepsilon_0\varepsilon_\infty^{\|} = 52$ (AlN) and 11 (GaN) GPa. The piezoelectric contribution modifies the elastic constant C_{33} by about 14 (AlN) or 3 (GaN)%. However, in the ratio (2.7) the residual effect of the piezoelectric polarization is much smaller. Hence, we omit this effect in the following discussion. Meanwhile, there are studies which show that also the pyroelectric effect $\sim P_{Tz}$ is weak [24]. Only in certain cases the polarization influence should become measurable.

2.3.2 Strained geometries and dielectric properties

Results of total energy optimizations using normconserving pseudopotentials (second version in Section 2.2.1) for structural parameters of GaN and AlN [69] are represented in Fig. 2.8 versus given biaxial strain. As a consequence of the Poisson effect one observes a more or less linear decrease of the lattice constant c with increasing tensile biaxial strain. The behavior of GaN and AlN is rather similar. Surprisingly the effect of a biaxial strain on the internal-cell

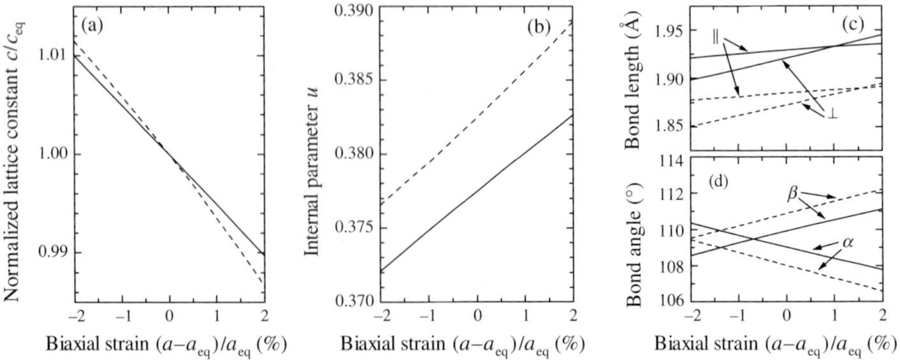

Fig. 2.8 Parameters of the atomic geometry of the wurtzite structure versus biaxial strain. GaN: solid line, AlN: dashed line. The lattice constant c (a), the internal-cell parameter u (b), the bond lengths (c), and the bond angles (d) are plotted. The bonds along the c-axis (forming angles of about 70° with the c-axis) are labeled by || (⊥). The label α (β) denotes the angle between the c-axis and the three bonds nearly perpendicular to it (between two bonds nearly perpendicular to the c-axis). (After ref. [69].)

parameter u and, hence, on the internal strain $(u-u_{eq})/u_{eq}$ is completely different. The u parameter increases with rising tensile biaxial strain.

The rather strong internal-strain effects dominate the variations of the bond lengths and bond angles [Figs 2.8(c) and (d)]. The length uc of the bonds parallel to the c-axis increases only slightly as a consequence of opposite variations of u and c. The length $\sqrt{(a^2/3) + (\frac{1}{2} - u)^2 c^2}$ of the bonds non-parallel to the c-axis increases much more strongly with a tensile biaxial strain. Both strain and internal strain give rise to a stretching of these bonds. The two different types of bond angles, α and β, behave differently versus biaxial strain. There is a tendency for the reduction of the deformation of the bonding tetrahedra towards compressive biaxial strain. The angles approach the value 109.47° of the ideal tetrahedra. In the opposite direction, i.e. for tensile biaxial strain, the bonding zig-zag chains in planes parallel to the c-axis are flattened. The bonding tetrahedra are compressed along the c-axis by shrinking the vertical distance $(\frac{1}{2} - u)$ within an X–N bilayer (X=Ga, Al) towards a planar structure. The angle α tends towards 90°, whereas the angles β tend towards 120°. This implies a tendency for dehybridization from ideal sp^3 hybrids towards sp^2 and p_z orbitals.

The Born effective charges as well as the high-frequency dielectric constants in Fig. 2.9 behave similarly (oppositely) versus biaxial and internal strain for the parallel (perpendicular) tensor components. This is understandable in the light of the discussion of the geometrical parameters. The strong dependence of the dielectric properties on the internal strain makes obvious, that beyond the macroscopic elasticity theory, also the changes on an atomic scale within the unit cell are important. Such changes are automatically included in the results of total-energy optimizations. All quantities considered in Fig. 2.9 increase with rising tensile biaxial strain. However, the parallel components of

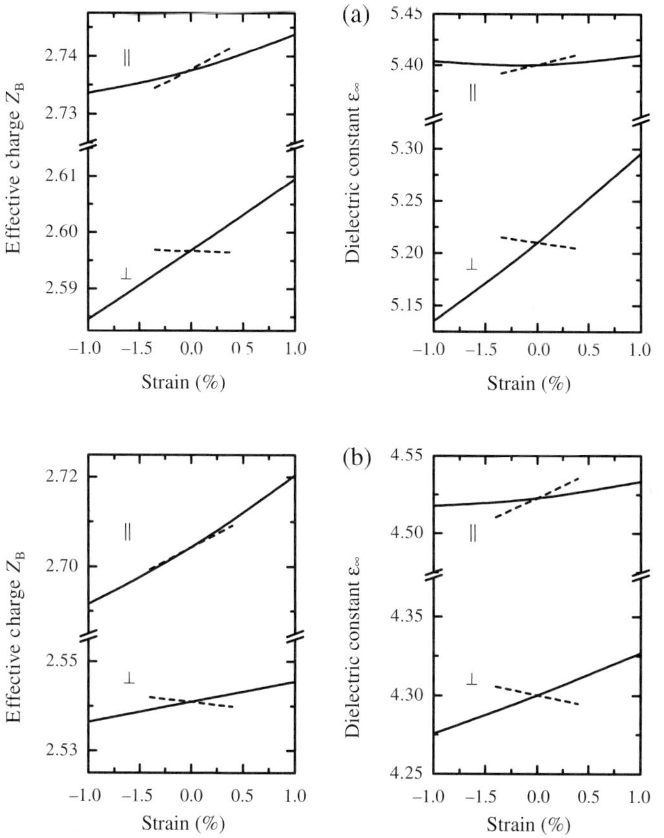

Fig. 2.9 Tensor components of the Born effective charge Z_B and the electronic dielectric constant ε_∞ versus biaxial strain (solid lines). In addition the dependence on a corresponding internal strain $(u - u_{eq})/u_{eq}$ is also plotted (dashed lines). (After ref. [70].)

the dielectric constants behave rather nonlinear. In the case of the dielectric constants, the general increase with the biaxial strain follows the dependence of the bond lengths in Fig. 2.8, since the electronic polarizability is nearly proportional to an average bond length [71]. The Born effective charge can approximately be described by the relation $Z_B = -1 + 4g + \frac{8}{3}g(1 - g^2)$ [71] with g as the charge-asymmetry coefficient introduced in Section 2.2.2 as a measure of the bond ionicity. The ionicity of the III–nitrides increases with increasing bond length [72]. This explains the increase of Z_B with increasing tensile biaxial strain in Fig. 2.9.

2.3.3 Deformation potentials of electrons and phonons

No splittings of the energy bands of the wurtzite nitrides occur in the presence of a strain perturbation that conserves the C_{6v}^4 space-group symmetry.

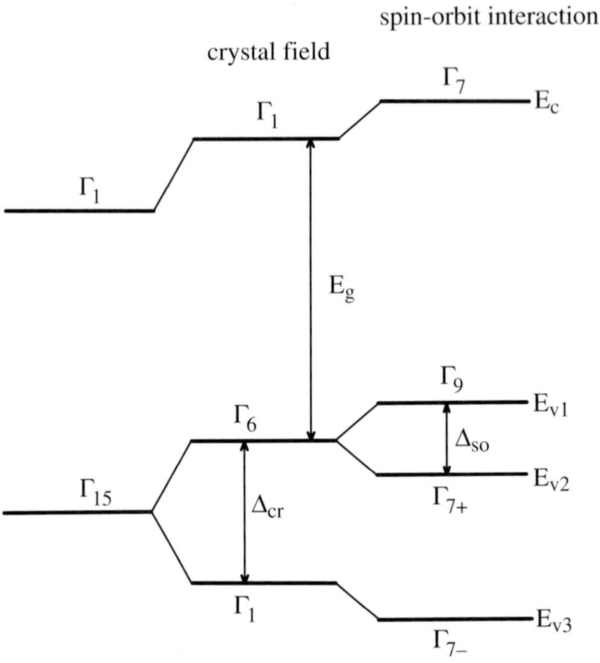

Fig. 2.10 Electronic energy bands (schematic) around the fundamental gap at the Γ point. The gap energy and the splittings of the top of the valence bands by the spin-orbit interaction and the crystal field are indicated.

Qualitatively the band structure is conserved as schematically indicated in Fig. 2.10 for the bands at the Γ point around the fundamental gap E_g. The splittings Δ_{so} and Δ_{cr} of the valence bands due to the spin-orbit interaction and the crystal field are indicated. They give rise to the heavy-hole band E_{v1} (Γ_9), the light-hole band E_{v2} (Γ_{7+}), and the crystal-field split-off hole band E_{v3} (Γ_{7-}). Figure 2.10 shows the situation of GaN with a positive crystal-field splitting and a small but positive spin-orbit parameter. The situation may be reversed for the other nitrides. In the limit of vanishing spin-orbit splitting only the Γ_6 and Γ_1 levels occur instead of Γ_9, Γ_{7+}, and Γ_{7-}. For instance, in the case of AlN the Γ_1 level is higher in energy than the Γ_6 level due to a negative crystal-field splitting [44, 73]. The *ab initio* methods described in Section 2.2.1 give Δ_{cr}=−210 (−244) meV (AlN), Δ_{cr}=30 (9) meV (GaN), and Δ_{cr}=15 (−) meV (InN) within the first (second) version. Characteristic band-structure parameters derived by another group are Δ_{cr}=36 meV (GaN) or −215 meV (AlN) and Δ_{so}=15 meV (GaN) or 20.4 meV (AlN) [74]. However, the values for GaN and InN depend remarkably on the inclusion of *d* electrons or not. Despite of the large atomic parameters in Table 2.1, there is a further reduction of the spin-orbit splitting Δ_{so} of the VBM. The fact that Δ_{so}=12.6 meV for InN is smaller than for either GaN (Δ_{so}=18.5 meV) or AlN (Δ_{so}=20 meV) [19] suggests a large negative contribution to this splitting due to the effect of the In4*d* electrons. In the case of a biaxial strain in the plane perpendicular to the *c*-axis the levels can be described by [75]

$$E_c = E_g + \Delta_{cr} + \frac{1}{3}\Delta_{so} + P_\epsilon,$$

$$E_{v1} = \Delta_{cr} + \frac{1}{3}\Delta_{so} + \Theta_\epsilon + \lambda_\epsilon, \tag{2.8}$$

$$E_{v2} = \frac{1}{2}\left(\Delta_{cr} - \frac{1}{3}\Delta_{so} + \Theta_\epsilon\right) + \lambda_\epsilon + \sqrt{\frac{1}{4}\left(\Delta_{cr} - \frac{1}{3}\Delta_{so} + \Theta_\epsilon\right)^2 + \frac{2}{9}\Delta_{so}^2},$$

$$E_{v3} = \frac{1}{2}\left(\Delta_{cr} - \frac{1}{3}\Delta_{so} + \Theta_\epsilon\right) + \lambda_\epsilon - \sqrt{\frac{1}{4}\left(\Delta_{cr} - \frac{1}{3}\Delta_{so} + \Theta_\epsilon\right)^2 + \frac{2}{9}\Delta_{so}^2},$$

with the hydrostatic energy shifts of the conduction band and the valence bands

$$P_\epsilon = a_{cz}\epsilon_{zz} + 2a_{cx}\epsilon_{xx},$$

$$\lambda_\epsilon = D_1\epsilon_{zz} + 2D_2\epsilon_{xx} \tag{2.9}$$

as well as the additional splitting of the valence bands

$$\Theta_\epsilon = D_3\epsilon_{zz} + 2D_4\epsilon_{xx}. \tag{2.10}$$

The character of the last quantity becomes evident in the quasicubic approximation, $D_1 - D_2 = -D_3 \approx 2D_4$. The quantities a_{cz}, a_{cx}, D_1, D_2, D_3 and D_4 are the deformation potential constants of the electron and hole bands.

There are also *first-principles* calculations of the deformation potentials in the wurtzite case [58]. Within the quasicubic approximation and neglecting the spin-orbit splitting the ($\Gamma_{1c}-\Gamma_{6v}$) transition, i.e. the (E_c-E_{v1}) gap and, hence, the ($\Gamma_{1c}-\Gamma_{15v}$) transition in the cubic limit, is shifted according to

$$\Delta E_g = a_V(2\epsilon_{xx} + \epsilon_{zz}) + \frac{d_5}{\sqrt{24}}(\epsilon_{xx} - \epsilon_{zz}). \tag{2.11}$$

For small uniform changes of the crystal volume V the variation of the fundamental transition energy E_g is defined by the volume deformation potential $dE_g/d\ln V|_{V=V_{eq}}$. It can be related to the deformation potentials in expression (2.8) by $a_V=(a_{cz}+2a_{cx})-(D_1+2D_2)$. Using the bulk modulus B (Table 2.2) the volume deformation potential is directly related to the linear pressure coefficient by $dE_g/dp|_{p=0} = -a_V/B$. The shear potential is given by

$d_5 = -\sqrt{(8/3)}[2(a_{cz} - a_{cx}) + (D_1 - D_2)]$. Only for zinc-blende nitrides is an extended study of the deformation potentials of the electronic bands available [40]. Characteristic values of the shear deformation potentials are d_5=−7.5 (AlN), −5.3 (GaN), and −3.8 (InN) eV [40].

Results of parameter-free calculations [26, 38, 40, 76] are listed in Table 2.7 for the volume deformation potentials and compared with available experimental data [77, 78]. The volume or pressure influence on the fundamental gaps depends rather weakly on the polytype. This may be explained by the similar orbital character of the band extrema participating in the corresponding optical transitions. Only in the case of AlN the deformation potentials differ due to the direct (indirect) character in wurtzite (zinc-blende) structure. In the case of GaN the calculation and measurements [26, 38, 40, 77] confirm that the pressure coefficient of the gap is slightly larger for the wurtzite type by about 1 meV/GPa. The total variation of the volume deformation potentials with the group-III cation is much stronger than the variation of this quantity among the other III–V compounds [79]. No clear chemical trend can be observed. Again, there are two opposite tendencies influencing the pressure coefficients. On the one hand, $dE_g/dp|_{p=0}$ decreases with the lattice constant. This explains the much smaller values of the nitrides in comparison to conventional III–V compounds. On the other hand, the bond ionicity acts in the opposite direction. As a result the coefficients exhibit a non-monotonous behavior with a maximum value for GaN.

The Raman frequencies of the strained wurtzite crystals are shifted or splitted with respect to the strain-free values. In the linear strain limit these shifts and splittings are related to the strain tensor by [80]

$$\begin{aligned}\Delta\omega(A_1) &= a(A_1)(\epsilon_{xx} + \epsilon_{yy}) + b(A_1)\epsilon_{zz},\\ \Delta\omega(E_{1/2}) &= a(E_{1/2})(\epsilon_{xx} + \epsilon_{yy}) + b(E_{1,2})\epsilon_{zz}\\ &\quad \pm c(E_{1/2})[(\epsilon_{xx} - \epsilon_{yy})^2 + 4\epsilon_{xy}^2]^{1/2}\end{aligned} \tag{2.12}$$

in dependence on the mode symmetry $j=A_1$ or $E_{1/2}$. The non-Raman-active B_1 modes follow a similar relation as the A_1 modes. The coefficients $a(j)$, $b(j)$, and

Table 2.7 Volume deformation potential $a_V = dE_g/d\ln V|_{V=V_{eq}}$ (in eV) and linear pressure coefficient $dE_g/dp|_{p=0}$ (in meV/GPa) of the fundamental gap. First value: zinc-blende, second value: wurtzite.

Coefficient	Reference	AlN		GaN		InN	
$-a_V$	calc. [26, 38]	–	9.3	10.2	11.1	3.2	2.2
	calc. [40]	0.4	9.0	6.4	6.9	3.0	2.8
	calc. [76]	0.4	8.8	7.4	7.8	2.2	4.1
$dE_g/dp\|_{p=0}$	calc. [26, 38]	–	46	56	57	24	17
	calc. [40]	2	44	32	33	22	19
	exper. [77]			43	44		
	exper. [78]					19	31

$c(j)$ are the corresponding phonon deformation potentials per unit strain. The combinations $[2a(j)+b(j)]$ and $[a(j)-b(j)]$ directly give the effects of hydrostatic strain or axial strain parallel to the c-axis. The effect of a trigonal strain is characterized by the deformation potential $c(j)$. In the case of a biaxial strain, $\epsilon_{xx}=\epsilon_{yy}$ and $\epsilon_{xy}=0$, eqn (2.12) changes over into

$$\Delta\omega(j) = 2a(j)\epsilon_{xx} + b(j)\epsilon_{zz}. \tag{2.13}$$

Using Hooke's law (2.2), i.e. neglecting the effect of the polarization fields, expression (2.13) can be converted into

$$\Delta\omega(j) = 2\tilde{a}(j)\sigma_{xx} + \tilde{b}(j)\sigma_{zz}. \tag{2.14}$$

Instead of the deformation potentials, $a(j)$ and $b(j)$, for a given unit strain, now the constants $\tilde{a}(j)$ and $\tilde{b}(j)$ for a given stress appear.

The dependence of the zone-center phonon frequencies on biaxial strain is represented in Fig. 2.11 for AlN and GaN. It is similar to that in the case of a negative hydrostatic pressure [72]. Apart from the lower E_2 mode, all other modes are softened with increasing biaxial strain. In contrast to the case of pure internal strain, this happens for all mode symmetries. The decrease versus biaxial strain in Fig. 2.11 is rather similar for the majority of mode frequencies. The opposite behavior of the E_2 modes is correlated with their mode softening in the presence of hydrostatic pressure [72]. It is interpreted to indicate the high-pressure phase transition of the group-III nitrides.

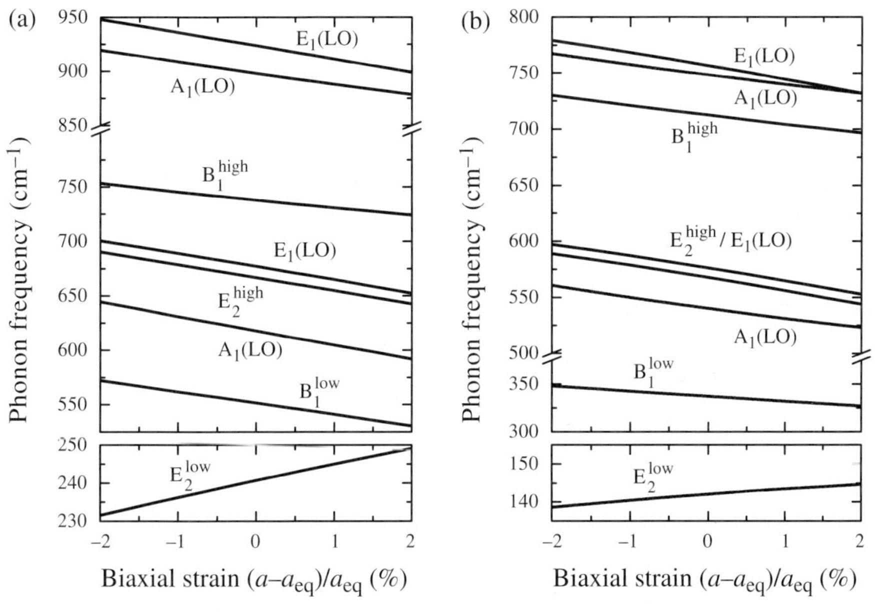

Fig. 2.11 Zone-center phonon frequencies versus biaxial strain for AlN (a) and GaN (b). (After ref. [70].)

The nearly linear slopes in Fig. 2.11 can be used to determine deformation potentials for given biaxial strain. The combination of the results for biaxial strain with the mode dependences on uniaxial strain allow the determination of both deformation potentials $a(j)$ and $b(j)$. Using calculated elastic stiffness constants [69] the deformation potentials $\tilde{a}(j)$ and $\tilde{b}(j)$ can also be derived [81]. The resulting phonon deformation potentials are listed in Tables 2.8 and 2.9 and, in the GaN case, are compared with values measured recently [82, 83]. In the case of the $a(j)$ and $b(j)$ deformation potentials, the agreement with the measured values is excellent, except for the lower E_2 mode where the strain influence is weak and the error bars become large. The agreement for GaN in the case of $\tilde{a}(j)$ and $\tilde{b}(j)$ is less satisfying. This is mainly a consequence of the elastic stiffness constants used by the experimentalists different from those used in the calculations. In general, the discrepancies between the

Table 2.8 Calculated phonon deformation potentials $a(j)$ and $b(j)$ (in cm^{-1}) as well as $\tilde{a}(j)$ and $\tilde{b}(j)$ (in cm^{-1}/GPa) for GaN [81]. They are compared with experimental data [82, 83].

Def.pot.	E_2^{low}	A_1(TO)	E_1(TO)	E_2^{high}	A_1(LO)	E_1(LO)
$a(j)$						
Calc. [81]	77	−639	−718	−740	−663	−776
Expt. [82]				−818	−685	
Expt. [83]	115	−630	−820	−850		
$b(j)$						
Calc. [81]	7	−690	−592	−727	−877	−704
Expt. [82]				−797	−997	
Expt. [83]	−80	−1290	−680	−920		
$\tilde{a}(j)$						
Calc. [81]	0.17	−1.01	−1.23	−1.20	−0.96	−1.29
Expt. [82]				−1.45	−0.4	
Expt. [83]	0.25	−0.7	−1.4	−1.35		
$\tilde{b}(j)$						
Calc. [81]	−0.06	−1.19	−0.80	−1.12	−1.66	−1.4
Expt. [82]				−1.27	−3.26	
Expt. [83]	−0.25	−2.66	−0.88	−1.47		

Table 2.9 Calculated phonon deformation potentials $a(j)$ and $b(j)$ (in cm^{-1}) as well as $\tilde{a}(j)$ and $\tilde{b}(j)$ (in cm^{-1}/GPa) for AlN.

Def.pot.	E_2^{low}	A_1(TO)	E_1(TO)	E_2^{high}	A_1(LO)	E_1(LO)
$a(j)$	150	−784	−832	−877	−743	−864
$b(j)$	−227	−390	−746	−911	−735	−809
$\tilde{a}(j)$	0.47	−1.42	−1.29	−1.28	−1.11	−1.32
$\tilde{b}(j)$	−0.89	−0.2	−1.21	−1.65	−1.35	−1.37

experimental and the theoretical values may also be due to the second set of experimental results needed for the calculation of the deformation potentials, which is taken from independent measurements. The deformation potential constants of the optical phonons in Tables 2.8 and 2.9 do not vary remarkably with the mode index. On average the deformation potentials are larger for AlN than for GaN. This is a consequence of the stronger covalent bonding in this nitride. In any case, it is inappropriate to use the deformation potential constants obtained for GaN also in the AlN case. Another important fact for AlN is the violation of the relation $C_{11}+C_{12}=C_{13}+C_{33}$ between the elastic stiffness constants, which is nearly fulfilled for GaN. As a consequence the deformation potentials cannot be directly related to the mode Grüneisen parameters by the relation $\gamma(j)=-[2a(j)+b(j)]/[3\omega(j)]$ as in the GaN case. Using the values from Tables 2.6 and 2.8, the Grüneisen parameters of GaN are underestimated by about 15% in comparison with other calculated values [40]. This is in contrast to conventional III–V semiconductors for which the DFPT gives excellent Grüneisen parameters.

2.4 Alloys

2.4.1 Thermodynamics

In order to investigate the stability and the properties of unstrained and strained pseudobinary $X_xGa_{1-x}N$ alloys (X=In, Al), one has to calculate the Helmholtz free energy $F(x, T)$ as a function of the X molar fraction and the absolute temperature. Since wurtzite and zinc-blende mixed crystals represent lattices of nearly tetrahedrally coordinated atoms, their thermodynamical behavior should be similar. Moreover, an applied biaxial strain, at least in a plane perpendicular to a cubic axis, can be simulated easily. For that reason, the calculations discussed in the following are restricted to cubic mixed crystals. The N atoms occupy one fcc sublattice, whereas the X and Ga atoms are randomly distributed over the other fcc sublattice.

The mixing free energy of a random alloy, $\Delta F(x, T)=\Delta U(x, T)-T\Delta S(x, T)$, can be described within a special cluster expansion approach [84, 85], the so-called generalized quasichemical approximation (GQCA) [86]. In this approximation the considered mixed crystal $X_xGa_{1-x}N$ is divided into an ensemble of clusters, each of which is taken to be independent—statistically and energetically—of the surrounding atomic configuration. As a random alloy the material is assumed to be spatially homogeneous everywhere on a macroscopic (or at least mesoscopic) length scale. On the cation sublattice one considers N sites on which the X=In, Al and Ga atoms take up some configuration. The numbers of X and Ga atoms are

$$\begin{aligned} N_X &= xN, \\ N_{Ga} &= (1-x)N, \\ N &= N_X + N_{Ga}. \end{aligned} \tag{2.15}$$

The system is first composed of M $2n$-atom clusters, ($nM = N$), which are treated independently. For a given alloy configuration the clusters are arranged in ($J+1$) classes with distinct total energies of the clusters, E_j with $j=0,1,\ldots,J$), which can be calculated within the framework of the DFT-LDA (first version, cf. sub-section 2.2.1) [87]. In each class there are M_j clusters with the same energy E_j. It holds $M = \sum_j M_j$. Introducing the fraction $x_j = M_j/M$ of clusters of energy E_j, the last equation can be written as

$$\sum_{j=0}^{J} x_j = 1. \tag{2.16}$$

Besides the energy the clusters differ with respect to the number n_j of X atoms and the number $(n-n_j)$ of Ga atoms, respectively. These numbers are subject to a constraint due to the given averaged composition x, i.e. the X molar fraction, as

$$\sum_{j=0}^{J} n_j x_j = nx. \tag{2.17}$$

According to the Connolly–Williams method [88], the constraints (2.16) and (2.17) can be interpreted as special cases of the representation of the configurationally averaged and hence composition-dependent (and in general also temperature-dependent) value

$$P(x,T) = \sum_{j=0}^{J} x_j P_j \tag{2.18}$$

of a property P of interest of the pseudobinary alloy with mainly compositional disorder. The considered property is different for each cluster and depends on the cluster class index j according to P_j. In the case of eqn (2.17) it holds $P_j = n_j/n$ and $P(x,T) = x$. The definition of the configurational average (2.18) allows the introduction of fluctuations around the mean values by considering the root-mean-square (rms) deviations

$$\Delta P(x,T) = \left\{ \sum_{j=0}^{J} x_j P_j^2 - \left[\sum_{j=0}^{J} x_j P_j \right]^2 \right\}^{1/2}. \tag{2.19}$$

The definition (2.18) is also applicable to the total energy. For the alloy configured with the set of clusters $M_0, M_1, \ldots, M_J$ the mixing internal energy divided by the number M of clusters reads as

$$\Delta U(x,T) = \sum_{j=0}^{J} x_j E_j - [(1-x)E_0 + xE_J]. \tag{2.20}$$

The configurational or mixing entropy can be calculated from the Boltzmann definition. $\Delta S(x, T) = k_B \ln W$, where W is the number of ways to configure the alloy with the set $M_0, M_1, \ldots, M_J$ of clusters [86]. This number W of configurations is given by the total number of ways $N!/(N_X! N_{Ga}!)$, in which N_{Ga} Ga atoms and N_X X atoms can be placed on N sites. First, this number has to be multiplied with the probability $M!/\prod_{j=0}^{J} M_j!$, i.e. the fraction of ways to arrange the different clusters $M_0, M_1, \ldots, M_J$. Second, with the assumption that the clusters are independent, one has to multiply with the joint probability of finding the set of clusters $M_0, M_1, \ldots, M_J$ in the alloy, $\prod_{j=0}^{J} (x_j^0)^{M_j}$, with

$$x_j^0 = g_j x^{n_j} (1 - x)^{n - n_j} \tag{2.21}$$

the fraction of clusters of type j for a given composition x in a regular solid solution. The degeneracy factor g_j is the number of ways of arranging the alloying cations in a cluster with energy E_j. Summarizing, with the expressions (2.15), the fraction $x_j = M_j/M$, and eqn (2.21) the mixing entropy divided by the number of clusters M follows as

$$\Delta S(x, T) = -k_{\rm B} \left\{ n [x \ln x + (1 - x) \ln (1 - x)] + \sum_{j=0}^{J} x_j \ln \left(\frac{x_j}{x_j^0} \right) \right\}. \tag{2.22}$$

We mention that for $x_j = x_j^0$ the GQCA expression (2.22) changes over into the well-known mixing entropy of an ideal solid solution [89].

The cluster fractions x_j are unknown. However, they can be interpreted as variational parameters. Then, by minimizing the free energy one finds $(\beta = 1/k_B T)$

$$x_j = \frac{g_j \eta^j e^{-\beta E_j}}{\sum_{j=0}^{J} g_j \eta^j e^{-\beta E_j}}. \tag{2.23}$$

The unknown parameter η has to be determined from the constraint (2.17). In the high-temperature limit $\beta \to 0$, it holds $\eta = x^{n_j}(1 - x)^{n - n_j}$ and the fractions x_j are equal to the probabilities of a random cluster distribution (2.21).

Within the GQCA the clusters with the energy E_j used in the calculation of the internal energy should have a reasonable size. The smallest clusters that consider local correlation are 16-bond clusters with a central N atom and four alloying atoms bound to the environment by 12 second-neighbor N atoms [89]. This corresponds to $n = 4$. There is another advantage using such 16-bond or 8-atom clusters. Since no two clusters share the same alloying atoms, in other words, since the cluster surfaces are chemically ordered, the clusters are statistically independent. In the absence of external stresses, there are five kinds $j = 0, \ldots, 4$ $(J = 4)$ of clusters distinguished by distinct cluster energies E_j. The degeneracy factors are $g_j = \binom{4}{j} = 4!/j!(4 - j)!$, i.e. $g_0 = 1$, $g_1 = 4$, $g_2 = 6$, $g_3 = 4$, and $g_4 = 1$. Each cluster occupies one cube in the crystal.

The atomic structure of each cluster is optimized for a given lattice constant a_j. The equilibrium lattice constants follow from the minimum cluster energies $E_j = E_j(a_j)$. In the presence of a biaxial strain parallel to a cubic axis the cluster energies are minimized with respect to two lattice constants, $a_{j\|}$ and $a_{j\perp}$. Pseudomorphic growth of the alloy is considered. The biaxial strain is modeled by fixing the in-plane lattice constant $a_{j\|} = a_{\|}$. This gives rise to an inhomogeneous strain distribution $\epsilon_{j\|} = (a_{\|} - a_j)/a_j$ over the clusters forming the mixed crystal. The strain varies locally on the characteristic length scale of the cluster sizes.

2.4.2 Miscibility and strain

In general, the cluster energies E_j do not depend linearly on j and, hence, give rise to an asymmetric dependence of the mixing internal energy and the mixing enthalpy. They cannot be described anymore by an expression such as $\Delta U = \Omega x(1-x)$ for the mixing enthalpy, as in the limit of the ideal solution model [90, 91]. Rather a composition-dependent interaction parameter Ω has to be considered, influencing the miscibility gap [87]. Mixing free energies of $In_xGa_{1-x}N$ resulting for different growth temperatures are plotted in Fig. 2.12 together with the accompanying temperature-composition phase diagram. The shapes of the ΔF curves versus the In molar fraction are asymmetric. The behavior of the excess free energy below the critical temperature at $T = 1295$ K in Fig. 2.12 indicates that there is a miscibility gap for x in the interval $x_1 < x < x_2$. Here x_1 and x_2 are the points at which the common tangent line touches the ΔF curve. There are two other special composition values, x_1' and x_2', called the spinodal points (the inflection points of ΔF). For $x_1 < x < x_1'$ or $x_2' < x < x_2$, the alloy is metastable against local decomposition, whereas for $x_1' < x < x_2'$, the mixed crystal is inherently unstable. It may decompose into homogeneous random alloys with the averaged compositions x_1 and x_2 on a length scale larger than the cluster sizes, i.e. larger than 1 nm.

By varying the temperature and tracing these miscibility gaps one can construct a $T-x$ phase diagram, which contains the binodal and spinodal curves (Fig. 2.12b). The spinodal curve marks the equilibrium solubility limit. Only for temperatures above this curve is a homogeneous random alloy with a defined average composition x predicted. The phase separation (below the binodal curve), which should occur at growth temperatures in $In_xGa_{1-x}N$, is driven by the internal strain due to the mixing of the two lattice-mismatched components InN and GaN. For $x \lesssim 0.5$ ($x \gtrsim 0.5$) indium (gallium) atoms are excluded from the $In_xGa_{1-x}N$ lattice to form alloys of distinct compositions, hence, locally the internal misfit strain is reduced, with a subsequent reduction of the strain energy of the system. Such a picture is in agreement with experimental findings [93, 94, 95, 96]. The phase diagram depicted in Fig. 2.12b is also similar to that obtained by other authors [90, 91]. However, from the *ab initio* calculations one obtains a larger range of miscibility for higher In molar fractions x. Moreover, the phase boundaries are much steeper for $x \to 0$ and $x \to 1$ than within the ideal solution model.

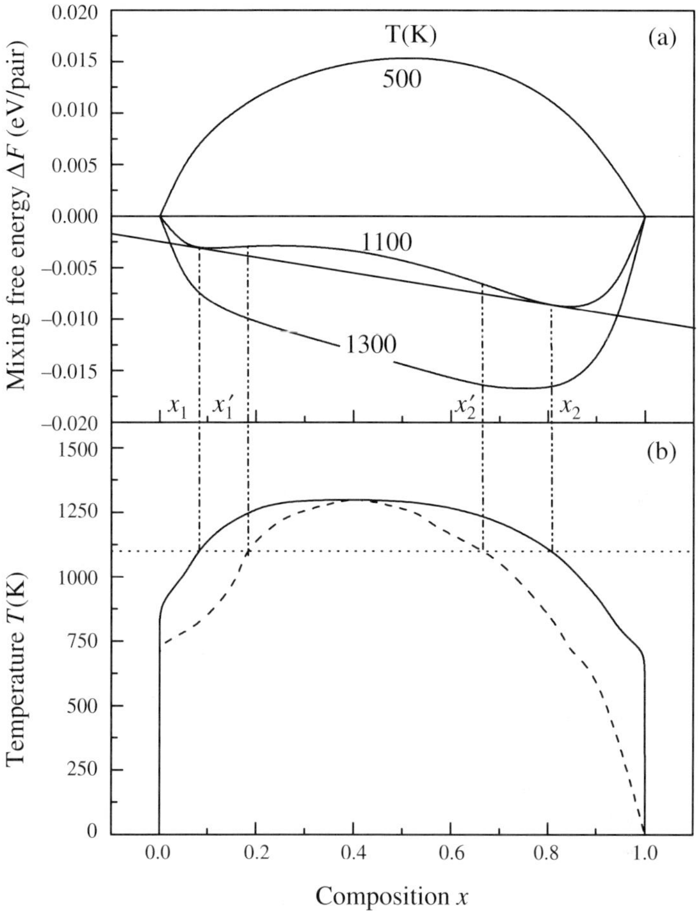

Fig. 2.12 (a) Mixing free energy $\Delta F(x, T)$ of $In_xGa_{1-x}N$ alloys versus composition. In the 1100 K case the common tangent is also plotted. (b) $T-x$ phase diagram of $In_xGa_{1-x}N$ solid line: binodal curve, dashed line: spinodal curve. (After ref. [92].)

The influence of a biaxial strain is demonstrated in Fig. 2.13. An inhomogeneous strain distribution is assumed in the alloy on a microscopic length scale in agreement with the assumption of pseudomorphic growth. Figure 2.13 shows the mixing free energy $\Delta F(x, T)$ for the $In_xGa_{1-x}N$ alloy as a function of composition calculated for the temperature range of interest. Results are plotted for alloys pseudomorphically grown on either a rigid GaN(001) buffer layer ($a_{\|}=a_{GaN}$), or an InN substrate ($a_{\|}=a_{InN}$). When the strong biaxial strain effects are introduced, ΔF becomes negative in the entire range of the alloy composition and exhibits a pronounced single minimum. The $T-x$ phase diagram undergoes a dramatic change leading to an almost zero critical temperature. Spinodal decomposition taking place in unstrained samples is predicted to be fully suppressed in the samples under biaxial strain, at least in the

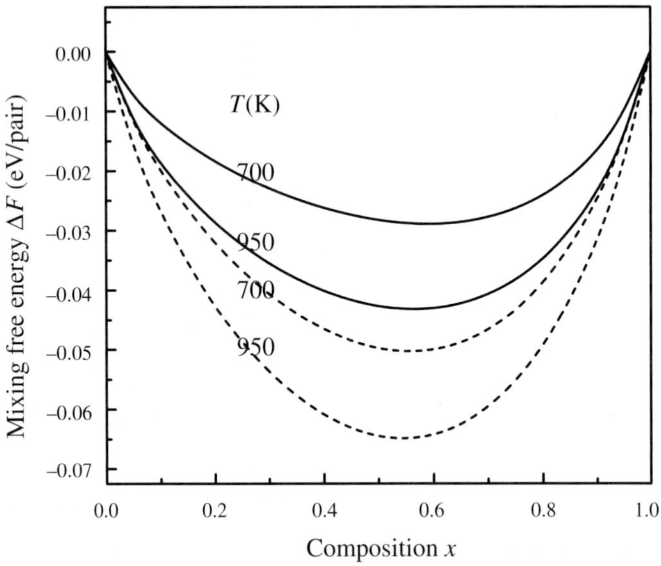

Fig. 2.13 Mixing free energy of inhomogeneously strained $In_xGa_{1-x}N$ alloys versus composition for two different temperatures $T=700$ and 950 K. The biaxial strain is defined by $a_{\|}=a_{GaN}$ (solid line) or $a_{\|}=a_{InN}$ (dashed line). (After ref. [92].)

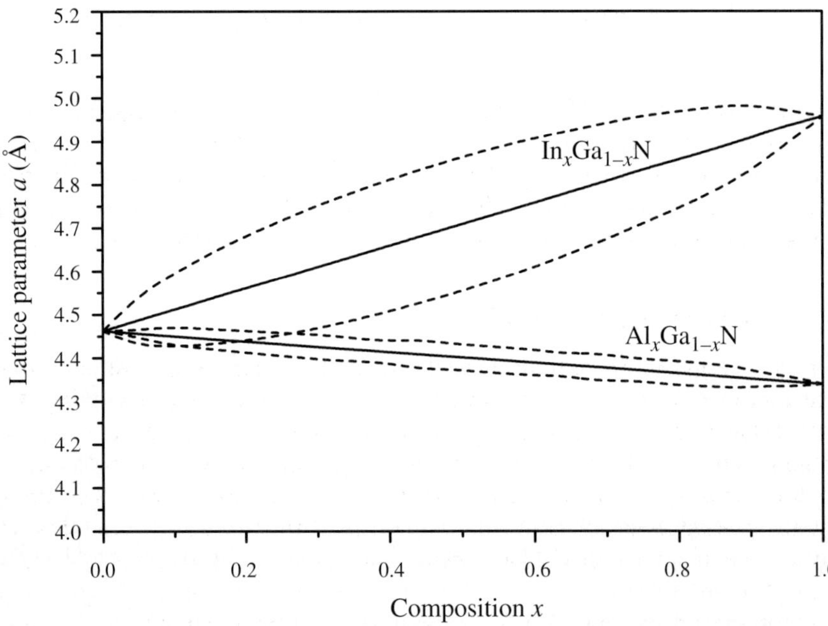

Fig. 2.14 Configurationally averaged lattice parameter a versus composition x. A growth temperature $T=950$ K is assumed. The range of the rms fluctuations in the lattice constant is indicated by dashed lines. (After ref. [87].)

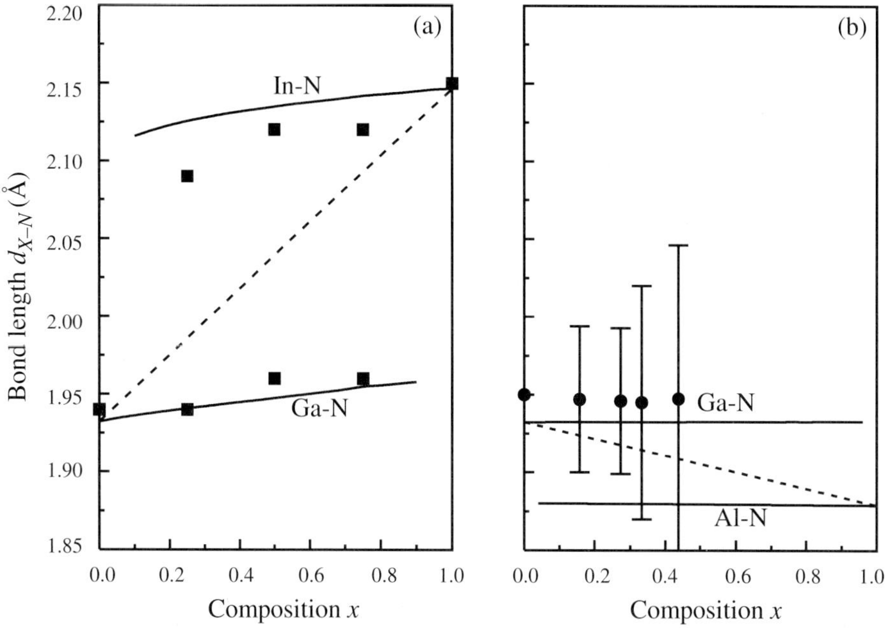

Fig. 2.15 Averaged bond lengths (solid lines) versus composition in $In_xGa_{1-x}N$ (a) and $Al_xGa_{1-x}N$ (b). The dashed line corresponds to the fictitious common bond length according to Vegard's rule. The filled squares and dots describe measured values [97, 98]. A growth temperature $T = 950$ K is assumed. (From ref. [87].)

extreme cases of about $|\epsilon_{\|}| = 0.1x$. However, also for smaller in-plane strains of about $|\epsilon_{\|}| = 0.01x$ remarkable changes happen [87]. The critical temperature is lowered and, simultaneously, the miscibility gap as well as the composition region of spontaneous decomposition are reduced. There is a clear tendency for suppression of the phase separation in $In_xGa_{1-x}N$ due to elastic strain.

2.4.3 Structure and composition fluctuations

The total-energy minimizations directly give the structural parameters as the lattice constant a, the bond lengths, as well as the second nearest-neighbour (2NN) distances for each cluster type. In Figs 2.14 and 2.15 the configurationally averaged values (2.18) are plotted for $In_xGa_{1-x}N$ and $Al_xGa_{1-x}N$. In the case of the lattice constants also their rms fluctuations (2.19) are shown. The configurationally averaged lattice constants a (Fig. 2.14) fulfil Vegard's law. However, there are mean-square deviations with a maximum value of about 3% in the $In_xGa_{1-x}N$ case.

The calculated average Ga-N, Al-N, and In-N bond lengths versus x are compared in Fig. 2.15 with Extended X-Ray Absorption Fine Structure (EXAFS) measurements [97, 98]. In agreement with experiment two different bond lengths d_{X-N} and d_{Ga-N} occur in each ternary $X_xGa_{1-x}N$ alloy. Apart from the small overbinding effect due to the used DFT-LDA and the fact that a possible strain in the samples influences the measured lengths, there is quantitative agreement with the experimental results. The same reasonable agreement can be stated with recent calculations performed within the Valence Force Field method, using extremely large clusters [99]. Figure 2.15 makes obvious that Vegard's rule fails for the bond lengths. The situation is similar for the 2NN distances [87]. These distances also do not follow Vegard's rule. The magnitude of the 2NN distances is more related to the covalent radii of the contributing atoms, although there is an influence of the surrounding cation distribution. Taking into account the overbinding tendency in DFT-LDA, again qualitative and quantitative agreement with experimental [97, 98] and other theoretical [99] predictions can be stated, in particular if the uncertainties due to the undetermined strain state of the samples, the length fluctuations, and the restrictions of the cluster-expansion method are taken into consideration.

Apart from the phase decomposition in mixed ternary crystals composition fluctuations Δx occur in each random alloy with a fixed average molar fraction x. Their characteristic length scale is given by the size of the considered clusters, i.e., $\lesssim 1\,\text{nm}$. The composition fluctuations follow from expression (2.19) for the mean-square deviations. Results are represented in Fig. 2.16. They basically show the same tendency as the fluctuations in an ideal solid solution, $\Delta x = \sqrt{x(1-x)/2}$. However, compared with $Al_xGa_{1-x}N$ they are larger in the $In_xGa_{1-x}N$ alloys indicating considerable deviations of these ternary nitrides

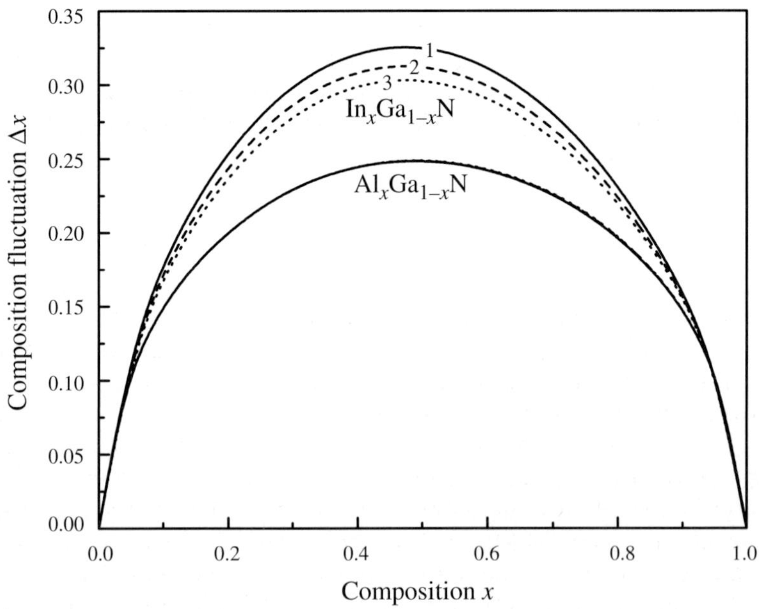

Fig. 2.16 Mean-square deviations Δx of the averaged composition x in $In_xGa_{1-x}N$ and $Al_xGa_{1-x}N$ alloys. Three different temperatures: $T=800$ (1), 950 (2), and 1100 (3) K are assumed. (After ref. [87].)

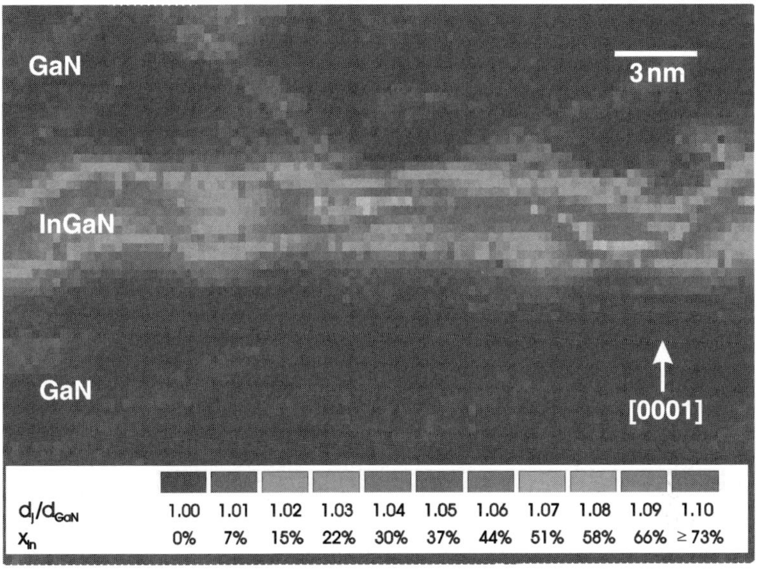

Fig. 2.17 DALI evaluation of a high-resolution transmission electron microscopy image to reveal the indium distribution in an $In_xGa_{1-x}N/GaN$ quantum-well structure. (After ref. [101] and private communication of D. Gerthsen.)

from ideal solutions. Moreover, the mirror symmetry of the Δx curves with respect to $x=0.5$ is destroyed. The maximum fluctuations occur for molar fractions $x<0.5$. Such composition fluctuations have been observed experimentally [100, 101] in $In_xGa_{1-x}N$ layers of $In_xGa_{1-x}N/GaN$ quantum-well structures as indicated in Fig. 2.17.

2.4.4 Gap bowing: influence of fluctuations and strain

In principle, a ternary alloy as $In_xGa_{1-x}N$ with chemical disorder and also partially structural disorder does not allow the definition of a band structure and, hence, of a fundamental energy gap. However, since the deviations from ordered systems are not too large, the cluster expansion method can be used to derive an average energy gap $E_g(x)$ and a mean-square deviation $\Delta E_g(x)$ from this value according to expressions (2.18) and (2.19). This cannot only be done in the strain-free case, but also in the presence of biaxial strain. Here, the method of Van de Walle [102] is used to determine the energy gap of a cubic system in the presence of strain. The strain dependence is governed by the hydrostatic deformation potentials $(a_c-a_v)=-6.92$ eV (GaN) and -3.14 eV (InN), the shear deformation potentials $b_v=-1.67$ eV (GaN) and -1.37 eV (InN), as well as the ratio of the elastic stiffness constants $c_{12}/c_{11}=0.54$ (GaN) and 0.69 (InN) [103]. These values are in very good agreement with those derived from other calculations [10, 40, 58, 76].

Results are summarized in Fig. 2.18. In the case of the average energy gap $E_g(x)$ the DFT-LDA values are corrected by a linear function in x that allows us to reproduce nearly the measured values for GaN and InN. These 'quasi-particle' gap corrections do not influence the bowing behavior of the gap function

$$E_g(x) = (1 - x)E_g(\text{GaN}) + xE_g(\text{InN}) - bx(1 - x). \qquad (2.24)$$

They are only used to guide the eyes. The use of the large value of the experimental InN gap [11] does not mean its acceptance. The bowing is still related only to the computed DFT-LDA values. Without strain one observes a downward bowing of the energy gap with a rather small bowing parameter $b = 0.68$ eV, a value being rather typical for the majority of DFT-LDA calculations [104]. As a consequence of a possible hole localization [105] the bowing parameter itself depends on composition.

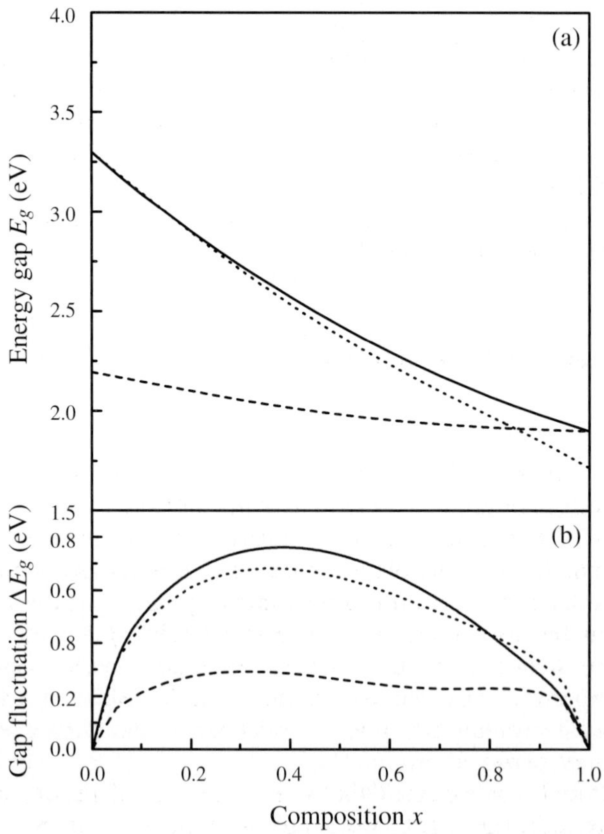

Fig. 2.18 Average energy gap (a) versus composition for $In_xGa_{1-x}N$ alloys. Solid line: unstrained situation, dashed line: $a_{\|} = a_{\text{GaN}}$, and dotted line: $a_{\|} = a_{\text{InN}}$. In the lower panel (b) the rms deviations ΔE_g of the gap are given. $T = 950$ K is assumed. (After ref. [103].)

The fluctuations of the energy gap, ΔE_g, in Fig. 2.18 play a very important role. Besides of an average energy gap they allow the definition of a minimum gap, $E_g - \Delta E_g$, with a completely different composition dependence and, hence, bowing behavior. Whereas E_g may be observed in absorption spectra, $E_g - \Delta E_g$ may give the positions of the emission lines. Consequently the gap fluctuations should give a remarkable contribution to the Stokes shift. Considering the unstrained situation, the bowing parameter of the minimum gap $(E_g - \Delta E_g)$ is much larger than that of E_g. The bowing of the minimum gap is again composition-dependent. Averaging over the interval $x = 0 \ldots 0.2$ one obtains a value of $b = 4.88$ [103]. It may explain the large values derived from measured spectra for $x < 0.2$, i.e. in the limit of low In molar fractions [106, 107]. In contrast to photoluminescence and photoreflection spectroscopy, other optical methods such as ellipsometry and transmission spectroscopy seem to detect indeed the larger average gap E_g with a much smaller bowing parameter b [108, 109]. The result depends on the determination of the gap, e.g. whether the onset or a value within the Urbach tail of the absorption is used to define the energy gap. Moreover, the bowing parameter depends on the chosen energy gap of InN.

Another important influence on the gap bowing may be due to biaxial strain. One observes that for both tensile and compressive strains, the curves $E_g = E_g(x)$ in Fig. 2.18 become more linear in comparison to the unstrained situation, resulting in a smaller value for the bowing parameter b. This is in agreement with the findings of other authors, that the bowing parameter measured for strained ternary alloys has to be increased during the extrapolation to the unstrained situation [106, 107].

2.5 Spontaneous polarization fields

2.5.1 Pyroelectricity

As materials with partially ionic bonds the group-III nitrides exhibit the piezoelectric effect. Piezoelectric polarization fields are induced for internal displacements of the group-III atoms relative to the nitrogen atoms in an elementary cell. In the zinc-blende case (T_d^2) this happens only for shear strains. For wurtzite crystals with lower symmetry (C_{6v}^4) piezoelectric polarization fields can be also induced by diagonal strains. However, even in the equilibrium the lower-symmetry wurtzite polytypes are expected to possess already an intrinsic spontaneous polarization field (cf. eqn (2.3)), since the four tetrahedral bonds around one atom are no longer equivalent, so that wurtzite crystals represent pyroelectrics.

In bulk material it is assumed that rearrangement of surface charges nullifies a spatially uniform spontaneous polarization. The effect of the spontaneous polarization fields can therefore only be observed in layered structures such as heterostructures consisting of wurtzite crystals of different compounds. In general, in heterostructures or inhomogeneous alloy layers variations in the composition are expected to create nonvanishing polarization fields. However, the same effect should also occur in systems with variations of the crystal structure, e.g. in heterocrystalline structures consisting of nitride layers of both the wurtzite and zinc-blende polytype.

Indeed, the influence of such intrinsic polarization fields has been observed via the quantum-confined Stark effect in quantum-well structures based mainly on the $Al_xGa_{1-x}N/GaN$ system [23]. The quantum-confined Stark effect in $In_xGa_{1-x}N/GaN$ heterostructures seems to be governed by piezoelectric fields [110]. In the case of field-effect transistors based on $Al_xGa_{1-x}N/GaN$ heterostructures the spontaneous and piezoelectric polarization fields can play the role of additional design parameters [67]. These fields give rise to high sheet carrier concentrations at the interfaces also in unintentionally doped samples. Meanwhile, there are also predictions [24] that according to the relations (2.2)–(2.4) the polarization fields give rise to small atomic displacements of the atoms in heterostructures with respect to the equilibrium positions in bulk crystals. In infinite strained GaN/AlN superlattices the effect should be observable by shifts of the Raman frequencies of a few cm^{-1}. However, the electronic properties are also influenced by the polarization fields [111].

2.5.2 Microscopic approach

The macroscopic polarization in an electronic system is related to the charge density by its definition as electric dipole moment per unit volume. Fortunately, in the *ab initio* theories one has direct access to the charge density. The electron density is even the central quantity of the DFT. Unfortunately, for an infinite crystal without center of inversion symmetry and using periodic boundary conditions in the computations, the polarization is an ill-defined quantity. In explicit calculations one has to eliminate truncation and surface effects. In the literature such calculations have been performed within the Berry phase approach [112] or a superlattice treatment [26].

The polarization field has only a non-vanishing component parallel to the *c*-axis of the wurtzite structure, which is indicated by the *z*-axis in the following. In the superlattice approximation the macroscopic termination-dependent longitudinal polarization P_{Lz} in the presence of a non-vanishing depolarization field [68, 113], i.e. the total polarization (2.3) in the wurtzite layers in the absence of external charges and a piezoelectric polarization, can be derived from *first principles* as the difference of the polarizations in a cubic (3C) reference material and in the wurtzite (2H) structure. For symmetry reasons the polarization field should vanish in a real cubic crystal. However, the accompanying electric field is nonzero in the cubic layers leading to a sawtooth potential in a superlattice arrangement of zinc-blende and wurtzite layers. The spontaneous polarization in SiC has been studied [114] using heterocrystalline structures (more strictly: superlattices) with supercells consisting of both polytypes, 3C and 2H. The macroscopic polarization P_{Lz} is related to homogeneous electric fields in each polytype region and, hence, to the gradient of the spatially averaged total one-electron potential by

$$P_{Lz} = -\varepsilon_0[E_z^{2H} - E_z^{3C}] = \varepsilon_0\left[\frac{d}{dz}\overline{V}_{\text{tot}}(z)|_{2H} - \frac{d}{dz}\overline{V}_{\text{tot}}(z)|_{3C}\right]. \quad (2.25)$$

The total single-particle potential occurs in the Kohn–Sham equation of the DFT-LDA. It is represented in Fig. 2.19 for a $(3C)_6(2H)_6$ superlattice containing 30 [111]- and [0001]-oriented III–N bilayers. The lattice constant of the cubic system is adapted to the in-plane constant of the wurtzite structure to avoid piezoelectric fields. The potential averaged over the *xy*-plane perpendicular to the *c*-axis shows the typical atomic oscillations. The additional spatial average in *z*-direction then gives the mentioned macroscopic saw-tooth potential (cf. Fig. 2.19) related to the macroscopic electric fields.

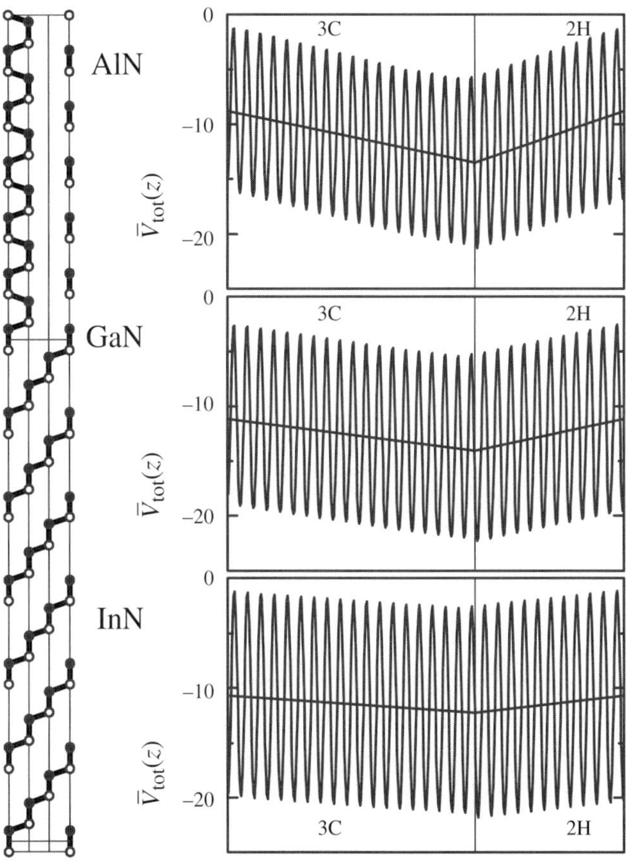

Fig. 2.19 Structure of a $(3C)_6(2H)_6$ superlattice cell and averaged total one-electron potentials. The vertical lines separate cubic and hexagonal stacking regions in AlN, GaN, and InN. (After ref. [26].)

2.5.3 Field strengths and origin

The polarization field strengths P_{Lz} derived from expression (2.25) are listed in Table 2.10. In order to determine the macroscopic zero-field or transverse polarization P_{Tz} [68, 113], the polarization induced in the electronic system by the electric fields in the hexagonal and cubic regions have to be taken into account. This effect is proportional to the electronic susceptibility. It holds

$$P_{Tz} = \bar{\varepsilon}_{\infty} P_{Lz} \tag{2.26}$$

with an effective electronic dielectric constant of the considered superlattice in z-direction

$$\bar{\varepsilon}_{\infty} = \frac{d_{3C}\varepsilon_{\infty}^{\parallel} + d_{2H}\varepsilon_{\infty}}{d_{3C} + d_{2H}}, \tag{2.27}$$

where the thicknesses, d_{2H} and d_{3C}, as well as the components of the dielectric tensor parallel to the c-axis, $\varepsilon_{\infty}^{\parallel}$ and ε_{∞}, appear. From the *ab initio* calculations we take $\bar{\varepsilon}_{\infty} = 4.52$ (AlN) [30], 5.41 (GaN) [31], and 8.49 (InN) [68]. The values resulting for P_{Tz} are also given in Table 2.10 and compared with results of the Berry-phase approach [112].

Interpreting the charge-asymmetry coefficients $g_{\parallel}$ and $g_{\perp}$ [37] (cf. Section 2.2.1) as the ionicities of the bonds parallel to the c-axis or with an angle of about 70° to the c-axis, one has to deal with bond dipoles $\mathbf{p}_{\parallel/\perp} = eg_{\parallel/\perp}\tau_{\parallel/\perp}$, where $\tau_{\parallel/\perp}$ denote vectors of the deformed tetrahedra in the wurtzite structure. The summation over the eight bonds per 2H unit cell with the volume Ω_0 leads to the expression

$$P_{Tz} \approx -\frac{2ec}{\Omega_0}\left[4g_{\perp}\left(\frac{3}{8} - u\right) + (g_{\perp} - g_{\parallel})u\right]. \tag{2.28}$$

There are two contributions to the spontaneous polarization field parallel to the c-axis. The first contribution $\sim(\frac{3}{8} - u)$ is directly related to the nonideality of the tetrahedra in the wurtzite structure. The second term is proportional to the difference of the ionicities of the two inequivalent bonds in the structure. The values $g_{\parallel}$ and $g_{\perp}$ are almost identical with those g of the zinc-blende

Table 2.10 Calculated spontaneous polarization fields (in units of 10^{-3} C/m^3) of wurtzite group-III nitrides.

Polarization	Approach	AlN	GaN	InN
P_{Lz}	superlattice [26]	–26	–14	–6
P_{Tz}	superlattice [26]	–120	–74	–50
P_{Tz}	Berry phase [112]	–81	–29	–32
P_{Tz}	model [26]	–51	–50	–22

structure (cf. Sec. 2.2.1). For AlN it holds $g_{\perp}=g(1-0.0025)$ and $g_{\parallel}=g(1+0.0259)$ [30]. Despite of the small variations with the bond direction, the second term is not negligible, in particular for nitrides with geometry parameters u and c/a close to the ideal values. Results using the geometry parameters in Table 2.2, the g-values from ref. [37], and the model described above [eqn (2.28)] are also given in Table 2.10.

The two *ab initio* calculations and the estimate in the framework of the bond-orbital model give rise to the same order of magnitude of the intrinsic polarization (cf. Table 2.10). Also the only experimental value $P_{Tz}=-0.055$ C/m^2 available for AlN [115] is in the range of the calculated ones. Physically relevant are only the polarization differences at the interfaces. They are in reasonable agreement for the combination AlN-GaN but not for GaN-InN. The variation of the absolute numbers between the two *first-principles* calculations is mainly a consequence of the extreme accuracy of the structural optimization that is required for polarization calculations. Small changes in the geometry parameters may induce big variations in the polarization fields. For instance, the large deviations in the results for GaN may be related to the difference in the calculated u parameters, $u=0.381$ [26] and $u=0.376$ [112]. Because of the accuracy requirements for the atomic structure one should be sceptical concerning the possibility of highly accurate calculations of the spontaneous polarization fields. The accuracy of the determination of the equilibrium atomic structure parameters within standard DFT-LDA calculations is perhaps not sufficient enough. The calculations should be really converged with respect to the number of plane waves, the density of the **k**-point mesh, and the precision of the minimization procedure. This holds in particular for the determination of the internal-cell parameter u, but also for the ratio of the lattice constants c and a.

2.6 Concluding remarks

In this chapter some ideas have been presented as to how solid-state properties can be predicted from *first-principles* calculations. They are based on the density functional theory in its local version, pseudopotentials, and a plane-wave expansion of the eigenfunctions. We have seen that the modern theory of solids can be successfully applied to the study of the energetical, structural, and dynamical properties of the group-III nitrides including their polytypic structures, strained layers, and alloys. After inclusion of many-body effects, as for instance the quasiparticle self-energy shifts, even electronic excitation energies can be predicted. There is a high degree of agreement between different calculations as well as between theory and experiment. Encouraged by this success *ab initio* methods will gain further importance for the understanding of unsolved problems in the field of III–nitride physics and technology.

The microscopic processes occurring during molecular beam epitaxy (MBE), metal-organic vapor phase epitaxy (MOVPE), and others will be explored from atomistic calculations. This requires not only the knowledge of the thermodynamics of the III-nitride surfaces [116, 117] but also knowledge about the chemical reactions between the substrate surface and the atoms or molecules used as reactants or precursors. This knowledge concerns the details of the processes occurring on the growing surface, such as diffusion, nucleation, and attachment or detachment of adatoms to existing islands and steps [118]. In recent years there was considerable progress in the understanding of surface chemical reactions using *ab initio* methods, in particular in the case of conventional III–V compounds. The dynamics of individual reactive species on the surface has been modeled in detail; either by integrating the classical trajectory of a molecule/atom over a few hundred femtoseconds or by the use of quantum-mechanic scattering techniques.

Another application of *ab initio* methods with increasing importance is the description of complex properties which can be measured after excitation of a III-nitride sample. One example is the measurement of one-electron excitation energies, where single electrons or holes are added to the system. It is necessary to dress the added electron or hole with the polarization cloud of the surrounding electrons. Quasiparticle shifts [44, 45, 49, 50, 51] have to be taken into account to obtain a band structure that can be compared with experimental values. The situation is more extreme for the case of optical properties. Here, photons interact with the system to produce electron–hole pairs. In addition to interacting with their respective polarization clouds, the excited electrons and holes interact strongly with each other. These excitonic effects can be now modeled also for nitride crystals to calculate their entire absorption spectra [119]. In the future these many-body effects will also be included in the description of the optical properties of more complex III-nitride systems as alloys and nanostructures.

Future research in the field of group-III nitrides will be focused on nanostructures based on different nitrides or alloys. Two classes of structures are particularly fruitful and important objects for *ab initio* studies of preparation and properties, quantum dot arrays and superlattices or multiquantum well structures. Their physical properties make them extremely attractive for a wide range of device applications. Meanwhile, self-assembled growth of nanometer-scale dots has been reported for GaN/AlGaN and InGaN/AlGaN systems. The Stranski–Krastanov growth mode as well as surfactant-mediated growth has been used. An atomic picture of the preparation processes and the properties of the dots is still missing. The same is true for the superlattices. GaN/AlGaN superlattices have been already successfully grown and characterized [67, 120]. However, there are only preliminary *first-principles* studies of the structural and dynamical properties of these layered structures [38, 121].

Note added in proof:

Meanwhile, the predicted small fundamental energy gap of InN seems to be confirmed experimentally. Absorption, photoluminescence and photoluminescence excitation spectroscopies [V.Y. Davydov, A.A. Klochikhin, R.P. Seisyan, *et al.*

(2002) *Phys. Stat. Sol. B* **229**, R1] indicate a gap energy of about 0.9 eV for high-quality MBE layers.

Acknowledgements

I thank M. Ferhat, J. Furthmüller, U. Grossner, W.G. Schmidt, L.K. Teles, and J.-M. Wagner for collaboration and/or reading of the manuscript. This work was supported by the Deutsche Forschungsgemeinschaft (Schwerpunktprogramm Gruppe-III–Nitride, Grant No. Be 1346/8-5).

References

1. Hohenberg, P. and Kohn, W. (1964). Inhomogeneous electron gas. *Phys. Rev.* **136**, B864–71.
2. Kohn, W. and Sham, L. J. (1965). Self-consistent equations including exchange and correlation effects. *Phys. Rev.* **140**, A1133–8.
3. Ceperley, D. M. and Alder, B. J. (1980). Ground state of the electron gas by stochastic method. *Phys. Rev. Lett.* **45**, 566–9.
4. Perdew, J. P. and Zunger, A. (1981). Self-interaction correction to density-functional approximations for many-electron systems. *Phys. Rev. B* **23**, 5048–79.
5. von Barth, U. and Hedin, L. (1972). A local exchange-correlation potential for spin polarized case. *J. Phys. C: Solid State Phys.* **5**, 1629–42.
6. Bechstedt, F. (1992). Quasiparticle corrections for energy gaps in semiconductors. *Adv. Solid State Phys.* **32**, 161–77.
7. Hedin, L. and Lundqvist, S. (1969). Effect of electron–electron and electron–phonon interactions on the one-electron states of solids. *Solid State Phys.* **23**, 1–181.
8. Hybertsen, M. S. and Louie, S. G. (1986). Electron correlation in semiconductors and insulators: Band gaps and quasiparticle energies. *Phys. Rev. B* **34**, 5390–413.
9. Gianozzi, P., de Gironcoli, S., Pavone, P., and Barone, S. (1991). Ab initio calculation of phonon dispersions in semiconductors. *Phys. Rev. B* **43**, 7231–42.
10. Wright, A. F. (1997). Elastic properties of zinc-blende and wurtzite AlN, GaN, and InN. *J. Appl. Phys.* **82**, 2833–9.
11. Tansley, T. L. and Foley, C. P. (1986). Optical band gap of indium nitride. *J. Appl. Phys.* **59**, 3241–4.
12. Tabata, A., Lima, A. P., Teles, L. K., *et al.* (1999). Structural properties and Raman modes of zinc-blende InN epitaxial layers. *Appl. Phys. Lett.* **74**, 362–4.
13. Fiorentini, V., Methfessel, M., and Scheffler, M. (1993). Electronic and structural properties of GaN by the full-potential linear muffin-tin orbitals method: The role of d electrons. *Phys. Rev. B* **47**, 13353–62.
14. Wright, A. F. and Nelson, J. S. (1994). Explicit treatment of the Gallium 3*d* electrons in GaN using the plane wave pseudopotential method. *Phys. Rev. B* **50**, 2159–69.
15. Wright, A. F. and Nelson, J. S. (1995). Consistent structural properties for AlN, GaN, and InN. *Phys. Rev. B* **51**, 7866–9.
16. Kresse, G. and Furthmüller, J. (1996). Efficient iterative schemes for *ab initio* total-energy calculations using a plane-wave basis set. *Phys. Rev. B* **54**, 1169–86.
17. Kresse, G. and Furthmüller, J. (1996). Efficiency of ab-initio total energy calculations for metals and semiconductors using a plane-wave basis set. *Computat. Materials Science* **6**, 15–50.

18. Sargent-Welch (1980). *Table of periodic properties of the elements*. Sargent-Welch Scientific Co., Skokie, Illinois.
19. Cardona, M. and Christensen, N. E. (2000). Spin-orbit splittings in AlN, GaN, and InN. *Solid State Commun.* **116**, 421–5.
20. Yeh, C.-Y., Lu, Z. W., Froyen, S., and Zunger, A. (1992). Zinc-blende-wurtzite polytypism in semiconductors. *Phys. Rev. B* **46**, 10085–97.
21. Nakamura, S. and Fasol, G. (1997). *The blue laser diode*. Springer, Heidelberg.
22. Bernardini, F. and Fiorentini, V. (2000). Polarization fields in nitride nanostructures: 10 points to think about. *Appl. Surf. Sci.* **166**, 23–29.
23. Leroux, M., Grandjean, N., Laügt, M., *et al.* (1998). Quantum confined Stark effect due to built-in internal polarization fields in (Al,Ga)N/GaN quantum wells. *Phys. Rev. B* **58**, R13371–4.
24. Gleize, J., Frandon, J., Renucci, M. A., and Bechstedt, F. (2001). Influence of spontaneous and piezoelectronic polarizations on the phonon frequencies in strained GaN/AlN superlattices. *Phys. Rev. B* **63**, 3308–11.
25. Grossner, U., Furthmüller, J., and Bechstedt, F. (1998). Bond-rotation versus bond-contraction relaxation of (110) surfaces of group-III nitrides. *Phys. Rev. B* **58**, R1722–5.
26. Bechstedt, F., Grossner, U., and Furthmüller, J. (2000). Dynamics and polarization of III-nitride lattices: A *first-principles* study. *Phys. Rev. B* **62**, 8003–11.
27. Furthmüller, J., Käckell, P., Bechstedt, F., and Kresse, G. (2000). Extreme softening of Vanderbilt pseudopotentials: General rules and case studies of first-row and d-electron elements. *Phys. Rev. B* **61**, 4576–87.
28. Chadi, D. J. and Cohen, M. L.. (1973). Special points in the Brillouin zone. *Phys. Rev. B* **8**, 5747–53.
29. Monkhorst, H. J. and Pack, J. D. (1976). Special points for Brillouin-zone integrations. *Phys. Rev. B* **13**, 5188–92.
30. Karch, K. and Bechstedt, F. (1997). *Ab initio* lattice dynamics of BN and AlN: Covalent versus ionic forces. *Phys. Rev. B* **56**, 7404–15.
31. Karch, K., Wagner, J.-M., and Bechstedt, F. (1998). *Ab initio* study of structural, dielectric, and dynamical porperties of GaN. *Phys. Rev. B* **57**, 7043–9.
32. Troullier, N. and Martins, J. L. (1991). Efficient pseudopotentials for plane-wave calculations. *Phys. Rev. B* **43**, 1993–2006.
33. Karch, K., Bechstedt, F., and Pletl, T. (1997). Lattice dynamics of GaN: Effects of 3d electrons. *Phys. Rev. B* **56**, 3560–3.
34. Louie, S. G., Froyen, S., and Cohen, M. L. (1982). Nonlinear ionic pseudopotentials in spin-density-functional calculations. *Phys. Rev. B* **26**, 1738–42.
35. Stampfl, C. and Van de Walle, C. G. (1999). Density-functional calculations for III–V nitrides using local-density approximation and generalized gradient approximation. *Phys. Rev. B* **59**, 5521–35.
36. Lawaetz, P. (1972). Stability of the wurtzite structure. *Phys. Rev. B* **5**, 4039–45.
37. Garcia, A. and Cohen, M. L. (1993). First-principles ionicity scales. I. Charge asymmetry in the solid state. *Phys. Rev. B* **47**, 4215–20.
38. Grossner, U., Furthmüller, J., and Bechstedt, F. (2000). Dynamics and polarization of group-III nitrides. *IPAP Conf. Series* **1**, 56–9.
39. Vogel, D., Krüger, P., and Pollmann, J. (1997). Structural and electronic properties of group-III nitrides. *Phys. Rev. B* **55**, 12836–9.
40. Kim, K., Lambrecht, W. R. L., and Segall, B. (1996). Elastic and related properties of tetrahedrally bonded BN, AlN, GaN, and InN. *Phys. Rev. B* **53**, 16310–26.
41. Lambrecht, W. R. L., Segall, B., Strite, S., *et al.* (1994). X-ray photoelectron spectroscopy of the valence band and semicore Ga 3*d* states in GaN. *Phys. Rev. B* **50**, 14155–60.

42. Nakadeira, A. and Tanaka, H. (1998). Metalorganic vapor-phase epitaxial growth and characterization of cubic $Al_xGa_{1-x}N$ alloy. *Jpn. J. Appl. Phys*. **37**, 1449–52.
43. Koizumi, T., Okumura, H., Balakrishnan, K. *et al.* (1999). Growth and characterization of cubic AlGaN and AlN epilayers by RF-plasma assisted MBE. *J. Cryst. Growth* **201/202**, 341–5.
44. Rubio, A., Corkill, J. L., Cohen, M. L., Shirley, E. L., and Louie, S.G. (1993). Quasiparticle band structure of AlN and GaN. *Phys. Rev. B* **48**, 11810–16.
45. Palummo, M., Del Sole, R., Reining, L., Bechstedt, F., and Cappellini, G. (1995). Screening models and simplified GW approaches: Si & GaN as test cases. *Solid State Commun*. **59**, 393–8.
46. Perry, P. B. and Rutz, R. F. (1978). The optical absorption edge of single-crystal AlN prepared by a close-space vapor process. *Appl. Phys. Lett*. **33**, 319–21.
47. Hong, H., Pavlidis, D., Brown, S. W., and Rand, S. C. (1995). Photoluminescence investigation of GaN films grown by metalorganic chemical vapor deposition on (100)GaAs. *J. Appl. Phys*. **77**, 1705–9.
48. Monemar, B. (1974). Fundamental energy gap of GaN from photoluminescence excitation spectra. *Phys. Rev. B* **10**, 676–81.
49. Bechstedt, F. and Del Sole, R. (1988). Analytical treatment of band-gap underestimates in the local-density approximation. *Phys. Rev. B* **38**, 7710–6.
50. Persson, C., Sernelius, B. E., Ferreira da Silva, A., Ahuja, R., and Johansson, B. (2001). Effective electron and hole masses in intrinsic and heavily doped GaN and AlN. *J. Phys.: Condensed Matter* **13**, 8915–22.
51. Persson, C., Ahuja, R., Ferreira da Silva, A., and Johansson, B. (2001). First-principles calculations of dielectric function in cubic and wurtzite InN. *J. Phys.: Condensed Matter* **13**, 8545–50.
52. O'Donnell, K. P. O. (2001). A mystery wrapped in an enigma: Optical properties of InGaN alloys. *Phys. Stat. Sol.* (a) **183**, 117–24.
53. Ruf, T., Serrano, J., Cardona, M., *et al.* (2000). Phonon dispersion curves in wurtzite-structure GaN determined by inelastic X-ray scattering. *Phys. Rev. Lett*. **86**, 906–9.
54. Bungaro, C., Rapcewicz, K., and Bernholc, J. (2000). *Ab initio* phonon dispersions of wurtzite AlN, GaN, and InN. *Phys. Rev. B* **61**, 6720–5.
55. Harima, H., Inoue, T., Nakashima, S., *et al.* (1999). Raman studies on phonon modes in cubic AlGaN alloy. *Appl. Phys. Lett*. **74**, 191–3.
56. Siegle, H., Kaczmarczyk, G., Fillipidis, L., *et al.* (1997). Zone-boundary phonons in hexagonal and cubic GaN. *Phys. Rev.* B **55**, 7000–4.
57. Tabata, A., Enderlein, R., Leite, J. R., *et al.* (1996). Comparative Raman studies of cubic and hexagonal GaN epitaxial layers. *J. Appl. Phys*. **79**, 4137–40.
58. Shimada, K., Sota, T., and Suzuki, K. (1998). First principles study on electronic and elastic properties of BN, AlN, and GaN. *J. Appl. Phys.* **84**, 4951–8.
59. Dyck, J. S., Kim, K., Limpijumnong, S., *et al.* (2000). Identification of Raman-active phonon modes in oriented platelets of InN and polycrystalline InN. *Solid State Commun*. **114**, 355–60.
60. Davydov, V. Y., Kitaev, Y. E., Goncharuk, I. N., *et al.* (1998). Phonon spectrum of wurtzite GaN and AlN. Experiment and theory. *J. Cryst. Growth* **189/190**, 656–60.
61. Davydov, V. Y., Emtsev, V. V., Goncharuk, I. N., *et al.* (1999). Experimental and theoretical studies of phonons in hexagonal InN. *Appl. Phys. Lett.* **75**, 3297–9.
62. Filipidis, L., Siegle, H., Hoffmann, A., *et al.* (1996). Raman frequencies and angular dispersion of polar modes in aluminum nitride and gallium nitride. *Phys. Stat. Sol.* (b) **198**, 621–7.
63. Jain, S.C., Willander, M., Narayan, J., and Van Overstraeten, R. (2000). III-nitrides: Growth, characterization, and properties. *J. Appl. Phys.* **87**, 965–1006.

64. Steude, G., Meyer, B. K., Göldner, A., *et al.* (1999). Strain modification of GaN in AlGaN/GaN epitaxial films. *Jpn. J. Appl. Phys.* **38**, L498–500.
65. Gleize, J., Demangeot, F., Frandon, J., *et al.* (1999). Phonons in a strained hexagonal GaN-AlN superlattice. *Appl. Phys. Lett.* **74**, 703–5.
66. Dieulesaint, E. and Roger, D. (1980). *Elastic waves in solids, applications to signal processing.* Wiley, New York.
67. Yu, E. T., Dang, X. Z., Asbeck, P. M., Lau, S. S., and Sullivan, G. J. (1999). Spontaneous and piezoelectric polarization effects in III–V nitride heterostructures. *J. Vac. Sci. Technol. B* **17**, 1742–9.
68. Bernardini, F., Fiorentini, V., and Vanderbilt, D. (1997). Polarization-based calculation of the dielectric tensor of polar crystals. *Phys. Rev. Lett.* **79**, 3958–61.
69. Wagner, J.-M. and Bechstedt, F. (2002). Properties of strained GaN and AlN layers: *Ab initio* studies. *Phys. Rev. B* (resubmitted).
70. Wagner, J.-M. and Bechstedt, F. (1999). Strain influence on III–Nitrides: *Ab initio* studies of structural, lattice-dynamical, and dielectric properties. *Phys. Stat. Sol.* (b) **216**, 793–8.
71. Harrison, W. A. (1989). *Electronic structure and the properties of solids*. Dover Publications, Mineola.
72. Wagner, J.-M. and Bechstedt, F. (2000). Pressure dependence of the dielectric and lattice-dynamical properties of GaN and AlN. *Phys. Rev.* B **62**, 4526–34.
73. Suzuki, M., Uenoyama, T., and Yanase, A. (1995). First-principles calculations of the effective-mass parameters of AlN and GaN. *Phys. Rev. B* **52**, 8132–9.
74. Kim, K., Lambrecht, W. R. L., Segall, B., and van Schilfgaarde, M. (1997). Effective masses and valence-band splittings in GaN and AlN. *Phys. Rev. B* **56**, 7363–75.
75. Chuang, S. L. and Chang, C. S. (1996). **K.p** method for strained wurtzite semiconductors. *Phys. Rev. B* **54**, 2491–504.
76. Gorczyca, I. and Christensen, N. E. (1994). Optical and structural properties of III–V nitrides under pressure. *Phys. Rev. B* **50**, 4397–415.
77. Reimann, K., Steube, M., Brandt, O., Yang, H., and Ploog, K. H. (1998). Direct comparison of the pressure-induced band-gap shifts in cubic and hexagonal GaN. *J. Appl. Phys*. **84**, 2971–3.
78. Weinstein, B. A., Perlin, P., Christensen, N. E., *et al.* (1998). Cubic InN inclusions: Proposed explanation for the small pressure-shift anomaly of the luminescence in InGaN-based quantum wells. *Solid State Commun.* **106**, 567–71.
79. Blacha, A., Presting, H., and Cardona, M. (1984). Deformation potentials of $k = 0$ states of tetrahedral semiconductors. *Phys. Stat. Sol.* (b) **126**, 11–36.
80. Briggs, R. J. and Ramdas, A. K. (1976). Piezospectroscopic study of the Raman spectrum of cadmium sulfide. *Phys. Rev. B* **13**, 5518–29.
81. Wagner, J.-M. and Bechstedt, F. (2000). Phonon deformation potentials of α-GaN and -AlN: An *ab initio* calculation. *Appl. Phys. Lett.* **77**, 346–8.
82. Demangeot, F., Frandon, J., Renucci, M. A., *et al.* (1996). Raman determination of phonon deformation potentials in α-GaN. *Solid State Commun.* **100**, 207–10.
83. Davydov, V. Y., Aveskiev, N. S., Goncharuk, I. N., *et al.* (1997). Raman and photoluminescence studies of biaxial strain in GaN epitaxial layers grown on 6H-SiC. *J. Appl. Phys.* **82**, 5097–102.
84. Wei, S. H., Ferreira, L. G., and Zunger, A. (1990). First-principles calculation of temperature-composition phase diagrams of semiconductor alloys. *Phys. Rev. B* **41**, 8240–69.
85. Zunger, A. (1994). First-principles statistical mechanics of semiconductor alloys and intermetallic compounds. In *Statics and dynamics of alloy phase transformations*, pp. 361–419. (Eds P.E.A. Turchi and A. Gonis). Plenum Press, New York.

86. Sher, A., Van Schilfgaarde, M., Chen, A-B., and Chen, W. (1987). Quasichemical approximation in binary alloys. *Phys. Rev. B* **36**, 4279–95.
87. Teles, L. K., Furthmüller, J., Scolfaro, L. M. R., Leite, J. R., and Bechstedt, F. (2000). First-principles calculations of the thermodynamic and structural properties of strained $In_xGa_{1-x}N$ and $Al_xGa_{1-x}N$ alloys. *Phys. Rev. B* **62**, 2475–85.
88. Conolly, J. W. D. and Williams, A. R. (1983). Density-functional theory applied to phase transformations in transition-metal alloys. *Phys. Rev. B* **27**, 5169–72.
89. Chen, A.-B. and Cher, A. (1995). *Semiconductor alloys*. Plenum Press, New York.
90. Ho, I. and Stringfellow, G.B. (1996). Solid phase immiscibility in InGaN. *Appl. Phys. Lett.* **69**, 2701–3.
91. Ho, I. and Stringfellow, G. B. (1997). Solubility of nitrogen in binary III–V systems. *J. Cryst. Growth* **178**, 1–7.
92. Teles, L. K., Scolfaro, L. M. R., Furthmüller, J., Leite, J. R., and Bechstedt, F. (2000). Influence of biaxial strain on thermodynamic, structural and electronic properties of $In_xGa_{1-x}N$ alloys. *IPAP Conf. Series* **1**, 689–92.
93. Doppalapudi, D., Basu, S. N., Ludwig, K. F., and Moustakas, T. D. (1998). Phase separation and ordering in InGaN alloys grown by molecular beam epitaxy. *J. Appl. Phys.* **84**, 1389–95.
94. El-Masry, N. A., Piner, E. L., Liu, S. X., and Bedair, S. M. (1998). Phase separation in InGaN grown by metalorganic chemical vapor deposition. *Appl. Phys. Lett.* **72**, 40–2.
95. Lemos, V., Silveira, E., Leite, J. R., *et al.* (2000). Blue-green photoluminescence from phase-separated quantum dots in cubic $In_xGa_{1-x}N$ epitaxial layers. *Phys. Rev. Lett.* **84**, 3666–9.
96. Kaschner, A., Hoffmann, A., Thomsen, C., *et al.* (2000). Evidence for phase separation in InGaN by resonant Raman scattering. *Phys. Stat. Sol.* (a) **175**, R4–6.
97. Jeffs, N. J., Blant, A. V., Cheng, T. S., *et al.* (1998). EXAFS studies of group III-nitrides. *MRS Proc.* **512**, 519–24.
98. Miyano, K. E., Woicik, J. C., Robins, L. H., Bouldin, C. E., and Wickenden, D. K. (1997). Extended X-ray absorption fine structure study of $Al_xGa_{1-x}N$ films. *Appl. Phys. Lett.* **70**, 2108–10.
99. Mattila, T. and Zunger, A. (1999). Predicted bond length variation in wurtzite and zinc-blende InGaN and AlGaN alloys. *J. Appl. Phys.* **85**, 160–7.
100. Wieser, N., Ambacher, O., Felsl, H.-P., Görgens, L., and Stutzmann, M. (1999). Compositional fluctuations in InGaN/GaN double heterostructures investigated by selectively excited photoluminescence and Raman spectroscopy. *Appl. Phys. Lett.* **74**, 3981–3.
101. Gerthsen, D., Hahn, E., Neubauer, B., *et al.* (2000). Composition fluctuations in InGaN analyzed by transmission electron microscopy. *Phys. Stat. Sol.* (a) **177**, 145–55.
102. Van de Walle, C. G. (1989). Band lineups and deformation potentials in the model solid theory. *Phys. Rev. B* **39**, 1871–83.
103. Teles, L. K., Furthmüller, J., Scolfaro, L. M. R., Leite, J. R., and Bechstedt, F. (2001). Influence of composition fluctuations and strain on gap bowing in $In_xGa_{1-x}N$. *Phys. Rev. B* **63**, 5204–8.
104. Wright, A. F. and Nelson, J. S. (1995). Bowing parameters for zinc-blende $Al_{1-x}Ga_xN$ and $Ga_{1-x}In_xN$. *Appl. Phys. Lett.* **66**, 3051–3.
105. Bellaiche, L., Mattila, T., Wang, L.-W., Wei, S.-W., and Zunger, A. (1999). Resonant hole localization and anomalous optical bowing in InGaN alloys. *Appl. Phys. Lett.* **74**, 1842–5.
106. Wetzel, C., Takeuchi, T., Yamaguchi, S., *et al.* (1998). Optical band gap in $Ga_{1-x}In_xN$ ($0 < x < 0.2$) on GaN by photoreflection. *Appl. Phys. Lett.* **73**, 1994–6.

107. Van de Walle, C. G., McCluskey, M. D., Master, C. P., Romano, L. T., and Johnson, N.M. (1999). Large and composition-dependent band-gap bowing in $In_xGa_{1-x}N$ alloys. *Materials Science Eng. B* **59**, 274–8.
108. Goldhahn, R., Scheiner, J., Shokhovets, S., *et al.* (1999). Determination of optical constants for cubic $In_xGa_{1-x}N$ layers. *Phys. Stat. Sol.* (b) **216**, 265–8.
109. Ramakrishnan, A., Wagner, J., Kunzer, M., *et al.* (2000). Piezoelectric field and confinement effects on the dielectric function spectrum of InGaN/GaN quantum wells. *Appl. Phys. Lett.* **76**, 79–81.
110. Takeuchi, T., Wetzel, C., Yamaguchi, S., *et al.* (1998). Determination of piezoelectric fields in strained GaInN quantum wells using the quantum-confined stark effect. *Appl. Phys. Lett.* **73**, 1691–3.
111. Park, S.-W. and Chuang, S.-L. (2000). Comparison of zinc-blende and wurtzite GaN semiconductors with spontaneous polarization and piezoelectric field effects. *J. Appl. Phys.* **87**, 353–64.
112. Bernardini, F., Fiorentini, V., and Vanderbilt, D. (1997). Spontaneous polarization and piezoelectric constants of III–V nitrides. *Phys. Rev. B* **56**, R10024–7.
113. Posternak, M., Baldereschi, A., Catellani, A., and Resta, R. (1990). *Ab initio* study of spontaneous polarization of pyroelectric BeO. *Phys. Rev. Lett.* **64**, 1777–80.
114. Qteish, A., Heine, V., and Needs, R. J. (1992). Polarization, band lineups, and stability of SiC polytypes. *Phys. Rev. B* **45**, 6534–42.
115. Albrecht, M., Christiansen, S., Remmele, T., *et al.* (1999). Spontaneous polarization of AlN measured by holography. *Institute of Physics Conference Series*, No. **164**, 361–6.
116. Grossner, U., Furthmüller, J., and Bechstedt, F. (1999). Initial stages of III-nitride growth. *Appl. Phys. Lett.* **74**, 3851–3.
117. Munkholm, A., Stephenson, G.B., Eastman, J.A., *et al.* (1999). Surface structure of GaN(0001) in the chemical vapor deposition environment. *Phys. Rev. Lett.* **83**, 741–4.
118. Zywietz, T., Neugebauer, J., and Scheffler, M. (1998). Adatom diffusion at GaN(0001) and $(000\bar{1})$ surfaces. *Appl. Phys. Lett.* **73**, 487–9.
119. Benedict, L. X. and Shirley, E. L. (1999). *Ab initio* calculation of $\varepsilon_2(\omega)$ including the electron-hole interaction: Application to GaN and CaF_2. *Phys. Rev. B* **59**, 5441–51.
120. Davydov, V. Y., Klochikhin, A. A., Goncharuk, I. N., *et al.* (2000). Raman studies of hexagonal $GaN/Al_xGa_{1-x}N$ multilayered structures. *IPAP Conference Series* **1**, 665–68.
121. Wagner, J. M., Gleize, J., and Bechstedt, F. (2000). Strain effects on structural, dielectric and lattice-dynamical properties of short-period GaN/AlN superlattices. *IPAP Conference Series* **1**, 669–72.

3 The homoepitaxial challenge for low-dimensional nitrides

Izabella Grzegory and Sylwester Porowski

3.1 Introduction

GaN and its alloys with Al and In recently became the basic materials for short-wavelength optoelectronics. This was due mainly to their direct energy gaps covering the whole visible spectrum and a large part of the UV range (6.2, 3.4, and 1.9 eV for AlN, GaN, and InN respectively). At present, high brightness blue and green light-emitting diodes (LEDs) and low-power blue laser diodes (LDs) are commercially available [1]. On the other hand, however, the development of GaN-based technology was and still is, strongly limited by difficulties in obtaining large, high quality crystals which could be used as substrates for epitaxial deposition of multilayer quantum structures necessary for devices. This is a direct consequence of thermodynamical properties of GaN [2] (and also AlN [3]) in particular, the melting conditions of which are so extreme, that the application of the common growth methods from stoichiometric liquids is technically impossible.

The spectacular development of GaN-based optoelectronic devices was achieved due to elaboration of the two-step Metal Organic Chemical Vapor Deposition (MOCVD) processes for the growth of both n- and p-type (GaAlIn)N epitaxial structures on highly mismatched sapphire substrates [1, 4]. In these structures dislocation densities are as high as 10^8–10^{10} cm^{-2}, but nevertheless, very efficient luminescence is possible especially if the structures contain InGaN [5]. Thus, it was suggested that GaN and its alloys are much more 'tolerant' of the presence of structural defects that the conventional III–V systems, due to the much lower minority carriers' diffusion length and relative immobility of structural defects in these strongly bonded crystals [6]. It is, however, well established [7] that structural defects present in heteroepitaxial nitrides influence the optical and electrical properties of GaN-based epitaxial layers and quantum structures, which lowers the efficiency and lifetime of some important classes of devices.

The dominant extended defects in the typical heteroepitaxial nitrides are threading dislocations (TD) and, associated with them, so-called V-defects, being the perturbation of the planar growth of InGaN and AlGaN due to the formation of low energy {10–11] planes around dislocations. The role of the threading dislocations in the non-radiative recombination processes in GaN and InGaN has been revealed by, for example, the cathodoluminescence (CL) mapping correlated with the TEM imaging [8]. These findings were supported by theoretical calculations [9] showing that some types of threading dislocations in nitrides are electrically and optically active.

The extremely high efficiency of the luminescence observed for InGaN, despite a very large number of dislocations, is often explained by the strong localization of carriers in deep potential wells caused by compositional fluctuations in InGaN alloys, which eliminates interaction of the carriers with dislocations [5]. The size and the nature of these fluctuations are still not precisely defined. This is difficult, since the incorporation of indium into the GaN lattice is a very unstable process (due to the instability of InN at high temperatures and the miscibility gap in GaN-InN system), very sensitive to the growth mechanisms and conditions. Thus InGaN of very different microstructures is grown in different laboratories.

At high excitations, however, even for In-containing structures, dislocations acting as the non-radiative recombination centers [5] quench the luminescence, and the efficiency of light-emitting devices is significantly limited. Therefore, in contrast to the InGaN-based LEDs, for the laser diodes (LD) dislocations limit both the efficiency and the lifetime of devices, and this is the most important obstacle in the construction of high-power blue lasers.

Homoepitaxy is the deposition of a thin layer of a crystalline material on the crystalline substrate of the same material having therefore exactly the same crystal structure and thermal expansion coefficients. If the substrate is perfect, the homoepitaxial layer should be also a perfect crystal. To approach this result experimentally, a very good coupling between the substrate and the growing layer is necessary. This means excellent preparation of the substrate surface and the choice of the conditions of epitaxy assuring stable crystallization of the layer. Such a stable growth can proceed by a two-dimensional layer by layer mechanism (mostly at low temperatures and/or on low indices exactly oriented surfaces) or by the atomic step flow (on surfaces vicinal to the low indices ones). From the point of view of the uniformity of the resulting layer, the step flow growth mode is the optimum one. This is because the incorporation of unintentional and intentional impurities is often very sensitive on the microscopic situation on the growing surface. The growth by propagation of an array of the well determined steps should lead therefore to the most uniform distribution of the impurities and minimization of the strains in the homoepitaxial layer.

The epitaxial growth techniques for III–V nitrides are already very well developed and it seems that the real technological challenge at present is to obtain large, high quality GaN and/or AlN crystals. For this reason, the methods for obtaining low dislocation density substrates for nitrides are reviewed in the first part of this chapter.

In the following, some examples of dislocation-free epitaxial GaN-based structures are analyzed to demonstrate the role of dislocations in both In-free and In-containing layers.

3.2 Low dislocation density substrates for low dislocation density GaN-based structures

It is already commonly accepted that the elimination of dislocations from structures based on GaN and its alloys will result in further important developments of both optoelectronic and electronic devices. Therefore considerable effort is being applied to overcome the problem of a defect-free, lattice-matched substrate for GaN.

Table 3.1 compares melting temperatures and pressures of most typical semiconductor materials like Si, GaAs, GaP, and the III–V nitrides.

The melting temperatures and the corresponding equilibrium pressure at melting of GaN and other III–N compounds are so extreme that they have not even been measured. The melting temperatures have been calculated by the use of Van Vechten's Quantum Dielectric Theory of Chemical Bonding [10]. The corresponding pressure follows from the extrapolation of the experimental equilibrium p_{N2}–T data. The table shows that both temperature and pressure at melting of GaN are much higher that for typical semiconductors. They are rather similar to the conditions used for high pressure synthesis of diamond. Due to these extreme melting conditions, GaN (like AlN and InN) cannot be grown from its stoichiometric melt by the Czochralski or Bridgman methods commonly used for typical semiconductors; it has to be crystallized by methods which allow lower temperatures and pressures.

Most spectacular and relatively widely applied, are methods based on controllable variations of the relative normal and lateral growth rates of GaN epitaxial layers deposited on foreign substrates patterned in different ways. These methods, called ELOG (Epitaxial Lateral Over Growth)[14], give quasi bulk large diameter GaN substrates which are locally almost dislocation-free. Usually the resulting substrates consist of a few micrometers wide near dislocation-free stripes divided by a few micrometers wide stripes with dislocation density of about $10^7 cm^{-2}$.

Table 3.1 Melting conditions of semiconductors.

Crystal	T^M, °C	p^M, atm.
Si	1400	<1
GaAs	1250	15
GaP	1465	30
GaN	**2500** [10]	**45 000** [11]
AlN	**2800** [10]	**>100** [12]
InN	**2200** [10]	**>60 000** [13]
diamond (synthesis)	1600	60 000

It has been demonstrated that, indeed, the reduction of defect density in the device structures deposited on the ELOG-type substrates leads to significant improvement of their performance. For instance, the external quantum efficiency of In-free GaN-based UV LEDs on the ELOG substrates increased fewer times by comparison with the diodes grown by the conventional MOCVD on sapphire [5]. The ELOG-type substrates were very important for the development of violet InGaN-based lasers, since only the application of these substrates allowed an increase in the lifetime of the devices over 10 000 h and their further commercialization.

Despite great advantages of the ELOG methods, there are some drawbacks, like strong bending of the substrates or quite complicated processing necessary for fabrication of both the substrates themselves and the devices on them.

Another 'epitaxial' approach to the low dislocation density substrate problem which does not require complicated processing, is concerned with methods based on the introduction of highly Silicon-doped interlayers during the epitaxial growth of GaN [15] or AlGaN [16] on SiC substrates. This allows for lower dislocation densities in the wafers by 2–3 orders of magnitude. It is suggested that this improvement is related to the anti-surfactant effect of Si leading to the self-organized formation of 3D-nano-masks and further growth by similar mechanisms as it is in the case of ELOG. Despite still quite high defect density (10^6–10^7 cm^{-2}), the substrates grown by the above methods can be as big as 3 inches (the current size of SiC substrates), and this is their great advantage.

Obviously, the most direct solution to the problem is to grow perfect bulk GaN and/or AlN crystals. There are several groups trying to get this badly needed result. The simplest (ideologically, not technically) methods allowing the growth of nitrides at temperatures and pressures much less extreme than it is required for melting, is the growth by sublimation–re-condensation and the growth from solutions.

Large (up to 14 mm in diameter) AlN boules have been grown [17] by the sublimation–re-condensation method based on the pioneering results of G. Slack [3]. A possibility to use these crystals for preparation of 1 cm^2 substrates with a very low density of low angle grain boundaries has been reported [17].

The solution growth methods are used mainly for crystallization of GaN. There are two techniques which give promising results. The low pressure–low temperature top seeded pulling from Ga-containing liquids [18] is very attractive in terms of growth rates achieved (1–2 mm/h) and the resulting crystals were boules up to 25 mm in diameter and 12 mm in length. The material is still polycrystalline but it contains single crystalline domains as large as 5–7 mm as it was reported recently [19].

The best structural quality GaN crystals (see Section 3.1.2) are grown from solutions of atomic nitrogen in liquid gallium under high N_2 pressure. This technique has been extensively developed at the High Pressure Research Center, Warsaw Poland [] and also recently, at the Japan Energy Corporation and Yamaguchi University [21]. The disadvantage of the method is that it produces individual platelets of the relatively small size of 1 cm^2. It is believed that further scaling of the crystal size by the use of bigger volume crucibles is possible. There are also promising results of directional crystallization with the use of single crystalline GaN seeds [22] (see also Section 3.1).

In this paper, the advantages of the use of dislocation-free GaN substrates for quantum structures will be demonstrated mostly by examples of the structures deposited on the substrates grown in Warsaw. This material is briefly characterized below.

3.3 GaN single crystalline substrates by the HNPS method

3.3.1 Morphology and size

The GaN crystals grown by the High Nitrogen Pressure Solution (see for example, ref. [2]) method are of wurzite structure, mainly in the form of hexagonal platelets (Fig. 3.1). The large hexagonal surfaces correspond to {0001} polar crystallographic planes. The polarity of the crystal surfaces was identified by etching in hot alkali water solutions, since the Ga-polar surface is inert to etching whereas the N-polar one is chemically active and can be etched quite easily [23]. The method was calibrated by CBED [24] and XPS [23] measurements. The side faces of the crystals are mainly the polar {10-11} and also nonpolar {10-10} planes. The habit of the crystals does not change for solutions containing up to about 1 at.% of Mg or Be.

The crystals in the form of hexagonal platelets growing slowly, with a rate <0.1 mm/h into {10$\underline{1}$0} directions (perpendicular to the *c*-axis), are usually single crystals of perfect morphology, suggesting stable layer-by-layer growth. They are transparent, with flat mirror-like faces. Quite often, however, the

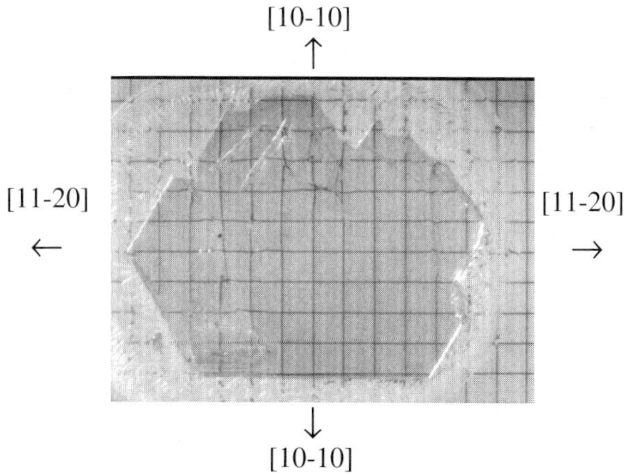

Fig. 3.1 Typical GaN crystal grown from solution of atomic nitrogen in liquid Ga at high N_2 pressure. The distance between grid lines, 1mm.

features of the unstable growth (macrosteps at the edges, periodic cellular structures) and also three dimensional, hexagonal growth figures grown during cooling of the system after crystallization in temperature gradient, are observed on the crystal surfaces. Thus polishing is necessary before using the crystals as substrates for epitaxial growth.

The average lateral size of crystals grown without an intentional seeding scales with the diameter of the high pressure reactor and at present is of about 10 mm.

From the analysis of the growth of GaN crystals without an intentional seeding it was deduced, that the growth into ⟨10-10⟩–directions, at the conditions corresponding to the growth rate of 0.05–0.1 mm/h, is stable and that, at the same conditions, the growth into directions ⟨11-20⟩ is also stable and even faster. Therefore the size of the stable platelets should be a function of the growth time (and volume of the solution) and further scaling of the crystal size with time and the volume of the crucible should be possible. It is also clear that the seeded growth into directions perpendicular to the *c*-axis should occur fast and in a stable way, especially into directions ⟨11-20⟩ [Fig. 3.2(a)]. This should lead to better control of the shape of the crystals and their distribution in the growth solution.

Seeded growth into directions parallel to the *c*-axis seems to be much more challenging since the growth rates observed in spontaneous crystallization are small (1–2 μm/h) and the growth shows strong tendencies to be unstable. Directional crystallization in vertical configuration, in stabilizing temperature gradient perpendicular to the growth front gives promising results in terms of the growth stability and the quality of the material [Fig. 3.2(b)] [25]. The method has to be optimized towards the increase of the growth rates by establishing even more uniform supersaturation at the crystallization front.

Since the hexagonal platelets grown without an intentional seeding have been used as substrates for epitaxy, the characterization given below is limited to them.

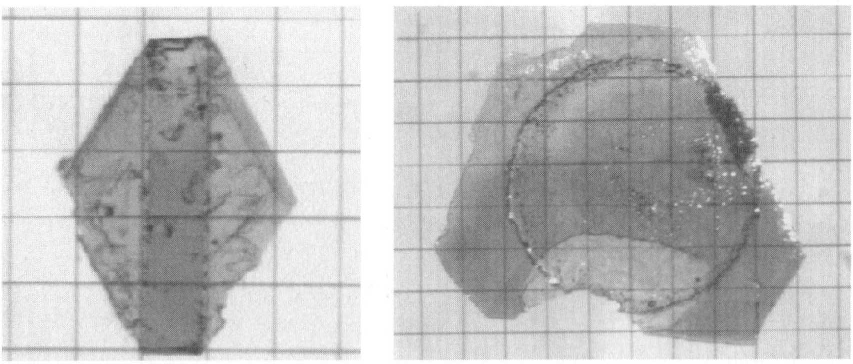

(a) (b)

Fig. 3.2 Seeded growth of GaN: (a) in directions ⟨11–20⟩, (b) in direction [0001]. The thickness of the new material grown during 50 h. directional crystallization process is about 0.5 mm.

3.3.2 Physical properties

The synthesis of GaN from its constituents is possibly due to the dissociative chemisorption of N_2 molecule on the Ga surface [26]. Nitrogen molecule dissociates on the Ga surface only if it overcomes the potential barrier of about 3.5 eV. For oxygen interacting with Ga there is no potential barrier for dissociation [27] and, therefore, even traces of this impurity in the growth system are the source of the unintentional oxygen doping of GaN. Therefore the crystals are strongly n-type with free electron concentration of about $5 \times 10^{19}\,cm^{-3}$ (metallic conductivity) and mobility of about 60 cm^2/Vs [28]. These free carriers can be fully eliminated by Mg or Be acceptors added into the growth solution. The resistivity of the crystals then becomes as high as 10^4–$10^6\,\Omega\,cm$ at 300 K. Usually the GaN:Mg crystals become p-type with an activation energy of 150 meV, at temperatures slightly exceeding 300 K [29]. The GaN:Be crystals are also highly resistive with the constant activation energy of 1.46 eV at temperatures up to at least 1000 K [30].

The structure of the pressure-grown GaN crystals has been studied by X-ray diffraction (XRD) [31], transmission electron microscopy (TEM) [i.e. 24, 32, 33, 34], defect selective etching (DSE) [35], and atomic force microscopy (AFM) of the homoepitaxial layers [36].

For the conductive crystals, the shape of the X-ray rocking curves ((0002) CuKα reflection) depends on the size of the crystal. The full widths at half maximum (FWHM) are 20–30 arcsec for 1mm crystals and 30–40 arcsec for 1-3 mm ones. For larger platelets the rocking curves often split into a few ~30–40 arcsec peaks showing a presence of low angle (1–3 arcmin) boundaries separating grains of a few millimeters in size. Misorientation between grains increases monotonically from one end to another end of the crystal [31]. It has been suggested that this is related to the polar character of the platelets growth. It was shown [25, 37, 38] that part of the platelet grown with N-polarity slightly differs in its physical properties from the part grown with Ga-polarity due to different content of both native and impurity related point defects. This can cause some strain and its subsequent relaxation through the formation of the low angle boundaries.

It was shown by Z. Liliental-Weber et al. [24], by TEM examination that the N-polar $(000\bar{1})$ surface of the n-type pressure grown GaN crystals (especially for the smaller ones) is often atomically flat (2–3 monolayer steps present) and that the crystals under this surface are practically free of extended defects. Under the opposite, rough surface a number of extended defects like stacking faults, dislocation loops, and Ga microprecipitates were observed. The relative thickness of this part usually consists of 10% of the entire thickness of the platelet. It seems that the presence of these defects is related to the growth instabilities often observed on the Ga-polar surface of the crystals grown without an intentional doping.

The detailed TEM studies of Mg-doped crystals are reported in ref. [32]. In particular it is shown that the introduction of Mg induces the formation of set of specific extended defects, being mainly the stacking faults situated at the {0001} polar surfaces of the crystals investigated.

For the use of the crystals as substrates for epitaxy, the near surface part of the material, being often the result of the unstable growth, has to be removed by mechanical polishing and the subsequent reactive ion etching (Ga-side) [40] or mechano-chemical polishing (N-side) [39].

As a rule the extended defects are not observed by TEM in the crystals used as substrates for both N and Ga polarity epitaxial growth. Therefore, if the epitaxy is properly performed both the substrate and the layer quantum structure do not contain dislocations. Figure 3.3 shows the AFM image of the Ga-polar (0001) surface of GaN substrate prepared for epitaxy by mechanical polishing, reactive ion etching, and annealing in the MOCVD system just before the epitaxial deposition. The array of straight monoatomic steps is the proof of both perfect substrate structure and good surface preparation.

Unfortunately, the TEM and AFM techniques allow analysis of only very small area samples. Therefore in order to measure dislocation densities in GaN, the Defect Selective Etching methods have been developed [35]. It was shown that etching in molten KOH-NaOH eutectics reveals dislocations in both GaN heteroepitaxial layers and GaN pressure-grown single crystals. The examples are presented in Fig. 3.4.

Figure 3.4a shows the result of defect selective etching of a typical GaN heteroepitaxial layer grown by MOCVD on sapphire substrate. High density of the etch pits is clearly visible. The same method applied for bulk crystals gives a very small (10–100 cm^{-2}) number of the etch pits. The typical pit observed on both heteroepitaxial layers and crystals is shown in Fig. 3.4(c). The pattern

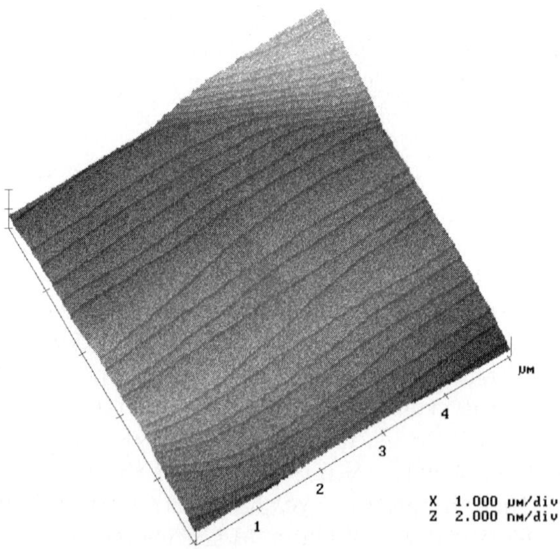

Fig. 3.3 Morphology of Ga-polar surface of pressure-grown GaN substrates (AFM, by G. Nowak) after annealing in gas mixture consisting of ammonia, hydrogen, and nitrogen with flows of 1000 : 1000 : 3300 sccm at atmospheric pressure. The sample was annealed for 10 min at 1050 °C. The surface before annealing has been subjected to the mechanical polishing and the reactive ion etching.

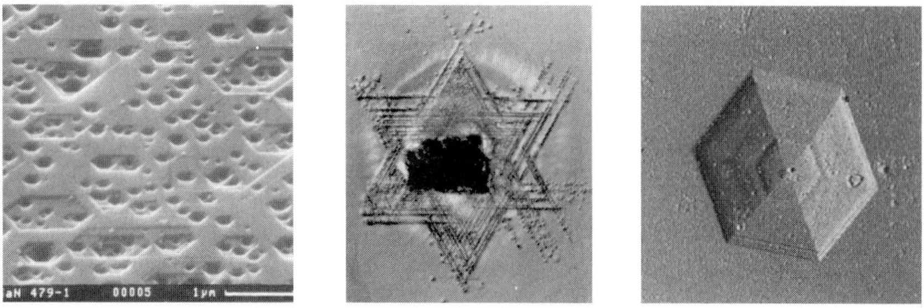

(a) (b) (c)

Fig. 3.4 Defect selective etching of GaN: (a) GaN/sapphire heteroepitaxial layer after etching in molten KOH-NaOH eutectics [35], (b) GaN pressure-grown single crystal after indentation with diamond and etching in molten KOH-NaOH eutectics. The average size of the star-like pattern is 100 μm, (c) etch pit (EP) on the GaN crystal surface after etching in molten KOH-NaOH eutectics, observed density of EP is $10–10^2 cm^2$.

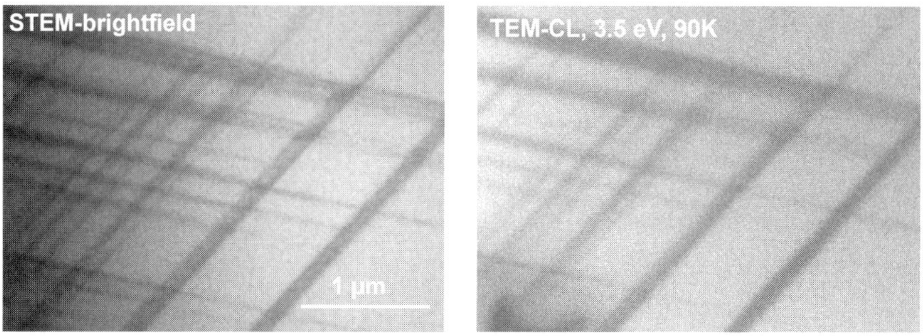

(a) (b)

Fig. 3.5 Dislocations generated in pressure-grown GaN crystals as non-radiative recombination centers: (a) TEM image of dislocations generated in GaN by indentation [see also Fig. 3.4 (b)], (b) near band-edge CL mapping of the area shown in Fig. 3.5(a). (Courtesy of M. Albrecht.)

in the Fig. 3.4(b) is the result of DSE of GaN single crystal with dislocations generated intentionally by indentation with diamond [40]. The etch pits surround the imprint of the diamond showing the area where dislocations are present and that the remaining material is dislocation-free. The CL studies [34] of these intentionally generated defects revealed that many of them are non-radiative recombination centers for band-edge emission of GaN. In Fig. 3.5, we have shown one of the monochromatic (3.5 cV) CL mappings from the indented area of the n-type pressure-grown GaN crystal.

3.4 GaN homoepitaxy on GaN pressure-grown substrates

If the surface preparation and the conditions of the epitaxial growth are right, the structure of GaN homoepitaxial layers, in terms of dislocation density,

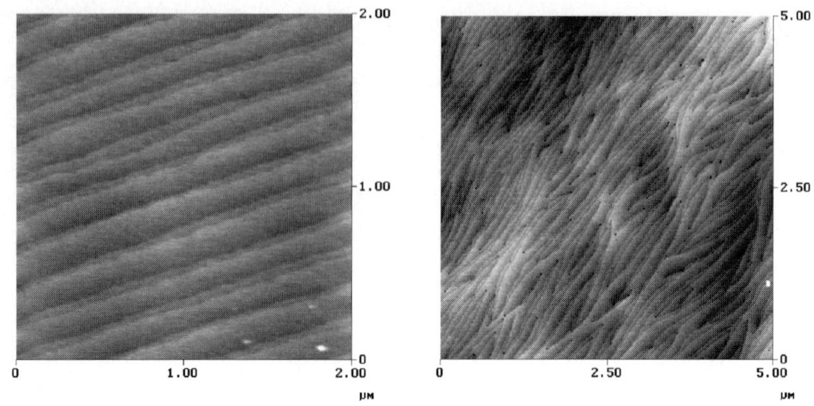

(a) (b)

Fig. 3.6 Surface morphology of GaN epitaxial layers deposited by MOCVD [41] on different substrates: (a) GaN pressure-grown crystal, (b) sapphire.

follows the structure of the GaN substrates. Figure 3.6 compares surface morphology of GaN epitaxial layers deposited by MOCVD on GaN crystal and on GaN/sapphire substrate, in the same run [41]. The atomic step flow on the surface of the homoepitaxial layer is not perturbed as it is observed to be for the GaN layer containing threading dislocations. If a GaN homoepitaxial layer is near dislocation-free, its optical properties depend mostly on the concentration and distribution of point defects incorporated during the epitaxial growth. It is a function of the purity of the growth system, the growth conditions, and the orientation of the substrate.

Due to the lack of strains and the high degree of homogeneity of the homoepitaxial GaN, the exciton related peaks in low temperature PL spectra are usually very narrow. The FWHM of the bound exciton lines less than 0.5 meV is observed for layers grown by both MOCVD (i.e. ref [42, 43]) and MBE (i.e. ref. [44], and [45]) on the Ga-polar surfaces of the GaN substrates (Fig. 3.7). For layers deposited on the N-polar surfaces the spectra are usually much wider due to the enhanced and non-uniform incorporation of unintentional impurities on this chemically active side of GaN crystal. This can be suppressed by the use of surfaces vicinal to the (000-1) as was shown by Zauner *et al.* [46], who obtained narrow excitonic spectra for GaN layers grown on misoriented N-polar surfaces of the GaN substrates (Fig. 3.7b).

For GaN layers grown by MBE with NH_3 nitrogen source, it was shown [45] that the band-edge PL integrated intensity at RT for the homoepitaxial GaN was about 60 times larger than for GaN grown on sapphire in the same growth conditions. Also the decay of the PL in these homoepitaxial layers was much slower than for corresponding (grown in similar conditions) heteroepitaxial GaN [45]. This indicated a substantial suppression of the non-radiative recombination by elimination of threading dislocations from the layers.

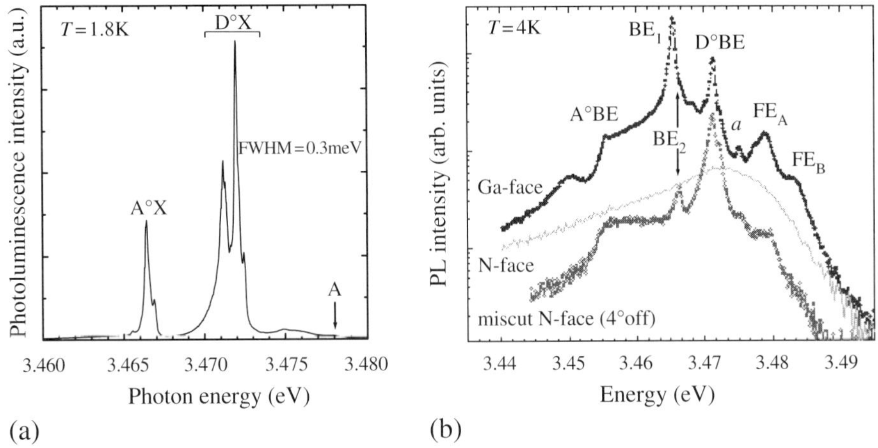

Fig. 3.7 Low temperature PL spectra of GaN homoepitaxial layers: (a) grown by NH_3 MBE on the Ga-polar surface of GaN substrate [45], (b) grown on N-polar surfaces (exactly oriented and vicinal) of GaN substrates [46].

The PL decay has been also studied at high excitations [47] for GaN layers grown by MOCVD on sapphire and pressure-grown GaN substrates. The decay time (at RT) of 450 ps measured for homoepitaxial material was five times longer than for corresponding heteroepitaxial layer grown at the same conditions. Figure 3.8 shows the decay of spontaneous luminescence measured at high excitation, close to the stimulated emission threshold for homo- and heteroepitaxial GaN.

The optically pumped LT stimulated emission [48] experiments confirmed that the near dislocation-free GaN is a much more efficient source of light than GaN-containing dislocations. The stimulated emission in GaN homoepitaxial layer grown by MOCVD started at the excitation power density lower than 1 MW/cm^2 whereas for GaN grown on SiC only at 10 MW/cm^2. The stimulated emission peak for the homoepitaxial layer was as narrow as about 5 meV (about 40 meV for GaN/SiC).

3.5 Quantum structures

3.5.1 Problems

As mentioned in the Introduction, there are also 'defects' increasing the optical efficiency of nitrides. It is often suggested that these 'defects' are the short-range potential fluctuations in InGaN alloys leading to localization of carriers and therefore to limitation of the influence of dislocations as the non-radiative recombination centers. In this model, the influence of dislocations on the optical efficiency of InGaN can be eliminated if the distance between dislocations exceeds the size of the potential fluctuations. However, for high injection currents

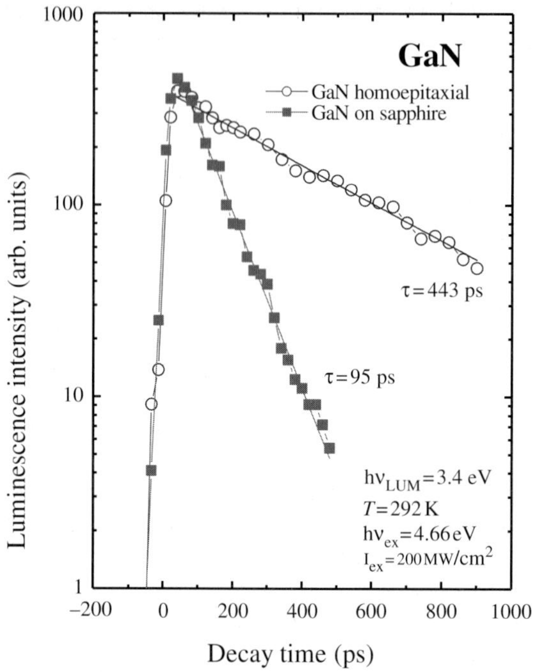

Fig. 3.8 Spontaneous luminescence transients for GaN films grown on sapphire (solid squares) and on GaN substrates (open circles) [48].

(i.e. in LDs), the potential minima are too shallow to contain all the injected carriers and therefore, for lasers, the reduction of dislocation density in the material is much more important than for LEDs. The above model explains in a consistent way the following characteristics of Nakamura's (the unquestionable leader in the field of GaN-based technology and optoelectronics) optoelectronic devices:

- The nitride LEDs containing In in their active structures are much more efficient than LEDs with pure GaN active layers (not alloyed with In).
- InGaN LEDs efficiency does not depend on dislocation density (if Al_2O_3 or ELOG are used) in contrast to the efficiency of In-free UV diodes.
- At high injection currents, the high dislocation density lowers the efficiency even for InGaN LEDs.

Therefore, one can expect that the elimination of dislocations from the structures should lead to:

- High efficiency In-free UV LEDs.
- Higher efficiency and higher power of both UV and visible laser diodes than is possible with dislocated structures.

In the following both In-free and In-containing structures are considered and the results supporting the above expectations are presented.

3.5.2 GaN/AlGaN—structures without indium

The growth of GaN/AlGaN structures on GaN pressure grown crystals has been studied mainly by MBE [33, 49, 50, 51]. Near dislocation-free GaN/AlGaN multi-quantum wells and structures have been grown on the N-polar surfaces of GaN crystals by RF plasma assisted MBE (PA MBE) [33, 49] and on the Ga-polar surfaces by MBE with NH_3 nitrogen source (R MBE) [45, 50]. As a rule, the optical properties of the structures without dislocations were much better than for similar structures grown in the same conditions, on sapphire.

For example, in Fig. 3.9, the integrated intensities of the PL coming from similar structures (8 ML GaN SQW with the $Al_{0.1}Ga_{0.9}N$ 50 nm barriers) grown by R MBE [45] on sapphire and bulk pressure grown GaN substrates are compared. It is shown that for the structure deposited on GaN, the PL is much stronger, especially at RT. For the structure grown on GaN substrate, the PL intensity starts to decrease with temperature only at about 100 K mainly due to the thermal escape of carriers from the quantum well towards the AlGaN barriers, as was observed for classical III–V QW heterostructures with dislocations densities lower than $10^3 cm^{-2}$ [51]. For heteroepitaxial structure, the presence of dislocations strongly influences the non-radiative recombination processes. The PL intensity starts to decrease at much lower

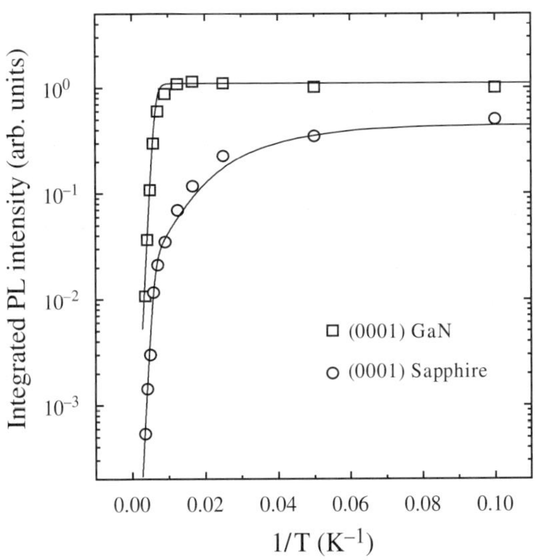

Fig. 3.9 Temperature dependence of the integrated PL intensity of homoepitaxial (squares) and heteroepitaxial (circles) $GaN/Al_{0.1}Ga_{0.9}N$ QWs [45].

temperatures due to delocalization of excitons and their further interaction with dislocations.

For devices on dislocation-free substrates, it is especially important that further increase of the emission efficiency can be achieved by increasing of the Al content in the barriers. This opens the possibility for the construction of highly efficient UV LEDs and lasers.

The semi-insulating GaN:Mg substrates have been used for growth GaN/AlGaN heterostructures with two-dimensional electron gas (2DEG) by R MBE [50]. The Hall mobility for 2DEG as high as 60 100 cm^2/Vs at 1.5 K (one of the highest reported for GaN) has been measured. The Shubnikov–de Haas oscillations starting at 1.8T and well defined QHE indicated the high quality of the MBE grown heterostructure. From the analysis of the scattering processes it was suggested that in such near dislocation-free structures the ionized impurity scattering was the dominant low-temperature mobility limiting mechanism.

3.5.3 InGaN containing structures

The InGaN near dislocation-free epitaxial layers and quantum wells were grown on the pressure-grown crystals mainly by MOCVD by Leszczynski *et al.* [36], in the High Pressure Research Center Unipress, Warsaw, Poland. The standard 470 nm LED structure with a single InGaN quantum well has also been grown by S. Nakamura [52] on the Unipress pressure grown crystal. In Fig. 3.10 the TEM images of one of the InGaN MQWs grown in Warsaw is shown to demonstrate that the structures are dislocation-free.

The compositional homogeneity of these materials was extensively studied by M. Albrecht *et al.* [53] using TEM and cathodoluminescence (CL) methods. It was shown that the InGaN microstructure is very sensitive to the growth conditions (temperature, growth rate, microscopic growth mode) and a

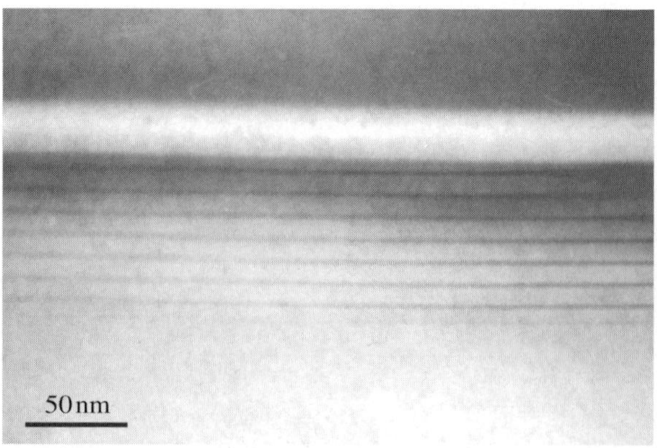

Fig. 3.10 InGaN MQW grown on the Ga-polar (0001) surface of GaN substrate by MOCVD [36], TEM. (Courtesy of M. Albrecht).

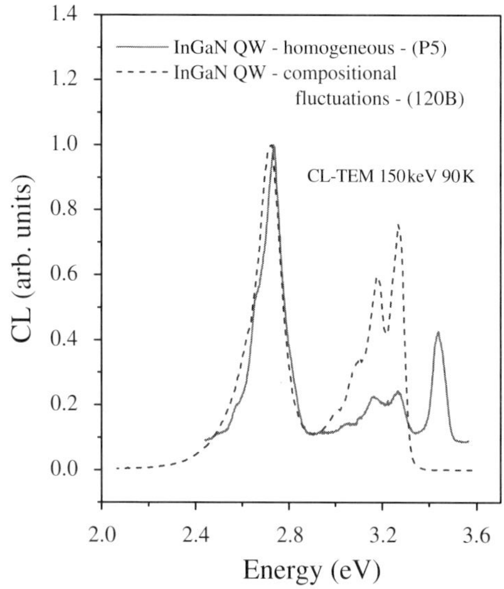

Fig. 3.11 Cathodoluminescence spectra coming from InGaN MQW grown on GaN substrate with large fluctuations of In content and from QW with homogeneous In distribution in InGaN. The average In fractions were the same in both structures. (Courtesy of M. Albrecht.)

different InGaN microstructure can be observed even for samples of the same average In content. For example, for InGaN QWs with 15% of In, the HRTEM tetragonal distortion analysis has shown either perfectly homogeneous distribution of In or 8–20% fluctuations of the In content with dimensions of a few nanometers. Nevertheless the optical properties of these QWs, with very different microstructures, were strikingly similar. This is shown in Fig. 3.11 where the cathodoluminescence spectra coming from these QWs are compared. The figure shows that the main CL emission lines coming from QW grown on GaN substrates with large (8–20%), nanometer scale In compositional fluctuations has a very similar energy position and FWHM than the line from the homogeneous QW, but with the same average In fraction in the well.

Figure 3.12 compares the electroluminescence spectra from LEDs with $In_{0.15}Ga_{0.85}N$ SQW grown on different substrates. And again the spectra are independently very similar on the substrate used.

Therefore, it can be suggested that the nanometer scale compositional fluctuations often observed in InGaN, are not responsible for the high efficiency InGaN based LEDs, which was the one of the most commonly accepted explanations. This suggestion is consistent with the observation of Leszczynski *et al.* [41] that the PL peaks energy position for InGaN QWs grown on various substrates (sapphire, SiC and GaN) depends only on the average In content (measured by X-ray and RBS), not on dislocation density.

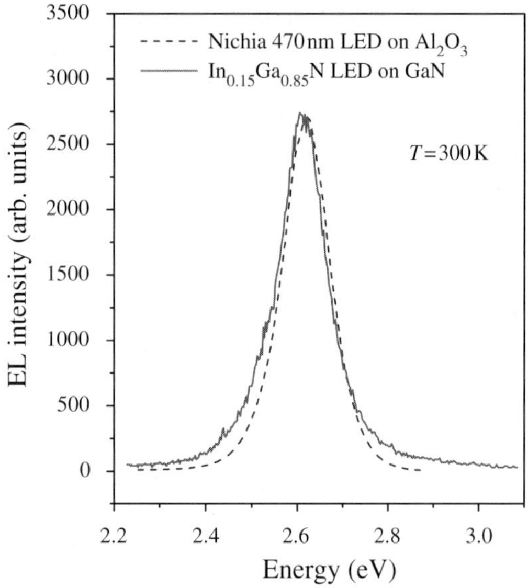

Fig. 3.12 Electroluminescence spectra from LEDs with InGaN SQW grown on different substrates. The average In fractions are the same in both structures.

The last observation and the dependence obtained is in agreement with the data of O'Donnell *et al.* [54] who found a linear relationship between PL band peak energy and measured In fraction in InGaN (for In content ranging from 0 to 40%). The data made it possible to suggest that localization in InGaN is due to non-random distribution of In atoms in InGaN in the atomic scale.

Thus, it seems that indeed:

- The strong localization of carriers in InGaN is not due to detectable compositional fluctuations because the emission from the homogeneous InGaN is the same as from InGaN with the large fluctuations detected, with similar average In content. I.
- The presence of centers localizing the carriers is the intrinsic property of InGaN on the atomic level. The additional support for this statement is that the emission peaks coming from InGaN of the same average composition are relatively broad (a few tenths of meV even for homogeneous, dislocation-free material) and of the same FWHM independently of the presence of the detectable In content fluctuations.

On the other hand, however, quite significant differences between the $In_{0.09}Ga_{0.81}N$ 0.2 μm-thick layers deposited on the pressure-grown GaN, SiC and sapphire substrates in the same run, have been found [55]. Despite slight differences (up to 50 meV, at 9 K) in the positions of the PL peaks (most probably caused by differences in thermal conductivity of the substrates), differences in the peak widths has been also observed. The FWHM of the peaks were 48, 80, and 90 meV for GaN, SiC and sapphire substrates respectively. The analysis of the spectral diffusion (shift of the peak position with time) of the PL of these three samples allowed the conclusion that the bandgap fluctuations in the InGaN layer grown on GaN substrate were much smaller (lower than 3 meV) than in the layers deposited on both sapphire (40meV) and SiC (90 meV) substrates. This indicates, for this particular growth, the strong influence of dislocations on the In segregation processes often observed by other authors (as, for example, in ref. [7]).

It is obvious that for dislocation-free InGaN structures, especially those to be used for lasers, the optimum is the homogeneous distribution of In in the quantum wells. Any fluctuations whatever their nature (i.e. caused by composition or thickness inhomogeneities), are no longer necessary. To achieve a high degree of homogeneity in the InGaN layers the best solution seems to be the growth on slightly misoriented substrates (like it was for the incorporation of unintentional impurities in the N-polar homoepitaxial GaN [46]). Then the continuous presence of the well determined steps on the surface allows to limit the growth mechanism to the step flow mode eliminating the necessity of two-dimensional nucleation which could lead to both the segregation of In and less perfect interfaces between quantum wells and barriers.

3.5.4 More complicated structures

Dislocations in the structures grown on the almost defect-free substrates can obviously appear as a result of the lattice mismatch between GaN and its ternaries InGaN and AlGaN. This was analyzed by Leszczynski *et al.* [36] by the X-ray measurements of lattice parameters of various GaN-based epitaxial layers deposited on GaN substrates in the function of InGaN (AlGaN) composition and thickness. It was shown that the boundary between fully strained and relaxed InGaN (AlGaN) layers is at a safe distance from mismatch–thickness area for layers necessary for a typical blue laser structure. These results are collected on the diagram in Fig. 3.13a. Figure 3.13(b) shows an example of the multilayer structure similar to the full structure of blue laser. The example confirms that no mismatch dislocations are generated if the structure is grown on dislocation-free GaN substrate.

The pressure-grown GaN crystal have been tested by S. Nakamura [52] as substrate for the InGaN 405 nm laser. The cw 30 mW devices with a lifetime exceeding 3000 h were constructed. The 30 mW power has been achieved at a current of 62 mA. For typical InGaN MQW LDs, at a current of 62 mA the output power was about 15 mW. This increase in the efficiency confirmed that dislocations are the main limiting factors for high power GaN-based lasers.

If the substrate is dislocation-free, the strain relaxation processes in low misfit heteroepitaxy is unaffected by the presence of threading dislocations. Both elastic and plastic relaxation has been studied [33] for GaN/AlGaN multilayer structures grown by plasma assisted MBE on GaN pressure grown substrates. It was shown that the crystal growth and relaxation in III-nitrides is not different from conventional III–V systems. The elastic strain relaxation by sinusoidal undulation of the AlGaN layers has been observed above

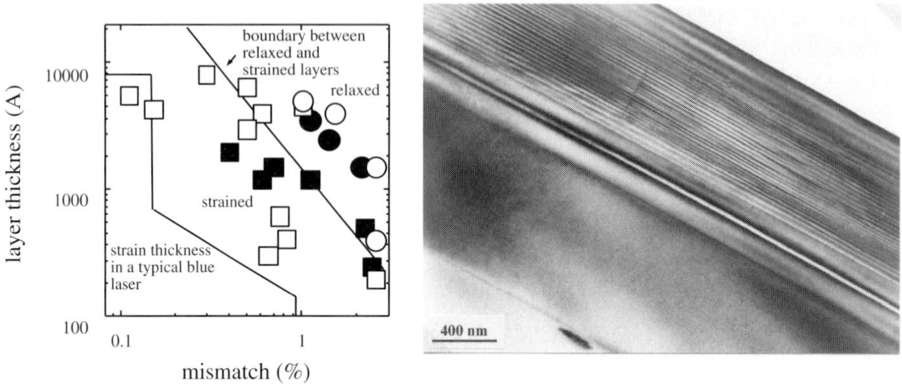

(a) (b)

Fig. 3.13 Critical conditions for III–N ternaries: (a) X-ray data: open circles–AlGaN relaxed, open squares–AlGaN strained, filled squares–InGaN strained, open squares–InGaN relaxed, (b) TEM image of the multilayer structure deposited on GaN substrate by MOCVD, the sequence of layers from the lower left corner: n-GaN, n-$Al_{0.11}Ga_{0.89}N$/n-GaN superlattice, n-GaN, $In_{0.09}Ga_{0.91}N$, p-GaN, p-$Al_{0.14}Ga_{0.86}N$/p-GaN superlattice, p-GaN.

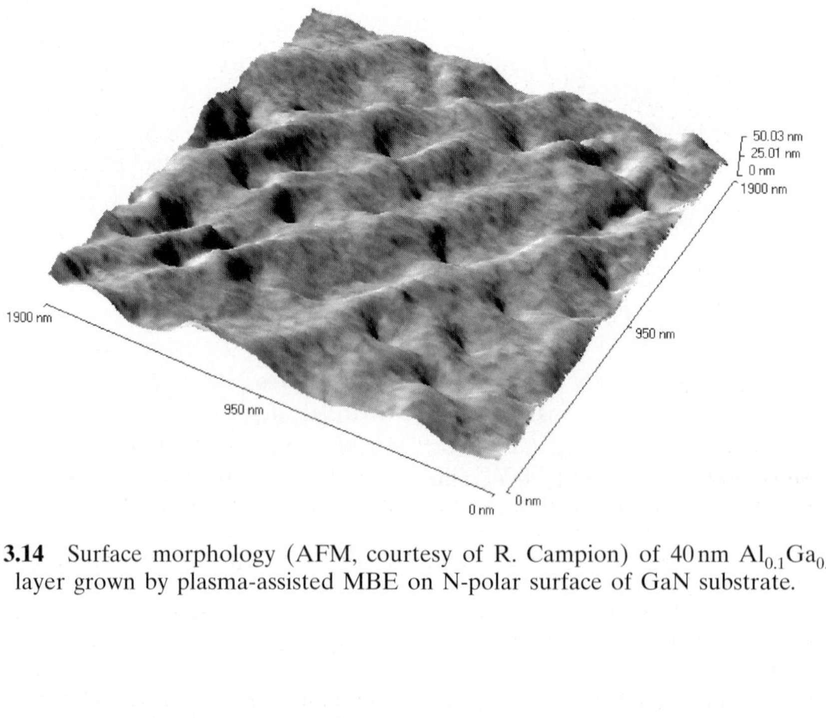

Fig. 3.14 Surface morphology (AFM, courtesy of R. Campion) of 40 nm $Al_{0.1}Ga_{0.9}N$ layer grown by plasma-assisted MBE on N-polar surface of GaN substrate.

a critical thickness depending on the Al content. An example of the AlGaN surface morphology with the sinusoidal undulation is shown in Fig. 3.14. The results obtained in ref. [33] indicate new possibilities for designing low misfit heteroepitaxial growth including self-organized formation of quantum structures.

3.6 Conclusions

The perfect single crystalline lattice-matched substrates for III nitrides will provide new technological possibilities and will stimulate further development of III–N physics and technology. The epitaxial growth on such substrates, especially on their surfaces vicinal to the low index crystallographic planes, should result in much more uniform layers regarding composition of ternary or quaternary compounds and both intentional and unintentional doping. This should also lead to the reduction of thickness fluctuations of thin layers constituting a quantum structure. Perfect interfaces are especially important for nitrides grown in polar (0001) directions, since then the fluctuations of the polarization induced internal electric fields can be minimized. The availability of bulk GaN and AlN substrates oriented in non-polar directions will open a new way for determination of the role of polarization-related effects in nitrides. The progress in both AlN [17] and GaN [22] bulk crystallization suggests that this possibility will appear in the very near future.

The application of dislocation free GaN substrates allows limitation of dislocation densities in the epitaxial GaN-based structures to very low values (lower than $10^2\,cm^{-2}$). In particular the full structures of InGaN-based blue lasers can be grown without creation of misfit dislocations. Thus the important centers of non-radiative recombination can be eliminated from both In-free and In-containing quantum structures, which is of crucial importance for efficient UV-emitting devices and for both UV and visible GaN-based high power lasers.

The absence of threading dislocations in the substrates should allow a bandgap engineering based on the typical elastic and plastic strain relaxation processes such as for classical low misfit heteroepitaxial systems.

3.7 References

1. Nakamura, S. and Fasol, G. (1997). Room temperature CW operation of InGaN MQW LDs. *The blue laser diode*, pp. 277–300. Springer-Verlag, Berlin.
2. Karpinski, J. and Porowski, S. (1984). High pressure thermodynamics of GaN. *J. Cryst. Growth* **66**(1), 11–20.
3. Slack, G. A. and McNelly, T. F. (1976). Growth of high purity AlN crystals. *J. Cryst. Growth* **34**, 263–79.
4. Nakamura, S., Mukai, T., Senoh, M., and Iwasa, N. (1992). Thermal annealing effects on p-type Mg-doped GaN films. *J. Appl. Phys.* (Part 2–Letters), **31**(2B), L139–42.
5. Nakamura, S. (2000). UV/blue/green InGaN based LEDs and laser diodes grown on epitaxially laterally overgrown GaN. *IEICE Trans.* **E83-C**(4), 529–35.

6. Speck, J. S. and Rosner, S. J. (1999). The role of threading dislocations in the physical properties of GaN and its alloys. *Physica B* 273–4, 24–32.
7. Sugahara, T. and Sakai, S. (2000) Role of dislocations in InGaN/GaN quantum wells grown on bulk GaN and sapphire substrates. *IEICE Trans.* **E83-C**(4), 598–604.
8. Rosner, S. J., Carr, E. C., Ludowise, M. J., Girolami, G., and Erikson, H. I. (1997). Correlation of cathodoluminescence inhomogeneity with microstructural defects in epitaxial GaN grown by metalorganic Chemica-vapor deposition. *Appl. Phys. Lett.* **70**, 420–6.
9. Elsner, J., Jones, R., Heggie, M. I. *et al.* (1998). Deep acceptors trapped atthreading-edge dislocations in GaN. *Phys. Rev. B* **58**, 12571–4.
10. Van Vechten, J. A. (1973). Quantum dielectric theory of electronegativity in covalent systems. III. Pressure-temperature phase diagrams, heats of mixing, and distribution coefficients. *Phys. Rev. B* **7**, 1479–507.
11. Karpinski, J., Jun, J., and Porowski, S. (1984). Equilibrium pressure of N/sub 2/over GaN and high pressure solution growth of GaN. *J. Cryst. Growth* **66**, 1–10.
12. Class, W. (1968). *Contract Rep.* NASA-Cr–1171.
13. Grzegory, I. (1996). Thermodynamics and crystal growth of III–N compounds at N2 pressure up to 2 GPa. In *Proc. Joint XV AIRAPT and XXXV EHPRG Conference* (Ed. W. Trzeciakowski), pp. 14–21.World Scientific, Singapore.
14. Hiramatsu, K. and Usui, A. (1999). Selective area growth and epitaxial lateral overgrowth of GaN. *EMIS Datareview Series* No. 23, published by INSPEC, The Institution of Electrical Engineers, London, 440–4.
15. Tanaka, S., Takeuchi, M., and Aoyagi, Y. (2000). Anti-surfactant in III-nitride epitaxy-quantum dot formation and dislocation termination. *Jpn. J. Appl. Phys.* (Part 2–Letters), **39**(8B), 15, L831–4.
16. Ainoya, M., Hirayama, H., Kinoshita, A., Hirata, A., and Aoyagi, Y. (2000). Presented at MRS 2000 Fall Meeting, Boston 2000.
17. Rojo, J. C., Slack, G. A., Morgan, K., Schowalter, L. J., Yakovlev, N. L., and Shusterman, Y. (2000). Growth of self-seeded aluminum nitride by sublimation-recondensation and substrate preparation. Presented at MRS 2000 Fall Meeting, Boston 2000.
18. Ivantsov, V. A., Sukhoveev, V. A., and Dmitriev, V. A. (1997). GaN crystals grown from a liquid phase at reduced pressure. *Gallium Nitride and Related Materials II. Symposium. Materials Research Society*, 143–8. Pittsburgh, PA, USA.
19. Albrecht, M., Nerding, M., Strunk, H. P., Ivantsov, V. A., Sukhoveev, V., and Dmitriev, V. A. (2000). Structure and optical properties of GaN bulk crystals grown from the liquid phase. Presented at MRS 2000 Fall Meeting, Boston 2000.
20. Grzegory, I. and Porowski, S. (1999). Growth of bulk GaN. *EMIS Datareview Series* No. 23, published by INSPEC, The Institution of Electrical Engineers, London, 440–4.
21. Inoue, T., Seki, Y., Oda, O., Kurai, S., Yamada, Y., and Taguchi, T. (2000). Growth of bulk GaN single crystals by the pressure controlled solution method. *Jpn J. Appl. Phys.* **39**, 2394–8.
22. Grzegory, I., Boc'kowski, M., Lucznik, B., Wróblewski, M., Teisseyre, H., Borysiuk, J., and Porowski, S. (2001). Seeded growth of GaN at high N_2 pressure on (0001) polar surfaces of GaN single crystalline substrates. To be presented at the International Conference on Materials for Advanced Technologies, 1–6 July 2001, Singapore.
23. Seelmann-Eggebert, M., Weyher, J. L., Obloh, H., Zimmermann, H., Rar, A., and Porowski, S. (1997). Polarity of (00.1) GaN epilayers grown on a (00.1) sapphire. *Appl. Phys. Lett.* **71**(18), 2635–7.

24. Liliental-Weber, Z., Kisielowski, C., Ruvimov, S., Chen, Y., Washburn, J., Grzegory, I. *et al.* (1996). Structural characterization of bulk GaN crystals grown under high hydrostatic pressure. *J. Electron. Mater.* **25**(9), 1545–50.
25. Grzegory, I. (2001). High pressure growth of bulk GaN from solutions in gallium. *J. Phys. Condensed Matter* **13**, 6875–92.
26. Krukowski, S., Romanowski, Z., Grzegory, I., and Porowski, S. (1999). Interaction of N2 molecule with liquid Ga surface – quantum mechanical calculations (DFT). *J. Cryst. Growth* **189/190**, 159–162.
27. Krukowski, S., Bockowski, M., Lucznik, B., Grzegory, I., Porowski, S., Suski, T., and Romanowski, Z. (2001). High-Nitrogen-Pressure growth of GaN single crystals: doping and physical properties. *J. Phys.: Condensed Matter* **13**, 1–11.L
28. Perlin, P., Camasel, J., Knap, W. *et al.* (1995). Investigation of longitudinal-opticalphonon-plasmon coupled modes in highly conducting bulk GaN. *Appl. Phys. Lett.* **67**, 2524–6.
29. Litwin-Staszewska, E., Suski, T., Piotrzkowski, R. *et al.* (2001). Temperature dependence of electrical properties of gallium nitride bulk single crystals doped with Mg and their evolution with annealing. *J. Appl. Phys.* **89**(12), 7960–5.
30. Suski, T., Litwin-Staszewska, E., and Perlin, P. (2000). Optical and electrical properties of GaN:Be grown under high nitrogen pressure. To be published in the *Proceedings of EGW-4, Nottingham 2000.*
31. Leszczynski, M., Grzegory, I., Teisseyre, H. *et al.* (1996). The microstructure of gallium nitride monocrystals grown at high pressure. *J. Cryst. Growth* **169**(2), 235–42.
32. Liliental-Weber, Z., Benamara, M., Swider, W. *et al.* (1999). Ordering in bulk GaN:Mg samples: defects caused by Mg doping. *Physica B* 273–4, 124–9.
33. Christiansen, S. H., Albrecht, M., Strunk, H. P. *et al.* (1999). Relaxation processes of AlGaN/GaN heterostructures grown onto single crystal GaN(0001) substrates. *Proc. ICNS'99, Monpellier, France.* In *Physica Status Solidi* (a), **176**(1), 285–9.
34. Weyher, J. L., Albrecht, M., Wosinski, T., Nowak, G., Strunk, H. P., and Porowski, S. (2001). Study of individual grown-in and indentation-induced dislocations in GaN by defect-selective etching and transmission electron microscopy. *Materials Science & Engineering B–Solid State Materials for Advanced Technology*, vol. B80, no. 1–3, 22 March 2001, pp. 318–21, Elsevier.
35. Weyher, J. L., Brown, P. D., Rouviere, J-L., Wosinski, T., Zauner, A. R. A., and Grzegory, I. (2000). Recent advances in defect-selective etching of GaN. *J. Cryst. Growth* **210**(1–3), 151–6.
36. Leszczynski, M., Prystawko, P., Czernecki, R. *et al.* (2000). Epitaxy on GaN bulk crystals. To be published in *Proc. International Conference on Solid State Crystals, Zakopane, Poland*, Oct. 9–13, 2000.
37. Suski, T., Jun, J., Leszczynski, M. *et al.* (1998). High pressure fabrication and processing of GaN:Mg. *Mater. Science and Engineering B* **59**, 1–10.
38. Frayssinet, E. (2000). Elaboration et etude d'heterojonctions GaN/AlGaN deposees sur GaN massif. Ph.D. thesis, Universite Montpellier II.
39. Weyher, J. L., Müller, S., Grzegory, I., and Porowski, S. (1997). Chemical polishing of bulk and epitaxial GaN. *J. Cryst. Growth* **182**, 17–23.
40. Karouta, F., Weyher, J. L., Jacobs, B., Nowak, G., Presz, A., Grzegory, I., and Kaufmann, L. M. F. (1999). Final polishing of Ga-polar GaN substrates using reactive ion etching. *J. Electron. Mater.* **28**(12), 1448–51.
41. Leszczynski, M., Prystawko, P., and Nowak, G. (1999). Private communication.
42. Stepniewski, R., Potemski, M., Wysmolek *et al.* (1999). Symmetry of excitons in GaN. *Phys. Rev. B–Condensed Matter* **60**(7), 15, 4438–41.

43. Kirchner, C., Schwegler, V., Eberhard, F. *et al.* (1999). Homoepitaxial growth of GaN by metalorganic vapor phase epitaxy: A benchmark for GaN technology. *Appl. Phys. Lett.* **75**, 1098.
44. Teisseyre, H., Nowak, G., Leszczynski, M. *et al.* (1996). MRS Internet *J. Nitride Semicond. Res.* **1**, 13.
45. Grandjean, N., Damilano, B., Massies, J., Neu, G., Teissere, M., Grzegory, I. *et al.* (2000). Optical properties of GaN epilayers and GaN/AlGaN quantum wells grown by molecular beam epitaxy on GaN(0001) single crystal substrate. *J. Appl. Phys.* **88**(1,1), 183–7.
46. Zauner, A. R. A., Weyher, J. L., Plomb, M. *et al.* (2000). Homo-epitaxial GaN growth on exact and misoriented single crystals: suppression of hillock formation. *J. Cryst. Growth* **210**, 435–43.
47. Jursenas, S., Kurilčik, N., Kurilčik, G. *et al.* (2001). Decay of stimulated and spontaneous emission in highly excited homoepitaxial GaN. *Appl. Phys. Lett.* **78**(24), 3776–8.
48. Holst, J. Chr., Hoffman, A., Leszczynski, M., and Porowski, S. Private communication.
49. Foxon, C. T., Davis, C. S., Novikov, S. V. *et al.* (1999). *Proc. ICNS'99, Montpellier, France*, in *Phys. Stat. Sol.* (a), **176**(723).
50. Frayssinet, E., Knap, W. P., and Lorenzini, N. (2000). High electron mobility in AlGaN/GaN heterostructures grown on bulk GaN substrates. *Appl. Phys. Lett.* **77**(16), 2551–3.
51. Lambkin, J. D., Dunstan, D. J., Homewood, K. P., Howard, L. K., and Emeny, M. T. (1990). Thermal quenching of the photoluminescence of InGaAs/GaAs and InGaAs/AlGaAs strained-layer quantum wells. *Appl. Phys. Lett.* **57**, 1986–8.
52. Nakamura, S. (1999). Private communication.
53. Albrecht, M. Private communication.
54. O'Donnell K. P., Martin, R. W., and Middleton, P. G. (1999). Origin of luminescence from InGaN diodes. [journal paper] *Phys. Rev. Lett.* **82**(1,4), 237–40.
55. Korona, K. P., Kuhl, J., Prystawko, P., Leszczynski, M., Perlin, P., and Suski, T. (2001). Influence of substrate on the photoluminescence dynamics in InGaN epilayers. Submitted to *Appl. Phys. Lett.*

4 HVPE-GaN quasi-substrates for nitride device structures

Tanya Paskova and Bo Monemar

4.1 Introduction

The development of III-nitride semiconductor devices for optoelectronic and electronic applications is driven by improvements in the epitaxial growth of these semiconductor materials. Device structures have traditionally been obtained using either MOCVD or MBE in many semiconductor material systems. The present device applications for III-nitrides have used MOCVD (particularly in the commercially work at Nichia Chemical [1] and Cree Research [2]) and MBE growth has lagged behind for optoelectronic device applications. However, recent developments in the MBE technology result in improved material quality and make the competition between them sharper. Moreover, the best quality MBE layers have been obtained using MOCVD-GaN templates [3], indicating that a combination of different growth techniques may be a way for producing the next generation III-nitride materials.

Despite remarkable achievements, the nitride materials still suffer from a very high defect density due to the lattice mismatch between the nitrides and all the available foreign substrates. Since bulk GaN single crystals are still limited to very small crystals, the HVPE growth technique attracts great attention, mainly due to its ability to produce thick layers of high quality with high growth rates at low cost. Also, recently developed techniques for substrate removal give the opportunity to develop quasi-substrate material of good quality [3–7]. Development of the HVPE technique combined with other techniques for producing GaN templates may become the key to resolve the high-density defect issue in the III-nitride device technology.

A high quality thick HVPE-GaN layer may be used either as a template, residing on the widely used sapphire substrate, or as a free-standing quasi-substrate for consequently grown device structures. In this work we review the current knowledge of the key growth-related properties of thick HVPE-GaN films relevant for the quasi-substrate application. We describe the HVPE growth process specific to III-nitride materials in order to gain insight into the fundamental aspects of the process and its relation to the material quality. The discussion is focused on three areas of critical importance for HVPE-GaN quasi-substrate application and consequently for the device performance.

First of all, the crystal quality of the thick layers and the possibilities for defect reduction and improvement of optical and electrical properties are reviewed. Second, the substrate separation processes are discussed. Third, physical properties of the free-standing films after separation and regrown films on the free-standing quasi-substrate films are briefly summarized with emphasis on the critical issues of further exploration in crystal growth improvement.

4.2 Thermodynamic basis

4.2.1 Thermodynamic analysis of HVPE growth of GaN

Gallium nitride is characterized by a high bond energy of 2.24 eV/bond and the estimated values of its melting temperature (2790 °C) and decomposition N_2 pressure (>30 kbar) [9] make melt growth impossible. One choice is therefore to obtain GaN single crystals from solution. However, the solubility of N_2 in liquid Ga metal also suffers from the high bond energy of N_2 (226 kcal/mole) and requires rather difficult conditions like a temperature of ~1500 °C and a pressure close to 15 kbar. The HVPE technique is another option to grow thick quasi-substrates due to its high growth rate up to 800 μm/h [10] although typically used values are much lower, 50–120 μm/h.

An accurate thermodynamical analysis of the HVPE growth process needs a postulation of the chemical reaction pathway. All practical methods for synthesis of GaN at high temperature utilize an active source of nitrogen, such as ammonia, NH_3, and a vapour source of Ga. In the HVPE growth, Ga precursors can be obtained by two reactions:

$$\mathrm{Ga(l) + HCl(g) \rightarrow GaCl(g) + 1/2H_2(g)} \tag{4.1a}$$

$$\mathrm{GaCl(l) + 2HCl(g) \rightarrow GaCl_3(g) + H_2(g),} \tag{4.1b}$$

Further, two reaction pathways yelding GaN deposition can exist [3]:

$$\mathrm{GaCl(g) + NH_3(g) \rightarrow GaN(s) + HCl(g) + H_2(g)} \tag{4.2a}$$

$$\mathrm{3GaCl(g) + 2NH_3(g) \rightarrow 2GaN(s) + GaCl(g) + 3H_2(g),} \tag{4.2b}$$

where l=liquid, g=gas, s=solid.

The GaN deposition is determined by the efficiency of the two [eqns (4.1) and (4.2)] chemical reactions. The efficiency of the first step [eqn (4.1)] was found to be up to 99% [11] depending on the temperature, the position of the HCl inlet, the carrier gas ambient and the liquid Ga surface exposed to the HCl gas. The second chemical reaction step depends on the reversion of the extent of ammonia decomposition into nitrogen and hydrogen, because GaN cannot be deposited by direct reaction between GaCl and N_2. It is known that ammonia is a thermodynamically unstable gas at temperatures usually employed in the GaN growth. However, the thermal decomposition of NH_3 is a very slow reaction and when no catalyst is present, no more than about 3–4% of the NH_3 was decomposed at temperatures higher than 950 °C [11].

The main gaseous species implied in the HVPE growth of GaN are GaCl, $GaCl_3$, HCl, NH_3, H_2, and the carrier gas. Thermodynamical functions of the material, the enthalpy $H(T)$, entropy $S(T)$ and specific heat $C_p(T)$, of all the species involved in the growth have been calculated by Przhevalski *et al.* [12] for the standard pressure of 1 atm and arbitrary temperature using a polynomial approximation of the Gibbs free energy. Detailed thermodynamical calculations of HVPE growth of GaN has been published by Ban *et al.* [11], Seifert *et al.* [10], and recently by Koukitu *et al.* [13] and Aujol *et al.* [14]. Based on their analysis, several important features can be pointed out, different from the growth of other III–V materials, and should be taken into account for successful HVPE-GaN growth. (i) An inert ambient atmosphere is needed and is more effective for the GaN growth than hydrogen atmosphere which is favorable for other III–V systems, because in the GaAs growth, for example, hydrogen is a reactant while it is a reaction product in the GaN growth; (ii) In HVPE-GaN growth, NH_3 should be used as the source of the group V element rather than a nitrogen halide (NCl_3), which is highly explosive. Moreover, the thermal dissociation of NH_3 results in the formation of N_2 molecules which are extremely stable and unreactive at the temperatures used while the thermal dissociation of $AsCl_3$, in GaAs growth, results in the formation of As_2 and As_4 molecules, which typically remain volatile and chemically reactive in next step of the growth reaction; (iii) Another problem comes from the fact that the GaN reaction tends to create huge amounts of extraneous products of NH_3Cl, $GaCl_3$ and $GaCl_3{\cdot}NH_3$, leading to a condensation and undesirable deposition of GaN on the quartz.

4.2.2 Kinetic analysis of HVPE growth of GaN

The kinetics of the process and growth mechanisms occurring at the solid/vapor interface during HVPE growth of GaN is at a low level of understanding, although it is widely admitted that the initial nucleation is of critical importance for good-quality growth. The available models [14, 15] assume a surface process involving the following steps: (i) adsorption of NH_3 molecules, (ii) adsorption of N atoms coming from ammonia decomposition, (iii) adsorption of GaCl molecules on the N atoms forming NGaCl, (iv) decomposition of the NGaCl via different desorption mechanisms (Fig. 4.1). Two of them were suggested in analogy with the GaAs model: desorption forming HCl and desorption forming $GaCl_3$ [15]. One more $GaCl_2$ desorption mechanism was also recently suggested based on the experimental results [16]. Statistical treatment of the dynamic equilibrium between the adsorbed and gas

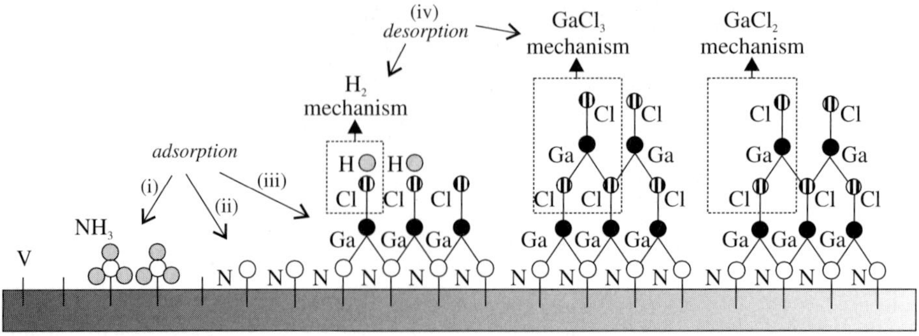

Fig. 4.1 Schematic drawing of possible reaction steps. Adsorption (i–NH_3 adsorption, ii–N adsorption, iii–GaCl adsorption) and desorption (H_2 desorption [15], $GaCl_3$ desorption [15] and $GaCl_2$ desorption [16]) mechanisms of the surface growth process of HVPE growth of GaN.

phase species allowed explicit expressions of the growth rates via the different pathways.

The growth rate in HVPE-GaN growth is the most interesting parameter, especially for its quasi-bulk application. Although several papers have dealt with the dependence of the growth rate on the growth conditions, the information obtained is not sufficient for a complete picture of determining growth mechanisms in the HVPE-GaN growth. The three dimensional (3D) nucleation growth is a common growth mechanism for heteroepitaxy of systems with high lattice mismatch between the substrate and the layer, and islanding driven by an increasing strain component in the total system free energy usually leads to relaxation of the system [17]. However, the study of microstructural evolution in the early stages of HVPE-GaN growth revealed a very fast and abrupt full coalescence [18], in contrast to slower island coalescence in MOCVD-GaN growth [19]. A 2D-multilayer growth is a realistic mechanism ensuring typically smooth morphology [20, 21] with a presence of steps with 30–80 nm wide and 5–6 Å height [Fig. 4.2(a, b)]. Additionally, a contribution of a spiral growth mechanism was also evidenced at a growth rate higher than 80 μm/h [Fig. 4.2(c, d)] [21]. So, the simultaneous operation of both mechanisms: 2D-multilayer as a base and spiral growth which provides steps and enhances nucleation sites in the upper part of the films might explain the high growth rates approached in the HVPE growth of GaN.

4.3 HVPE-GaN thick layers—heteroepitaxial growth

Although considerable progress has been made in the crystal quality of thick GaN layers grown by HVPE over the last years, many defect-related issues of thick films are still unsolved or subject of controversy. Different types of extended defects are common in heteroepitaxial GaN growth because of

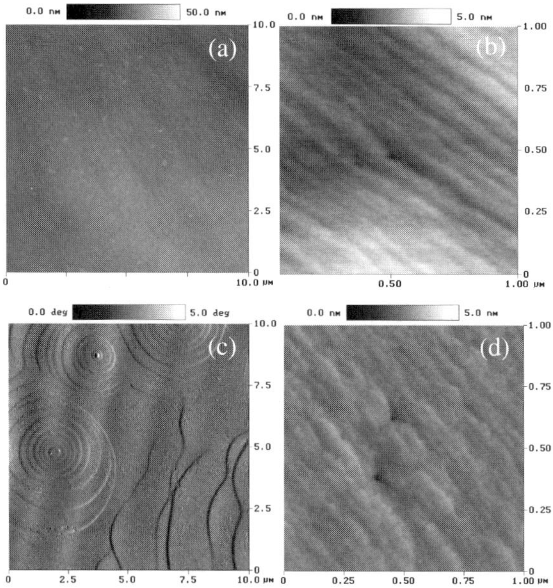

Fig. 4.2 AFM micrographs of the surface of 70-μm-thick GaN layers grown at growth rates of: (a) and (b) 30 μm/h; and (c) and (d) 80 μm/h.

the dissimilarities between the lattice parameters and thermal expansion coefficients of GaN and the substrate. Since most of the structural defects can propagate into the subsequently grown structure, the understanding of the fundamental processes leading to their formation is very important, in order to optimize the nucleation and growth stages to reduce the high defect density in the layers.

4.3.1 Specific characteristics of HVPE-GaN

In HVPE-GaN layers grown on different substrates, domain structure near the substrate interface was observed by cathodoluminescence (CL) imaging, as reported by several groups [22–24]. We have studied in detail the elongated domains of an inverted pyramidal shape at the interface of GaN grown on sapphire without a buffer [25–27]. Bright domains of similar shape and smaller dimensions are also reported at the GaN/sapphire interface with ZnO buffer [28], and at the interface of GaN grown on SiC [24]. Recently, similar observations were reported in very thick epitaxial lateral overgrown HVPE-GaN layers [29] in the coalescence regions. They seem to be inherent defects in highly mismatched GaN heteroepitaxy at high growth rates.

An overall view of the cross-section of a representative sample obtained by CL imaging shows well-defined bright domains close to the sapphire interface [see Fig. 4.3(a)]. The strong emission contrast can be attributed to difference in radiative lifetimes most probably related to increased impurity incorporation

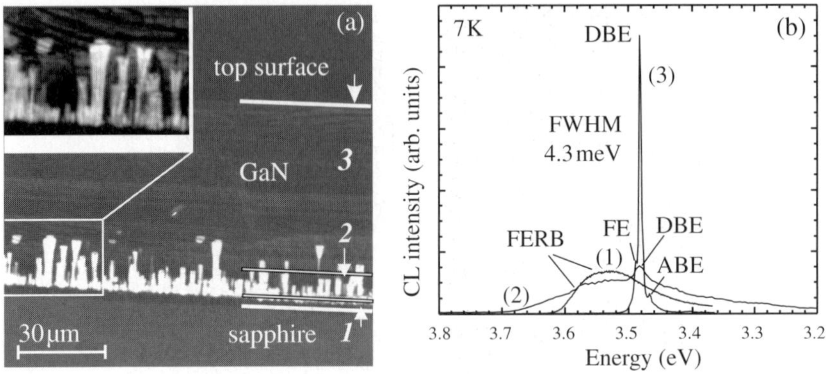

Fig. 4.3 (a) Panchromatic CL image of cross-section of 75-μm-thick GaN layer with three specific layer regions. (b) Local CL spectra taken in the respective regions in cross-section of the film.

into the bright domains. It is confirmed by the CL spectra [Fig. 4.3(b)] taken from areas aligned along the [0001] direction as marked in the panchromatic CL image. The spectrum from the interface region (1) shows a broad blueshifted emission typical of highly doped material ($>10^{18} cm^{-3}$) found to be due to free-electron recombination (FERB) [30]. The spectrum from the columnar region (2) shows a simultaneous presence of excitonic and FERB CL emissions. This broad emission band decreases with increasing thickness, disappears completely at a certain distance from the sapphire, and the luminescence becomes excitonic over the whole remaining part (3) of the film typical for good-quality material with a free carrier concentration lower than $10^{17} cm^{-3}$.

It was found that the columnar domain presence has a strong impact on the electrical properties of the material [30]. That is why avoiding such domains is of crucial importance. A study of their evolution showed that most of the domains preserve a shape of an elongated inverted pyramid, and are terminated within a sublayer with a thickness varying in the range of 1–15 μm. Although the thickness of the sublayer can be reduced by changing the growth conditions, mainly the growth rate [26], it cannot be avoided completely without using buffer layers. Very few of the domains become larger in size, propagate through the whole layer and terminate at the layer surface with a hexagonal inverted pyramid-like pit (15–20 μm in diameter) dependent on the layer thickness and/or the growth rate. Some of them are empty, others are filled with overgrown material [25]. The pits are regularly oriented, with the characteristic hexagonal shape defined by the boundary lines perpendicular to the [1–100] directions, and inclined walls along polar planes {1–101}.

The TEM images taken from the same interface region, observed by the CL, show isolated domains of a similar shape, elongated in the [0001] direction [Fig. 4.4(a)]. They are about 100–300 nm wide at the interface and up to a few micrometers when terminated. Good-quality material and a few dislocations

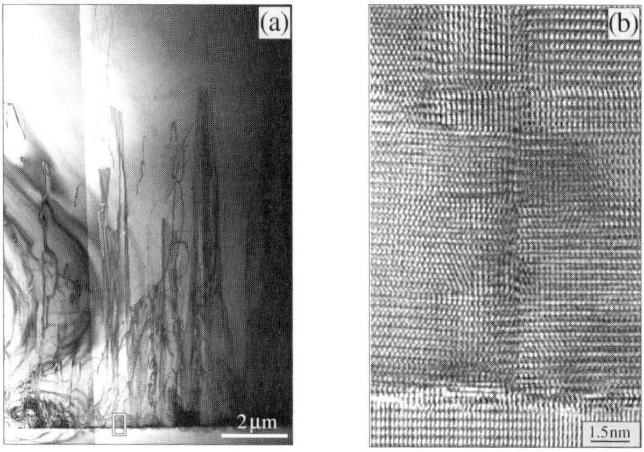

Fig. 4.4 Cross-sectional TEM (a) and HRTEM (b) images of GaN/sapphire interface. The box indicates the spot for the HRTEM image.

are observed between them. Two of the six limiting boundaries enclosing the domains are seen in the TEM image. An inclination angle of ~4° is estimated between the boundaries and the normal to the GaN (0001) plane. A higher magnification image of a region at the layer/sapphire interface [marked by a square in Fig. 4.4(a)] reveals that the domain boundary is associated with a substrate *c*/3 step formed along the {11–20} Al_2O_3 planes [Fig. 4.4(b)]. The image shows that the domain boundary arises from the coalescence of two crystallites of nucleation islands that grow on the two adjacent terraces on both sides of the step. The two crystallites are slightly tilted (~1°) towards the boundary plane. In Fig. 4.4(b) one can also see that the lattices from the two sides of the boundary are displaced. Following the alignment of the lattice planes it seems that there is a shift along [0001] axis on one side of the boundary introduced by the step, and this forms a stacking mismatch boundary.

4.3.2 Thickness improvement

The occurrence of the basal plane stacking faults and stacking mismatch boundaries during the early stages of growth could help the stress relaxation mechanism in such heteroepitaxial systems. Micro-Raman spectroscopy measurements performed in cross-section along a cleaved edge of a sample can quantify the stress distribution with growth thickness via the E_2 mode position sensitivity to the strain. The relaxation effect is clearly revealed by the observed strain gradient along the layer thickness. After the initial strain decrease during the first micrometers, the remainder of the layer is left almost constantly compressively stressed in the (0001) plane with the stress amount of about 0.3 Gpa [27]. The steep strain decrease mostly within the columnar domain region gives evidence that the lattice can elastically relax the built-in

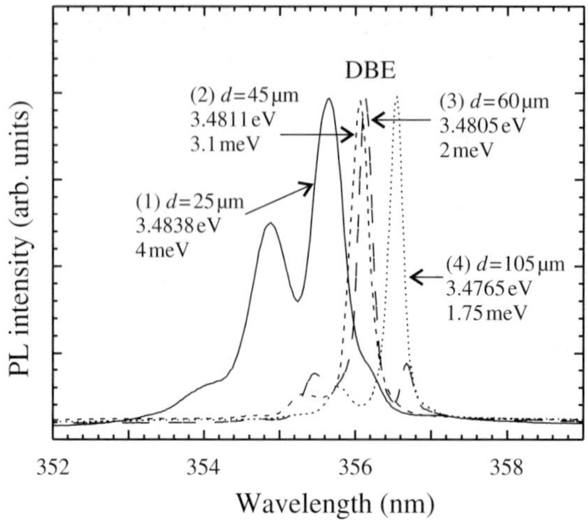

Fig. 4.5 Near-band-gap photoluminescence spectra (2 K) of GaN layers grown directly on sapphire with different thicknesses.

strain by the domain formation in the initial stage of growth. Thus, an explanation for formation of features with a specific symmetry structure during initial growth of strained films is based on the need of minimisation of the total film energy.

The tendency of strain decrease with layer thickness is confirmed by an overall macroscopic estimation of the strain relaxation effect done by measuring photoluminescence (PL) spectra from the top surface of layers with different thicknesses (Fig. 4.5). The near band gap PL spectra are typically dominated by exciton emission with well pronounced and sharp peaks attributed to free excitons (FX), and donor (DBE) and acceptor (ABE) bound excitons [31]. We performed a systematic study of the PL peak positions of thick HVPE layers with thicknesses from 15 to 130 µm and compared to the DBE energy position of the strain-free homoepitaxial grown GaN (3.471 eV [32]). The PL peak shift and the tendency of stress decrease with increasing thickness have been obtained in all samples studied.

Another specific feature of the thick GaN layers is a great reduction of the FWHM of the strongest DBE exciton line (Fig. 4.5) down to ~1.5 meV indicative of improvement of the layers with increasing thickness, as reported by many authors [33–38] (FWHM of 1 meV is reported for a 300-µm-thick HVPE-GaN film). That correlates with the dislocation density reduction of more than several orders of magnitude (from $10^{10}\,cm^{-2}$ to $10^{6}\,cm^{-2}$) within a hundred of micrometers in HVPE-GaN films [34–36]. Improvement of electrical parameters with increasing thickness is also a specific feature of such layers. Especially strong is the difference in the free carrier concentrations between the interface highly defective region and the rest of the film, resulting in the necessity of a two-layer model for electrical parameter analysis [37, 38]. Götz *et al.* [37] directly measured values of ~10^{17} at the top surface and ~10^{20} at the interface by mechanical thinning of a 13-µm-thick film.

4.4 Growth optimizations and material improvements

The optimization of the HVPE-GaN growth is impeded by the typical complications of the crystal growth technology like the multidisciplinary nature and complex multiparameter character of the process [39]. The latter obstacle is really very complicated because the large number of the growth parameters, typically including geometrical, thermal, chemical, supersaturation, etc., have not only to be optimized, they have also to be compromised between opposing requirements. For example, the growth rate should be low for reducing the thickness of the interface region of columnar growth [26], but it should be also high to achieve time-efficient growth. The need for heteroepitaxial nucleation is another issue that complicates the GaN growth. Several recently suggested techniques resulted in significant improvements (although they are still under development), such as different single buffers or multiple interlayers, and also different modifications of forced growth in horizontal in-plane directions, as: (i) lateral epitaxial overgrowth (ELOG) around masks of different materials and shapes; (ii) ELOG around grooved substrates and (iii) mass-transport during annealing of patterned GaN.

4.4.1 Buffers for HVPE growth of GaN

4.4.1.1 *Low temperature HVPE-GaN buffers*

Although early work on low temperature GaN buffers in the same HVPE chamber was not successful in the HVPE technology, a recently developed two-step process, employing a nucleation at ~550 °C [40] or 900 °C [41], has been shown to result in a good quality material with very promising characteristics.

4.4.1.2 *ZnO buffer*

HVPE-GaN layers grown on ZnO buffers have shown one of the best parameters for HVPE-GaN material up to now—a mobility of 880 cm^2/Vs with free carrier concentration of $(6–7)\times10^{16}\,cm^{-3}$; a dislocation density down to $(6–7)\times10^{8}\ cm^{-2}$ for 74-µm-thick layer; a thermal conductivity of 1.9 W/cm-K; a good optical quality characterized with FWHM of the dominant DBE line of (1.8–2) meV; a good crystal quality with FWHM of the rocking curve of the (0002) reflection of (4.2–7) arcsec. These layers are also characterized by a highly defective interface layer with a thickness of about 200 nm [34] and thickness improvement of all properties [37, 38]. A detailed description of such layers can be found in ref. [42].

The ZnO buffer has been suggested as a buffer for HVPE growth of thick GaN films (up to 300 μm thickness) by Detchprohm *et al.* [43] and has been used for substrate delamination due to its easy chemical etching. Further, the ZnO buffer has been developed and optimized at Lincoln laboratory during the last years and now it is one of the best working buffers for growth of thick GaN of good quality, also demonstrated in specific device applications as avalanche photodiodes [44].

4.4.1.3 *MOCVD-GaN templates*

The use of MOCVD thin films as templates for improving the GaN growth on sapphire has been firstly shown in MBE growth. We have successfully employed the MOCVD-GaN templates in the HVPE growth [45–48]. Two types of MOCVD templates, intentionally doped with Si with a free carrier concentration of $2\times10^{18}\,cm^{-3}$, and nominally undoped GaN layers with a free carrier concentration of $2\times10^{17}\,cm^{-3}$ were used in order to elucidate the effect of Si doping of the MOCVD templates on the defect arrangement of the overgrown films, since the Si doping of MOCVD-GaN is known to play a role in reduction of the biaxial stress and the density of vertical screw and edge dislocations in the layers [49].

We found that the growth on templates helps to: (1) eliminate the large scale structural defects at the sapphire/GaN interface [Fig. 4.6(a, b)]; (2) reduce the concentration and non-uniform distribution of residual carriers; and (3) relieve partly the compressive stress in the HVPE-GaN layers. We also found that the overgrown HVPE layer inherits the character of threading dislocation distribution of the templates [Fig. 4.7(a, c)]. Two different types of dislocation arrangements have been observed in hydride vapor phase epitaxial GaN films grown on sapphire substrates depending on undoped or Si-doped GaN templates: (i) predominantly straight threading dislocations parallel to the [0001] direction in the layer grown on an undoped template [Fig. 4.7(a)]; (ii) a network of interacting dislocations of edge, screw and mixed character in the layer grown on a Si-doped template [Fig. 4.7(c)]. The two types of defect distribution result in different surface morphologies, respectively: (i) low-angle grain boundaries formed by pure edge dislocations around spiral grown

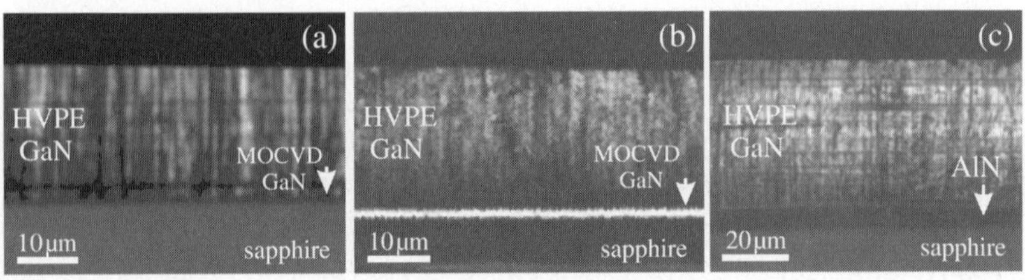

Fig. 4.6 Low temperature (4.6 K) panchromatic CL images of HVPE-GaN grown on: (a) undoped MOCVD-GaN template; (b) Si-doped MOCVD-GaN template; an (c) high temperature reactively sputtered AlN buffer.

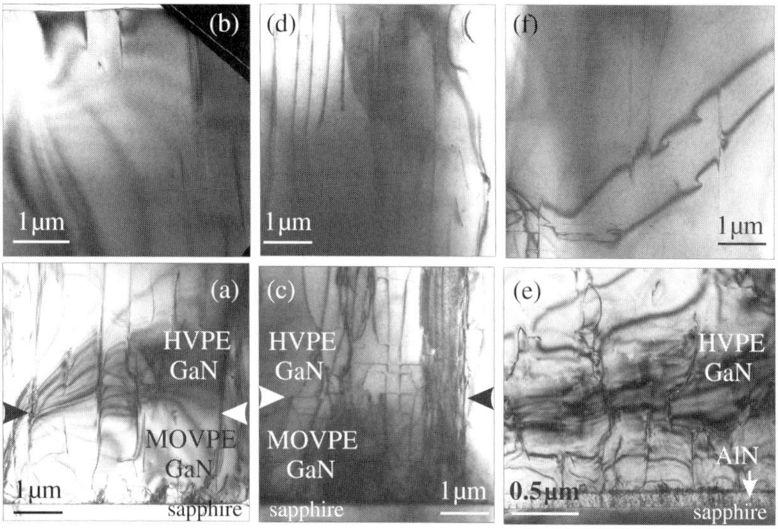

Fig. 4.7 TEM images in cross-section of HVPE-GaN grown on: (a, b) undoped MOCVD-GaN template; (c, d) Si-doped MOCVD-GaN template; and (e, f) high temperature reactively sputtered AlN buffer, taken at the interface (a, c, e) and top surface regions (b, d, f), respectively. Arrows mark the interface positions.

hillocks and (ii) a smooth surface intersected by randomly distributed dislocations. The Si doping of the GaN templates was found to enhance defect interaction in the templates and to enable a reduction of the dislocation density in the overgrown thick GaN films, although it does not lead to an improvement of the overall structural properties of the material. A reduction of dislocation density to $\sim 7 \times 10^7\,cm^{-2}$ at the top 100 nm of a 25-μm-thick film grown on the Si-doped template was obtained as a result of promoted interactions and thus annihilation of some of them. The total dislocation density in a 25-μm-thick film grown on the undoped template was determined to be $\sim 4 \times 10^8\,cm^{-2}$, keeping this value almost constant throughout the entire HVPE-GaN film. We conclude that the quality of the thick HVPE-GaN layers is significantly improved and they are more homogeneous and more relaxed (Fig. 4.8) as a result of eliminating the large-scale structural defects using growth on MOCVD-GaN templates.

4.4.1.4 *High temperature reactively sputtered AlN buffers*

High temperature (HT) reactively sputtered AlN buffer layers were also found to be effective in suppressing defect formation [Fig. 4.6(c)] in the GaN nucleation region. This buffer approach also permits reducing of tensile stress during GaN growth and leads to optimised film structure and properties [50, 51].

Cross-sectional TEM images reveal two zones in the GaN layer, contrary to the three zones reported for MOCVD-GaN growth utilizing a LT AlN buffer

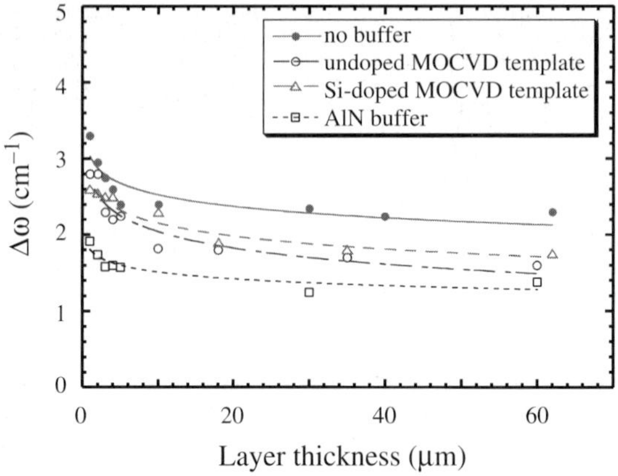

Fig. 4.8 Frequency shift of the E_2 mode as a function of the distance from the sapphire substrate of: HVPE-GaN on sapphire (black dots); and HVPE-GaN on undoped MOCVD-GaN template (circles); HVPE-GaN on Si-doped MOCVD-GaN template (triangles); and HVPE-GaN on AlN buffer (squares).

[Fig. 4.7(e)]. The highly faulted nucleation zone is not found here. HRTEM study of the interface structure shows atomically abrupt interfaces with a good match between the respective lattices [51]. The AlN comprises well-defined slightly misoriented domains and accommodates most of the misfit defects usually generated in the interface region.

Photoluminescence and Raman scattering spectra of the GaN films grown on AlN buffers show a decrease of the stress compared to that of layers grown directly on sapphire. There is a clear shift of the exciton positions to lower energies as expected for less strained GaN material. The position of the E_2 Raman mode (Fig. 4.8) was also found to be closer to the frequency position (566.2 cm^{-1}) of the strain-free bulk material [52], indicating a mechanism for misfit defect and strain relaxation. Using the HT AlN buffer layers, the highly defective interface regions responsible for non-uniform free carrier distribution in thick GaN films are dramatically reduced. This improved structure leads to a lower free carrier concentration in the range of 10^{17} cm^3 as determined by Hall-effect measurements. Additionally, micro-Raman scattering study in cross-sections confirms the homogeneity of the electrical parameters in the films based on the absence of the phonon–plasmon coupled mode and the feature at 650 cm^{-1} attributed to high doping [53].

4.4.2 Epitaxial lateral overgrowth

High dislocation and point defect densities remain as typical problems in thick films even with the aid of buffer layers. Significant reduction of the structural defects has been achieved through the use of epitaxial lateral overgrowth techniques in thin MOCVD-GaN films. This approach was applied in HVPE-GaN films as well [54–57].

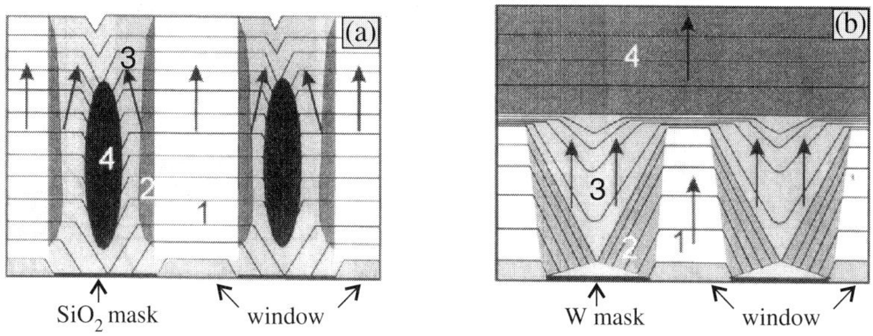

Fig. 4.9 Two different models of facet tilting (marked by arrows) and luminescence of ELOG HVPE-GaN using: (a) SiO_2 and (b) W masks. Numbers denote regions with: *1*—excitonic emission; *2(a)*—no emission; *3(a)* and *4(a)*, and *2(b)* and *3(b)*—emission due to high electron concentration; *4(b)*—excitonic emission. (after Hiramatsu *et al.* [56]).

The growth procedure developed by Usui *et al.* [54] consists principally of a HVPE selective homoepitaxial growth on a thin MOCVD grown GaN layer. The MOCVD growth was performed using the conventional two-step procedure, consisting of a 20-nm-thick *LT* buffer layer and high-temperature main overlayer. A SiO_2 layer was deposited by chemical vapor deposition and then fabricated into mask/window stripes with a period of about 7 μm, using conventional photolithographic techniques. After surface cleaning this structure was used for subsequent main HVPE growth. Such layers showed a dramatic reduction in the threading dislocation density to $2 \times 10^7 cm^2$ [54, 55]. At the same time a complex distribution of impurity incorporation and facet tilting around the masks was observed [56]. A schematic drawing of such a structure is shown in Fig. 4.9(a). Depending on the pattern orientation and dimensions, the defect structure and properties of the grown films have been found to differ. Parillaud *et al.* [57] used several mask geometries with a star-shaped pattern, an array of hexagonal holes, or parallel stripes with different widths and periods to determine the optimal stripe orientation and differences in HVPE lateral growth rates.

An alternative to using W mask instead of SiO_2 has been demonstrated by Hiramatsu *et al.* [56]. The W films were deposited by the RF sputtering method on a 4-μm-thick MOCVD-GaN layer and standard photolithography with a chemical etching was used to fabricate the stripe mask pattern. The lateral overgrowth rate on the W mask was found to be higher than that on the SiO_2, related to different growth faceting [Fig. 4.9(b)]. The ELOG-GaN layers on the W mask have better surface morphology and crystalline quality than that with SiO_2 mask, although the defect structure exhibited a large structural defect as a void in the initial coalescence region.

4.4.3 Mass transport

The mass-transport technique has been suggested very recently as a simple solution for avoiding mask deposition, which is a source of undesired impurity incorporation as well as strain pattern. Nitta *et al.* [58] has suggested the new technique of annealing the grooved GaN surface in a MOCVD chamber. As a result an overgrowth into the trenches occurred. Due to the lateral growth there is also a bending of the dislocations and a great reduction of the dislocation density. The results on mass-transport of HVPE-GaN in a HVPE reactor agree with that finding, although a spatially resolved study of optical properties revealed an increase of the donor-acceptor pair emission in the mass-transport regions [59] indicating some preferable impurity incorporation and/or redistribution due to the different growth mode.

4.5 Free-standing GaN films

4.5.1 Substrate removal

There are several approaches for GaN separation from the substrate depending mainly on the type of the substrate used.

The easiest way for substrate removal is a chemical etching in case of using Si as a substrate. Silicon is inexpensive and industrially established which makes the Si substrate application for thick quasi-bulk GaN growth very attractive. There are first results appearing on successful growth of thick GaN layers on Si substrate using the ELOG technique [60].

Another approach has been taken by Melnik *et al.* [61], who removed SiC substrates by reactive ion etching, producing free-standing GaN with 30 mm diameter. GaN epitaxial layers up to 200 μm in thickness have been directly grown on the (0001)Si face of SiC substrates without any buffer layer. After the growth, the SiC substrate was removed from the GaN film by reactively ion etching in SF_6 containing gas mixture at an etch rate of about 0.2 μm/min. The biggest obstacle in this approach, i.e. growing thick crack-free GaN layers on SiC remains unsolved, however.

If sapphire is used as a substrate, the mechanical polishing of HVPE thick GaN layers seems to be a very promising approach because of its simplicity, although a very long time is typically required for the procedure to minimize additional crack appearance. Kim *et al.* [62] have demonstrated sapphire removal of a 300–400 μm crack-free GaN layer. After the growth, the sapphire substrate was mechanically removed using diamond-coated cloths. Finally, both sides of the free-standing samples were mechanically polished to the thickness of about 350 μm using diamond slurry to a surface roughness of 10–17 Å.

Using a similar approach of polishing and/or cutting away, the sapphire Nakamura *et al.* attained improvements in the lifetime of laser diodes with cleaved facets [63]. In order to avoid cracking problems and to reduce the defect density, they used HVPE overgrowth on a 20-μm-thick MOCVD-ELOG structure substrate [8]. After HVPE growth of a 200-μm-thick film, the sapphire substrate together with the ELOG and the first 50–60 μm of the more defective part of the HVPE-GaN layer was removed [Fig. 4.10(a)] in order to obtain a good GaN substrate with a thickness of approximately 150 μm.

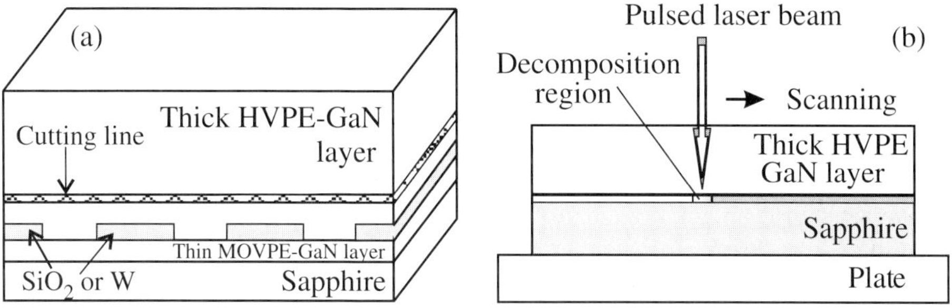

Fig. 4.10 Schematic diagrams of substrate removal by: (a) polishing of sapphire together with ~50-μm-thick interface region; (b) laser lift-off process for removal of the sapphire substrate from thick HVPE grown GaN films.

Laser-induced lift-off processes to separate GaN from the sapphire using different pulsed UV lasers have been demonstrated by Wong *et al.* [4] and Kelly *et al.* [5]. The process works by irradiating the sapphire/GaN interface with intense laser pulses just at or above the absorption edge of GaN [Fig. 4.10(b)]. It leads to a fast and strong heating which causes thermal decomposition in the interface regions of the film, yielding metallic Ga and nitrogen gas effusion. As it was suggested that the absorption takes place predominantly in the defective interface regions of the GaN films, our findings show that a highly defective interface region localized in a very thin sublayer close to the sapphire is of importance for an easy delamination. We have separated HVPE-GaN layers with thickness in the range of 10–120 μm grown either directly on sapphire or using HT AlN buffers and MOCVD-GaN templates [64]. The samples grown on AlN and on Si-doped MOCVD-GaN buffers were delaminated from the substrate just in small pieces, while the layers grown without a buffer or using undoped MOCVD templates were easily separated without cracking. Cross-sectional transmission electron microscopy images reveal that the highly defective nucleation regions of the GaN layers grown without a buffer and on undoped MOCVD templates are localized in very thin interfacial layers of about 800 nm [Fig. 4.5(a)] and 400 nm [Fig. 4.7(a)], respectively. The Si doping leads to bending the dislocations into the basal planes and thus defects become more randomly and uniformly distributed in a thicker interface region [Fig. 4.7(c)]. So, the highly defective interface region is not so strongly localized in size and density. Similar effects were found in the layers grown on HT AlN buffers. Thus we suggest that the thin defective layer absorbs more efficiently the laser power and this leads to an easier thermal decomposition of this interface region. Free-standing GaN wafers up to 2" size obtained by the lift-off technique are nowadays demonstrated in the research development stage [6, 7] but much more detailed work, including optimisation of the wafer temperature, power, size and scanning speed of the laser beam, etc., is needed before an optimised procedure for substrate delamination can be obtained

4.5.2 Properties of the free-standing films

The convergent beam electron diffraction technique was employed to determine the polarity of free-standing HVPE-GaN films [65]. The epitaxial as-grown surface was confirmed to be Ga-face while the side juxtaposed to the substrate before separation was confirmed to be of N polarity.

The structural and electrical properties of the free-standing HVPE-GaN with Ga- or N-faces showed a large difference. The dislocation density was found $>10^{10}\,cm^{-2}$ at the N-face while it was of 10^5–$10^6\,cm^{-2}$ at the Ga-face [65] and the donor concentration was about $10^{20}\,cm^{-3}$ at the interface side and as low as $3\times10^{15}\,cm^{-3}$ at the Ga-face surface of a 300-μm-thick free-standing layer [66]. The results, apparently, agree with those shown for thick GaN films residing on substrates.

A comparison between the PL spectra taken from the Ga- and N-faces shows similar emission bands, typically, near-band-gap emission at ~3.47 eV, DAP emission at ~3.29 eV and red emission at ~1.8 eV (Fig. 4.11). The Ga-face is fully exciton dominated with well distinguished exciton peaks while the N-face was found to show a broad emission (FERB) together with exciton emission in the near-band-gap region in full agreement with the CL results in cross-sections of the as-grown films. After polishing the defective interface region, the N-face spectrum becomes exciton dominated (Fig. 4.11 inset) and the exciton PL emission intensity from the polished N-face appeared to exceed that from the Ga-face about 10 times (Fig. 4.11). Additionally, Yun et al. [67] found that chemical etching in H_3PO_4 improves the N-face surface and resulted in a PL spectrum with linewidths similar to that from the Ga-face. Further detailed investigations need to be done to fully clarify the relation

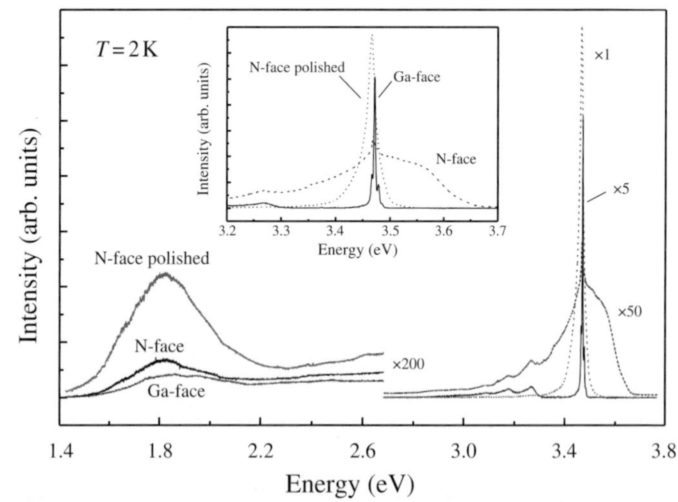

Fig. 4.11 Photoluminescence spectra of free-standing GaN film from both Ga- and N-faces. The expanded views of the near-band-edge are shown inset.

between the defect distribution and emission properties in both faces of HVPE-GaN quasi-substrates.

Due to the difference of thermal expansion coefficients between GaN and sapphire, the as-grown thick GaN films always exhibit a strong convex curvature [6, 7]. The radius of curvature was found to depend on the thicknesses of both sapphire substrate and GaN film. Interestingly, it decreased with increasing sapphire thickness and increased with thickness of the GaN film [7]. For 300-μm-thick films the radius was reported to be in the range of 0.4–0.6 m. After the delamination, the curvature was found to decrease with a radius greater than 2 m [6], and also to occur with a concave shape [7] (Fig. 4.12).

There is also a clear improvement of the crystalline quality after the sapphire separation as viewed by FWHM of the X-ray rocking curves for (0002) reflection. Typical values were measured to be 250–600 arcsec before lift-off and less than 200 arcsec after the lift-off [6, 7]. This fact is believed to be due to the reduction of the curvature, i.e. the increasing of the radius. On the other hand, the N-side (interface) of the free-standing films was always observed as having worse crystalline quality compared to the Ga-side. The latter is strongly related to the different defect density at the both sides of the HVPE-GaN quasi-substrate, and also reflects the stress still present in the free-standing films.

We performed a systematic study of the PL peak positions of thick HVPE layers with different thicknesses before and after the lift-off separation process [64, 68], and the results are shown in Fig. 4.13. It can be seen that the shift of the PL peaks is more pronounced for relatively thin films with a thickness of about 15 μm, while the shift is rather small ~1 meV in thicker films and remains almost constant independent of increasing thickness. There is no

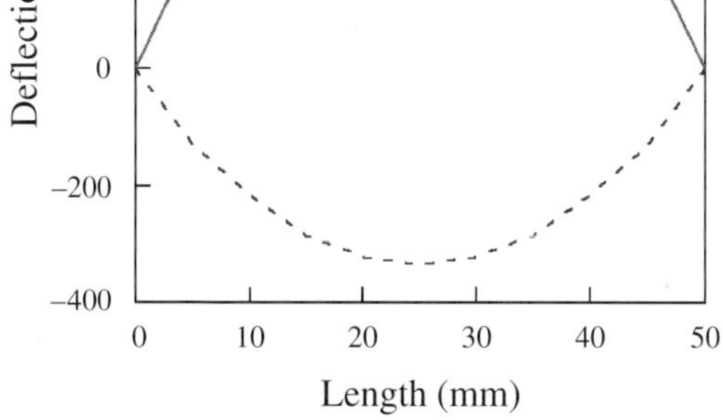

Fig. 4.12 Bowing of as-grown and free-standing GaN films viewed from the Ga surface (After Park *et al.* [7].)

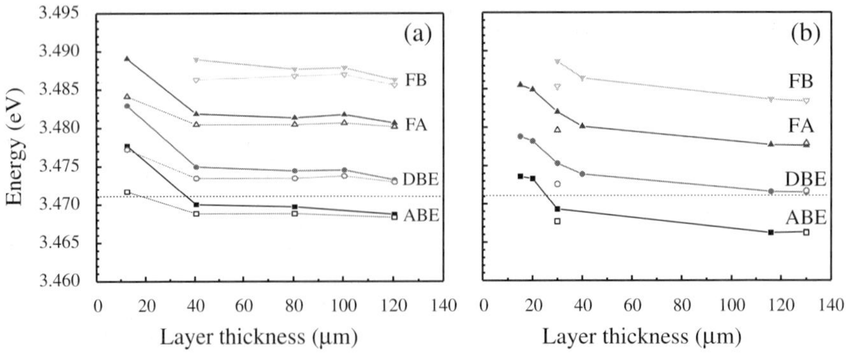

Fig. 4.13 Photoluminescence peak positions of thick HVPE-GaN films, before and after separation, with different thicknesses grown directly on sapphire (a) and grown on sapphire with a MOCVD-GaN template (b). The full symbols present the peak positions in as-grown layers, the open symbols present the respective peak positions in the same films after substrate separation.

essential shift of the donor-bound exciton position of the free-standing films compared to those of the thick as-grown films. Such a shift would be expected if the main stress in the as-grown films is due to substrate-induced compressive strain, and its small value may be explained by a relaxation present in the thick (more than 40 μm) films even before the lift-off. However, there is a small shift of about 2–3 meV in all samples studied with respect to the energy position (3.471 eV) line of the homoepitaxial strain-free material indicating that the small amount of stress present in the free-standing films is also related to defect induced strain. Similar behavior of PL shifting [Fig. 4.13(b)] and the tendency of stress decrease with increasing thickness has been obtained in a series of separated and as-grown samples when the initial film was grown on sapphire using MOCVD templates, which exhibit a lower defect density. In such films, the DBE peak positions are closer to the reported value for unstrained GaN [32].

4.6 Homoepitaxial regrowth on HVPE-GaN quasi-substrates

The Ga-face and the N-face of the free-standing HVPE-GaN quasi-substrates exhibited a wide variation in the chemical stability in addition to the huge variation of their defect density and crystal quality. Thus, very different substrate preparation is required for the two faces of the substrates. The mechanical polishing of the Ga-face needs to be combined either with dry etching or photo-assisted chemical etching in order to ensure epi-ready surfaces. Only a few reports on GaN surface preparation are presented so far [69, 70]. Despite the uncertainty in the preparation procedure, successful regrowths on the Ga-face of the GaN quasi-substrates have been demonstrated even without any polishing steps.

4.6.1 HVPE homoepitaxial regrowth

We have studied HVPE-GaN overgrowths with different thicknesses on HVPE-GaN quasi-substrates with different defect density [64]. There is a clear improvement of the surface morphology in all overgrown layers as revealed by AFM imaging. Figure 4.14 shows two images of a 40-μm-thick GaN layer as-grown on sapphire and of an overgrown layer with a thickness of about 150 μm on the same free-standing layer separated from the sapphire. The surface of the as-grown layer on sapphire is quite smooth consisting of an array of terraces with a width of about 120 nm separated by 5–6 Å steps. However, one can also see that some of the steps are terminated tangentially at the edges of surface depressions due to dislocations of mixed type [71]. The overgrown film on the free-standing GaN film has a lower surface roughness compared to that of the as-grown layer. No terrace depressions were observed and the density of dislocations intersecting the surface is two times less, indicating a decrease of threading dislocations and a change in the dominant growth mechanism.

A photoluminescence spectrum of an as-grown 40-μm-thick GaN layer on sapphire together with a spectrum of the same film after separation and two spectra of the overgrown free-standing GaN films (up to thicknesses of 100 μm and 200 μm) are shown in Fig. 4.15(a). As can be seen the narrow (~2 meV) exciton peaks in the spectrum of the as-grown film are fully reproduced in the overgrown films indicating their good crystalline quality. These rather small variations in the PL peak positions correspond to a small variation in the bandgap and the corresponding lattice parameters in the films, and reflect the replication of the defect distribution in the material (both dislocations and point defects).

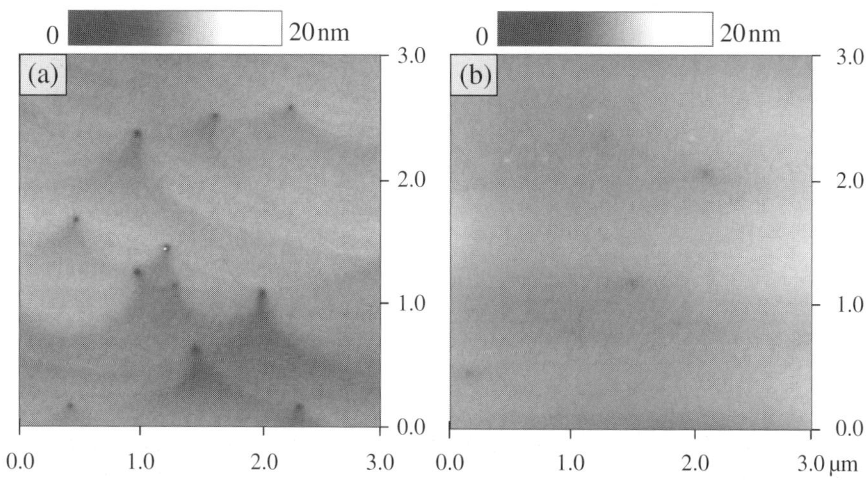

Fig. 4.14 AFM images of as-grown GaN film directly on sapphire (a) and regrown film on free-standing GaN (b).

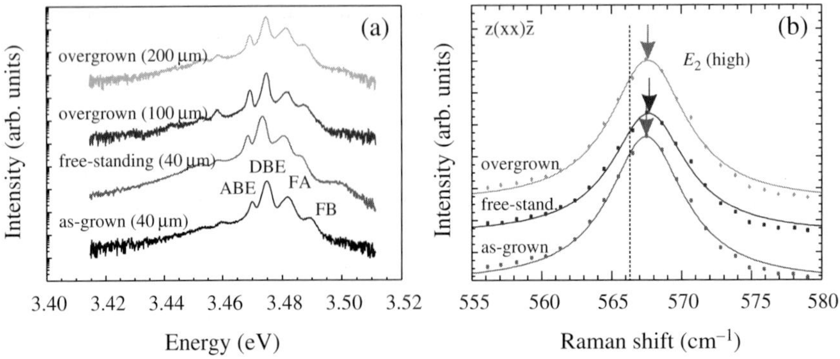

Fig. 4.15 (a) Low temperature (2 K) photoluminescence spectra, and (b) Raman scattering spectra of HVPE-GaN films: as-grown; free-standing; and overgrown HVPE-GaN films on the free-standing GaN films.

The stress in the overgrown GaN layers was analysed by Raman scattering measurements. Three Raman spectra in the region of the E_2 mode are shown in Fig. 4.15(b). The samples are the same as those shown in Fig. 4.15(a). The solid lines represent calculated spectra derived from a fitting assuming a Lorentzian line shape. The peak positions of the as-grown, the separated, and the overgrown films are observed to be very close to each other indicating a very small change in the strain. The frequency shift to higher wave numbers with respect to the reference value of 566.2 cm^{-1} [52] for homoepitaxial and bulk unstressed material, verifies the presence of some compressive strain in the layers in agreement with the assignment of the PL line shift.

Based on both PL and RS experimental data, it is clear that all samples studied (crack-free with a thickness less than 50 μm) are under a small compressive strain and the removing of the substrate does not lead to a complete removal of the stress. Thus, two possible explanations can be suggested, in principle. If the main strain in thick as-grown GaN layers on sapphire is due to the substrate induced biaxial compressive stress it should persist after the substrate separation showing plasticity. The second option is an assumption that the main strain in thick HVPE-GaN films is hydrostatic strain due to internal defects.

The residual strain along the *c*-axis was also studied by high resolution XRD. The full width at half maximum (FWHM) values of the 2θ-ω rocking curves were found to be less than 46 arcsec for all samples. The linewidths of the (0002) reflection in the overgrown layers showed a small broadening which increases with increasing thickness together with some asymmetry indicative of a small stress gradient in these thicker films probably due to an internal defect redistribution [64]. The lattice constants measurements show a higher *c* lattice constant while the *a* parameter is smaller in the as-grown layers and is higher in the free-standing and regrown layers than the 'ideal parameters' $c = 5.1856$ Å and $a = 3.188$ Å [72] reported for unstrained material. It leads to the conclusion that in as-grown crack-free HVPE-GaN films with thickness of about 40 μm the main compressive strain is due to the hydrostatic compressive stress with a small contribution from the biaxial compressive stress. The free-standing and the overgrown HVPE films are mainly hydrostatically compressed. The residual stress in thick GaN layers was tentatively attributed to the dislocation-related inhomogeneously distributed stress along the growth axis. Since the overgrown layers replicate mainly the defect distribution character and persist the stress of the underlying films, the defect and stress control in the as-grown HVPE-GaN films is of significant importance.

4.6.2 MOCVD homoepitaxial growth on HVPE-GaN quasi-substrates

Results on MOCVD homoepitaxial GaN growth on HVPE-GaN were recently shown by Vaudo *et al.* [6] and Miskys *et al.* [73]. A very smooth surface with a lower density of pit terminations was achieved similarly as in HVPE regrowth. No increase of the impurity concentration and defect density at the interface between the HVPE-GaN and MOCVD-GaN was observed by cross-sectional SIMS and TEM studies. However, a slight broadening of the linewidth of (0002) reflection in the DCXRC spectra was found, although it was relatively narrow with a FWHM of about 73 arcsec.

A comparison was made [6] between identical light emitting diode structures grown on three substrates, sapphire with AlN buffer, 10 μm-thick HVPE-GaN on sapphire and free-standing HVPE-GaN quasi-substrate. The structures consisted of a GaN layer doped with Si, an InGaN layer and a GaN layer doped with Mg, and with Ti/Al contacts on the GaN:Si layer and Ni/Au

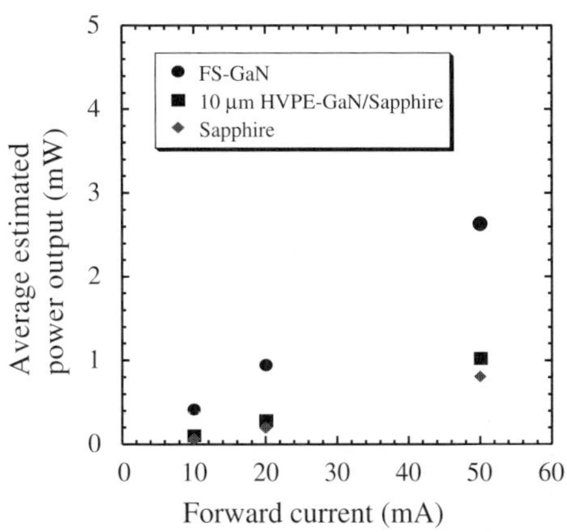

Fig. 4.16 Comparison of relative light output power versus current from a GaN/InGaN LEDs fabricated on: free-standing GaN; 10-μm-thick GaN layer residing on sapphire; and directly on sapphire. (After Vaudo *et al.* [6].)

contacts on the GaN:Mg layer. The output power for LEDs on the free-standing GaN substrates was observed to be four times greater compared to that on the other two substrates [6] (Fig. 4.16) indicating superior performance of this device and the suitability of the HVPE-GaN quasi-substrates for nitride device structures.

4.7 Conclusions

The current crystalline quality of nitride materials is still not sufficient for the realization of their full potentials. The high-density threading dislocation problem and strain-related phenomena in the nitride heterostructures remain to be clarified. The homoepitaxial growth on GaN substrates is the only reliable direction for growth of high quality materials and the HVPE-GaN quasi-substrate development is the most promising approach at present. Therefore, further improvements in the crystal quality of HVPE-GaN quasi-substrates will lead to a better defect control of the entire device structures.

Acknowledgements

We would like to thank all our collaborators P. P. Paskov, E. Valcheva, E. M. Goldys, B. Arnaudov, S. Evtimova, M. Abrashev, T. Shubina, and V. Ratnikov for their contribution to the characterization of the films and fruitful discussions. J. Birch and S. Tungasmita, and R. Beccard and M. Heuken are acknowledged for supplying the AlN and the MOCVD-GaN buffers. We also wish to thank all colleagues providing the non-owned figures used in this work.

References

1. Nakamura, S. (2000). UV/blue/green InGaN-based LEDs and laser diodes grown on epitaxially laterally overgrown GaN. *IEICE Trans. Electron.* **E83-C**, 529–35.
2. Sheppard, S.T., Doverspike, K., Leonard, M., Pribble, W.L., Allen, S.T., and Palmour, J.W. (2000). Improved 10 GHz operation of GaN/AlGaN HEMTs on silicon carbide. *Material Science Forum* **338–42**, 1643–6.
3. Speck, J. S., Heying, B., Elsass, C., Fini, P., Green, D., Haus, E. *et al.* (2001). Progress in the MBE growth of GaN and its alloys. *Materials Research Society Symposium Proceedings*, 2000. Boston, MA, G7.4.
4. Wong, W. S., Sands, T., and Cheung, N. W. (1998). Damage-free separation of GaN thin films from sapphire substrates. *Appl. Phys. Lett.* **72**, 599–601.
5. Kelly, M. K., Vaudo, R. P., Phanse, V. M., Görgens, L., Ambacher, O., and Stutzmann, M. (1999). Large free-standing GaN substrates by hydride vapor phase epitaxy and laser induced liftoff. *Jpn. Appl. Phys.* **38**, L217–19.
6. Vaudo, R. P., Brandes, G. R., Flynn, J. S. *et al.* (2000). Synthesis and properties of HVPE nitride substrates. Proc IWN2000, Nagoya, Jpn, *IPAP Conference Series* **C1**, 15–8.

7. Park, S. S., Park, I. W., and Choh, S. H. (2000). Bowing of GaN substrates by hydride vapor phase epitaxy. Proc. IWN2000, Nagoya, Jpn, *IPAP Conference Series* **C1**, 60–3.
8. Usui, A., Sunakawa, H., Kobayashi, K., Watanabe, H., Yamaguchi, A. A., and Mizuta, M. (2001). Characteristics of FIELO-GaN grown by hydride vapor phase epitaxy. *Materials Research Society Symposium Proceedings* 2000, Boston, MA, G5.6.
9. Krukowski, S. (1997). Thermodynamic and high/pressure growth of (Al, Ga, In)N single crystals. *Diamond and Related Materials* **6**, 1515–23.
10. Seifert, W., Fitzl, G., and Butter, E. (1981). Study of the growth rate in VPE of GaN. *J. Cryst. Growth* **52**, 257–62.
11. Ban, V.S. (1972). Mass spectrometric studies of vapor phase crystal growth: II. GaN. *J. Electrochem. Soc.* **119**, 761–5.
12. Przhevalskii, I.N., Yu, S., Karpov, S., and Makarov, Y. N. (1998). Thermodynamic properties of group-III nitrides and related species. *MRS Internet J. Nitride Semicond. Res.* **3**, art 30.
13. Koukitu, A., Hama, S., Taki, T., and Seki, H. (1998). Thermodynamic analysis of hydride vapor phase epitaxy of GaN. *Jpn. J. Appl. Phys.* **37**, 762–5.
14. Aujol, E., Napierala, J., Trassoudaine, A., Gil-Lafon, E., and Cadoret, R. (2001). Hydrogen and nitrogen ambient effects on epitaxial growth of GaN by hydride vapor phase epitaxy. *J. Cryst. Growth* **222**, 538–48.
15. Cadoret, R. (1999). Growth mechanisms of (00.1)GaN substrates in the hydride-vapor-phase method: surface diffusion, spiral growth, H_2 and $GaCl_3$ mechanisms. *J. Cryst. Growth* **205**, 123–35.
16. Trassaudaine, A., Aujol, E., Cadoret, R., Paskova, T., and Monemar, B. (2001). A new mechanism in the growth process of GaN by HVPE. *Materials Research Society Symposium Proceedings* 2000, Boston, G.3.2.
17. Greene, J.E. (1993). Physics of film growth from the vapor phase. In *Multicomponent and multilayered thin films for advanced technologies: techniques, fundamentals, and devices* (ed. O. Auciello), pp. 39–56. Kluwer Academic Publishers, Boston, Mass.
18. Golan, Y., Wu, X. H., Speck, J. S., Vaudo, R. P., and Phanse, V. M. (1998). Morphology and microstructural evolution of early stages of hydride vapor phase epitaxy of GaN on sapphire. *Appl. Phys. Lett.* **73**, 3090–3.
19. Akasaki, I., Amano, H., Koide, Y., Haramatsu, K., and Sawaki, N. (1989). Effects of AlN buffer layer on crystallographic structure and on electrical and optical properties of GaN and $Ga_{1-x}Al_xN$ ($0<x\leq0.4$) films grown on sapphire substrate by MOVPE. *J. Cryst. Growth* **98**, 209–19.
20. Molnar, R. J., Götz, W., Romano, L. T., and Johnson, N. M. (1997). Growth of gallium nitride by hydride vapor phase epitaxy. *J. Cryst. Growth* **178**, 147–56.
21. Paskova, T., Svedberg, E. B., Madsen, L. D. *et al.* (1999). Domain structure of thick GaN layers grown by hydride vapor phase epitaxy. *Materials Research Society Symposium Proceedings* **537**, G3.16.
22. Goldys, E. M., Paskova, T., Ivanov, I. G., Arnaudov, B., and Monemar, B. (1998). Direct observation of large-scale nonuniformities in hydride vapor phase epitaxy grown gallium nitride by cathodoluminescence. *Appl. Phys. Lett.* **73**, 3583–5.
23. Kim, H. M., Oh, J. E., and Kang, T. W. (2000). Cathodoluminescence characterization of thick GaN films grown by HVPE. Proc.IWN2000, Nagoya, Jpn. *IPAP Conference Series* **C1**, 49–52.
24. Salviati, G., Albrecht, M., Zanotti-Fregonara, C. *et al.* (1999). Cathodoluminescence and transmission electron microscopy study of the influence of crystal defects on optical transitions in GaN. *Phys. Stat. Sol. (a)* **171**, 325–39.

25. Paskova, T., Goldys, E. M., and Monemar, B. (1999). Hydride vapor-phase epitaxy growth and cathodoluminescence characterization of thick GaN films. *J. Cryst. Growth* **200**, 1–11.
26. Paskova, T., Goldys, E. M., Yakimova, R., Svedberg, E. B., Henry, A., and Monemar, B. (2000). Influence of growth rate on the structure of thick GaN layers grown by HVPE. *J. Cryst. Growth* **208**, 18–26.
27. Valcheva, E., Paskova, T., Persson, P. A. O., Hultman, L., and Monemar, B. (2002). Misfit defect formation in hydride vapor-phase epitaxy GaN grown on sapphire. *Appl. Phys. Lett.* **80**, (Issue 9, March 4).
28. Bertram, F., Srinivasan, S., Ponce, F. A., Riemann, T., Christen, J., and Molnar, R. J. (2001). Spatial variation of luminescence in thick GaN films. *Appl. Phys. Lett.* **78**, 1222–4.
29. Kim, C., Yi, J., Yang, M., *et al.* (2000). Inversion domains triggering recovery of lumnescence uniformity in epitaxially lateral overgrown thick GaN film. *Appl. Phys. Lett.* **77**, 4319–21.
30. Arnaudov, B., Paskova, T., Goldys, E. M., *et al.* (1999). Contribution of free-electronrecombination to the luminescence spectra of thick GaN films grown by hydride vapor phase epitaxy. *J. Appl. Phys.* **85**, 7888–92.
31. Pozina, G., Bergman, P. J., Paskova, T., and Monemar, B. (1999). Bound exciton dynamics in GaN grown by hydride vapor-phase epitaxy. *Appl. Phys. Lett.* **75**, 4124–6.
32. Kornitzer, K., Ebner, T., Thonke, K. *et al.* (1999). Photoluminescence and reflectancespectroscopy of excitonic transitions in high-quality homoepitaxial GaN films. *Phys. Rev. B* **60**, 1471–3.
33. Chernyak, L., Osinky, A., Nootz, G., *et al.* (2000). Electron beam and optical depth profiling of quasi-bulk GaN. *Appl. Phys. Lett.* **77**, 2695–7.
34. Romano, L. T., Krusor, B. S., and Molnar, R. J. (1997). Structure of GaN films grown by hydride vapor phase epitaxy. *Appl. Phys. Lett.* **71**, 2283–5.
35. Albrecht, M., Nikitina, I. P., Nikolaev, A. E., Melnik, Y. V., Dmitriev, V. A., and Strunk, H. P. (1999). Dislocation reduction in AlN and GaN bulk crystals grown by HVPE. *Phys. Stat. Sol. (a)* **176**, 453–8.
36. Lee, K. and Auh, K. (2001). Dislocation density of GaN grown by hydride vapor phase epitaxy. *MRS Internet J. Nitride Semicond. Res.* **6**, art. 9.
37. Götz, W., Walker, J., Romano, L. T., Johnson, N. M., and Molnar, R.J. (1997). Thickness dependence of electronic properties of GaN epi-layers. *Material Research Society Symposium Proceedings* **449**, 525–30.
38. Look, D. C., and Molnar, R. J. (1997). Degenerate layer at GaN/sapphire interface: influence on Hall-effect measurements. *Appl. Phys. Lett.* **70**, 3377–9.
39. Scheel, H. J. (2000). Historical aspects of crystal growth technology. *J. Cryst. Growth* **211**, 1–12.
40. Tavernier, P. R., Etzkorn, E. V., Wang, Y., and Clarke, D. R. (2000). Two-step growth of high quality GaN by hydride vapor phase epitaxy. *Appl. Phys. Lett.* **77**, 1804–6.
41. Zhang, W., Alves, H. R., Bläsing, J., Hofmann, D. M., Krost, A., and Meyer, B. K. (2000). Effect of high temperature GaN buffer layer on the growth of thick GaN by hydride vapor phase epitaxy. Proc IWN2000, Nagoya, *Jpn. IPAP Conference Series* **C1**, 46–8.
42. Molnar, R. J. (1999). Hydride vapor phase epitaxial growth of III-V nitrides. In *Gallium Nitride (GaN) II*, (eds J. I. Pankove, and T. D. Moustakas), pp. 1–31. Academic Press, San Diego, CA.
43. Detchprohm, T., Amano, H., Hiramatsu, K., and Akasaki, I. (1993). The growth of thick GaN film on sapphire substrate by using ZnO buffer layers. *J. Cryst. Growth* **128**, 384–90.
44. McIntosh, K. A., Molnar, R. J., Mahoney, L. J., *et al.* (1999). GaN avalanche photodiodes grown by hydride vapor phase epitaxy. *Appl. Phys. Lett.* **75**, 3485–7.

45. Paskova, T., Tungasmita, S., Valcheva, E., *et al.* (2000). Hydride vapor phase homoepitaxial growth of GaN on MOCVD-grown templates. *Materials Research Society Symposium Proceedings* **5S1**, W3.14.
46. Valcheva, E., Paskova, T., Birch, J. *et al.* (2001). Elimination of nonuniformities in thick GaN films using metalorganic chemical vapor deposited GaN templates. *J. Appl. Phys.* **90**, 6011–16.
47. Paskova, T., Valcheva, E., Birch, J. *et al.* (2000). Effect of Si doping of metalorganic chemical vapor deposition-GaN templates on the defect arrangement in hydride vapor phase epitaxy-GaN overgrown layers. *J. Appl. Phys.* **88**, 5729–32.
48. Ratnikov, V., Kyutt, R., Shubina, T., Paskova, T., Valcheva, E., and Monemar, B. (2000). Brag and Laue diffraction study of dislocations in thick hydride vapor phase epitaxy GaN films. *J. Appl. Phys.* **88**, 6252–9.
49. Romano, L. T., Van de Walle, C. G., Ager III J. W., Götz, W., and Kern, R. S. (2000). Effect of Si-doping on strain, cracking and microstructure in GaN films grown by metalorganic chemical vapor deposition. *J. Appl. Phys.* **87**, 7745–52.
50. Paskova, T., Valcheva, E., Birch, J. *et al.* (2001). Defect and stress relaxation in HVPE-GaN films using high temperature sputtered AlN buffers. Proc. of EGW-4, Nottingham, July 4–7, 2000, *J. Cryst. Growth* **230**, 381–6.
51. Valcheva, E., Paskova, T., Tungasmita, S. *et al.* (2000). Interface structure of hydride vapor phase epitaxial GaN grown with high-temperature reactively sputtered AlN buffer. *Appl. Phys. Lett.* **76**, 1860–2.
52. Kisielowski, C., Kruger, J., Ruvimov, S. *et al.* (1996). Strain-related phenomena in GaN thin films. *Phys. Rev. B* **54**, 17745–53.
53. Kaschner, A., Hoffman, A., Thomsen, C. *et al.* (2000). Micro-Raman and cathodoluminescence studies of epitaxial laterally overgrown GaN with tungsten masks: a method to map the free-carrier concentration of thick GaN samples. *Appl. Phys. Lett.* **76**, 3418–20.
54. Usui, A. (1998). Bulk GaN crystal with low defect density grown by hydride vapor phase epitaxy. *Materials Research Society Symposium Proceedings* **482**, 233–44.
55. Sakai, A., Sunakawa, H., and Usui, A. (1999). Defect structure in selectively grown GaN films with low threading dislocation density. *Appl. Phys. Lett.* **71**, 2259–61.
56. Hiramatsu, K., Motogaito, A., Miyake, H. *et al.* (2000). Crystalline and optical properties of ELO GaN by HVPE using tungsten mask. *IEICE Trans. Electron.* **E83-C**, 620–5.
57. Porillaud, O., Wagner, V., Buehlmann, H. J., Ilegems, M. (1998). Localized epitaxy of GaN by HVPE on patterned substrates. *MRS Internet J. Nitride Semicond. Res.* **3**, art. 40.
58. Nitta, S., Kariya, M., Kashima, T., Yamaguchi, S., Amano, H., and Akasaki, I. (2000). Mass transport and reduction of threading dislocation in GaN. *Appl. Surf. Science* **159–60**, 421–6.
59. Paskova, T., Goldys, E. M., Paskov, P. P. *et al.* (2001). Mass-transport growth and optical emission properties of hydride vapor phase epitaxy GaN. *Appl. Phys. Lett.* **78**, 4130-2.
60. Gu, S., Zhang, L., Zhang, R., Wicks, G. W., and Kuech, T. F. (2000). Epitaxial lateral overgrowth of GaN on Si substrates by hydride vapor phase epitaxy. Proc IWN2000, Nagoya, Jpn. *IPAP Conference Series* **C1**, 23–6.
61. Melnik, Y. V., Nikolaev, A., Nikitina, I., Vassilevski, K. V., and Dmitriev, V. A. (1998). Properties of free-standing GaN bulk crystals grown by HVPE. *Materials Research Society Symposium Proceedings* **482**, 269–74.
62. Kim, S. T., Lee, Y. J., Moon, D. C., Hong, C. H., and Yoo, T. K. (1998). Preparation and properties of free-standing HVPE grown GaN substrates. *J. Cryst. Growth* **194**, 37–42.

63. Nakamura, S. (1999). InGaN-based violet laser diodes. *Semicond. Sci. Technol.* **14**, R27–40.
64. Paskova, T., Paskov, P. P., Birch, J. *et al.* (2000). HVPE regrowth on free-standing GaN quasi-substrates. Proc IWN2000, Nagoya, Jpn, *IPAP Conference Series* **C1**, 19–22.
65. Jasinski, J., Swider, W., Liliental-Weber, Z. *et al.* (2001). Characterization of free-standing hydride vapor phase epitaxy GaN. *Appl. Phys. Lett.* **78**, 2297–9.
66. Saxler, A., Look, D. C., Elhamri, S. *et al.* (2001). High mobility of n-type GaN substrates. *Appl. Phys. Lett.* **78**, 1873–5.
67. Yun, F., Reshchikov, M. A., Jones, K. *et al.* (2000). Electrical, structural, and optical characterization of free-standing GaN template grown by hydride vapor phase epitaxy. *Solid State Electron.* **44**, 2225–32.
68. Paskova, T., Paskov, P. P., Darakchieva, V., Tungasmita, S., Birch, J., and Monemar, B. (2001). Defect reduction in HVPE growth of GaN and related optical spectra. *Phys. Stat. Sol. (a)* **183**, 197–201.
69. Nagahama, S., Iwasa, N., Senoh, M. *et al.* (2000). High-power and long-lifetime InGaN multi-quantum-well laser diodes grown on low-dislocation density GaN substrates. *Jpn. J. Appl. Phys.* **39**, L647–50.
70. Visconti, P., Jones, K. M., Reshchikov, M. A., Cingolani, R., Morkoc, H., and Molnar, R. J. (2000). Characteristics of free-standing hydride vapor phase epitaxy grown GaN with very low defect concentration. *Appl. Phys. Lett.* **77**, 3743–5.
71. Heying, B., Tarsa, E. J., Elsass, C. R., Fini, P., DenBaars, S. P., and Speck, J. S. (1999). Dislocation mediated surface morphology of GaN. *J. Appl. Phys.* **85**, 6470–6.
72. Leszczynski, M., Suski, T., Perlin, P. *et al.* (1995). Lattice constants, thermal expansion and compressibility of gallium nitride. *J. Phys. D: Appl. Phys.* **28**, A149–53.
73. Miskys, C. R., Kelly, M. K., Ambacher, O., Martinez-Criado, G., and Stutzmann, M. (2000). GaN homoepitaxy by metalorganic chemical vapor deposition on free-standing GaN substrates. *Appl. Phys. Lett.* **77**, 1858–60.

5 Growth optimization of low-dimensional nitrides by metalorganic vapor phase epitaxy and dislocation control

Hiroshi Amano, Satoshi Kamiyama and Isamu Akasaki

Recent progress in the growth technology which uses metalorganic vapor phase epitaxy has led to the realization of nitride-based bright-blue/green light-emitting diodes, violet laser diodes, UV photodiodes, and high-electron-mobility transistors. Low-dimensional structures have been used in the active region of each device, thereby significantly improving this performance. In this section, first, development of the growth method of group III nitrides is reviewed. Next, artificial fabrication of low-dimensional structures and their luminescence and electrical properties are discussed. Finally, recent progress in the growth of low-dislocation nitrides by lateral epitaxy is reviewed.

5.1 Growth of group III nitrides on sapphire (low-temperature deposited buffer layer)

GaN was first synthesized almost 70 years ago by a direct reaction between metallic Ga and ammonia [1]. The first single crystalline GaN was grown in 1969 using hydride vapor phase epitaxy (HVPE) on sapphire substrate [2]. In 1971, a greenish blue metal-insulator-semiconductor (mis)-type LED based on GaN using Zn-related luminescence was fabricated by HVPE [3]. The following are the major disadvantages in fabricating mis-type GaN LEDs by HVPE: (1) generation of numerous cracks and pits, (2) poor controllability of the thickness of the insulating layer which governs the operating voltage, and (3) high concentration of residual donors in nominally undoped GaN. In order to reduce the concentration of residual donor, thick films had been indispensable [4]. Due to the tensile stress caused by the mosaicity of GaN at the initial growth stage on the sapphire substrate, however, many cracks are generated at the interface between the GaN and the sapphire when the thickness of the GaN film exceeds several tens of microns. Another issue is the thickness control. In case of the mis-type structures, the operating voltage strongly depends on the thickness of the acceptor-doped insulating layer. Due to the poor controllability of thickness by HVPE, it was difficult to reduce the operating voltage less than 10 [V].

Metalorganic vapor phase epitacy (MOVPE) was expected to solve these problems. Manasevit *et al.* synthesized GaN and AlN on 6H-SiC and sapphire substrates by MOVPE for the first time [5]. They used triethylgallium and ammonia as source gases. They obtained highly c-axis oriented poly-crystalline films. Their films were composed of small grains with sizes of sub-microns.

The use of a low-temperature deposited (LT-) buffer layer was firstly accomplished in 1986 [6]. The flatness was greatly improved by using the LT-buffer. High-resolution X-ray diffraction showed that mosaicity was remarkably reduced. Figure 5.1 schematically shows the mosaicity of the crystal and the profile of X-ray diffraction from (0002) and $(10\bar{1}0)$ plane. Mosaicity can be characterized by measuring both tilting and twisting which is estimated from the X-ray diffraction profile of (0002) and $(10\bar{1}0)$ plane, respectively by rocking the sample as indicated in the figure, for each diffraction. It should be emphasized that not only the tilting but also twisting is reduced by using LT-buffer. Table 5.1 summarizes the FWHM of the X-ray diffractions from (0002), $(10\bar{1}0)$ and $(20\bar{2}4)$ planes of two different GaN samples with (A) and without (B) LT-buffer. Tilting is very small in sample B, but the twisting is huge, resulting in the very poor PL property.

In the PL spectrum of the undoped GaN grown by LT-buffer at cryogenic temperature, strong free exciton lines corresponding to the valence band splitting can be clearly observed in addition to the donor- and acceptor-bound exciton lines. From these experiments, valence band structure of GaN was precisely determined [7]. The free exciton emission dominates in the PL spectrum even at room temperature [8, 9]. By using the LT-buffer, the residual donor

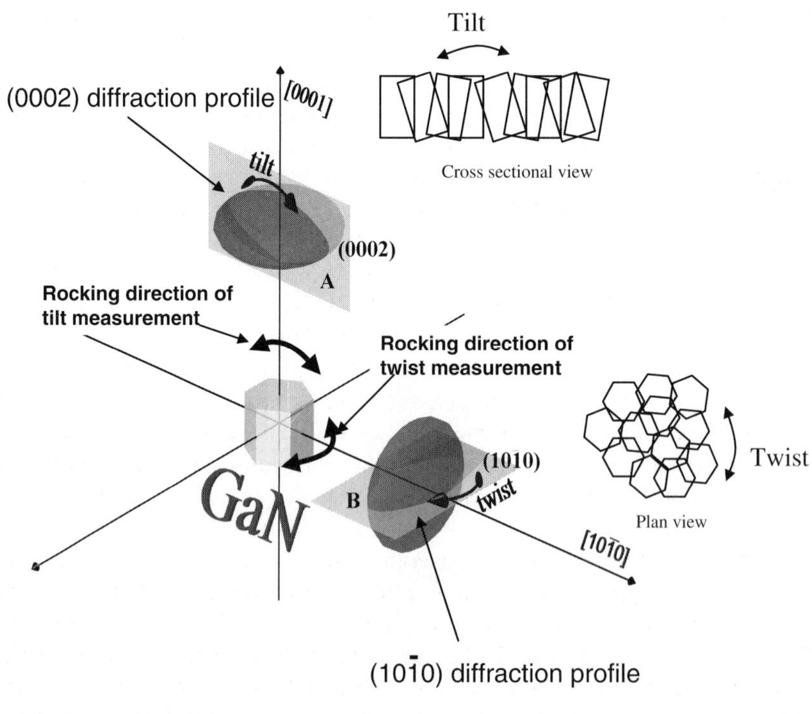

Fig. 5.1 Schematic drawing of the mosaicity of group-III nitride crystals and the corresponding X-ray diffraction profile. X-ray intensity profile of ω-mode for (0002) and $(10\bar{1}0)$ correspond to tilt and twist distribution, respectively. Rocking directions for measuring tilt and twist distributions are also shown in the figure.

Table 5.1 FWHMs of ω-mode with open slit and $2\theta/\omega$-mode with channel-cut analyzer crystal of (0002), $(10\bar{1}0)$ and $(20\bar{2}4)$ diffractions from 1.7 [μm] thick GaN. PL of each GaN is also shown. $\Delta\omega_c$; tilt, Δc; fluctuation of lattice constant c, $\Delta\omega_a$; twist, Δa; fluctuation of lattice constant a.

[arc sec.]

	(0002)		$(10\bar{1}0)$		$(20\bar{2}4)$	Ref. PL
	$\Delta\omega_c$	Δc	$\Delta\omega_a$	Δa	$\Delta\omega$	
			[arc sec.]			
Sample A	406	23	599	55	222	Intense
Sample B	48	33	5500	Too weak	284	Not detectable

concentration is reduced more than two orders of magnitude while the mobility is increased by an order of magnitude at room temperature. These results show that not only the crystalline quality, but also the luminescence as well as electrical properties are significantly improved as a result of using the LT-buffer method. It is pointed out that such high-quality GaN layers which are sufficiently thin to avoid crack generation can be realized.

Compared to GaN, it is difficult to grow high-quality and thick AlGaN. The strong reaction of the alkyl source of Al such as tri-methyl-aluminum and ammonia in vapor phase prohibited the precise control of the alloy composition of AlGaN. A high flow rate or low-pressure operation in a MOVPE system was preferable for the control of the alloy composition of AlGaN [10, 11]. The crystalline quality of AlGaN on sapphire was improved by using the LT-buffer layer, although it progressively worsened with the increase of the AlN molar fraction. The crystalline quality was significantly improved when AlGaN was grown on high-quality GaN, but the fracture of the film, which originates from the grown-in tensile stress induced by the lattice mismatch between AlGaN and GaN, occurs if the thickness of AlGaN exceeds a critical value [12, 13]. The fracture problem in an AlGaN on GaN heterostructure was solved by utilizing another LT-AlN layer. Figure 5.2 schematically shows how to use another LT-AlN layer. The LT-AlN layer is inserted between the underlying GaN layer and the upper AlGaN layer, and is called the 'LT-interlayer' [14]. The major effect of the LT-interlayer is the reduction of tensile stress during growth and reduction of the threading dislocations which have screw components [15].

Polycrystalline GaInN was first synthesized by Osamura *et al.* using sputtering [16]. Nagatomo *et al.* [17] and Yoshimoto *et al.* [18] grew single-crystalline GaInN by MOVPE. Compared to AlGaN, the composition of the GaInN is strongly dependent on the growth temperature. Figure 5.3 shows the dependence of the InN molar fraction in GaInN as a function of the growth temperature for a constant vapor composition. The GaInN growth process is affected by the sublimation of GaInN and/or a change of the sticking coefficient of the In precursor. Therefore, precise control of growth temperature is essential for the growth of GaInN having the desired composition.

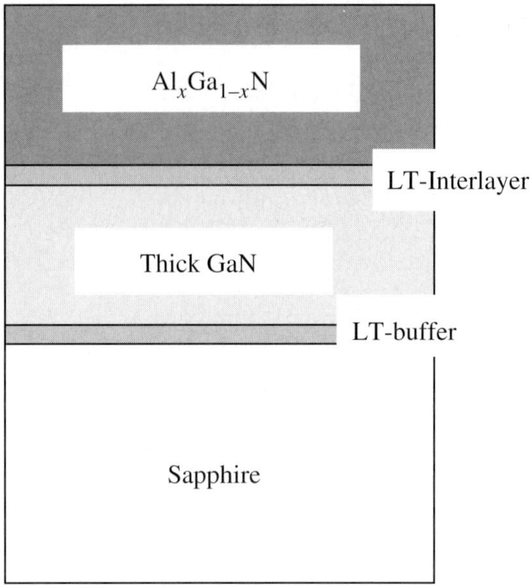

Fig. 5.2 Schematic drawing of the low temperature interlayer.

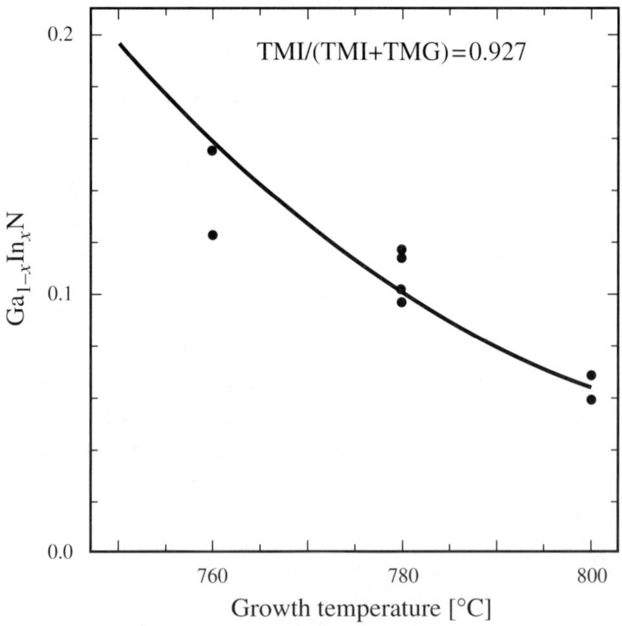

Fig. 5.3 Dependence of the solid composition of GaInN at a constant flow rate of TMGa and TMIn as a function of growth temperature.

5.2 Conductivity control of n-type nitrides

The resistivity of nominally undoped GaN grown using the LT-buffer layer is much higher than that of HVPE-grown GaN or MOVPE-grown GaN without the LT-buffer layer due to the decrease of the residual donor concentration, which is $10^{15} cm^{-3}$ or less. In order to control the conductivity of n-type GaN and other nitrides, doping with a donor impurity is indispensable. In 1990, SiH_4 diluted with H_2 or N_2 was found to be suitable for Si doping during MOVPE growth of GaN [19]. The electron concentration at room temperature can be controlled from the undoped level to higher than $10^{19} cm^{-3}$. Si_2H_6 has also been used as a Si precursor [20]. The activation energy of Si as a donor in GaN is around 30 [meV] [21, 22]. It increases with the increase of AlN molar fraction in $Al_xGa_{1-x}N$ alloys [23]. The conductivity of n-type GaInN is also controlled by Si-doping [24]. Nowadays, Si doping in MOVPE growth with a LT-buffer layer is widely adopted for conductivity control of *n*-type nitrides, i.e. GaN, AlGaN and GaInN. Oxygen was found to act as a donor in GaN, which was thought to be the unintentionally doped donor in the HVPE growth of GaN in the 70s [25]. Compared to silicon, oxygen tends to form deep levels when it is doped with AlGaN with a high AlN molar fraction [26]. Ge and Se donors were also reported [27, 28].

5.3 Realization of p-type nitrides

Many groups have attempted to produce p-type GaN. No group, however, succeeded until 1989. Good controllability of Mg concentration during MOVPE growth of GaN was reported using bis-cyclopentadienylmagnesium as the Mg precursor [29]. The added Mg was mostly inactive by hydrogen passivation when as-grown. A low-energy electron beam irradiation treatment was used to activate the Mg, and then to yield p-type GaN and p-n-junction-type LEDs for the first time [30]. In 1991, p-type $Al_xGa_{1-x}N$ was obtained in the same manner [31]. Subsequently, p-type GaN was also achieved by thermal annealing [32]. P-type $Ga_{1-x}In_xN$ was also achieved in 1995 [33]. The activation energy of Mg in group III nitrides depends on their bandgap. It decreases with the increase of InN molar fraction in $Ga_{1-x}In_xN$ alloys, while it increases with increase of the AlN molar fraction in $Al_xGa_{1-x}N$ alloys [34]. In other words, it is difficult to achieve highly conductive p-type $Al_xGa_{1-x}N$ alloys.

5.4 Fabrication of low-dimensional nitrides

Breakthrough of crystal growth of GaN using the LT-buffer in addition to conductivity control, and in particular, the realization of p-type GaN mentioned above, have led to the fabrication of low-dimensional structures, and thus to the achievement of high-performance blue and green light-emitting diodes (LEDs) and violet laser diodes (LDs).

A MOVPE-grown low-dimensional AlGaN based multilayered structure was first fabricated in 1991 [12]. Clear satellite peaks in the X-ray diffraction profile were reported. A high-electron mobility two-dimensional electron gas was observed at the interface between AlGaN and GaN which was also grown on a sapphire substrate covered with the LT-buffer layer [35].

The specific feature of this structure is the coherent growth of AlGaN on GaN and the corresponding piezoelectric field inside the AlGaN. In the case of wurtzite structures, if a mismatched heterostructure is grown along the [0001] direction, stress caused by the lattice mismatch cannot be relaxed by the screw dislocation with Burger's vector $\vec{b} = [0001]$ or the edge dislocation with Burger's vector $\vec{b} = [2\bar{1}\bar{1}0]$, because there are no inclined planes in these dislocations. Therefore, wurtzite crystal can be coherently grown without lattice relaxation to be much thicker than the zinc-blende crystal. Figure 5.4 shows the reciprocal space mapping of X-ray diffraction around the (20$\bar{2}$4) plane of the AlGaN/GaN heterostructure. The horizontal axis and vertical axis correspond to the reciprocal in-plane and plane normal lattice constants. As shown in the figure, the horizontal axis of the two peaks are aligned, which means that the in-plane lattice constants of AlGaN and GaN are the same, this is referred to as coherent growth.

In 1993, a low-dimensional GaInN-based multi-layered structure was fabricated [36]. In 1995, as shown in Figure 5.5, the PL efficiency of GaInN-based MQW was found to be strongly dependent on the thickness

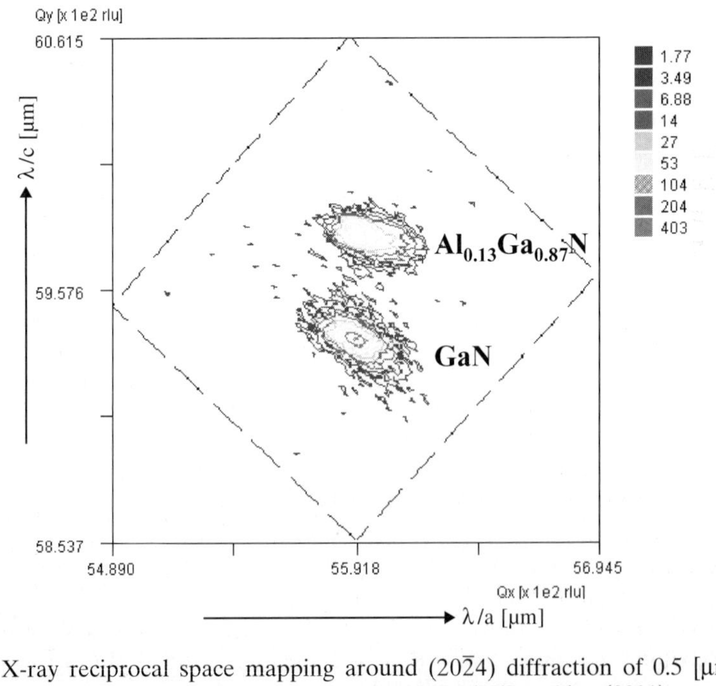

Fig. 5.4 X-ray reciprocal space mapping around (20$\bar{2}$4) diffraction of 0.5 [μm] thick $Al_{0.13}Ga_{0.87}N$ directly grown on GaN/LT-AlN/Sapphire (0001).

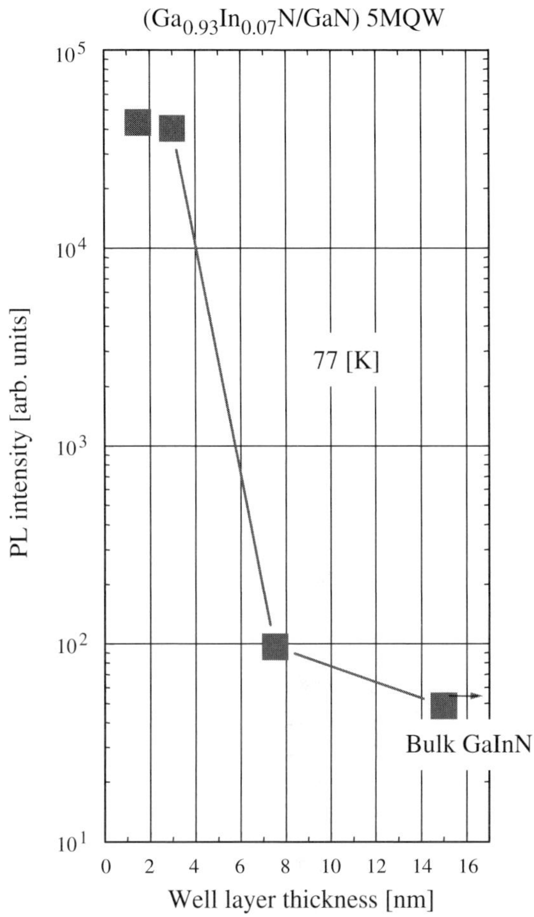

Fig. 5.5 77 [K] PL intensity of $Ga_{0.93}In_{0.07}N$/GaN 5 QWs grown on GaN/LT-AlN/Sapphire (0001) as a function of thickness of $Ga_{0.93}In_{0.07}N$ well layer.

of the well layer [37]. This finding led to the realization of highly efficient blue LEDs and violet LDs. The comprehension of the crystal structure of a GaInN/GaN multi-layered structure on high-quality GaN obtained by the X-ray reciprocal space mapping and transmission electron microscopy helps to understand their luminescence properties. Figure 5.6 shows reciprocal space mapping of X-ray diffraction around the $(20\bar{2}4)$ plane of the $Ga_{0.93}In_{0.07}N$/GaN heterostructure. Similar to AlGaN, GaInN also tends to be coherently grown on the underlying GaN. Precise control of thickness and alloy composition in the growth of GaInN/GaN MQW by MOVPE was demonstrated. Figures 5.7 and 5.8 show the X-ray diffraction profiles of GaInN/GaN MQW, in which thickness and alloy composition of the GaInN well were designed to be 11 monolayers and an InN fraction of

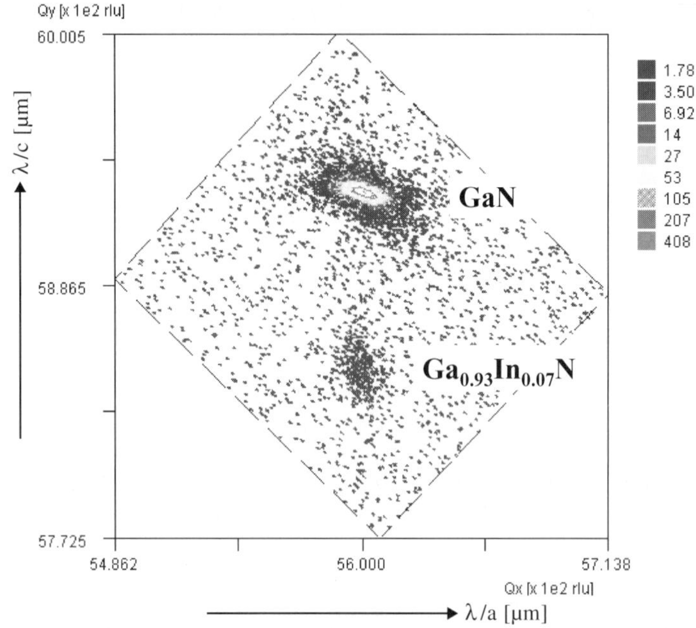

Fig. 5.6 X-ray reciprocal space mapping around $(20\bar{2}4)$ diffraction of 0.05 [μm] thick $Ga_{0.93}In_{0.07}N$ directly grown on GaN/LT-AlN/(0001) Sapphire.

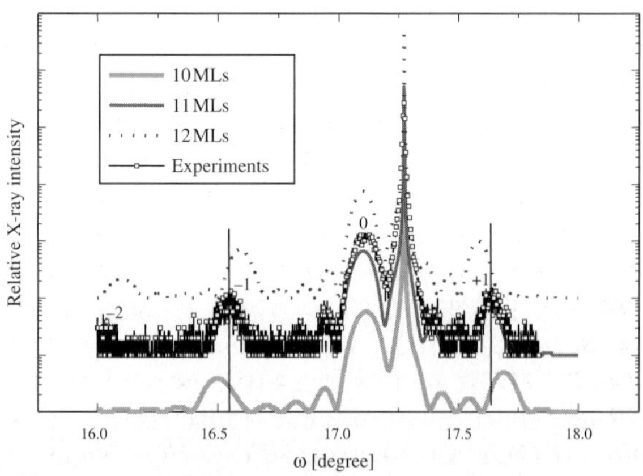

Fig. 5.7 2θ/ω-mode X-ray diffraction profile from (0002) plane of $Ga_{0.93}In_{0.07}N$/GaN MQW. Simulation profiles of different MQWs having well thickness of 10, 11, and 12 monolayers are also shown for comparison. A distinct difference is observed for 10 and 12 monolayers especially at the -1th order satellite peak.

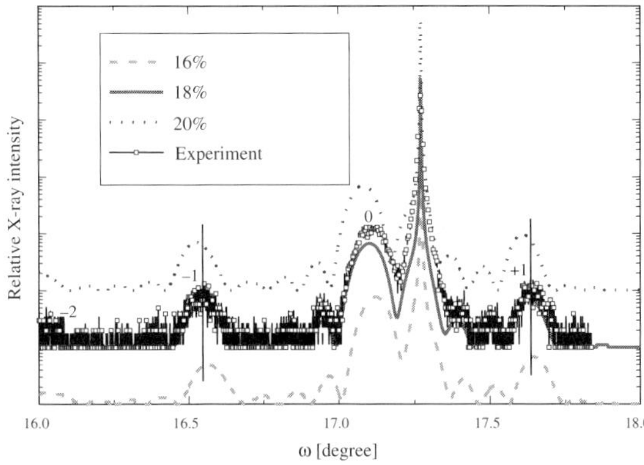

Fig. 5.8 $2\theta/\omega$-mode X-ray diffraction profile from (0002) plane of $Ga_{0.93}In_{0.07}N$/GaN MQW. Simulation profiles of different MQWs having different InN molar fractions of 16, 18 and 20% in the well layer are also shown for comparison. A distinct difference is observed for 16 and 20% especially at the -1th order satellite peak.

18%, respectively. The simulation results given in the figures clearly show that thickness and alloy composition can be controlled within + – one monolayer and 1%, respectively [38].

5.5 Quantum confined Stark effect and piezoelectricity

Not only the quantum size effect, but also the internal electric field mainly caused by the large piezoelectricity of nitrides and the spatial fluctuation of composition is found to greatly affect the luminescence properties of low-dimensional structures. These findings lead to the optimization of the structure of GaInN-based quantum wells. As shown in the previous section, it is found that [0001]-oriented thin GaInN layers can be coherently grown on a thick GaN layer, in other words, the GaInN layers were under biaxial compressive stress. In such a case, a piezoelectric field is induced since group III nitrides have large piezoelectric constants with [0001] orientation [39–41]. Therefore, optical properties of [0001]-oriented GaInN strained QWs should be strongly affected due to the quantum-confined Stark effect (QCSE). The internal fields due to the piezoelectric effect in the GaInN strained layer on thick GaN are theoretically calculated. In the calculation, a thin GaInN layer is assumed to be coherently grown on a thick GaN layer along the [0001] direction. In other words, the GaInN layer has the same *a*-axis lattice constant as the GaN layer, and it is under biaxial compressive stress perpendicular to [0001]. The relationship between piezoelectric polarization and strain in the case of a wurtzite structure is given by

$$\begin{pmatrix} P_x \\ P_y \\ P_z \end{pmatrix} = \begin{pmatrix} 0 & 0 & 0 & 0 & e_{15} & 0 \\ 0 & 0 & 0 & e_{15} & 0 & 0 \\ e_{31} & e_{31} & e_{33} & 0 & 0 & 0 \end{pmatrix} \begin{pmatrix} \varepsilon_{xx} \\ \varepsilon_{yy} \\ \varepsilon_{zz} \\ \varepsilon_{yz} \\ \varepsilon_{zx} \\ \varepsilon_{xy} \end{pmatrix}$$

where P_i, e_{ij} and ε_{ij} are the electric polarization, the piezoelectric constant and strain in the strained layer, respectively, and the z-direction is set parallel to the [0001] axis. The strain elements in the case of a wurtzite structure are

$$\varepsilon_{xx} = \varepsilon_{yy} = \frac{a_s - a_e}{a_e},$$

$$\varepsilon_{zz} = -\frac{2\,c_{13}}{c_{33}} \cdot \varepsilon_{xx},$$

$$\varepsilon_{yz} = \varepsilon_{zx} = \varepsilon_{xy} = 0,$$

where a_e, a_s and c_{ij} are the free-standing *a*-axis lattice constants of the epitaxial layer (GaInN) and the substrate (GaN), and the elastic stiffness constant for the epitaxial layer, respectively. The electric field along [0001] in the strained layer due to the piezoelectric effect is given by

$$E_z = -\frac{P_z}{\varepsilon_r \cdot \varepsilon_o},$$

where ε_r and ε_o are the dielectric constant of the material and the permittivity in free space, respectively. A piezoelectric field as large as a 1–2 [MV/cm] is estimated. Figure 5.9 shows the QCSE of the GaInN/GaN MQW sandwiched by the p-GaN and n-GaN. The large offset clearly reveals the existence of piezoelectric field [39]. As shown in the inset in Figure 5.9, the growth polarity of MOVPE-grown GaN on sapphire using a LT-buffer layer was found to be the Ga-face, which was also confirmed by the X-ray crystal truncation rod analysis [42].

5.6 Enhancement of impurity activation by the use of a low-dimensional structure

A low-dimensional structure is also effective in improving the electrical properties. Modulation doping in the $Al_xGa_{1-x}N$/GaN super-lattice is found to decrease the effective activation energy of Mg, thus enhancing the hole concentration [43]. The piezoelectric field is thought to play an important role

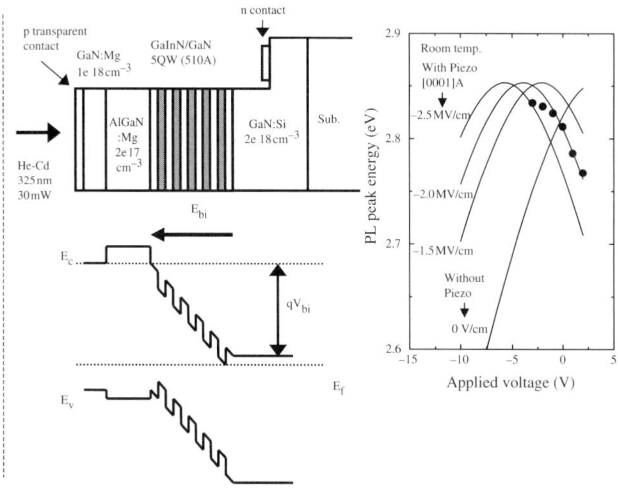

Fig. 5.9 Sample structure for p–i–n diode and its PL peak shift as a function of bias voltage. Simulation curve with and without a piezoelectric field is also shown. Piezoelectric fields in two different directions are considered.

in reducing the effective activation energy. Modulation doping in the GaN/ $Ga_{1-x}In_xN$ superlattice was also found to be effective [44].

5.7 Control of threading dislocations

Group III nitrides grown using a LT-buffer layer on the sapphire substrate contain mixed-type, pure-edge type and screw-type dislocations on the order of 10^9 [cm^{-2}] or more. Several attempts have been made to reduce the density of threading dislocations. All the methods, 'Epitaxial Lateral Overgrowth' or 'Lateral Epitaxial Overgrowth', 'Pendeo Epitaxy', 'Facet Initiated Epitaxial Lateral Overgrowth' or 'Facet Controlled Epitaxial Lateral Overgrowth' and 'Air-Bridged Lateral Epitaxial Growth' [45–49] use lateral seeding epitaxy as the modification of micro-channel epitaxy [50], thereby partly or entirely suppressing the propagation of threading dislocations over the upper layer. In case of nitrides, the bending of threading dislocations at the inclined plane during growth, which is caused by the strong stress, helps to reduce dislocation density.

A low-temperature deposited layer inserted between high-temperature grown nitrides is also found to be effective in reducing threading dislocations [14]. This technique can be applied to grow not only GaN, but also to grow $Al_xGa_{1-x}N$ covering the entire compositional range.

These techniques were used for the improvement of the performance of nitride-based devices. It was reported that the lifetime of violet LD was significantly improved. Leakage current of the p-n junction under a reverse voltage is markedly reduced, therefore a UV detector showing extremely low dark current was fabricated. PL efficiency of GaN-based and AlGaN-based

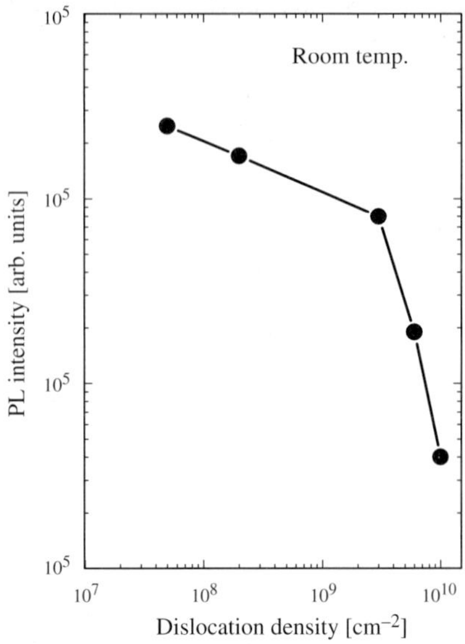

Fig. 5.10 PL intensity of $Al_{0.08}Ga_{0.92}N$:Si(8 [nm])/GaN(3 [nm])$_5$ MQWs at room temperature as a function of threading dislocation density. Inclined lateral epitaxy and LT-interlayer was used to grow samples.

MQWs is found to be greatly affected by the density of threading dislocations. Figure 5.10 shows the PL intensity of $Al_{0.08}Ga_{0.92}$ N:Si(8 [nm])/GaN (3 [nm])5 MQWs as a function of threading dislocation density. In growing samples, inclined lateral epitaxy and LT-interlayer was used. As shown in the figure, in order to achieve highly efficient UV-LEDs and UV-LDs, it is very important to reduce threading dislocations. High-performance UV-LEDs are fabricated on the low-dislocation density AlGaN.

5.8 Summary

The recent evolution of the fabrication of low-dimensional nitrides by MOVPE on highly mismatched substrates is briefly reviewed. The performance of several nitride-based devices has rapidly improved, thus the market size of nitride-based devices is rapidly expanding. In the next century, nitrides will play a major role in the optoelectronics industry.

References

1. Johnson, W. C., Parson, J. C., and Crew, M. C. (1932). Nitrogen compounds of gallium III, gallium nitride, *J. Phys. Chem.* **36**, 2561.
2. Maruska, H. P. and Tietjen, J. J. (1969). The preparation and properties of vapor-deposited single-crystal-line GaN, *Appl. Phys. Lett.* **15**, 327.

3. Pankove, J. I., Miller, E. A., Richman, D., and Berkeyheiser, J. E. (1971). Electroluminescence in GaN, *J. Lumin.* **4**, 63.
4. Ilegems, M. and Montgomery, H. C. (1973). Electrical properties of n-type vapor-grown gallium nitride, *J. Phys. Chem. Solids* **34**, 885.
5. Manasevit, H. M., Erdmann, F. M., and Simpson, W. I. (1971). The use of metalorganics in the preparation of semiconductor materials. IV. The nitrides of aluminum and gallium, *J. Electrochem. Soc.* **118**, 1864.
6. Amano, H., Sawaki, N., Akasaki, I., and Toyoda, Y. (1986). Metalorganic vapor phase epitaxial growth of a high quality GaN film using an AlN buffer layer, *Appl. Phys. Lett.*, **48**, 353.
7. For example, *Group III nitride semiconductor compounds*, (ed. Bernard Gil) (1998). Oxford University Press.
8. Bergman, J. P., Monemar, B., Amano, H., Akasaki, I., Hiramatsu, K., Sawaki, N., and Detchprohm, T. (1995). Exciton dynamics in GaN, *Conf. Ser. Inst. Phys.*, **142** , 931.
9. Chichibu, S., Azuhata, T., Sota T., and Nakamura, S. (1996). Excitonic emissions from hexagonal GaN epitaxial layers, *J. Appl. Phys.* **79**, 2784.
10. Koide, Y., Itoh, H., Sawaki, N., and Akasaki, I. (1986). Epitaxial growth and properties of $Al_xGa_{1-x}N$ by MOVPE, *J. Electrochem. Soc.* **133**, 1956.
11. Khan, M. A., Skogman, R. A., Schulze, R. G., and Gershenzon, M. (1983). Electrical properties and ion implantation of epitaxial GaN, grown by low pressure metalorganic chemical vapor deposition, *Appl. Phys. Lett.* **42**, 430.
12. Itoh, K., Kawamoto, T., Amano H., Hiramatsu K., and Akasaki, I. (1991). Metalorganic vapor phase epitaxial growth and properties of $GaN/Al_{0.1}Ga_{0.9}N$ layered structures, *Jpn. J. Appl. Phys.* **30**, 1924.
13. Ito, K., Hiramatsu, K., Amano, H., and Akasaki, I. (1990). Preparation of $Al_xGa_{1-x}N/GaN$ heterostructures by MOVPE, *J. Cryst. Growth* **104**, 533.
14. Amano, H., Iwaya, M., Kashima, T., Katsuragawa, M., Akasaki, I., Han, J. *et al.* (1998). Stress and defect control in GaN using low temperature interlayers, *Jpn. J. Appl. Phys*. **37**, L 1540.
15. Kashima, T., Nakamura, R., Iwaya, M., Katoh, H., Yamaguchi, S., Amano, H., and Akasaki I. (1999). Microscopic investigation of $Al_{0.43}Ga_{0.57}N$ on sapphire, *Jpn. J. Appl. Phys*. **38**, L 1515.
16. Osamura, K., Nakajima N., and Murakami, Y. (1972). Fundamental absorption edge in GaN, InN and their alloys, *Solid State Commun.* **11**, 617.
17. Nagatomo, T., Kuboyama, T., Minamino H., and Omoto, O. (1989). Properties of $Ga_{1-x}In_xN$ films prepared by MOVPE, *Jpn. J. Appl. Phys.* **28**, L1334.
18. Yoshimoto, N., Matsuoka, T., Sasaki, T., and Katsui, A. (1991). Photoluminescence of InGaN films grown at high temperature by metalorganic vapor phase epitaxy, *Appl. Phys. Lett.* **59**, 2251.
19. Amano, H. and Akasaki, I. (1990). Fabrication and properties of GaN p-n Junction LED, *Mater. Res. Soc. Ext. Abst.* **EA-21**, 165.
20. Rowland, L. B., Doverspike, K., and Gaskill, D. K. (1995). Silicon doping of GaN using disilane, *Appl. Phys. Lett.* **66**, 1495.
21. Meyer, B. K., Volm, D., Graber, A., Alt, H. C., Detchprohm, T., Amano, H., and Akasaki, I. (1995) Shallow donors in GaN—the binding energy and the electron effective mass, *Solid State Commun.* **95**, 597.
22. Goetz, W., Johnson, N. M., Chen, C., Liu, H., Kuo, C., and Imler, W. (1996). Activation energies of Si donors in GaN, *Appl. Phys. Lett.* **68**, 3144.
23. Steude, G., Hofmann, D. M., Meyer, B. K., Amano, H., and Akasaki, I. (1998). The Residual Donor Binding Energy in AlCaN Epitaxial Layers, *Phys. Stat. Sol.* (a)**165**, R3.
24. Nakamura, S., Mukai, T., and Senoh, M. (1993). Si-doped InGaN films grown on GaN films, *Jpn. J. Appl. Phys.* **32**, L16.

25. Van de Walle, C. G. and Neugebauer, J. (1997). Defects and doping in III–V nitrides, *Mater. Sci. Forum* **258/263**, 19.
26. Wetzel, C., Suski, T., Ager, J. W., Weber, E. R., Hailer, E. E., *et al.* (1997). Pressure induced deep gap state of oxygen in GaN, *Phys. Rev. Lett.* **78**, 3923.
27. Nakamura, S., Mukai, T., and Senoh, M. (1992). Si- and Ge doped GaN films grown with GaN buffers layers, *Jpn. J. Appl. Phys.* **31**, 2883.
28. Guo, J. D., Feng, M. S., and Pan, F. M. (1995). Se-doped GaN films grown by low-pressure metalorganic chemical vapor deposition, *Preventive Medicine* **24**, 5510.
29. Amano, H., Kitoh, M., Hiramatsu, K., and Akasaki, I. (1990). Growth and luminescence properties of Mg-doped GaN prepared by MOVPE, *J. Electrochem. Soc.* **137**, 1639.
30. Amano, H., Kito, M., Hiramatsu, K., Sawaki, N., and Akasaki, I. (1989), p-type conduction in Mg-doped GaN treated with low-energy electron beam irradiation (LEEBI), *Jpn. J. Appl. Phys.* **28**, L2112.
31. Tanaka, T., Watanabe, A., Amano, H., Kobayashi, Y., Akasaki, I., Yamazaki, S., and Koike M. (1994). p-type conduction in Mg-doped GaN and $Al_{0.08}Ga_{0.92}N$ grown by metalorganic vapor phase epitaxy, *Appl. Phys. Lett.* **65**, 593.
32. Nakamura, S., Senoh, M., and Mukai, T. (1991). Highly p-typed Mg-doped GaN films grown with GaN buffer layer, *Jpn. J. Appl. Phys.* **30**, L1708.
33. Yamasaki, S., Asami S., Shibata, N., Koike, M., Manabe, K., Tanaka, T., *et al.* (1995). p-type conduction in Mg-doped $Ga_{0.91}In_{0.09}N$ grown by metalorganic vapor-phase epitaxy, *Appl. Phys. Lett.* **66**, 1112.
34. Katsuragawa, M., Sota S., Komori, M., Anbe, C., Takeuchi, T., Sakai, H., Amano H., and Akasaki, I. (1998) Thermal ionization energy of Si and Mg in AlGaN, *J. Cryst. Growth* **189–190**, 528.
35. Khan Asif, M., Van Hove, J. M., Kuznia, J. N., and Olson, D. T. (1991). High electron mobility $GaN/Al_xGa_{1-x}N$ heterostructures grown by low-pressure metalorganic chemical vapor deposition, *Appl. Phys. Lett.* **58**, 2408.
36. Nakamura, S., Senoh, M., Iwasa, N., and Nagahama, S. (1995). High-brightness InGaN blue, green and yellow light-emitting diodes with quantum well structures, *Jpn. J. Appl. Phys.* **34**, L797.
37. Amano, H., and Akasaki, I. (1995). Fabrication and properties of GaN-based quantum well structure for short wavelength light emitter, Ext. Abstr. 1995 *Int. Conf. on Solid State Devices & Mater* 683.
38. Amano, H., Takeuchi, T., Yamaguchi, S., Nitta, S., Kariya, M., *et al.* (1998). Structural properties of nitrides grown by OMVPE on sapphire substrate, *Mater. Res. Soc. Symp. Proc.* 482, 479.
39. Takeuchi, T., Sota, S., Katsuragawa, M., Komori, M., Takeuchi, H., Amano, H., and Akasaki, I. (1997). Quantum-confined Stark effect due to piezoelectric fields in GaInN strained quantum wells, *Jpn. J. Appl. Phys.* **36**, L 382.
40. Bernardini, F., Fiorentini, V., and Vanderbilt, D. (1997). Spontaneous polarization and piezoelectric constants of III–V nitrides, *Phys. Rev. B* **56**, R10 024.
41. Wetzel, C., Takeuchi, T., Amano, H., and Akasaki, I. (1999). Piezoelectric Franz–Keldysh effect in strained GalnN/GaN heterostructures, *J. Appl. Phys.* **85**, 3786.
42. Tabuchi, M., Takeda, Y., Matsumoto, N., Amano, H., and Akasaki, I. (1999). X-Ray interference and crystal truncation rod observation of GaN and GaInN layers grown on sapphire with AIN buffer layer, *Jpn. J. Appl. Phys.—Supplement* **38**, 281.
43. Nakamura, S., Senoh, M., Nagahama, S., Iwasa, N., Yamada, T., Matsushita, T., *et al.* (1998). InGaN/GaN/AlGaN-based laser diodes with modulation-doped strained-layer superlattices grown on an epitaxially laterally overgrown GaN substrate, *Appl. Phys. Lett.* **72**, 211.

44. Kumakura, K., Makimoto, T. and Kobayashi, N. (2000). Efficient hole generation above 10^{19} cm^{-3} in Mg-doped InGaN/GaN superlattices at room temperature, *Jpn. J. Appl. Phys.* **39**, L195.
45. Usui, A., Sunakawa, H., Sakai, A., and Yamaguchi, A. (1997). Thick GaN epitaxial growth with low dislocation density by hydride vapor phase epitaxy, *Jpn. J. Appl. Phys.* **36**, L899.
46. Gehrke, T., Linthicum, K. J., Thomson, D. B., Rajagopal, P., Batchelor, A. D., and Davis, R. F. (1999). Pendeo-epitaxy of gallium nitride and aluminum nitride films and heterostructures on silicon carbide substrate, MRS *Internet J. Nitride Semicond. Res.* **4S1**, G3.2.
47. Hiramatsu, K., Nishiyama, K., Onishi, M., Mizutani, H., Narukawa, M, Motogaito, A., *et al.* (2000). Fabrication and characterization of low defect density GaN using facet- controlled epitaxial lateral overgrowth (FACELO), *J. Crystal Growth* **221**, 316.
48. Sakai, A., Sunakawa, H., Kimura, A., and Usui, A. (2000). Self-organized propagation of dislocations in GaN films during epitaxial lateral overgrowth, *Appl. Phys. Lett.* **76**, 442.
49. Izumi, T., Inoue, K., Narukawa, Y., Okamoto, K., Kawakami, Y., Fujita, S., *et al.* (2000). Recombination dynamics in GaN and InGaN/GaN multiple quantum wells on air-bridged lateral epitaxial grown GaN layers, *IPAP Conference Series* **1**, 599.
50. Ujiie, Y. and Nishinaga, T. (1989). Epitaxial lateral overgrowth of GaAs on a Si substrate, *Jpn. J. Appl. Phys.* **28**, L337.

6 The growth of low-dimensional nitrides by molecular beam epitaxy

Nicolas Grandjean, Benjamin Damilano, and Jean Massies

6.1 Introduction

Molecular beam epitaxy (MBE) is the technique of choice for the growth of low-dimensional semiconductor heterostructures. The ultra-high vacuum environment of MBE allows the use of an *in situ* electron probe like reflection high-energy electron diffraction (RHEED). RHEED is a powerful tool for controlling the growth in real time [1]. It allows the precise measurement of the growth rate [2], the determination of ternary alloy composition [3], and the investigation of strain relaxation during heteroepitaxy [4, 5]. MBE is also well suited for achieving abrupt interfaces. This is the consequence of molecular flux regime, which authorizes an instantaneous flux switching (simply using a shutter), and growth conditions far from equilibrium that avoid strong intermixing at the interfaces. For instance, the substrate temperatures used in MBE are much lower than those currently reported for metal-organics chemical vapor deposition (MOCVD) growth.

The realization of well-controlled low-dimensional structures is of prime importance for optoelectronic devices but also for fundamental studies. In particular, in wurtzite phase III–V nitrides grown along the [0001] polar axis, a giant electric field takes place in quantum wells (QWs) [6–13] and quantum dots (QDs) [14, 15], which strongly modifies their optical properties by quantum confined Stark effect (QCSE) [16]. The huge values of the electric field are the consequence of the significant lattice-mismatch between the different binary compounds, the very large piezoelectric constants, and the presence of a spontaneous polarization [17]. Spontaneous polarization is inherent to III–V nitrides and exists even for zero strain. Bernardini *et al.* have predicted polarization fields (piezo+spontaneous) as high as 10 MV/cm and 15 MV/cm for GaN/AlN and InN/GaN QWs, respectively [17]. Such huge electric fields have a detrimental effect on the radiative efficiency of quantum heterostructures because of the spatial separation of the carriers along the growth axis, which reduces the oscillator strength. Therefore, the radiative lifetime of carriers becomes very long and the probability that they will recombine on non-radiative defects increases drastically. This is especially the case in heteroepitaxial nitride layers as the density of threading dislocations is 10^8–$10^{10}\,cm^{-2}$ [18]. How then can we explain the extremely high level of performance of nitrides-based light-emitting diodes (LEDs) [19]? It was proposed that alloy phase separation in the InGaN active layer produces In-rich clusters acting as QDs [20]. The single idea is once the carriers are trapped in deep potential well, they can no longer diffuse toward non-radiative centers. This was first shown by Gérard *et al.* in the case of InAs/GaAs QDs grown on silicon substrate [21]. In that case, QDs are in fact not used for their zero-dimensional quantum properties but for their spatial confinnig properties. In-rich clusters in InGaN alloys are thus expected to play the same role as InAs/GaAs QDs.

In this chapter, we will first discuss the growth by MBE of group-III nitrides. Nitrogen sources, substrates issues, and real-time growth control, will be presented as well as the GaN material properties. In a second part, we will report on GaN/AlGaN QWs. We will investigate the effects of polarization field through the optical properties of quantum heterostructures. In a third part, we will present the growth of nitride QDs and discuss their radiative efficiency compared to QWs. Finally, the properties of InGaN/GaN heterostructures will be discussed and the role of In composition fluctuations in InGaN/GaN QWs will be emphasized.

6.2 MBE growth of group-III nitrides: a résumé

6.2.1 Nitrogen source: ammonia versus nitrogen plasma

The weak point of the MBE growth of nitrides has been for a long time related to the absence of efficient nitrogen sources. The first generation of nitrogen plasma-sources suffered from low growth rates and ionic species, which are detrimental for the material quality. Nowadays, the commercial nitrogen plasma sources allow the growth of high-quality GaN epilayers with growth rates larger than 1 μm/h.

Nitrogen plasma is obtained either by radio-frequency (RF) at 13.56 MHz or by electron cyclotron resonance (ECR) at 2.45 GHz. ECR plasma produces more ionic species than RF plasma [22].

An alternative way for growing nitrides by MBE is to use ammonia, which is cracked on the growth surface [23–31]. The overall GaN material properties, apart from the surface morphology due to different growth kinetics, are comparable to those obtained by plasma-assisted MBE. Another difference of NH_3-MBE is related to the dependence of the NH_3 cracking efficiency on the temperature of the GaN (0001) surface (Fig. 6.1) [32]. For temperatures lower than 450 °C, the efficiency is nearly zero and it is then very difficult to grow GaN in this temperature range. Attempts to increase the NH_3 efficiency using a high-temperature pre-cracking cell have failed. Actually, the product of the decomposition of NH_3 in a high temperature cell is mainly N_2 and H_2 [27]. The efficiency of NH_3 increases slowly for surface temperatures larger than 500 °C. At the usual growth temperature of GaN (~800 °C), it is only ~ 4%. This means that a large amount of NH_3 is needed for growing GaN at growth rates larger than 1 μm/h. It should be pointed out that the pumping of the uncracked NH_3 molecules is ensured by their condensation onto the liquid-nitrogen-cooled

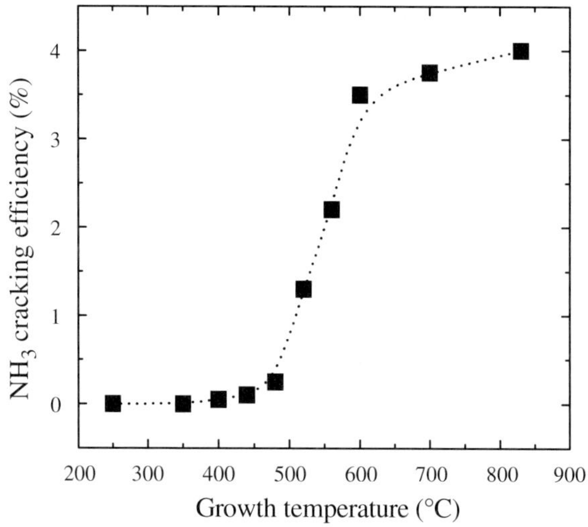

Fig. 6.1 Efficiency of NH_3 as a function of the substrate temperature.

cryopanels of the MBE chamber. During the growth, the pressure in the growth chamber is typically in the low 10^{-6} Torr range. However, special care must be taken when the cryopanels warms up because of the very large growth chamber pressure reached at the NH_3 desorption point. Particular attention must also be paid to parasitic reactions of NH_3 with Ga and Al molten charge that may lead to damage of the effusion cells. A double filament cell for Ga and a cold lip cell for Al can avoid severe problems.

The growth conditions of NH_3-MBE of III–V nitrides are in fact close to those currently used for III–V arsenides. GaN is grown with an excess of nitrogen active species, i.e. a V/III ratio larger than 1 (typically 5–10), considering a cracking efficiency of 4% [33]. Using these growth conditions, the growth is two-dimensional (2D) as indicated by a streaky RHEED pattern, and the GaN material properties are of high quality [33, 34]. On the other hand, faceting of the growing surface is observed for Ga-rich conditions. However, we have remarked that the presence of As (for instance in the case of the MBE system was previously used for the growth of III–V arsenides) allows the 2D growth to be sustained while being under Ga-rich conditions. On the contrary, a spotty RHEED pattern was observed for N-rich conditions in this case. Note that the presence of As on the growth front is indicated by a 2×2 surface reconstruction during growth instead of a 1×1.

The GaN growth conditions in plasma assisted MBE are different from those used in NH_3-MBE. It has indeed been reported that Ga-rich conditions are a prerequisite for achieving 2D growth and high-quality material [35–37]. This is ascribed to the high reactivity of atomic nitrogen with Ga at the growing surface. The surface diffusion length of Ga adatoms is quite low and not sufficient to ensure the crystalline quality. A lot of stacking faults are observed for N-rich conditions [37]. Several approaches have been developed to increase the surface diffusion length. One can mention migration enhanced epitaxy [38] and the use of In as surfactant [39, 40].

Contrary to NH_3 source, of which the efficiency depends on the substrate temperature, the plasma sources delivered a constant nitrogen active species flux. This is a key issue for the growth at low temperature of InN [41] and high In content InGaN alloys [42]. Plasma sources are also particularly well suited for the growth of cubic phase III–V nitrides.

6.2.2 Substrate issues

GaN single crystal substrates are nowadays still unavailable for nitride mass production because of their rarity and small size. They are used in basic research and are very important as the properties of homoepitaxial nitride layers serve as a benchmark [43].

6.2.2.1 *Sapphire*

$Al_2O_3(0001)$ substrate is currently used in MBE as in MOCVD. A nitridation of the sapphire surface is generally carried out before the growth of the buffer layer in order to improve the structural and optical properties of the GaN epilayers [44, 45]. The chemical nature of the nitrided sapphire surface depends on several parameters. For instance, Moustakas *et al.* have first reported that the exposure to an ECR nitrogen plasma of the sapphire substrate, heated at 850 °C, results in an AlN relaxed layer [44]. On the other hand, Uchida *et al.* (MOCVD) have shown that sapphire exposed to NH_3 at 1020 °C leads to an AlN_xO_{1-x} layer [46]. In fact, both AlN [44, 45, 47] and AlN_xO_{1-x} [48] have been observed in MBE depending on the nitrogen source and the nitridation conditions (temperature, duration, etc.). It must be pointed out that too long a nitridation time results in a degradation of the overall properties of the GaN epilayer. On the other hand, the nitridation of the sapphire substrate also plays on the polarity of the GaN epilayer. For instance, Ga-polarity is observed when an optimized nitridation is performed, while GaN epilayers grown directly on sapphire (without nitridation step) exhibit N-polarity [49].

The use of a buffer layer, GaN [28] or AlN [24], can improve the material quality. This was first demonstrated in 1983 by S. Yoshida *et al.* using AlN [24]. It should be pointed out that at low temperature (500 °C), it is necessary to increase substantially the NH_3 flux for the growth of the GaN buffer layer since the NH_3 efficiency is very low at that temperature [32]. Moreover, it has been reported that the growth temperature of the GaN buffer layer determines the overall properties of the overgrown GaN epilayer (crystal quality, surface roughness, residual doping, optical properties) [28]. In Fig. 6.2 are displayed cross-sectional transmission electron microscopy (TEM) images of 4 μm-thick GaN epilayers grown on top of a 250 Å-thick GaN buffer layer deposited either at 550 °C or 500 °C. One can see that the structural quality of the GaN epilayer is strongly affected by the growth temperature of the GaN

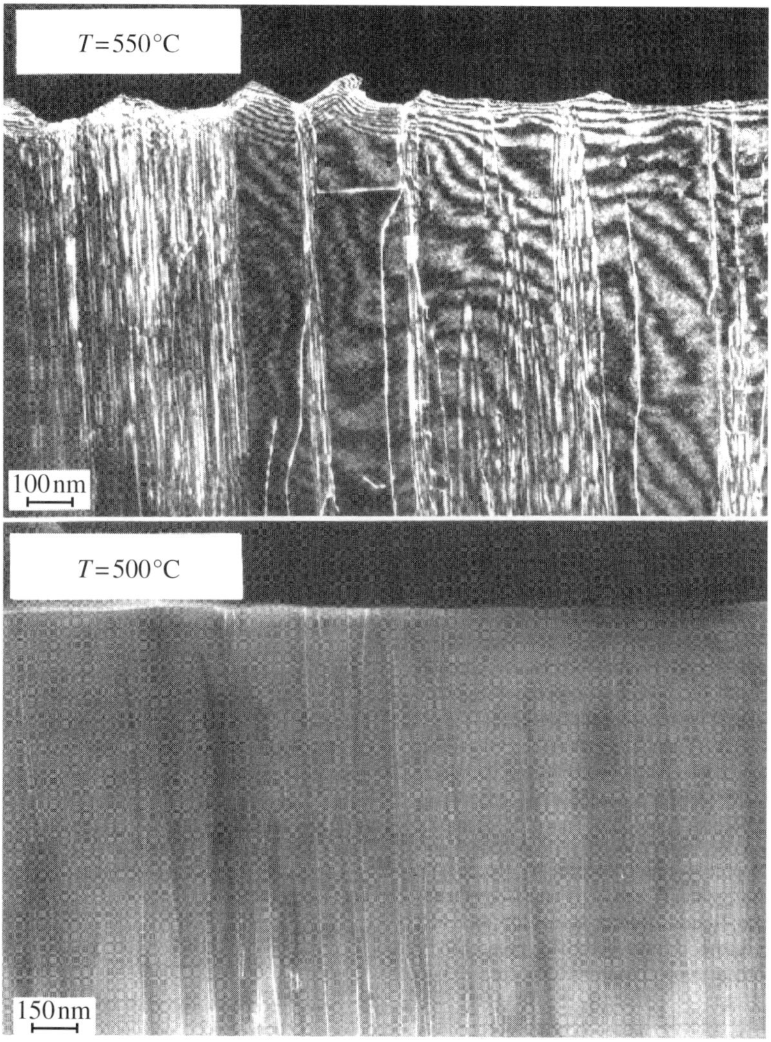

Fig. 6.2 TEM image of 4 μm-thick GaN epilayers grown on a 250 Å-thick GaN buffer deposited at 550 °C or 500 °C

buffer layer. For buffer grown at 550 °C, both the surface morphology and the crystal quality are poor. There is a lot of dislocations, stacking faults, and prismatic defects. Improvement of the material quality is observed by reducing the buffer layer growth temperature to 500 °C [28].

6.2.2.2 *Silicon carbide*

Silicon carbide is a competitive substrate for both optoelectronics, due to its electrical conductivity, and for high-power and high-frequency electronics because of its good thermal conductivity compared to sapphire. However, there are still only scarce reports concerning the MBE-growth of GaN on SiC substrates. The main reason is likely to be their expensive price, and the second reason is the difficulty of preparing good epi-ready surfaces. It has indeed been reported that the surface roughness of SiC substrates is crucial for the growth of high-quality nitrides [50–52] because steps are responsible for structural defects such as stacking faults and double positioning boundaries [51]. Nevertheless, substrate preparation procedures have been successfully developed [52, 53].

The mode of growth of GaN on SiC is Volmer–Weber, i.e. GaN does not wet the SiC surface [54, 55]. Consequently, dislocations are formed at the island edges. On the contrary, AlN wets the SiC surface [51, 54] and can therefore be used as a buffer layer for enhancing the GaN nucleation [52, 55]. However, AlN is not well suited for vertical electrical injection, and works have been devoted to the growth of high-quality GaN layers directly on SiC [31, 56, 57].

In the next coming years, SiC will certainly become a common substrate because of the rapid progresses in high-power and high-frequency electronics devices combined with the increasing numbers of substrate suppliers.

6.2.2.3 *Silicon*

Silicon is always considered as the potentially best substrate because of its low cost, large size, high quality, hardness, thermal conductivity, etc. However, the lattice-mismatch and the thermal expansion coefficient mismatch are severe drawbacks for the growth of high quality nitride layers. Nevertheless, LEDs and high electron mobility transistors (HEMT) have already been achieved on silicon substrates.

Si(111) substrates are well suited for wurtzite phase nitride compounds, while the (001) orientation is better for the cubic phase [58], although wurtzite GaN has been also observed on Si(001) [59, 60]. An AlN buffer layer is generally used before the growth of GaN because the latter does not wet silicon. With an appropriate growth procedure, the epitaxial GaN layers are as smooth as those obtained on sapphire substrate in the case of NH_3-MBE [61, 62]. The formation of cracks remains a key issue for the future of silicon substrates. The critical thickness for crack formation has been recently delayed to several microns by using strain-balanced engineering allowing to realize crack-free LEDs and HEMTs [63].

6.2.2.4 *Other substrates*

Despite their potential interest the following substrates have not lived up to this promise. ZnO was considered because of its small lattice-mismatch with GaN but no clear advantage compared to sapphire was reported. GaAs has been also used for both wurtzite and cubic phase nitrides despite a large lattice-mismatch (20%). The main problem is the thermal decomposition of GaAs at usual GaN growth temperature. Other substrates like $MgAl_2O_4$, $LiGaO_2$, and silica glass, have been also employed.

6.2.3 Real-time control of the growth

From the very beginning, the development of MBE has been closely associated with the use of RHEED to follow the growth in real-time, as demonstrated by the pioneering work of A.Y. Cho [1]. The power of RHEED as a growth control tool has latterly been considerably increased by the discovery of the specular-beam intensity oscillations and their use for growth [2] rate measurements as well as alloy composition determination [3]. A further important step has been the use of RHEED to follow the in-plane lattice parameter variation during epitaxial growth [4], which has been shown to be an extremely sensitive method to study relaxation phenomena in real time [5]. In group-III nitrides, there are only scarce reports concerning RHEED oscillation [64–67] because of the lack of suitable lattice-matched substrates, which generally results in 3D growth with a rough growth front.

However, RHEED intensity oscillations provide an unrivaled probe of the growth rate at the monolayer scale. Provided that the growth rate and the temperature correspond to the 2D layer-by-layer nucleation growth mode, well resolved RHEED oscillations can be observed for GaN and AlN [65] as well as for the ternaries $Al_xGa_{1-x}N$ [65] and $In_xGa_{1-x}N$ [67]. Figure 6.3 shows a set of typical RHEED intensity oscillations recorded in the case of GaN, $Al_xGa_{1-x}N$ and AlN growth on sapphire substrate. From such data, the alloy composition can easily be determined by comparing the oscillation frequencies associated with the binaries and the ternary alloy. The numerous RHEED intensity

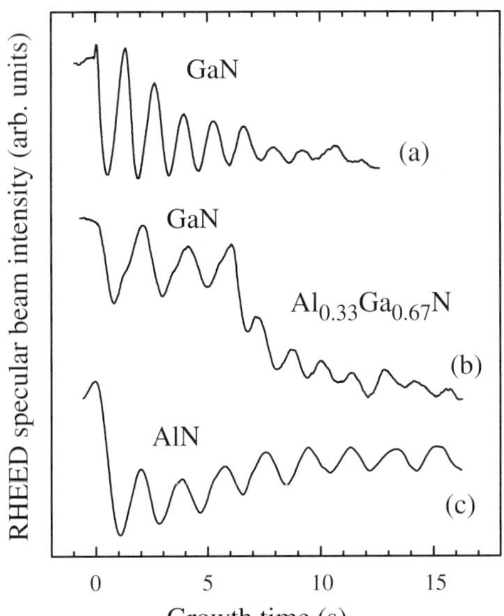

Fig. 6.3 RHEED specular beam intensity oscillations recorded during heteroepitaxial growth on sapphire of (a) GaN, (b) GaN and AlGaN, and (c) AlN.

oscillations displayed in Fig. 6.4 exemplify the smooth GaN surface morphology that produces homoepitaxial growth on GaN single crystal substrate.

RHEED can also be used to detect the onset of 3D island formation in the case of lattice-mismatch growth by recording the specular beam intensity and/or the Bragg spot intensity [67]. Figure 6.5 shows the Bragg intensity variation during the growth of $In_{0.17}Ga_{0.83}N$ on a GaN relaxed epilayer. The 2D–3D growth mode transition is indicated by the drastic increase of the Bragg intensity.

RHEED is also useful for controlling the plastic relaxation during the growth of GaN on AlN (or AlN on GaN). One can see in Fig. 6.6 that above 12 MLs the lattice constant of GaN starts to increase, while the RHEED pattern remains streaky. This is the signature of the generation of dislocations that release the accumulated elastic energy. Note that below 12 MLs, GaN is pseudomorphically strained onto AlN and the in-plane lattice constant of GaN is that of AlN.

RHEED can bring interesting information through the surface reconstruction observation. For Ga-polarity, the N-rich regime in plasma-assisted MBE growth is characterized by a 2×2 surface reconstruction [68]. In the case of NH_3-MBE, the surface reconstruction during growth is always 1×1, whatever the V/III ratio, at least if the growth chamber is free of As-containing species [69]. Otherwise, a 2×2 surface reconstruction is observed under Ga-rich conditions. The polarity of the GaN epilayer can also be deduced from the surface reconstruction. We have indeed checked by crossing RHEED observations and

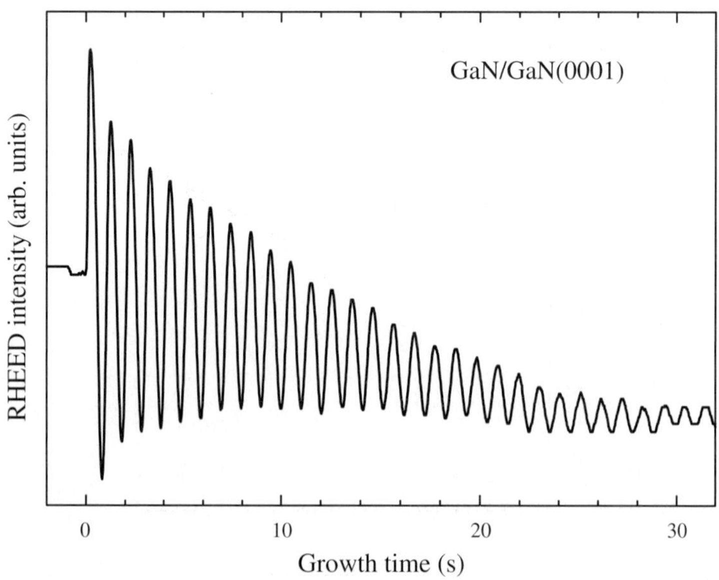

Fig. 6.4 RHEED specular beam intensity oscillations recorded during the homoepitaxial growth of GaN on GaN (0001) bulk single crystal.

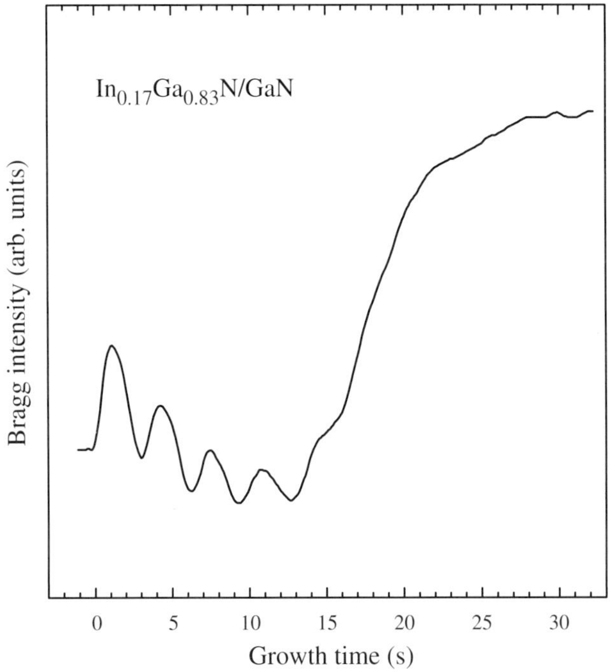

Fig. 6.5 RHEED Bragg intensity variation recorded during the growth of InGaN on GaN.

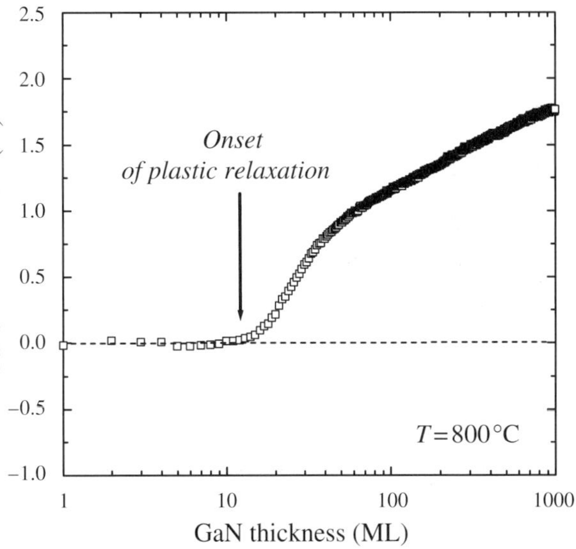

Fig. 6.6 In-plane lattice mismatch variation deduced from RHEED streak distance and recorded during the growth of GaN on AlN.

TEM convergent beam diffraction measurements that the Ga-polarity is indicated by a 2×2 surface reconstruction when cooling down below 500 °C the GaN surface under NH_3.

6.2.4 Overview of MBE grown GaN material properties

In Fig. 6.7 are reported low temperature PL spectra of GaN epilayers grown on (0001) GaN single crystal, (0001) sapphire, and (111) silicon substrates. On GaN bulk substrate, the PL linewidth of bound excitons is less than 0.5 meV. It is much larger on sapphire (~5 meV) and silicon (~8 meV) substrates because of inhomogeneous strain distribution. The energy position of A free-exciton depends on the strain state of the epilayer, and therefore, on the substrate [18]. Because of the expansion coefficient mismatch between GaN and silicon, the epilayers grown on Si(111) are under tensile strain and the GaN bandgap red-shifts. Moreover, for sufficiently high tensile strain, the A and B valence bands crossover [18], as experienced on the reflectivity spectrum (not shown).

Concerning the electrical properties, record electron mobility at 300 K (~1200 cm^2/Vs) has been reported for GaN epilayers grown by plasma-MBE on MOCVD template [70]. On sapphire substrate, an electron mobility

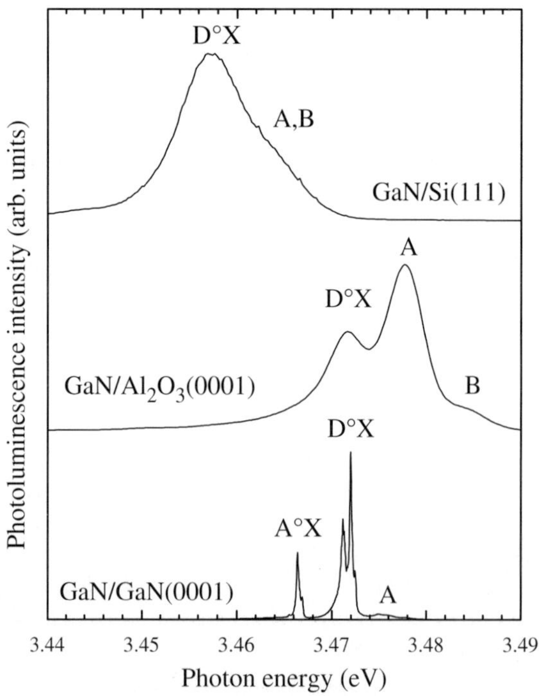

Fig. 6.7 Low temperature (2–10 K) PL spectra of GaN epilayers grown on Si(111), Al_2O_3 (0001), and GaN(0001).

of 560 cm^2/Vs has been achieved by NH_3-MBE on sapphire [30]. The smoothness of the MBE-grown surfaces is well exemplified by the high mobility of 2D electron gas at GaN/AlGaN interfaces. There are nowadays several groups that have reported mobility values as high as 60 000 cm^2/Vs at low temperature [71, 72]. The X-ray diffraction (XRD) rocking curve linewidth is 250–350 arcsec on sapphire [30] and 600 arcsec on silicon substrates [62]. The density of dislocations is in the low $10^9 cm^{-2}$ range on sapphire [30], in the middle of $10^9 cm^{-2}$ on silicon [62], and less than $10^5 cm^{-2}$ on GaN single crystal substrates [73].

6.3 AlGaN/GaN quantum well structures

6.3.1 Evidence of quantum confined Stark effect

The low temperature (2 K) PL spectrum of a 2 nm-thick GaN/$Al_{0.08}Ga_{0.92}N$ single (S) QW is reported in Fig. 6.8. The peak at 3.53 eV corresponds to the QW ground state transition energy. This is higher than the bandgap of GaN because of the additional quantum confinement energies of both the electron and hole. The peak at 3.64 eV is ascribed to the AlGaN barrier emission. The PL linewidth of the QW is only 9 meV that demonstrates the smoothness of the GaN/AlGaN interfaces. It should be noticed that the PL peak broadens as the Al composition in the barrier increases. Let us now consider a sample that

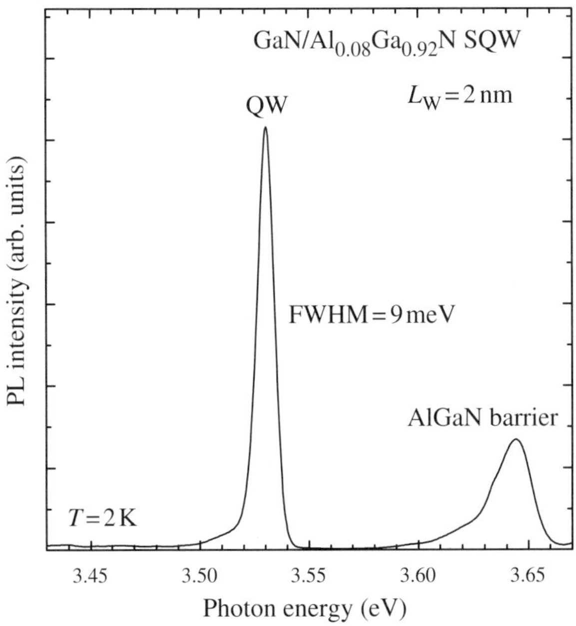

Fig. 6.8 Low temperature (2 K) PL spectrum of a GaN/$Al_{0.08}Ga_{0.92}N$ single SQW grown on sapphire by MBE.

contains four GaN/$Al_{0.13}Ga_{0.87}N$ QWs of 1, 2, 3, and 4.5 nm well thicknesses. The 10 K PL spectrum of this sample is displayed in Fig. 6.9. The salient feature is the transition energy of the wider well that is lower than the GaN bandgap emission (3.478 eV for A-free exciton in our GaN epilayers grown on sapphire). This is the consequence of the QCSE, which results from the giant polarization field inside the QW [16]. This electric field builds in wurtzite phase III–V nitrides heterostructures due to the existence of macroscopic polarization [17]: a monopolar charge appears at the interface between two nitride compounds and is equal to the difference of the total polarization (piezoelectric and spontaneous). The piezoelectric polarization depends on the strain whereas the spontaneous polarization does not. The magnitude of the built-in electric field can be estimated from the variation of the QW transition energy as a function of the well thickness [74]. The QW energy is estimated using simple envelope function calculations including the electric field as a fitting parameter. In Fig. 6.10 are reported the experimental PL energies and the calculated energies of GaN/$Al_{0.17}Ga_{0.83}N$ QWs taking into account an electric field of 710 kV/cm.

Let us come back to the relationship between the strain state of the QW structure and the internal electric field. High resolution (HR) XRD reciprocal space mapping around the (0002) peak indicates that the AlGaN barriers are pseudomorphically strained onto the GaN buffer layer, which is fully relaxed. Therefore, piezoelectric polarization occurs in the barriers and not in the well. This differs from the situation of GaInAs/GaAs QWs grown along [111] for which the piezoelectric polarization is in the well. However, since the electric field is the consequence of the formation of monopolar charges at the interfaces, which are created by the difference of the total macroscopic polarization

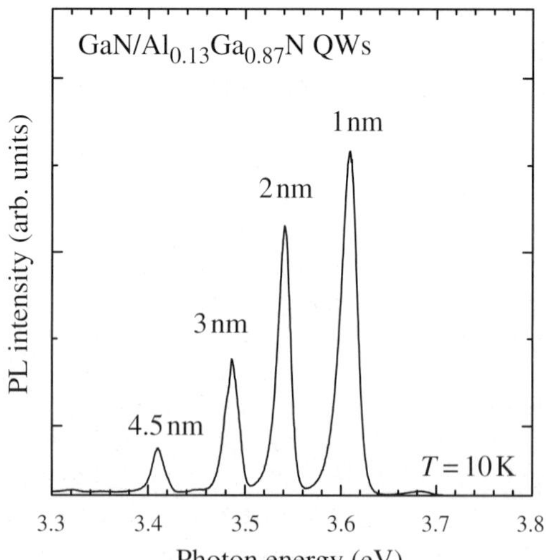

Fig. 6.9 Low temperature (10 K) PL spectrum of four GaN/$Al_{0.13}Ga_{0.87}N$ QWs. Of various thicknesses: 1, 2, 3, and 4.5 nm. The $Al_{0.13}Ga_{0.87}N$ barrier thickness is 10 nm.

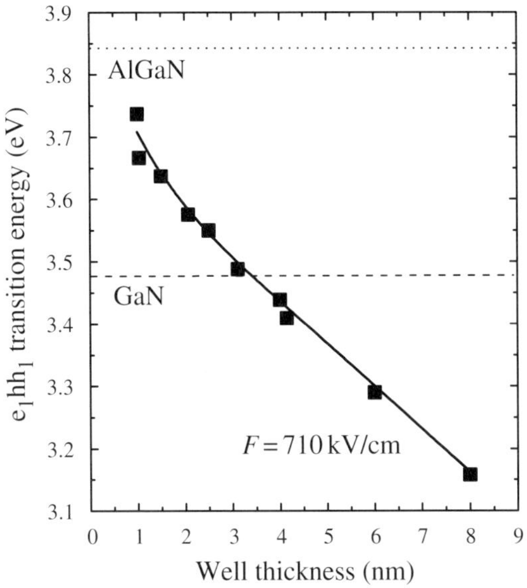

Fig. 6.10 Experimental (closed squares) and calculated e1-hh1 transition energies of $GaN/Al_{0.17}Ga_{0.83}N$ QWs (full line) including an electric field of 710 kV/cm.

between the well and the barrier, the electric field builds also in the quantum well in the case of nitride heterostructures. A counterintuitive remark is that the effect of a biaxial compressive strain in GaN well is the same as that of a biaxial tensile strain in AlGaN barriers (neglecting the deformation potential effects and the slight differences in piezoelectric constants).

Another important remark concerns the relevance of the electric field value we measured in Fig. 6.10 by PL experiments. The question is how the intrinsic electric field, given by the macroscopic polarization of nitrides, is screened by free-carriers or by photogenerated electron–hole pairs? The latter effect has been checked by reducing by two orders of magnitude the laser excitation power [74]. We did not observe any significant variation of the QW transition energy. One can then conclude that the electric field screening by laser excitation can be neglected, at least for the excitation power we used ($<1\,W/cm^2$) and for the considered samples, i.e. an electric field value lower than 1 MV/cm and well thicknesses lower than 5 nm. Indeed, the larger the electric field the larger the screening and the larger the well the larger the screening. The second origin of the electric field screening is the presence of residual donors in the barriers. This parameter has been investigated by intentionally Si-doping the AlGaN barriers to $1\times10^{18}\,cm^{-3}$ in a sample having exactly the same structure than the sample of Fig. 6.9. From the PL energy fitting, we deduced an electric field of 600 kV/cm instead of 710 kV/cm for undoped AlGaN barriers [75]. This screening can be evaluated following the approach proposed by Im *et al.* [76] that considers the depletion of the AlGaN barriers. For intentionally doped AlGaN barriers, we expect an electric field reduction of

180 kV/cm. This is larger than the value we measured (110 kV/cm). Actually, the discrepancy can be accounted for by assuming a residual doping in the undoped $Al_{0.17}Ga_{0.83}N$ layers of $4\times10^{17}\,cm^{-3}$. This means that the actual internal electric field is probably slightly higher than the experimental value of 710 kV/cm.

Finally, it should be pointed out that the previous value of 710 kV/cm was determined in a GaN/AlGaN MQW sample with an Al composition of 17%. Obviously, the electric field depends on the Al composition but also on the sample geometry, i.e. SQW or MQW.

6.3.2 Sample structure geometry and electric field

The sample geometry is a key parameter determining the electric field in heterostructures. We recall that the conservation of the electric displacement vector across a QW leads to

$$\varepsilon_w F_w - \varepsilon_b F_b = P_b - P_w \tag{6.1}$$

where ε_w (ε_b) is the well (barrier) dielectric constant, F_w (F_b) is the electric field in the well (barrier) and P_w (P_b) is the well (barrier) zero field total polarization. For a SQW (infinite barriers), the potential drop occurs in the QW and $F_w=(P_b-P_w)/\varepsilon_w$. In the case of a periodic MQW, the boundary condition $L_wF_w+L_bF_b=0$ (with L_w the well width and L_b the barrier thickness) gives:

$$F_w = L_b(P_b - P_w)/(\varepsilon_b L_w + \varepsilon_w L_b) \tag{6.2}$$

From eqn (6.2) one can see that the electric field is larger in a SQW than in a MQW with thin barriers. Nevertheless, the electric field in a MQW tends toward that of a SQW when the barriers become sufficiently thick compared to the well thickness.

Figure 6.11(a) shows the 10 K PL spectrum of a 4-nm-thick $GaN/Al_{0.17}Ga_{0.83}N$ SQW. With the aim of investigating the role of the sample geometry, two MQW structures containing four 4-nm-thick QWs have been grown with AlGaN barrier thicknesses of 5 nm [Fig. 6.11(b)] and 30 nm [Fig. 6.11(c)]. For the thinner barrier, the MQW energy transition exhibits a blue shift (35 meV) [Fig. 6.11(b)], while for a 30-nm-thick barrier the PL energy [Fig. 6.11(c)] is close to SQW transition energy [Fig. 6.11(a)]. This can be interpreted considering a reduction of the electric field in the first sample due to the redistribution of the huge polarization fields among wells and barriers [see eqn (6.2)] [77]. Such behavior has already been reported in the case of strained (111) grown (Ga,In)As/GaAs MQWs [78]. We have grown a set of $Al_{0.17}Ga_{0.83}N/GaN$ MQWs varying the barrier thickness. The electric fields width we deduced in these samples from PL experiments are reported in Fig. 6.12. Fitting these data with eqn (6.2) (full line) allows one to determine an electric field of 870 kV/cm in the case of a SQW with an Al composition of 17%.

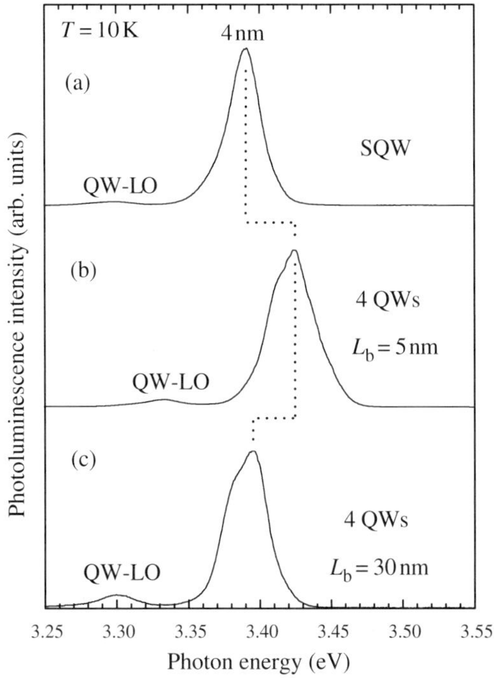

Fig. 6.11 Low temperature (10 K) PL spectra of (a) a $GaN/Al_{0.17}Ga_{0.83}N$ SQW, (b) a $GaN/Al_{0.17}Ga_{0.83}N$ MQW with a barrier thickness of 5 nm, and (c) a $GaN/Al_{0.17}Ga_{0.83}N$ MQW with a barrier thickness of 30 nm.

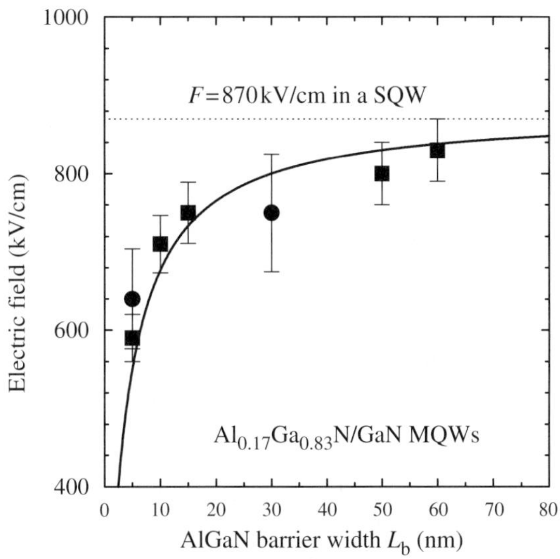

Fig. 6.12 Evolution of the electric field in a $GaN/Al_{0.17}Ga_{0.83}N$ MQW as a function of the barrier thickness.

6.3.3 Electric field versus Al composition

In order to investigate the dependence of the electric field on the Al composition, we have grown a set of GaN/$Al_xGa_{1-x}N$(x=0.08, 0.13, 0.17, and 0.27) QW samples in which the well thickness varies from 1 to 4 nm. The PL transition energy at 10 K are reported in Fig. 6.13. The thinner QWs (1 and 2 nm) exhibit a blue shift when increasing the Al composition. This is obviously the consequence of a stronger quantum carrier confinement due to higher barrier heights. In contrast, the energy of the wider wells (3 and 4 nm) decreases with the Al composition. This is the signature of a larger QCSE, which results from the augmentation of the internal electric field.

The value of the electric field in SQWs (closed squares) is reported in Fig. 6.14 as a function of the Al composition. One can see that the experimental values are lower than those calculated from the theoretical piezoelectric and spontaneous polarization coefficients [17]. At the moment, there is no clear explanation for this discrepancy.

6.4 GaN/AlN quantum dot structures

6.4.1 Growth of GaN/AlN QDs

GaN/AlN QDs have been realized taking advantage of the islanding phenomenon that results from the built-in strain due to the 2.5% lattice-mismatch

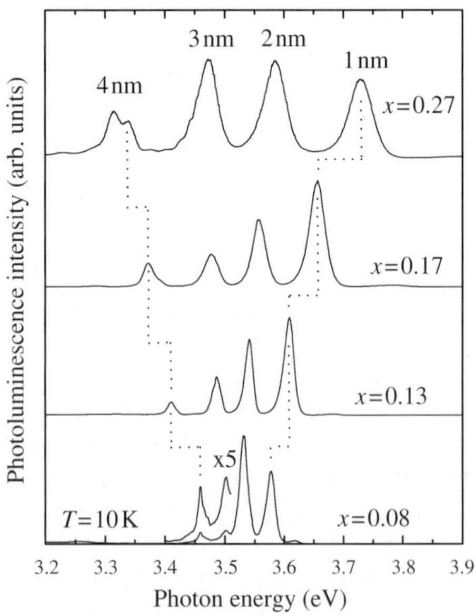

Fig. 6.13 Low temperature (10 K) PL spectra of GaN/AlGaN MQWs as a function of the Al composition.

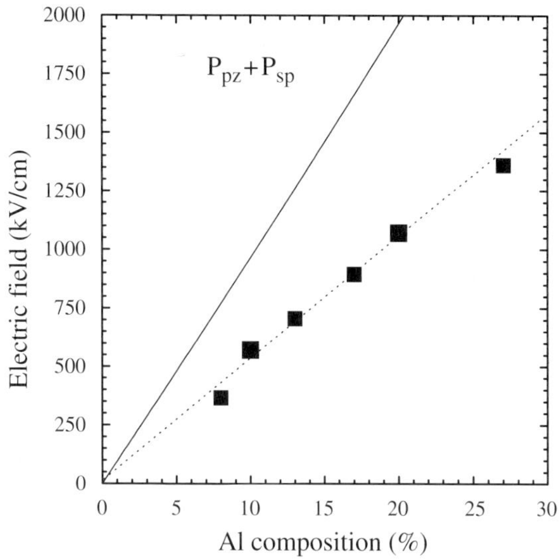

Fig. 6.14 Electric field value as a function of the Al composition in GaN/AlGaN SQW.

between GaN and AlN. It has been shown in plasma-assisted MBE that 3D islands self-form above a critical thickness of 3 MLs [14]. At this stage, the energy gain due to the elastic relaxation at the 3D island free-edges overcomes the increase of the surface free energy resulting from the 3D morphology. On the other hand, the mode of growth remains 2D during NH_3-MBE and the elastic energy is plastically relaxed at 12 MLs by dislocation generation [15]. Actually, we succeeded in forming 3D islands using an interruption of the NH_3 flux [15]. This can be attributed to the presence of hydrogen or NH_x species at the growth surface, which may play on the surface free energy. It has indeed been theoretically calculated that the adsorption of hydrogen on a GaN(0001) surface increases the surface free energy by a factor 4 [79].

In order to illustrate the formation of the GaN islands on AlN template, we show in Fig. 6.15 the RHEED patterns that correspond to the procedure we used. On a thick (>0.3 μm) AlN template, 5 MLs of GaN are deposited at 800 °C. The growth mode is 2D and the epilayer stable against a growth interruption under NH_3 [Fig. 6.15(a)]. Then, we turn off the NH_3 flux and instantaneously the RHEED pattern becomes spotty, indicating the formation of 3D islands [Fig. 6.15(b)]. Note that the formation of 3D islands supposes a sufficient mass transport, which is directly related to the surface diffusion length, and thus a sufficiently high temperature (770 °C in the present experiments). However, there is a limit for the upper temperatures as the decomposition of GaN under vacuum becomes very strong above 820 °C [80]. The 3D islands fabricated by this procedure have been characterized by atomic force microscopy (Fig. 6.16). Their density is very large (1 to 5×10^{11} cm^{-2}) compared to InAs/GaAs system. Finally, the 3D GaN islands are capped with AlN to form the QDs.

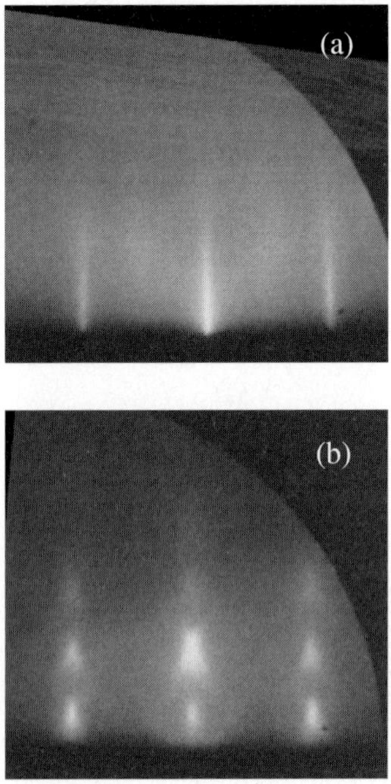

Fig. 6.15 RHEED patterns: (a) 5 MLs of GaN/AlN with NH_3 flux, and (b) 5 MLs of GaN/AlN without NH_3 flux.

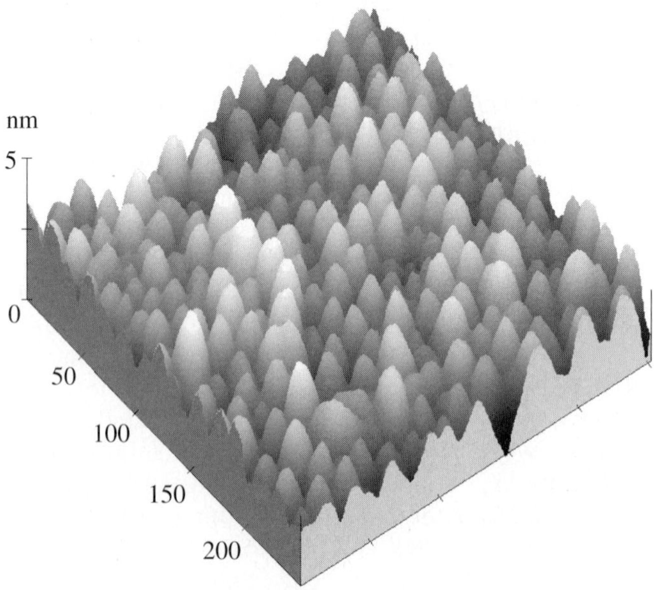

Fig. 6.16 AFM image of uncapped GaN islands formed on AlN by growth interruption without NH_3.

6.4.2 Electric field effects in QD structures

The optical properties of GaN/AlN QDs have been investigated by PL spectroscopy. A giant red-shift of the transition energy may occur as the dot height increases. This would be again a manifestation of the QCSE resulting from the internal electric field. Figure 6.17 displays the 300 K PL spectra of GaN/AlN QDs grown on Si(111) substrate. One can see that the PL energy maximum varies from 2.8 eV (blue) to 2.1 eV (orange). These energies are much lower than the bandgap of GaN forming the QDs. It should be recalled that GaN is pseudomorphically strained onto AlN and is thus under biaxial compression. A rough estimation leads to a 300 K GaN band-gap of 3.6 eV. Moreover, considering the additional quantum confinement energy arising from the very large bandgap discontinuity between GaN and AlN, one can then conclude to a giant QCSE. Theoretical calculations predict an electric field of 10 MV/cm in AlN/GaN heterostructures [17]. Such an amplitude can explain the experimental data, since it corresponds to an energy variation of 100 meV per Å.

The experimental electric field can be measured by plotting the QD PL energy as a function of the dot height measured by TEM. A value of 4 MV/cm can be estimated from the fit of the experimental data. This in reasonable agreement with previously published data [14]. However, we have to keep in mind that screening of the electric field by free carriers may occur [81]. The reduction of the optical power injection by four orders of magnitude leads to 150 meV red-shift of the QD PL energy for a dot height of about 4 nm.

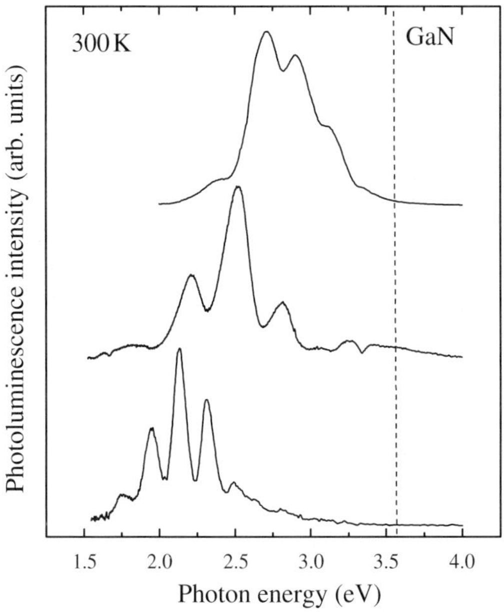

Fig. 6.17 300 K PL spectra of GaN/AlN QDs grown on (111) silicon substrate. The nominal GaN layer thicknesses are (a) 7, (b) 10, and (c) 12 MLs.

This corresponds to an increase of the electric field by about 400 kV/cm. Another source of the screening is the presence of residual donors in the barriers but it is difficult to evaluate this parameter in the present system.

We have just seen that the electric field is very large in GaN/AlN QDs, i.e. at least 4 MV/cm. Beside the giant red-shift of the PL emission energy, a second consequence is the drastic decrease of the oscillator strength. We have computed the square of the electron–hole wave function overlap in GaN/AlN QWs as a function of the well thickness. Because of the low aspect ratio (height/length) of GaN/AlN QDs, one can roughly consider that QDs behave like QWs as far as QCSE is concerned. The reduction of the oscillator strength is as high as seven orders of magnitude for a 6-nm wide QW (emission energy of 2.1 eV). On the other hand, this means that the radiative lifetime in QDs increases by several orders of magnitude since it is inversely proportional to the oscillator strength. At first sight, this would be highly detrimental to PL intensity. Indeed, if the radiative lifetime becomes much longer than the non-radiative lifetime, the non-radiative transitions will govern the carrier recombinations. Owing to the very high density of non-radiative recombination centers in nitrides, one can thus expect that the PL efficiency will rapidly decrease as the radiative lifetime increases via QCSE. Surprisingly, intense PL is observed at room temperature from GaN/AlN QDs grown on silicon substrate although the density of dislocations is greater than $10^{10}\,cm^{-2}$. Actually, such an apparent anomalous behavior can be understood considering the effect of carrier trapping in the dots. In that case, even if the radiative lifetime is very long, carriers are localized and therefore cannot diffuse toward non-radiative recombination centers like threading dislocations. To illustrate the effect of carrier trapping on the radiative efficiency, we compare the integrated PL intensity versus temperature of GaN/AlN QDs grown on silicon to that of a GaN epilayer grown on sapphire (Fig. 6.18). Whereas the density of dislocations is greater on silicon, the decrease of the PL intensity between 10 K and 300 K is very low ($<20\%$) for the QD system while it is huge in the case of the GaN epilayer.

6.5 InGaN/GaN quantum structures

6.5.1 QDs formed by Stranski–Krastanov growth mode transition

We have previously shown that use of QDs instead of homogeneous QWs or epilayers allows the radiative efficiency to increase by spatially confining the carriers. Obviously, this supposes that most of the QDs are free of defects, i.e. that their density is much more important than the density of threading dislocations. GaN/AlN QDs are attractive for efficient light emission over the whole visible range and even for white light. However, this can be obtained only by optical pumping and not by electrical injection since AlN p-n junction cannot be realized. On the other hand, InGaN/GaN QDs could be very useful for optoelectronic device applications such as LEDs. For instance, Narukawa

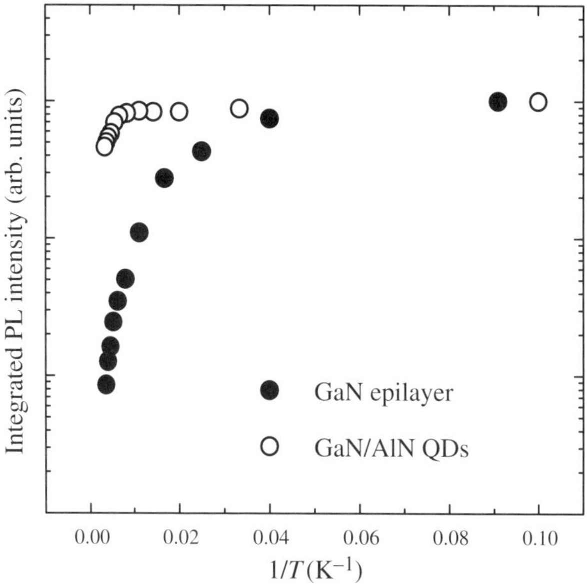

Fig. 6.18 Arrhenius plots of GaN/AlN QDs and GaN epilayer grown on (111) Si and (0001) Al_2O_3, respectively.

addressed that InGaN/GaN QDs, which are formed by InGaN phase separation during MOCVD growth process, would be responsible for the high efficiency of nitrides based optoelectronic devices [20].

MBE growth is well adapted for realizing InGaN/GaN QDs by taking advantage of the very high lattice-mismatch (11%) between GaN and InN. In that case, the strain induced by the lattice-mismatch provokes a SK growth mode regime: the formation of 3D islands above a critical thickness [67, 82]. As already mentioned for GaN/AlN QD, the driving force of the 3D islanding is the relaxation of the elastic energy at the 3D islands free-edges [5, 83] and the critical thickness is reached when the gain of elastic energy overpasses the surface free energy term [83]. The lattice-mismatch must also be sufficient, otherwise there is no 2D–3D transition because of a too low elastic energy term.

The critical In composition was measured *in situ* via RHEED intensity oscillations [67]. It is indeed well-known that the 2D–3D growth mode transition is associated with the rapid decrease of the specular beam intensity simultaneously to the increase of the Bragg spot intensity. We observed that 3D islands form only for In compositions larger than 12% (Fig. 6.19). It should be recalled that this critical In composition depends on the surface free-energy and/or on the kinetics (mass transport). For instance, a critical In composition of 18% was found in plasma-assisted MBE [82].

The optical properties of $In_{0.15}Ga_{0.85}N$/GaN QDs have been investigated by PL at room-temperature [84]. The PL energy maxima range from 3.1 eV to

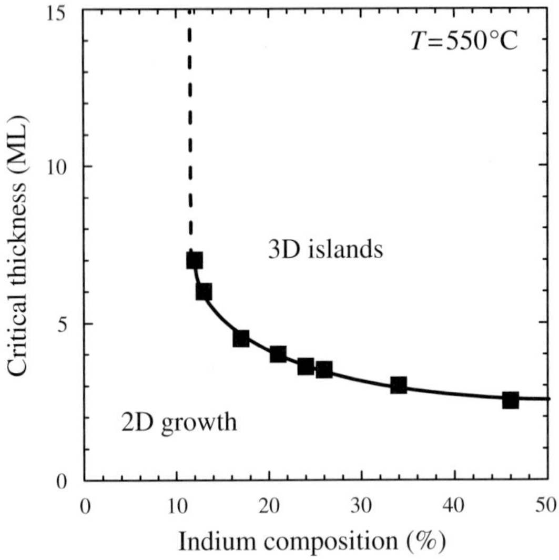

Fig. 6.19 Critical thickness for the 2D–3D growth mode transition during the growth of InGaN/GaN.

2.6 eV. The PL intensity rapidly vanishes for low-energy emission because in that case the dot height is too large and the oscillator strength is strongly reduced by QCSE. A larger In composition is thus required to get longer wavelength luminescence.

6.5.2 Surfactant effect of NH_3

In section 6.4.1, we have shown that NH_3 plays an important role on the surface free energy inhibiting the islanding phenomenon in the GaN/AlN lattice-mismatched system. In the case of InGaN growth, 3D islands form although NH_3 is supplied. Two reasons can be invoked to explain the difference between GaN and InGaN growth mode. First, because of the low growth temperature of InGaN alloys, the decomposition of NH_3 is weak, producing less surfactant active species (NH_x or H), which are considered to be responsible for the increase of the surface free energy. Second, it has been reported that an In wetting layer covers the InGaN growth front because of surface segregation effect (note that this phenomenon is well-known to occur during InGaAs growth) [85]. One can thus suspect that the surfactant active species are less bound to the surface and that the effect on the surface free-energy is thereby considerably reduced.

In order to counteract the lower efficiency of NH_3 in the case of InGaN growth, we have drastically increased the NH_3 flux. The result is exemplified in Fig. 6.20 where is reported the variation of the specular beam RHEED intensity recorded during the strained growth of $In_{0.13}Ga_{0.87}N$ on GaN for various NH_3 flow. One can clearly see that increasing the NH_3 flow delayed the 2D–3D growth mode transition (100 sccm) and can even suppress the

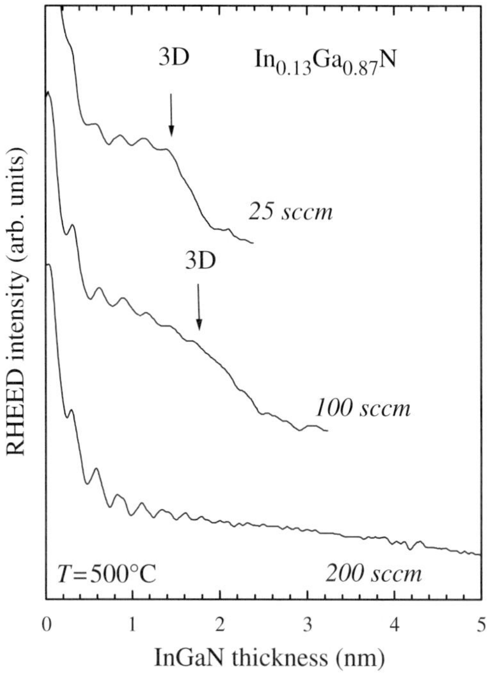

Fig. 6.20 RHEED intensity oscillations showing the delay of the 2D–3D growth mode transition when the NH_3 flux is increased.

islanding phenomenon (200 sccm). Regarding these results, NH_3 can be considered as a self-surfactant for the growth of InGaN on GaN.

We have grown $In_{0.2}Ga_{0.8}N$/GaN QWs using large NH_3 fluxes [85]. The first characterization was done by HRXRD. A spectrum recorded around the GaN (002) Bragg reflection of $5 \times In_{0.2}Ga_{0.8}N$(4 nm)/GaN(14 nm) MQWs is displayed in Fig. 6.21. The well resolved satellite peaks, up to the fourth-order fringe, are an indication of good interface quality and constant In composition from well to well. TEM in cross-section also confirms the abruptness of the InGaN/GaN interfaces (Fig. 6.22). Note that there is no evidence for InGaN alloy phase separation, although slight In composition fluctuations cannot be discarded.

6.5.3 QWs and the role of alloy composition fluctuations

PL measurements at low temperature have been carried out on $In_{0.2}Ga_{0.8}N$/GaN QW samples grown with a large NH_3 flow. PL spectra are displayed in Fig. 6.23 for various well thicknesses (from 3 nm to 5.5 nm). As a consequence of the QCSE, the PL energy range covers the whole visible spectrum. We have estimated the internal electric field to be as high as 2.5 MV/cm [87]. It should be pointed out that in the InGaN/GaN system most of the polarization field

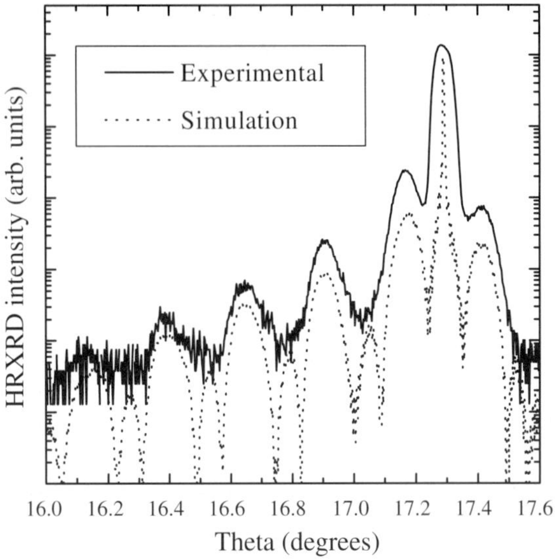

Fig. 6.21 HRXRD spectrum of a $5\times In_{0.2}Ga_{0.8}N$(4 nm)/GaN(14 nm) MQW.

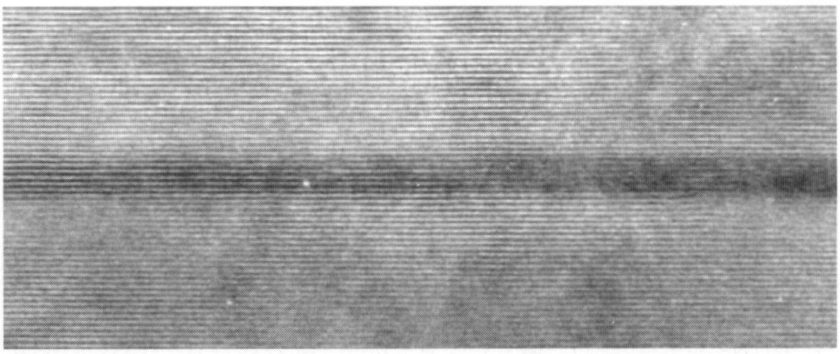

Fig. 6.22 TEM image of a 3.5-nm-thick $In_{0.2}Ga_{0.8}N$/GaN QW.

arises from the piezoelectric polarization and not from the spontaneous polarization because of the small difference between GaN and InN constants [17].

The PL linewidths at low temperature are 80±10 meV (150 meV at 300 K). These values are rather low if one considers the different factors contributing to the PL broadening in InGaN/GaN QW system. First of all, the very large bandgap bowing parameter [88] of the InGaN alloy is one of the main reasons. Indeed, only ±1% of In composition fluctuations lead to a bandgap variation of about 80 meV. Another source of the PL broadening is related to the giant QCSE. If we take into account an internal electric field of 2.5 MV/cm, a well thickness fluctuation of ±1 ML involves a PL energy variation of 130 meV. Moreover, an In composition fluctuation of ±1% induces an electric field variation of about ±125 kV/cm and consequently an energy variation of 50 meV.

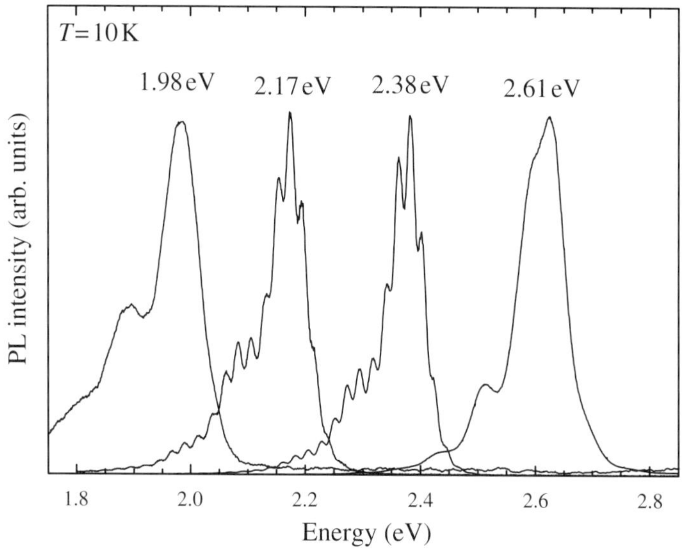

Fig. 6.23 Low temperature (10 K) PL spectra of $In_{0.2}Ga_{0.8}N$/GaN QWs of various thicknesses: 3, 3.5, 4.5, and 5.5 nm.

All these factors may produce a very large PL broadening even for low In composition and well thickness fluctuations. The experimental PL linewidths are thus a strong evidence for both In composition homogeneity and smooth QW interfaces.

Surprisingly, the PL efficiency of $In_xGa_{1-x}N(x>0.15)$/GaN QWs (closed circles) is much higher than that of GaN/AlGaN QWs (open circles), for same thicknesses, although the electric field is larger in the former (Fig. 6.24). By analogy with GaN/AlN QDs, one may suspect that localization effects occur in InGaN/GaN QWs. The degree of localization can be evaluated by the mobility-edge of excitons, i.e. the temperature at which excitons, which are localized at low temperature, become free. This behavior is generally experienced, when the localization energy is sufficiently high, by an S-shape of the PL energy variation as a function of the temperature [89]. We report in Fig. 6.25, the PL energy versus temperature for (a) GaN/$Al_{0.17}Ga_{0.83}N$ QW, and (b) $In_{0.2}Ga_{0.8}N$/GaN QW, both having the same well width of 2 nm. Also in Fig. 6.25 is plotted the semi-empirical Varshni's equation (full line), which roughly accounts for the temperature-dependent energy variation of free-exciton population. The mobility-edge temperature (arrows in Fig. 6.25) is reached as the experimental PL energy data coincide with Varshni's equation curve. It can be seen that the temperature is higher for the InGaN/GaN QW, in agreement with the higher PL efficiency. This indicates that carrier localization takes place in InGaN/GaN QWs probably because as small as the In composition fluctuations are, the combination of both the large bandgap bowing and the In composition-dependent electric field leads to deep potential

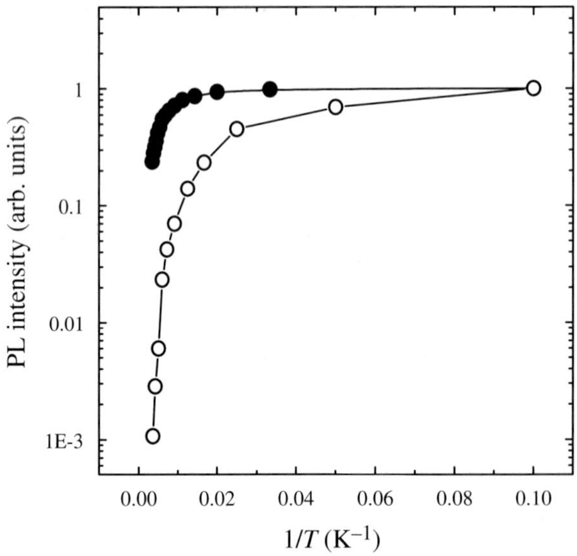

Fig. 6.24 Arrhenius plots of the PL intensity of InGaN/GaN (closed circles) and GaN/AlGaN (open circles) QWs.

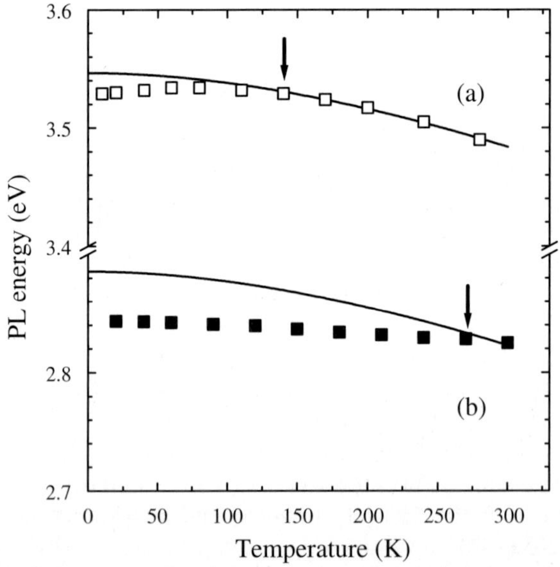

Fig. 6.25 PL energy of a GaN/AlGaN QW (open squares) and an InGaN/GaN QW (closed squares). Full lines correspond to Varshni's equation bandgap variation and arrows to the mobility edges.

fluctuations. It can then be proposed that it is not necessary to have InGaN alloy phase separation, forming In-rich QDs, to get high PL efficiency InGaN/GaN QWs.

6.6 Conclusion

In this chapter, we have demonstrated the unique capability of MBE for controlling *in situ* the growth of nitrides-based quantum heterostructures. These heterostructures present exciting features that are a giant QCSE and localization effects. The radiative efficiency of InGaN/GaN QWs is huge compared to equivalent GaN/AlGaN QWs. This explains why the InGaN/GaN QW-based LEDs are so efficient compared to UV LEDs based on GaN/AlGaN QW structures, as far as the heteroepitaxy is concerned.

Acknowledgements

The authors would like to thank F. Semond, F. Natali, S. Laügt, S. Vézian, P. Vennéguès, G. Neu, M. Leroux, J.-C. Gillon, and I. Koltchakian (CRHEA-CNRS), L. Siozade, and J. Leymarie (LASMEA-CNRS, Clermont-Fd University), B. Gil and P. Lefebvre (GES-CNRS, Montpellier University). This work is supported by BOQUANI and NANILUB contracts.

References

1. Cho, A. Y. (1969). *Surf. Sci.* **17**, 494.
2. Neave, J. H., Joyce, B. A., Robson, P. T., and Norton, N. (1983). *Appl. Phys. A* **31**, 1.
3. Massies, J., Turco, F., Saletes, A., and Contour, J. P. (1987). *J. Cryst. Growth* **80**, 307.
4. Price, G. L. (1991). *Phys. Rev. Lett.* **66**, 469.
5. Massies, J., and Grandjean, N. (1993). *Phys. Rev. Lett.* **71**, 1411.
6. Im, J. S., Kollmer, H., Off, J., Sohmer, A., Scholz, F., and Hangleiter, A. (1998). *Phys. Rev. B* **57**, R9435.
7. Leroux, M., Grandjean, N., Laügt, M., Massies, J., Gil, B., Lefebvre, P., and Bigenwald, P. (1998). *Phys. Rev. B* **58**, R13371.
8. Takeuchi, T., Wetzel, C., Yamaguchi, S., Sakai, H., Amano, H., Akasaki, I., *et al.* (1998) *Appl. Phys. Lett.* **73**, 1691.
9. Chichibu, S. F., Abare, A. C., Minsky, M. S., Keller, S., Fleischer, S. B., Bowers, J.E., *et al.* (1998). *Appl. Phys. Lett.* **73**, 2006.
10. Kim, H. S., Lin, J. Y., Jiang, H. X., Chow, W. W., Botchkarev, A., and Morkoç, H. (1999). *Appl. Phys. Lett.* **73**, 3426.
11. Lefebvre, P., Allègre, J., Gil, B., Mathieu, H., Grandjean, N., Leroux, M., *et al.* (1999). *Phys. Rev. B* **59**, 15363.
12. Simon, J., Langer, R., Barski, A., and Pelekanos, N. T. (2000). *Phys. Rev. B* **61**, 7211.
13. Berkovicz, E., Gershoni, D., Bahir, G., Lakin, E., Shilo, D., Zolotoyabko, E., *et al.* (2000). *Phys. Rev. B* **61**, 10994.
14. Widmann, F., Simon, J., Daudin, B., Feuillet, G., Rouvière, J. L., Pelekanos, N. T., and Fishman, G. (1998). *Phys. Rev. B* **58**, R15989.

15. Damilano, B., Grandjean, N., Semond, F., Massies, J., and Leroux, M. (1999). *Appl. Phys. Lett.* **75**, 962.
16. Bastard, G. (1988). Wave mechanics applied to semiconductors heterostructures, Edited by les Editions de Physique.
17. Bernardini, F., Fiorentini, V., and Vanderbilt, D. (1997). *Phys. Rev. B* **56**, R10024.
18. Gil, B. (ed.) (1998). *Group III nitride Semiconductor compounds.* Clarendon Press, Oxford.
19. Nakamura, S. (1999). *Semicond. Sci. Technol.* **14**, R27.
20. Narukawa, Y., Kawakami, Y., Funato, M., Shizuo Fujita, Shigeo Fujita, and Nakamura, S. (1997). *Appl. Phys. Lett.* **70**, 981.
21. Gérard, J. M., Cabrol, O., and Sermage, B. (1996). *Appl. Phys. Lett.* **68**, 3123.
22. Krtschil, A., Witte, H., Lisker, M., Christen, J., Birkle, U., Einfeldt, S., and Hommel, D. (1999). *Appl. Phys. Lett.* **74**, 2032.
23. Yoshida, S., Missawa, S., and Itoh, A. (1975). *Appl. Phys. Lett.* **26**, 461.
24. Yoshida, S., Misawa, S., and Gonda, S. (1983). *Appl. Phys. Lett.* **42**, 427.
25. Smith, M., Chen, G. D., Li, J. Z., Lin, J. Y., Jiang, H. X., Salvador, A., *et al.* (1995). *Appl. Phys. Lett.* **67**, 3387.
26. Yang, Z., Li, L. K., and Wang, W. I. (1995). *Appl. Phys. Lett.* **67**, 1686.
27. Kamp, M., Mayer, M., Pelzmann, A., Thies, A., Chung, H. Y., Sternschulte, H., *et al.* (1996). *Mater. Res. Soc. Symp. Proc.* **395**, 135.
28. Grandjean, N., Massies, J., Vennégués, P., Laügt, M., and Leroux, M. (1997). *Appl. Phys. Lett.* **70**, 643.
29. Jones, C. R., Lei, T., Kaspi, R., and Evans, K. R. (1996). *Mater. Res. Soc. Symp. Proc.* **395**, 141.
30. Tang, H., and Webb, J. B. (1999). *Appl. Phys. Lett.* **74**, 2373.
31. Thamm, A., Brandt, O., Takemura, Y., Trampert, A., and Ploog, K. H. (1999). *Appl. Phys. Lett.* **75**, 944.
32. Mesrine, M., Grandjean, N., and Massies, J. (1998). *Appl. Phys. Lett.* **72**, 350.
33. Grandjean, N., Leroux, M., Massies, J., Mesrine, M., and Laügt, M. (1999). *Jpn. J. Appl. Phys.* **38**, 618.
34. Grandjean, N., Massies, J., Vennégués, P., Leroux, M., Demangeot, F., Renucci, M., and Frandon, J. (1998). *J. Appl. Phys.* **83**, 1379.
35. Moustakas, T. D., and Molnar, R.J. (1993). *Mater. Res. Soc. Symp. Proc.* **281**, 753.
36. Botchkarev, A., Salvador, A., Sverdlov, B., Myoung, J., and Morkoç, H. (1995). *J. Appl. Phys.* **77**, 4455.
37. Tarsa, E. J., Heyring, B., Wu, X. H., Fini, P., DenBaars, S. P., and Speck, J. S. (1997). *J. Appl. Phys.* **82**, 5472.
38. Sugihara, D., Ikuchi, A., Kusakabe, K., Nakamura, S., Toyoura, Y., Yamada, T., and Kishino, K. (2001). *Jpn. J. Appl. Phys.* **39**, L197.
39. Widmann, F., Daudin, B., Feuillet, G., Pelekanos, N., and Rouvière, J. L. (1998). *Appl. Phys. Lett.* **73**, 2642.
40. Shen, X.-Q., and Aoyagi, Y., (1999). *Jpn. J. Appl. Phys.* **38**, L14.
41. Saito, Y., Teraguchi, N., Suzuki, A., Araki, T., and Nanishi, Y. (2001). *Jpn. J. Appl. Phys.* **40**, L91.
42. Ploog, K. H., Brandt, O., Muralidharan, R., Thamm, A., and Waltereit, P. (2000). *J. Vac. Sci. Technol. B* **18**, 2290.
43. Kirchner, C., Schnegler, V., Eberhard, F., Kamp, M., Ebeling, K. J., Konitzer, K., *et al.* (1999). *Appl. Phys. Lett.* **75**, 1098.
44. Moustakas, T. D., Molnar, R. J., Lei, T., Menon, G., and Eddy, Jr., C. R. (1992). *Mater. Res. Soc. Symp. Proc.* **242**, 427.
45. Grandjean, N., Massies, J., and Leroux, M. (1996). *Appl. Phys. Lett.* **69**, 2071.
46. Uchida, K., Watanabe, A., Yano, F., Kouguchi, M., Tanaka, T., and Minagawa, S. (1996). *J. Appl. Phys.* **79**, 3487.

47. Yeadon, M., Marshall, M. T., Hamdani, F., Pekin, S., Morkoç, H., and Gibson, J. M. (1998). *J. Appl. Phys.* **83**, 2847.
48. Widmann, F., Feuillet, G., Daudin, B., and Rouvière, J. L. (1999). *J. Appl. Phys.* **85**, 1550.
49. Tews, H., Averbeck, R., Graber, A., and Riechert, H. (1996). *Electron. Lett.* **32**, 2004.
50. Yang, B., Trampert, A., Jenichen, B., Brandt, O., and Ploog, K. H. (1998). *Appl. Phys. Lett.* **73**, 3869.
51. Tanaka, S., Scott, R.S., and Davis, R. F. (1995). *Appl. Phys. Lett.* **66**, 37.
52. Qizhen Xue, Xue, Q. K., Hasegawa, Y., Tsong, I. S., and Sakurai, T. (1999). *Appl. Phys. Lett.* **74**, 2468.
53. Zhou, L., Audurier, V., Pirouz, P., and Powell, J. A. (1997). *J. Electrochem. Soc.* **144**, L161.
54. Ferro, G., Okumura, H., and Yoshida, S. (1999). *Jpn. J. Appl. Phys.* **38**, 3634.
55. Waltereit, P., Brandt, O., Trampert, A., Ramsteiner, M., Reiche, M., Qi, M., and Ploog, K. H. (1999). *Appl. Phys. Lett.* **74**, 3660.
56. Yang, B., Trampert, A., Jenichen, B., Brandt, O., and Ploog, K. H. (1998). *Appl. Phys. Lett.* **73**, 3869.
57. Boscherini, F., Lantier, R., Rizzi, A., D'Acapito, F., and Mobilio, S. (1999). *Appl. Phys. Lett.* **74**, 3308.
58. Moustakas, T. D., and Molnar, R. J. (1993). *Physica B* **185**, 36.
59. Ohtani, A., Stevens, K. S., and Beresford, R. (1994). *Appl. Phys. Lett.* **65**, 61.
60. Yang, B., Trampert, A., Brandt, O., Jenichen, B., and Ploog, K. H. (1998). J. *Appl. Phys*. **83**, 3800.
61. Nikishin, S. A., Faleev, N. N., Antipov, V. G., Francoeur, S., Grave-de-Peralta, L., Seryogin, G.A. *et al.* (1999). *Appl. Phys. Lett.* **75**, 2073.
62. Semond, F., Lorenzini, P., Grandjean, N., and Massies, J. (2001). *Appl. Phys. Lett.* **78**, 335.
63. Semond, F., Grandjean, N., Damilano, B., Natali, F., and Massies, J. unpublished.
64. Li, L.K., Yang, Z., and Wang, W.I. (1995). *Electron. Lett.* **31**, 2127.
65. Grandjean, N., and Massies, J. (1997). *Appl. Phys. Lett.* **71**, 1816.
66. Daudin, B., and Widmann, F. (1997). *J. Cryst. Growth* **182**, 1.
67. Grandjean, N., and Massies, J. (1998). *Appl. Phys. Lett.* **72**, 1078.
68. Hacke, P., Feuillet, G., Okumura, H., and Yoshida, S. (1996). *Appl. Phys. Lett.* **69**, 2507.
69. Vézian, S. PhD thesis, Université de Nice, Sophia Antipolis.
70. Heying, B., Smorchkova, I., Poblenz, C., Elsass, C., Fini, P., and Den Baars, *et al.* (2000). *Appl. Phys. Lett.* **77**, 2885.
71. Elsass, C.R., Shlorchkova, I.P., Heying, B., Haus, E., Poblenz, C., Fini, P., *et al.* (2000). *Jpn. J. Appl. Phys.* **39**, L1023.
72. Frayssinet, E., Knap, W., Lorenzini, P., Grandjean, N., Massies, J., Skierbiszewski, C., *et al.* (2000). *Appl. Phys. Lett.* **77**, 2551.
73. Grandjean, N., Damilano, B., Massies, J., Neu, G., Teissere, M., Grzegory, I., *et al.* (2000). *J. Appl. Phys.* **88**, 183.
74. Grandjean, N., Damilano, B., Dalmasso, S., Leroux, M., Laügt, M., and Massies, J. (1999). *J. Appl. Phys.* **86**, 3714.
75. Grandjean, N., Damilano, B., Dalmasso, S., Leroux, M., Laügt, M., and Massies, J. (1999). *J. Phys. Stat. Solidi* (a) **176**, 219.
76. Im, J.S., Kollmer, H., Gfrorer, O., Off, J., Scholz, F., and Hangleiter, A. (1999). *MRS Internet Nitride Semicond. Res.* **4S1**, G6.20.
77. Leroux, M., Grandjean, N., Massies, J., Gil, B., Lefebvre, P., and Bigenwald, P. (1999). *Phys. Rev. B* **60**, 1496.
78. Pabla, A. S., Sanchez-Rojas, J. L., Woodhead, J., Grey, R., David, J. P. R., Rees, G.J., Hill, G. *et al.* (1993). *Appl. Phys. Lett.* **63**, 752.

79. Elsner, J., Haugk, M., Jungnickel, G., and Frauenheim, Th. (1998). *Solid State Commun.* **106**, 739.
80. Grandjean, N., Massies, J., Semond, F., Karpov, S.Y., and Talalaev, R.A. (1999). *Appl. Phys. Lett.*, **74**, 1854 and *Appl. Phys. Lett.* **75**, 3035.
81. Della Sala, F., Di Carlo, A., Lugli, P., Bernardini, F., Fiorentini, V., Scholz, R., and Jancu, J.-M. (1999). *Appl. Phys. Lett.* **74**, 2002.
82. Adelmann, C., Simon, J., Feuillet, G., Pelekanos, N. T., Daudin, B., and Fishman, G. (2000). *Appl. Phys. Lett.* **76**, 1570.
83. Snyder, C. W., Orr, B. G., Kessler, D., and Sander, L. M. (1991). *Phys. Rev. Lett.* **66**, 3032.
84. Damilano, B., Grandjean, N., Dalmasso, S., and Massies, J. (1999). *Appl. Phys. Lett.* **75,** 3751.
85. Chen, H., Smith, A. R., Feenstra, R. M., and Greve, D. W. (1999). *Mater. Res. Soc. Symp. Proc.* **537**, G9.5.
86. Damilano, B., Grandjean, N., Massies, J., Siozade, L., and Leymarie, J. (2000). *Appl. Phys. Lett.* **77**, 1268.
87. Lefebvre, P., Morel, A., Gallart, M., Taliercio, T., Allègre, J., Gil, B. *et al.* (2001). *Appl. Phys. Lett.* **78**, 1252.
88. McCluskey, M. D., Van de Walle, C. G., Master, C. P., Romano, L. T., and Johnson, N.M. (1998). *Appl. Phys. Lett.* **72**, 2725.
89. Yong-Hoon-Cho, Gainer, G. H., Fischer, A. J., Song, J .J., Keller, S., Mishra, U. K., and DenBaars, S. P. (1998). *Appl. Phys. Lett.* **73**, 1370.

7 Topological defects and low-dimensional nitride layers

Pierre Ruterana

7.1 Introduction

A large research effort is going on with wurtzite nitride semiconductors, due to their large direct bandgap which provides a possibility for optoelectronic devices from red to the ultraviolet range [1]. The bandgaps of $In_xGa_{1-x}N$ alloys cover a wide spectral range, from red (InN: 1.89 eV) to UV (GaN: 3.5 eV), and that of the AlGaN goes to deep UV(AlN: 6.2 eV). As expected, the most difficult part is InGaN, due mainly to the large mismatch (10%) between GaN and InN. In this case, ordering [2], phase separation [3], and growth instabilities have been reported [4]. For AlN and GaN, the misfit is smaller (≈2.5%) and one would expect more stable growth. The growth of quantum wells (QWs) appears to follow the 2D growth mode and that of high-quality quantum dots (QDs) has been achieved [5]. However, it was in this system that ordering along the *c*-axis was reported in AlGaN by Korakakis *et al.* [6], where the Ga and Al atoms separately occupy two simple hexagonal sublattices of the wurtzite system, giving rise to the lower symmetry P3m1 space group. Active areas in visible GaN-based electroluminescent diodes (ELDs) and laser diodes (LDs) consist mainly of InGaN/GaN multiple quantum wells. Results have been published concerning the phase separation which was mainly supposed to give rise to the formation of layers composed of InN and GaN patches. This was first reported in polycrystalline InGaN films that had been annealed at temperatures below 700 °C [7]. More recently, it was shown that a ternary layer grown by molecular beam epitaxy (MBE) with In fraction larger than 30% contains regions of pure In, but atomic In concentrations up to 80% could be incorporated without phase separation in GaN/InGaN/GaN heterostructures [3]. This was tentatively explained by the fact that the two compounds have a large mismatch of lattice parameters (~10%). In fact, the growth of InGaN layers may lead to the formation of the ternary alloy $In_xGa_{1-x}N$, InN and GaN, In, and finally in some cases to a ternary ordered alloy. Depending on the growth conditions one may obtain a combination of the various cases, along with the characteristic crystallographic defects.

As bulk crystals or wafers of these materials are not available, GaN has to be epitaxially grown on a large variety of substrates. The most commonly used substrates are sapphire and α-SiC which have hexagonal symmetry upon growth along the [0001] axis, and most of the work has been carried out on sapphire, which presents a large lattice-mismatch (16%). The active GaN layers contain large densities of crystallographic defects ($10^{10}\,nm^{-2}$), among which, those that may cross the whole epitaxial layer and be detrimental to the electro-optical properties are the threading dislocations [8, 9], the nanopipes [10], the inversion domains [11, 12], and the prismatic planar defects [13, 14]. As pointed out by many workers, the large majority of these defects are threading dislocations which originate from the particular growth mode of GaN on top of the (0001) sapphire or SiC substrates. This mosaic growth mode leads to islands which are rotated mostly around the *c*-axis and are bounded by **a** edge dislocations [15, 16]. These defects are present during the growth of the ternary QWs and QDs, they may take part in the strain relaxation [17] as well as in the diffusion processes which can result in phase segregation [18].

In what follows, the atomic structure of the above extended defects is reviewed, then the growth modes of the thin films are considered, with a special focus on the ternary InGaN. The various ordering types which have been reported for AlGaN and InGaN are then commented upon. A number of interesting results on the role of the edge threading dislocations on the nucleation of GaN/AlN QDs will also be discussed. The last part of this chapter deals with the complexity of the InGaN alloy. Using high-resolution electron microscopy, strain analysis, and finite elements modeling, it is shown that even in the best layers with low densities of V-shaped defects, the indium composition is not uniform. Inside 17% indium nominal composition QWs, clusters 2 nm high and 3 nm wide can reach 45% In composition.

7.2 Crystallographic considerations

The nitrides being studied (GaN, InN, AlN) crystallize in the cubic structure (sphalerite) or in the hexagonal structure (wurtzite).The latter is thermodynamically more stable and in conventional growth conditions on (0001) sapphire substrate, GaN is hexagonal (wurtzite : space group $P6_3mc$), with the lattice parameters $a=0.319\,nm$ and $c=0.518\,nm$. The anions (N^{3-}) form an hcp structure in which the cations (Ga^{3+}) occupy half of the tetrahedral sites.

The structure of a unit cell of GaN projected along [0001] is depicted schematically in Fig. 7.1. The open symbols represent γ sites which are occupied by nitrogen atoms, the Ga atoms are in the tetrahedral sites β. These latter sites can either be at height u above (β_1) or below (β_2) each N site, depending on the crystal polarity.

7.2.1 Epitaxial relationship

On top of the (0001) sapphire, the GaN layers have the well-known epitaxial relationship : $(0001)_{substrate}//(0001)_{GaN}$ and $[11\bar{2}0]_{substrate}//[0\bar{1}10]_{GaN}$.This relation leads to the continuation of the anion compact stacking (O^{2-} in the substrate,

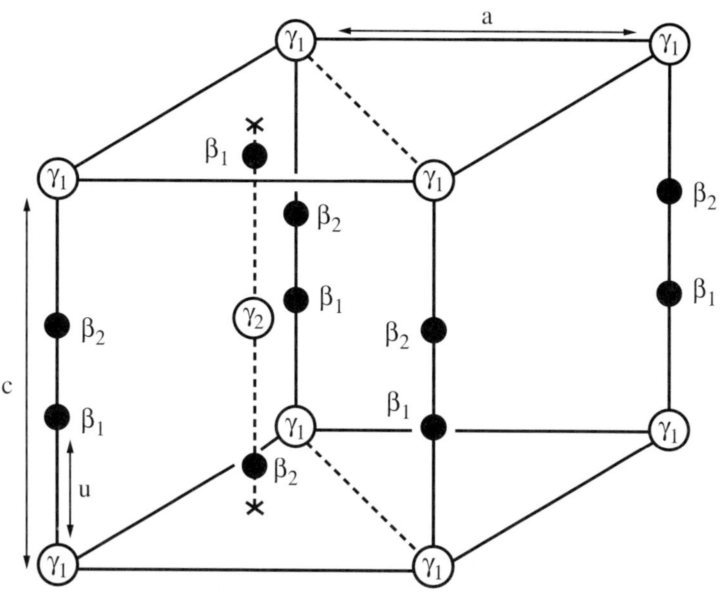

Fig. 7.1 The wurtzite unit cell with the β_1 and β_2 tetrahedral sites. If N occupies the γ sites, only one family of β sites can simultaneously be occupied by Ga atoms.

N^{3-} in the thin film).The apparent 30° rotation between the two crystals is due to the choice of the crystallographic origin (in the substrate, the origin is related to the Al^{3+} lattice).

At the interface, the cations switch from octahedral (Al^{3+} in Al_2O_3) to tetrahedral sites (Ga^{3+} in GaN). So, an important parameter could be the coordinence polyhedra of the cation that is between an oxygen plane and a nitrogen plane at the interface. There are three polyhedra to consider: each one corresponds to one of the three interstitial sites of the hcp structure (2 tetrahedral sites plus 1 octahedral site).

Furthermore, the hcp anion stacking does not have the same parameters in both materials. This leads to mismatch in the *a* and *c* directions of a hcp structure. In the basal plane, the mismatch, at room temperature, is equal to:

$$f_a = \frac{a^*_{\text{substrate}} - a_{\text{deposite}}}{a^*_{\text{substrate}}} = -16.09\,\%,$$

where $a^*_{\text{substrate}} = a_{\text{substrate}}/\sqrt{3}$, so as to respect the 30° rotation.

It was shown, after Bragg filtering of HREM images, that this lattice misfit is largely relaxed in the first monolayer of GaN, with a network of regularly spaced 60° dislocations. A residual stress of −2.1% was evaluated near the interface and related to the high density of threading defects present in the epitaxial layers [19]. In the *c* direction, the difference of lattice parameters leads to an even larger mismatch:

$$f_c = \frac{c^*_{\text{substrate}} - c_{\text{deposite}}}{c^*_{\text{substrate}}} = -19.7\%,$$

where $c^*_{\text{substrate}} = a_{\text{substrate}}/3$.

This mismatch may act at the substrate surface steps as the growth takes place at adjacent terraces. In the growth on 6H-SiC, there is no rotation of the lattice; the misfits are shown in Table 7.1.

A comprehensive description of the tetrahedrally coordinated structures is very useful for the investigation of the interface defects. The bonds describe a tetrahedron denoted T which possesses one atom species at each corner and the other atom species in its center [20]. The basal plane of this structure is defined by one face of the tetrahedron and the bond perpendicular to this latter defines the *c*-axis. A rotation of 180° around *c* produces a twinned variant denoted T′ [Fig. 7.2(a)]. The two variants are also related to each other by a mirror symmetry about a $\{1\bar{1}00\}$ plane. A tetrahedron can occupy one of the three possible positions in the basal plane and layers of tetrahedra are then denoted $T_1, T_2, T_3, T'_1, T'_2, T'_3$ [Fig. 7.2(b)]. The structure of these materials and their different polytypes can be completely described by stacking these six tetrahedra layers. Not all stacking sequences are possible, and two rules must

Table 7.1 Crystallograpic data on wurtzite GaN, AIN, InN and the substrates (sapphire and 6H-SiC)

Properties		AlN	GaN	InN	6H-SiC	Sapphire
Lattice parameter nm	a	0.3112	0.3189	0.35446	0.3081	0.4758
	c	0.4982	0.5185	0.57034	1.5092	1.2991
	c/a	1.6003	1.6252	1.609	1.633 (×3)	2.73
Thermal expansion	a	4.2	5.6	5.7	4.2	7.5
coefficient (10^{-6} C^{-1})	c	5.3	3.2	3.7	4.7	8.5
$d_{10\bar{1}0}$		0.2695	0.276	0.307	0.267	0.137
$d_{11\bar{2}0}$		0.1556	0.1593	0.1772	0.1541	0.2379
d_{0001}		0.249	0.259	0.285	0.2516	0.2165
$\Delta a/a_{SiC}$ (%)		1.04	3.54	–15.05		
$\Delta a/a_{sap}$(%)		–13.29	–16.09	–29		
$\Delta a/a_{GaN}$ (%)		2.34		–11.26		
$\Delta a/a_{AIN}$ (%)			–2.4	–13.9		
$\Delta c/c_{SiC}$ (%)		1.02	–2.9	–13.9		
$\Delta c/c_{Sap}$ (%)		–15	–19.6	12.19		
Polytype			2H		6H	
Spatial group				$P6_3mc$		$R\bar{3}c$

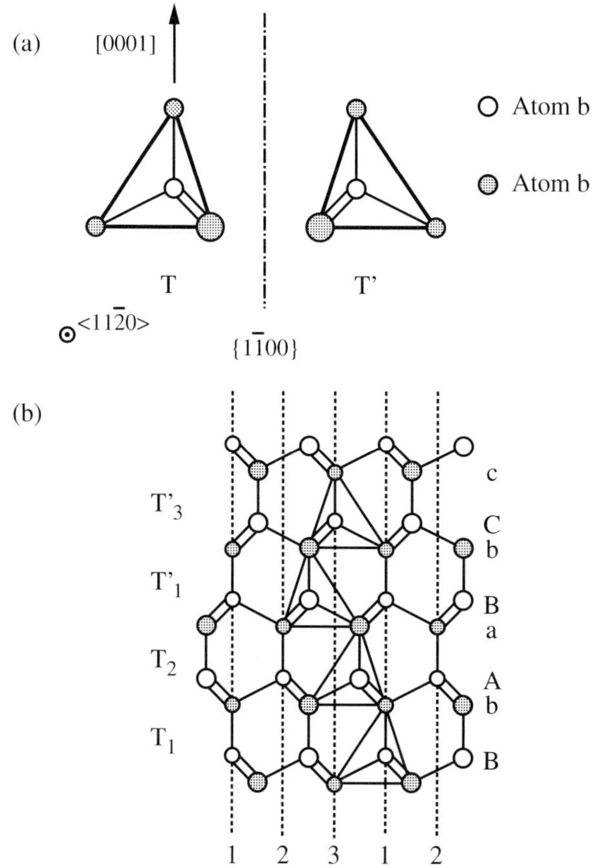

Fig. 7.2 Tetrahedron representation of the sphalerite and wurtzite materials. (a) The two twinned tetrahedra, (b) The rules for stacking the tetrahedra in the three fcc positions.

be respected to keep a corner sharing structure:

1. A tetrahedron T can be followed by another one of the same kind with the following subscript: T_1 T_2 T_3 and inversely for the twinned variant: T'_3 T'_2 T'_1.
2. A tetrahedron T_1 must be followed by the twinned variant of the preceding subscript: $T_1T'_3$ and inversely for its twinned variant: T'_1T_2.

For example, the structures of the different polytypes, under the present study, are described by the following sequences:

- T_1 T'_3 or T_2 T'_1 or T_3 T'_2 for the wurtzite structure which corresponds to the polytype denoted 2H in the Ramsdel notation;
- T_1 T_2 T'_1 T'_3 for the 4H polytype;
- T_1 T_2 T_3 T'_2 T'_1 T'_3 for the 6H polytype;
- T_1 T_2 T_3 or $T'_3T'_2$ T'_1 3C polytype.

7.2.2 Stacking faults in the hexagonal structure

The 2H polytype is the simplest one from which all the other polytypes can be described by introduction of periodic stacking faults (SFs). The tetrahedral notation is used in this paragraph to describe the structure of the three types of SFs (Fig. 7.3) existing in the wurtzite structure.

1. Intrinsic SF denoted I_1 : the following stacking sequence is $T_1\ T'_3\ \underline{T_1}\ \underline{T_2}$ T'1 T2 and the other possibility, I'_1 : $T'_3\ T_1\ \underline{T'_3}\ \underline{T'_2}\ T_3\ T'_2$, is related to I_1 by a rotation of 180° around the *c*-axis (the faulted sequence is underlined).
2. Intrinsic SF denoted I2: obtained by the dissociation of a 1/3 $\langle 11\bar{2}0\rangle$ dislocation into two Shockley partials. Its sequence is $T_1\ T'_3\ \underline{T_1}\ \underline{T_2}\ \underline{T_3}\ T'_2\ T_3$ and for the twinned variant I'_2: $T'_3\ T_1\ \underline{T'_3}\ \underline{T'_1}\ \underline{T'_2}\ T_3\ T'_2$.
3. Extrinsic SF denoted E: This defect corresponds to the insertion of a complete sequence of the sphalerite structure, e.g : $T_1T_2T_3$ or $T'_1\ T'_2\ T'_3$ in $T_1T'_3$ ^ $T_1T'_3$ leading to $T_1\ T'_3\ \underline{T_1}\ \underline{T_2}\ \underline{T_3}\ \underline{T_1}\ T'_3$, or $T_1\ \underline{T'_3}\ \underline{T'_1}\ \underline{T'_2}\ \underline{T'_3}\ T_1T'_3$

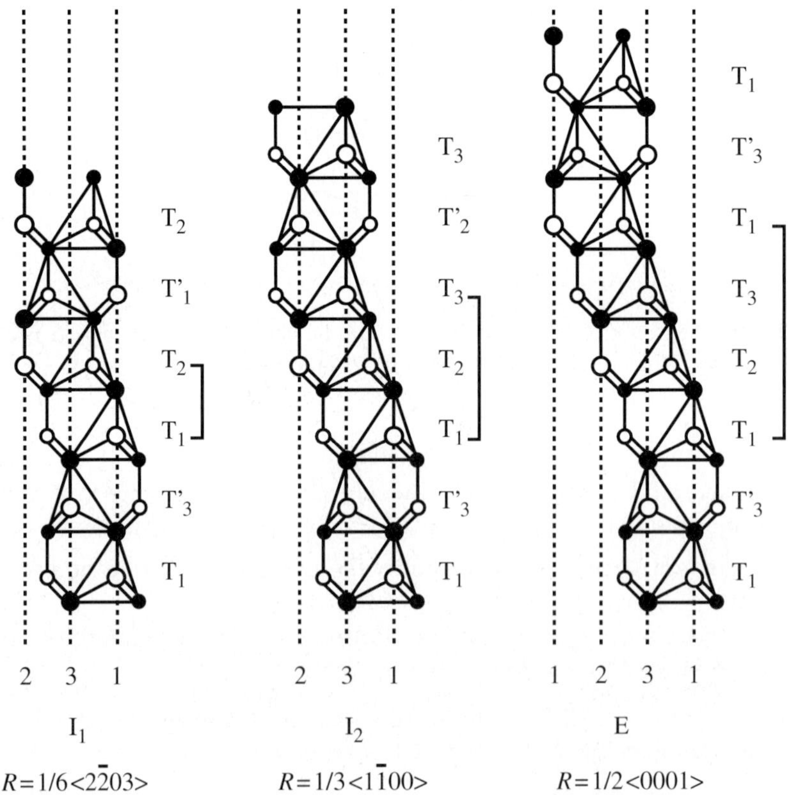

Fig. 7.3 The three basal stacking faults projected along $[11\bar{2}0]$ using the tetrahedron representation in case of GaN. (o) represent nitrogen, (•) gallium, (–) in-plane and (=) out of plane bonds.

which is the twinned variant. This defect is bounded by a Frank partial dislocation ($b = 1/2$ [0001]).

These three different sequences constitute the possible stacking faults in the wurtzite structure.

7.3 Extended defects in III-nitrides

7.3.1 Threading dislocations

Conventional TEM has shown that these dislocations originate at the interface and propagate to the sample surface. Their line is roughly parallel to the *c* growth axis and the large majority are edge type. Most of the results have been obtained in layers containing a huge density of threading dislocations (10^9–10^{10} nm^{-2}) following growth on sapphire (Fig. 7.4). The origin of this density has been suggested to be due to a particular growth mode which was named 'mosaic growth' [21]. Due to the large misfits, the adjacent islands are misoriented mainly by rotation about the *c*-axis and their boundaries are made of a pure edge dislocations. In general, they can be also tilted and *c* or $a+c$ dislocations may be formed [22]. The above growth mode may explain misorientation of 0.5 to 2° between adjacent grains [21]. It has been shown that *c* and $a+c$ dislocations tend to bend and annihilate; the density of those that reach the layer surface is only a few percent of that near the interface. Calculations have shown that the low optical activity of the edge threading dislocations may be explained by the formation of dimer bonds inside an eight

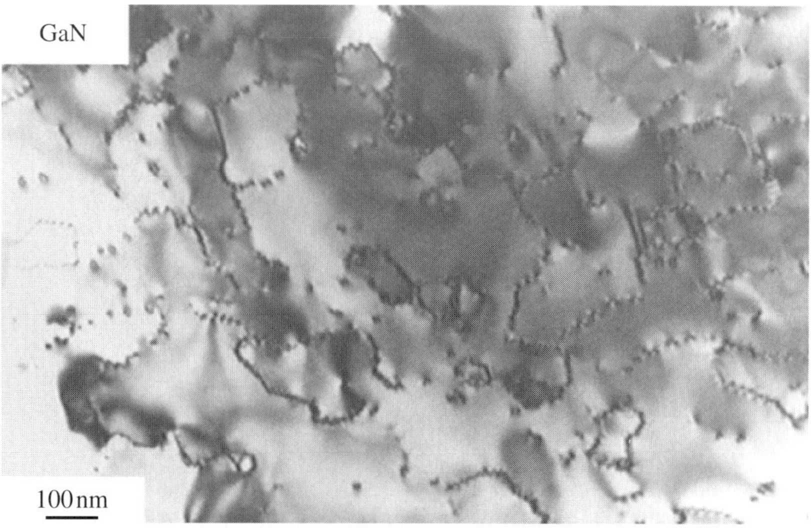

Fig. 7.4 A plan-view micrograph of a GaN layer viewed close to the [0001] zone axis, low-angle grain boundaries (closed domains in the image) are bounded by *a* edge threading dislocations.

atom ring core [23]. The most accurate observations published so far have shown that this is the case in MOCVD grown samples.

However, our HREM observations suggest that this may not be the case, at least in MBE grown GaN layers. In addition to the above atomic configuration, we have noticed typical contrast (Fig. 7.5) which is not explained by an eight atoms core. And from our observations, this configuration has more or less an equal frequency with the 8-atom ring core. Image simulations show that this core is compatible with a 5/7 atom ring configuration.

7.3.2 Nanopipes

Such defects are empty or filled holes which exhibit a dislocation character, they mostly have a regular shape. They can extend up to a few tens of nanometers and are usually limited by $\{10\bar{1}0\}$ planes. They have been called nanopipes in analogy to SiC in which such features can measure more than 10 μm and are called macropipes. From our experience they can be classified into two classes, those which have an edge component and those which are pure screw dislocations. As shown in Fig. 7.6, as large as 2*a* edge component may be exhibited by such defects. In our specimens, it was noticed that these nanopipes were confined inside the first 200 nm of the epitaxial layer [24]; they contain amorphous material in layers grown on 6H-SiC, whereas they are empty on top of sapphire [25]. Other authors have shown that the pure screw nanopipes cross the entire epitaxial layer [26]. However, a number of observations has been reported in which such defects either keep the same section or close and open up a few times on their way to the surface.

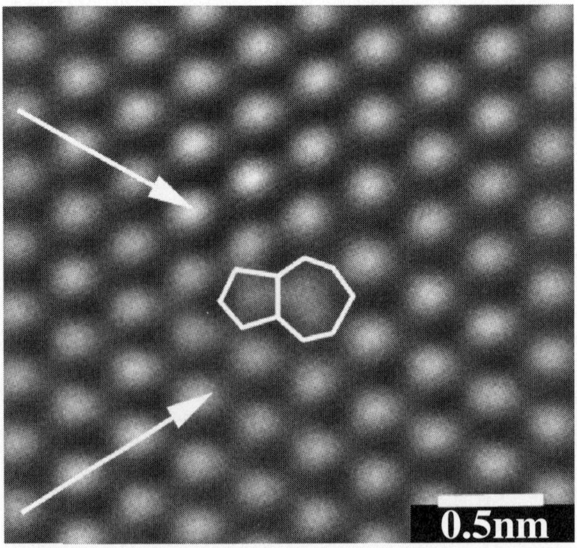

Fig. 7.5 A HREM image of a pure edge *a* threading dislocation, a 5/7 atom ring model is superimposed. The additional $\{10\bar{1}0\}$ lattice planes are shown (arrows).

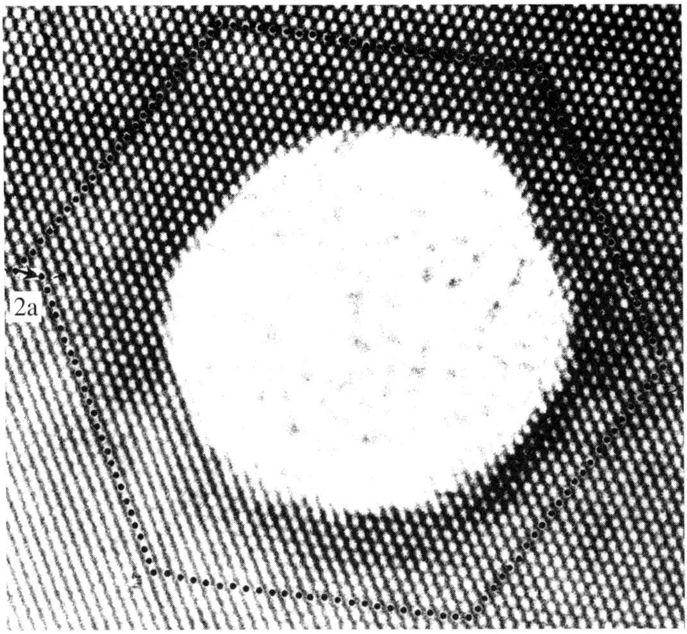

Fig. 7.6 A nanopipe which exhibits a $2a$ in-plane Burgers vector, it is filled with a amorphous material.

7.3.3 V-shape defects

Inverted pyramids can form at the surface of InGaN layers, they appear as soon as In is introduced and their density increases along with the indium content. These defects are empty and limited by $\{10\bar{1}1\}$ facets and in most reports they were shown to be connected to threading dialocations. Such dislocations are often seen to have a c component (Fig. 7.7). When they form inside QWs layers, the QW structure may grow on the pyramid walls of the open hexagonal inverted pyramids, thus leading to buried quantum lines on the $\{10\bar{1}1\}$ facets [27]. The formation of these defects may be kinetically controled as in normal InGaN/GaN MQW growth, the GaN barrier growth temperature is some 250 °C lower than used in high temperature GaN deposition in order to suppress re-evaporation of In. When such structures are capped later on by high temperature GaN growth which takes place at high surface diffusion rates, the V-defects are quickly planarized [27].

7.3.4 Stacking faults

7.3.4.1 *Basal stacking faults*

In bulk GaN grown under high nitrogen pressure, the three types of stacking faults were observed (I_1, I_2, and E), in GaN layers grown over sapphire, such faults are often bounded by Frank–Shockley partials. Recent work indicates

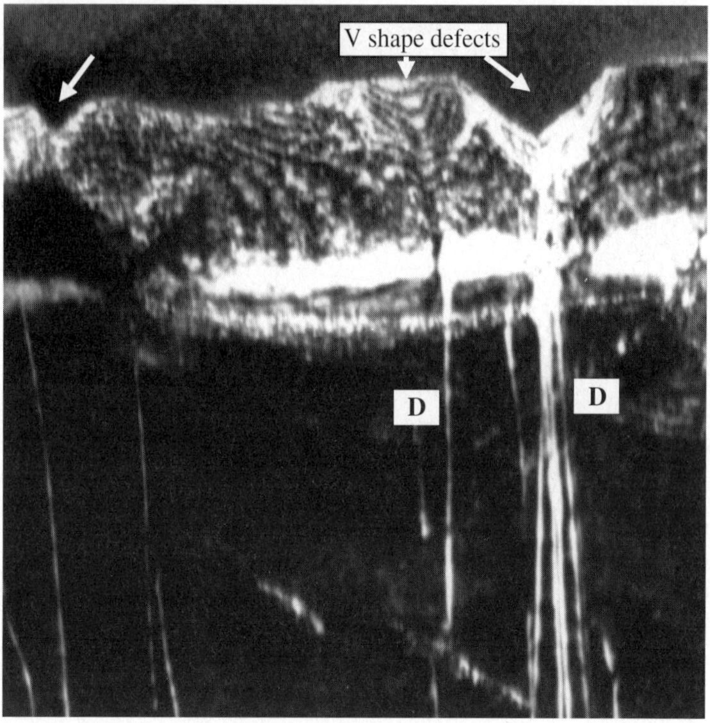

Fig. 7.7 V-shape defects at the top of a thick (200 nm) 10% InGaN layer, the connection to the threading dislocations is shown (D).

that stacking faults may not induce defect states in the bandgap, but that they can give rise to quantum-well-like regions of zinc-blende materials embedded in the wurtzite lattice that can bind electrons [28]. This was confirmed by photoluminescence and transmission electron microscopy experiments, indicating an excitonic transition at 3.4 eV due to the presence of stacking faults [29].

In the following, we present an analysis of basal stacking faults observed in the vicinity of the interface with the substrate. The AlN/GaN layers were grown on top of (0001) sapphire. At low magnification, we observed that a high density of defects is present near the interface, which are mostly dislocations and stacking faults. In the immediate vicinity with the sapphire substrate, the defects are long extended stacking faults whereas further from it (20 nm), we mainly observed short width faults with an extension less than a few ten $\{10\bar{1}0\}$ lattice spacings. The long extended stacking faults are mainly I_1 and I_2 faults; however, a few segments of E fault can also observed (Fig. 7.8). Inside the GaN layer at 15 nm from the interface substrate, there are narrow basal stacking faults (Fig. 7.9). Two additional half $\{10\bar{1}0\}$ planes are underlined, which are separated by about ten lattice planes. By referring to the perfect wurtzite structure, the wrong stacking sequences can be underlined and we determined that these faults are I_2. They may form by dissociation of a perfect

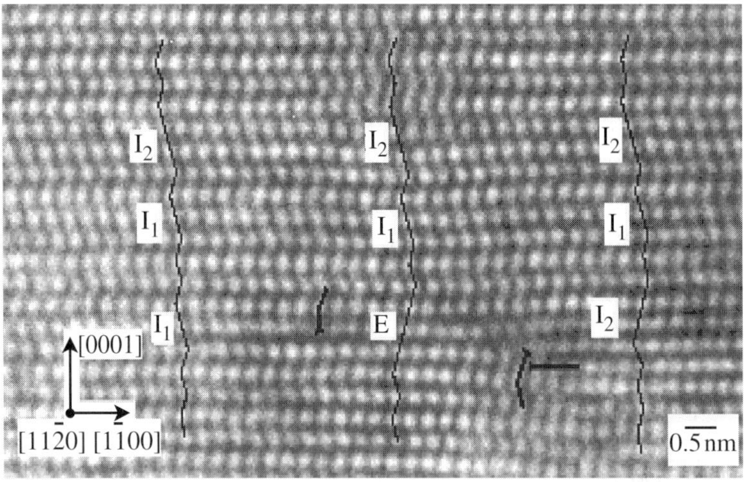

Fig. 7.8 Large basal stacking faults inside the GaN buffer layer close to the interface with sapphire, inside the buffer layer.

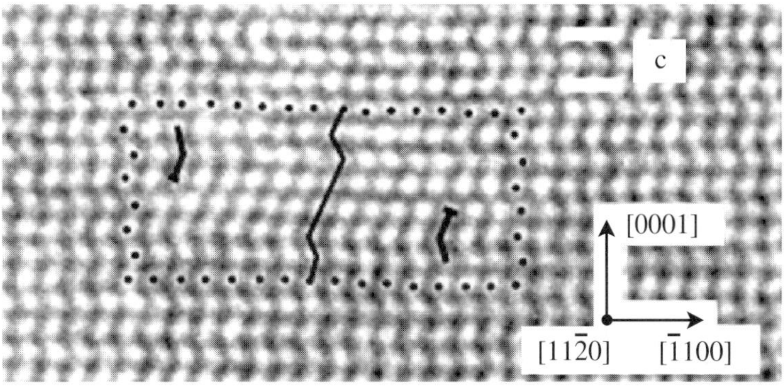

Fig. 7.9 HREM image of a small extension I2 loop often observed at the top of low-temperature buffer layers.

dislocation ($a = 1/3 <11\bar{2}0>$) into two Shockley partials. However, the Burgers circuit drawn around these faults did not exhibit any closure failure. The two partial dislocations bounding these basal faults have opposite Burgers vectors ($b = 1/3 <10\bar{1}0>$). These faults I_2 can be formed directly by shear and we note that their presence is related to the end of the buffer layer.

Besides the I_2 faults, I_1 faults are also present (Fig. 7.10) and the Burgers circuit drawn around shows up a closure failure equal to 1/3 [11$\bar{2}$0]. At the left of this fault, additional (0002) and {10$\bar{1}$0} half-planes may be shown, whereas on its right, it is one-half (0002) opposite plane. This I_1 fault corresponds to a climb-dissociated perfect dislocation into two Frank–Shockley partials,

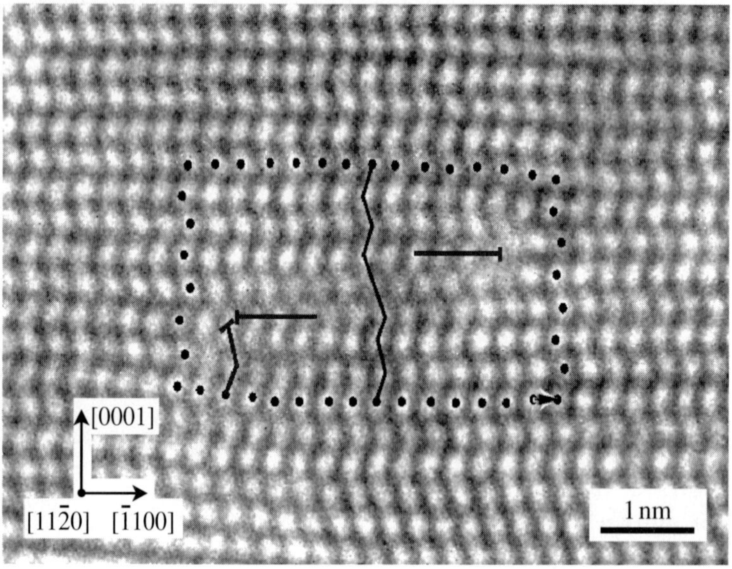

Fig. 7.10 Small I1 loop at the top of the low temperature GaN buffer layer.

following this reaction:

$$1/3 \langle 11\bar{2}0\rangle \rightarrow 1/6 \langle 20\bar{2}3\rangle + 1/6 \langle 02\bar{2}3\rangle \text{ or } a \rightarrow (p_1 + c_{/2}) + (p_2 - c_{/2}).$$

Previously, I_1 stacking faults were reported to exist in GaN layers grown on sapphire substrate by MOCVD [30, 31] whereas I_2 and E ones were not reported. The I_2 fault can form directly by shear or after a dissociation of a pure edge dislocation into two Shockley partials. The small faults are observed in the first 15 nm of the GaN layer, that corresponds to the thickness of the buffer layer. It was shown that the buffer layer is predominantly cubic and transformed in hexagonal GaN during heating [32]. Thus, these faults may be due to incomplete transformation from sphalerite to wurtzite structures. Besides, the formation of the extended faults may be due to the mosaïc growth mode of GaN/sapphire layers.

7.3.3.2 *Prismatic stacking faults*

In epitaxial layers, independent of the substrates, observations carried out in plan-view samples show defects which can either terminate by partial dislocations or form closed domains. Along the [0001] direction, it is difficult to see them edge on, due to the fact that they easily fold from the prismatic to the basal plane. For the closed domains, it is possible that these faults may limit inversion domains. Convergent beam analysis showed that they always have the same polarity as the matrix [33]. Two atomic models exist in the literature for the {11$\bar{2}$0} prismatic fault in wurtzite materials as originally characterized by conventional electron microscopy in the 60s by Blank *et al.* [34] [Fig. 7.11(a)] and Drum [35] [Fig. 7.11(b)], respectively. Using HREM characterization, in all the observations carried out along the [0001] direction, we found the Drum model atomic structure as shown in Fig. 7.12 along with the simulated image (inset).

In cross-section observations it was possible to characterize a prismatic fault linking two basal I_1 SFs which was found to exhibit a different contrast (Fig. 7.13). Inside the area defined by the projected image of the defect, the lattice fringes of the basal planes exhibit alternately a continuous fringe and high intensity spots (between points b–c on Fig. 7.13). This contrast matches the Blank model atomic configuration.

The displacement vector of these planar faults was confirmed in conventional as well as HREM experiments [36]. In the conventional weak beam technique, we made the g, ng experiment in order to extinguish the α fringes and we systematically found the 1/2 <$10\bar{1}1$> prismatic stacking fault corresponding to the Drum model [36].

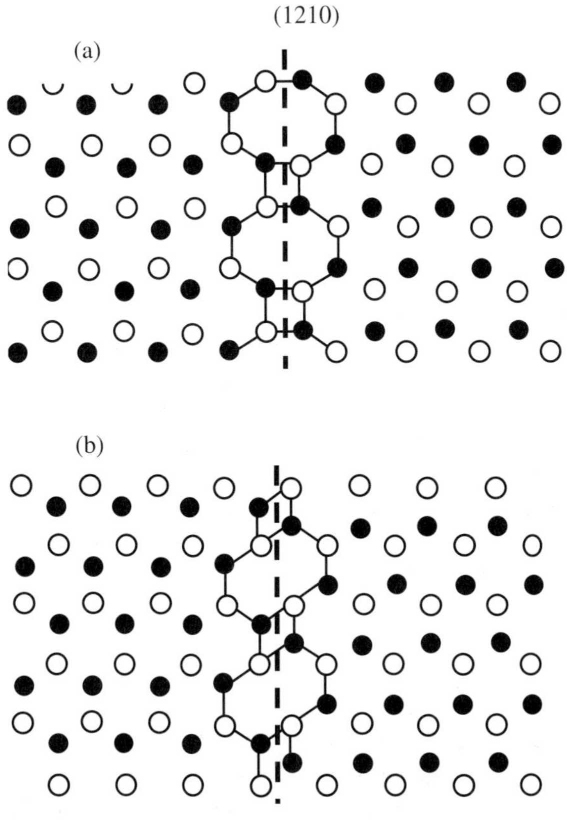

Fig. 7.11 The two models of the prismatic stacking fault projected along the [0001] axis; • Atoms are located at $c/8$ out of the projection plane: (a) the Drum model: 1/2 [$10\bar{1}1$], (b) the Blank model: 1/6 [$20\bar{2}3$].

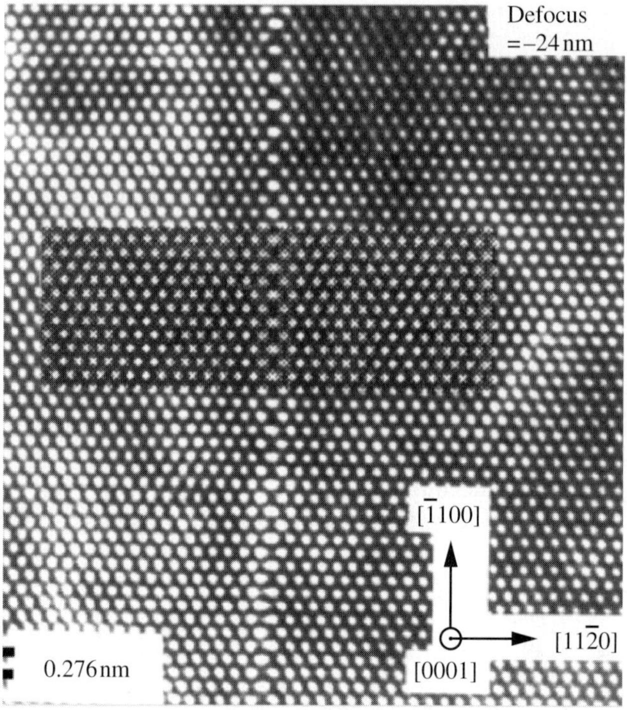

Fig. 7.12 Experimental image of the prismatic stacking fault recorded along the [0001] zone axis. A simulated image of the Drum model has been superimposed.

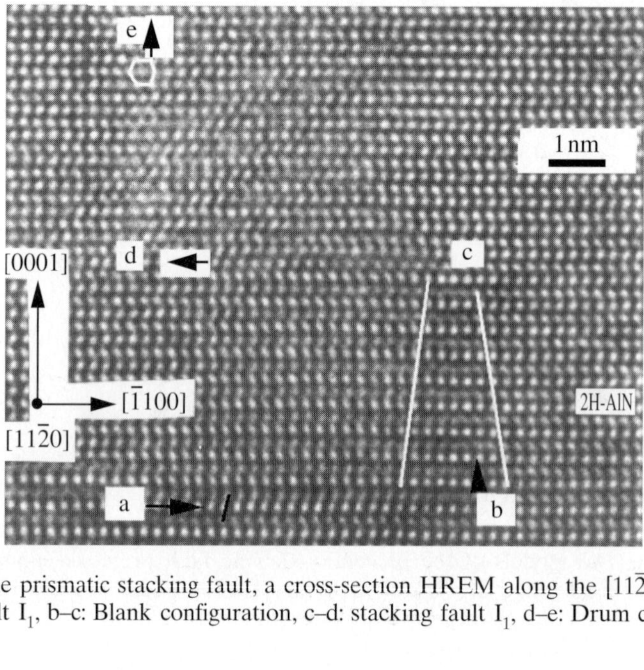

Fig. 7.13 The prismatic stacking fault, a cross-section HREM along the [11$\bar{2}$0], a–b: basal stacking fault I_1, b–c: Blank configuration, c–d: stacking fault I_1, d–e: Drum configuration.

7.3.4 Inversion domain boundaries

In wurtzite AlN and GaN which are non-centrosymmetric, defects due to polarity can form, such defects were reported for the first time by Aminoff and Broome [37]. The inversion domains which vertically cross the epitaxial layers are limited by $\{10\bar{1}0\}$ boundary planes. Such IDBs have been largely investigated theoretically as well as by several TEM methods. In our HREM investigation, we have shown that experimental micrographs recorded at specific defoci and thicknesses can be used in order to determine the atomic structure of the boundaries. The two possible models for the $\{10\bar{1}0\}$ IDBs are (i) the Holt model which contains wrong bond in the boundary [Fig. 7.14(a)], and (ii) the reconstructed V model [Fig. 14(b)] which can be deduced by a *c*/2 translation. We have used these models for image simulations and compared to experimental observations. For the V model, the stacking sequences are the same on both sides of the IDB, in contrast with the Holt model at –54 nm defocus. At –29 nm defocus, the tunnels are projected into white spots, they can be underlined for thicknesses ranging from 1 to 5.5 nm and from 10 to 15 nm. For the Holt model, the sequences are in the same direction whereas for the V model, they are opposite. Therefore, it is possible for to them distinguish [38].

For these IDBs, the boundary plane can be obtained by cutting two or one bond per atom leading to different atomic structures. The H models are characterized by the presence of wrong bonds at the boundary plane whereas the IDBs of the V model are composed of 4 and 8 atom rings. As seen above, the H and V models can be distinguished by considering the stacking sequences. In the experimental micrographs, we have found that the boundary

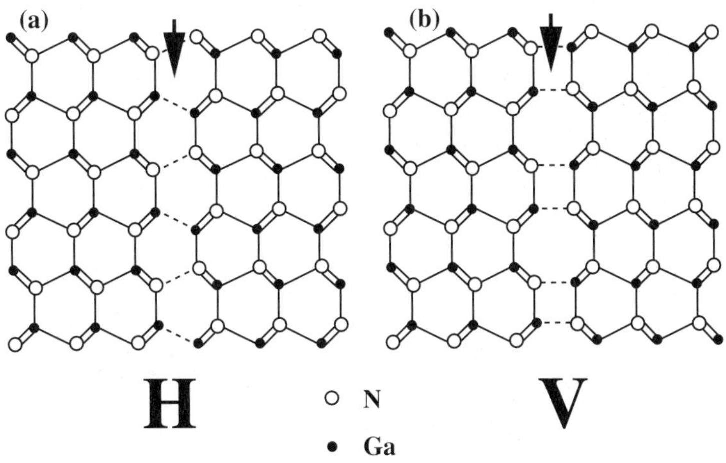

Fig. 7.14 The two models of the atomic structure of the $(10\bar{1}0)$ inversion domain boundary in wurtzite GaN, (a) The Holt model which has wrong bonds, (b) The reconstructed V or IDB* model.

plane moves from one configuration to the other of the same model in different samples depending on the growth conditions. In Fig. 7.15(a), the image of an IDB of one type is presented at –54 nm defocus. From A to B, the boundary plane is located between two rows of equivalent brighter spots indicating the Holt 1 (one wrong bond per atom) configuration, and as seen from B to C, only one row of brighter spots is present characterizing the Holt 2 configuration (two bonds per atom). In Fig. 7.15(b), an IDB from a different set of samples is shown for –29 nm defocus, from A to B, the two brighter spots show the V2 configuration and from B to C, we have a unique row of brighter spots characterizing the V1 configuration. For the two samples, the change from one configuration to another one is indicated by the inclined arrows.

7.3.5 The extended defects and active layers

In the devices, the active layers are deposited after growth of at least one-micron-thick GaN in which the density of extended defects is slightly reduced by more than an order of mangitude. The defects present inside the top layers are those that propagate vertically. They are the threading dislocations, the nanopipes, the inversion domains, and the prismatic faults. Inside the best MOCVD layers which are used for ELDs, the growth is controlably tuned to the Ga polarity, and such layers contain negligibly small densities of inversion domains. The prismatic stacking faults fold very easily to and from the basal plane where they were often found to terminate by an I_1 partial dislocation. It can easily be seen that the only basal stacking faults that will be present are those originating from the dissociation of threading dislocations.

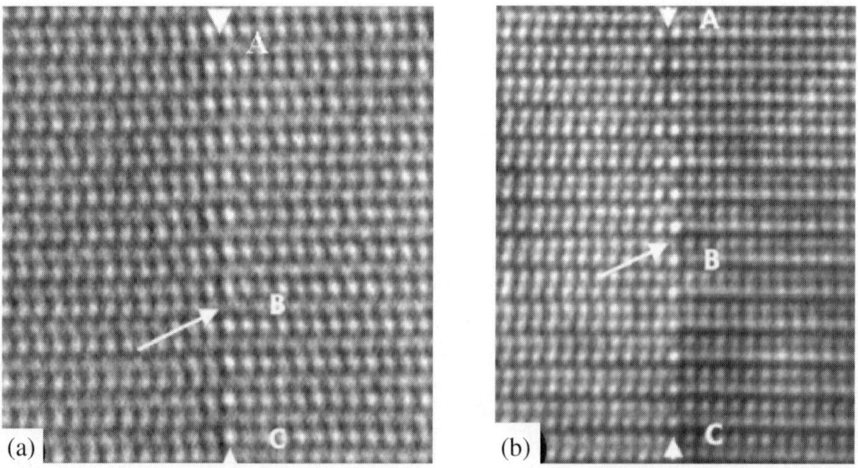

Fig. 7.15 Experimental images of the two cuts of the boundary plane: (a) the Holt model at 54 nm defocus , A–B: H1 configuration, B–C: H2 configuration; (b) V model at 29 nm defocus, A–B, V2 closely spaced planes and two bonds per atom inside the boundary, B–C, V1 largely spaced atomic planes and one atom bond inside the boundary plane.

As was shown fairly recently, such effects as silicon doping can induce this dissociation [39] and probably help in decreasing the density of the threading dislocations. Thus, the influence of planar defects on the performance of the devices (ELDs, LDs) is probably rather small, or at least can be controled by improving the growth conditions. The nanopipes are limited by mainly 6 $\{10\bar{1}0\}$ vertical facets that can cross the whole epitaxial layer; its presence inside the active layers will constitute areas of complex surface processes very detrimental to the device performance. However, the density of these defects is very low in the best layers ($< 10^5\,nm^{-2}$) in comparison to the size of the active areas ($1\,\mu m^2$) and diffusion lengths of the carriers (a few tens of nanometers). Therefore, the defects which play a critical role on the properties of the devices are the threading dislocations. Among these dislocations, it is now admitted that it is very difficult to decrease the pure edge dislocation density (a type) and they end by forming more than 90% of the defects that reach the surface. With densities in the range of 10^9–$10^{11}\,cm^{-2}$, the mean distance between two threading dislocations is 0.1 μm. These distances are comparable to the carriers' diffusion lengths and to the size of active layers.

In the following, we present the growth of the ternary alloys, and as the behavior is different for AlGaN and InGaN, the two cases will be dealt with separately.

7.4 Growth mechanisms in epitaxial layers

The growth of thin films has been categorized into three modes, namely the Frank–Van der Merwe (FVM) mode or layer-by-layer growth, the Stranski–Krastanov (SK) mode: the first layer remains smooth and then clusters form on top, and the Volmer–Weber (VW) mode, in which only clusters can form (Fig. 7.16). These growth modes are deduced from the energy balance between the surface, interface, and strain energies. Conventionally, the FVM growth mode is predominant in systems where the misfit is less than 1%, then the SK

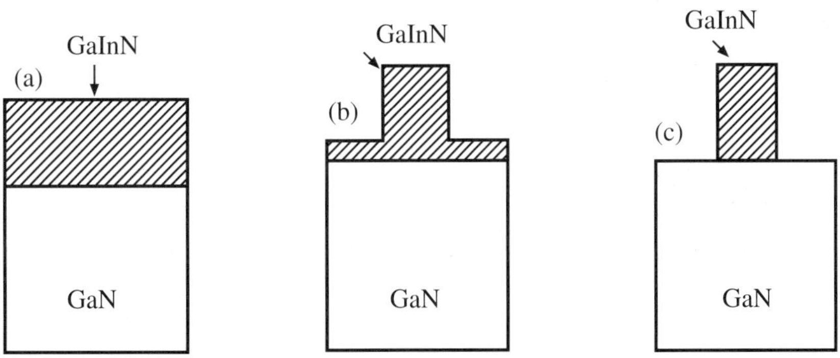

Fig. 7.16 Growth modes for thin epitaxial layers, (a) 2D Frank–Van der Merwe, (b) 2D–3D Stranski–Krastanov transition, (c) 3D Volmer–Weber growth.

mode takes over to about 4%, and then the three-dimensional VW growth dominates. Upon growth on the (0001) surface of GaN, the misfit is smaller for AlN and the growth mode should be SK. For InGaN, it is interesting to follow the development which was made recently by Nakajima *et al.* [40]. They determine the thickness–composition phase diagrams of the growth mode for III–V ternary systems by considering the energy balance between the strain, surface, and interface energies for the three growth modes. In their work, the FVM, SK, and VW modes can appear in the phase diagrams determined only by the calculation of the three kinds of energy. The effect of a buffer layer was considered because an AlN buffer layer can be effective in obtaining the layer-by-layer growth [41]. From the calculated phase diagrams, one explain the way to proceed in order to obtain the layer-by-layer growth or to grow self-assembled quantum dots.

7.4.1 Energy balance

In the calculation, the schematic geometries of a GaInN strained film on GaN substrate for the FVM mode, and a GaInN cluster on a GaInN wetting layer on a GaN substrate for the SK mode, are shown in Fig. 7.16. The volume of a cluster is taken to be equal to that of the wetting layer. The total volume of the strained film is equal to that of the cluster and wetting layer shown in Fig. 16(b). In the calculation, the strain energy is calculated on the basis of the rectangular shaped cluster, because an accurate stress can thus be estimated. A semispherical cluster was adopted to estimate the surface and interfacial energies because the semispherical cluster seems to be more real for estimating the surface and interfacial energies.

For the calculation of the strain energy, the method developed by Nakajima *et al.* [42] is used in order to calculate the precise stress distribution of an island on top of a substrate. In this method, the island is divided into many imaginary thin layers, and the face force and strain balance is considered over all these thin layers with coherent interfaces. Shear lag analysis [43] is then used to calculate the longitudinal stress distribution over the imaginary thin layers in the FVM, SK, and VW mode structures. The total strain energy of each structure G_{st} which includes the strain energy of the layer, cluster, buffer layer, and substrate is given by

$$G^{\mathrm{st}} = \sum_{i=1}^{m} U_i, = \sum_{i=1}^{m} \frac{A_i d_i \sigma_i^2}{2E_i}, \tag{7.1}$$

where U_i is the elastic strain energy in the ith imaginary thin layer, m is total number of imaginary thin layers which constitute each structure, and σ_i, E_i, A_i, and d_i are the stress, Young modulus, surface area, and thickness of the ith imaginary thin layer.

On the surface between the vapor and solid phases, the surface energy per unit area γ_s can be given approximately by:

$$\gamma_s = (1 - w/u)\Delta H_{v0} N_0^{2/3}, \tag{7.2}$$

where u is the number of nearest neighbors of an atom in the bulk of the crystal and w is the number of neighbors in the crystal of an atom on the surface in question. So, w/u means the number of bonds which connect a surface atom to atoms in the crystal, and $(1-w/u)$ means the number of dangling bonds of an atom on the surface. ΔH_{v0} is the enthalpy of evaporation of the material, and N_0 is the number of atoms per unit volume. The argument used is that the surface energy is the energy to break all of the nearest-neighbor bonds across a given plane. The number of atoms per unit surface area N_s can be related to N_0 as follows:

$$N_s = N_0^{2/3}.$$

For a III–V wurtzite-type compound, N_s is given by:

$$N_s = \frac{2}{\sqrt{3}a^2},$$

for the (0001) face, where a is the lattice constant of the III–V binary compound. For growth from vapor, ΔH_{v0} is given by the enthalpy of evaporation ΔH per mole as follows:

$$\Delta H_{v0} = \frac{\Delta H}{2N_A},$$

where N_A is Avogadro's number ($N_A = 6.023_* 10^{23}$). ΔH is estimated using Stringfellow's model [44] written by:

$$\Delta H = Ka^{-2.5}.$$

a is the lattice constant of the binary compound and the value of K is 1.15×10^7cal/mole Å$^{2.5}$. Therefore, the surface energy G^{su} can be written as follows:

$$G^{su} = (1 - \alpha)A_s\gamma_s = \frac{(1 - \alpha)A_s\Delta H}{4\sqrt{3}a^2 N_A} \tag{7.3}$$

for the (0001) face where w is 3 and u is 4, A_s is the surface area of each-structure, and α the reconstruction ratio of dangling bonds on the surface. A part of dangling bonds on the surface makes bonds with each other on the surface. The number of dangling bonds decreases as α increases.

The interface energy G^{if} between the layer and the substrate (FVM and SK) or between the cluster and the substrate (VW) was calculated by considering abrupt transition of bonding species at the heterointerface. The interface energy related to the abrupt transition of bonding species calculated as follows [45]:

$$G^{if} = A_i\gamma_i = A_i(1 - w/u)\Omega^s_{GaN-InN}x^2_{InN}\frac{N_i}{N_A} \tag{7.4}$$

where γ_i is the interface energy per unit area, $\Omega^s{}_{GaN-InN}$ is the interaction parameter of the GaInN compound, and N_i is the number of atoms per unit area at the interface. For the (0001) face, N_i can be given by:

$$N_i = \frac{2}{\sqrt{3}a_i^2},$$

where a_i is the strained-lattice constant of the layer or cluster at the interface.

The growth mode dependent part of the free energy G^f can be given as:

$$G^f = G^{st} + G^{su} + G^{st} + G^{if}$$

The free energies G^f for the FVM, SK, and VW growth mode structures are then calculated and the phase diagram of the growth is determined by comparing them.

7.4.2 Growth mode vs composition for the GaInN/GaN system

Equations (7.1–7.4) can be used to the FVM, SK, and VW modes energies. Table 7.2 gives the parameters used for the calculation of the strain energy. The shear modulus G is estimated using the Young modulus E and the Poisson ratio ν. The parameters of the ternary alloys are estimated from those of the binary compounds using the Vegard law, the value of the interaction parameter $\Omega^s{}_{GaN-InN}$ is 5980 cal/mole [46]. In this attempt, the surface energy was calculated on the basis of the assumption that the reconstruction did not occur on the surface. The strain energy of the FVM mode is the largest and that of the VW mode is the smallest. The strain energy increases as the InN fraction x increases or the lattice misfit increases. The surface energy of the FVM mode is the smallest and that of the VW mode is the largest. The surface energy slightly decreases as the InN fraction x increases because ΔH and $1/a^2$ decrease as x increases. The interface energy of the FVM mode is equal to that of the SK mode, and that of the VW mode is the smallest. The interface energy increases as the InN fraction increases because G^{if} increases as x increases. The strain energy becomes larger than the surface energy as the InN fraction increases. The interface energy is the smallest of the three kinds of energy. So, the relationship between the strain energy and the surface energy is very important in determining the stable mode. The strain, surface, and interface energies were calculated for the three kinds of structure with various numbers

Table 7.2 Materials properties for strain energy calculation [40].

	GaN	InN	AIN
Young modulus (GPa)	$2.38\ 10^4$	$1.25\ 10^4$	$2.97\ 10^4$
Poisson ratio	0.306	0.365	0.297
Shear modulus (GPa)	$9.11\ 10^3$	$4.58\ 10^3$	$1.145\ 10^4$

of lattice layers, and the free energy for each structure was derived by summing up these energies. The free energy of the FVM mode is the smallest between $x=0$ and 0.15, that of the SK mode between $x=0.15$ and 0.26, and that of the VW mode between $x=0.26$ and 1.0 when the number of lattice layers is two. That is, each growth mode is stable within the composition range with the smallest free energy. By comparing the free energy of each growth mode at various thicknesses, we can determine the thickness–composition phase diagram of the GaInN/GaN structure. Figure 7.17 shows the phase diagram of the GaInN/GaN structure. The phase diagram is expressed as a function of the number of lattice layers, L and the composition of InN, x. When the composition of InN or the lattice misfit is small, the FVM mode appears in the phase diagram. However, the VW mode appears in most regions of the phase diagram. The SK mode appears only in the small region between the FVM and VW modes. Therefore, it is difficult to obtain the layer-by-layer growth of GaInN with large InN fraction on top of GaN.

7.4.3 Effect of AlN buffer layer on the GaInN/GaN phase diagram

The surface energy of $Ga_{0.5}In_{0.5}N$ can be estimated to be $\gamma_s=4.92\times10^{-4}\,J/m^2$, and that of AlN can be estimated to be $\gamma_s=6.95\times10^{-4}\,J/m^2$ by using eqn (2). In order to know the effect of AlN buffer layer with a larger surface energy, the thickness–composition phase diagram was determined for the GaInN/AlN/GaN structure which has an AlN buffer layer between the GaInN

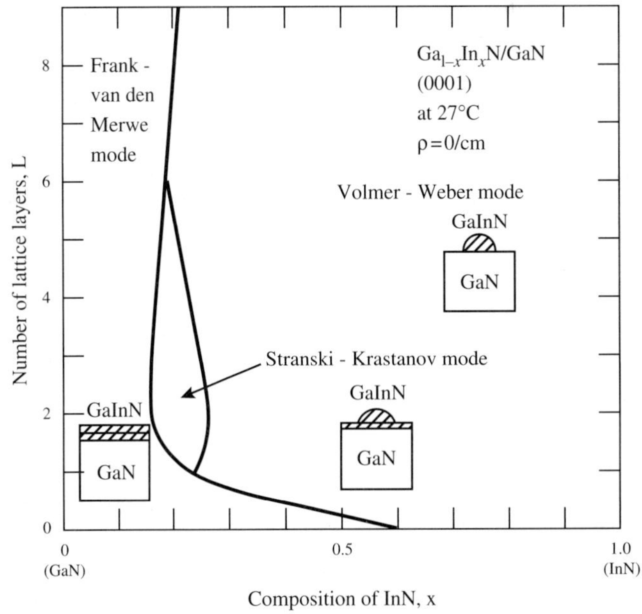

Fig. 7.17 Phase diagram of the growth modes of InGaN on top of a GaN surface versus indium composition [40].

layer and the GaN substrate. The structures consist of two lattice layers. When the InN fraction is smaller, the strain energy of the VW mode is the largest, and that of the SK mode is the smallest. The GaN substrate for the VW mode structure is more largely strained near the heterointerface by the AlN layer than that for the FVM and SK mode structure because the GaInN layer above the AlN layer reduces the strain in the GaN substrate. When the InN fraction is larger, the strain energy of the FVM mode is the largest and that of the VW mode is the smallest. The strain energy increases as the InN fraction increases. The surface energy of the FVM mode is the smallest and that of the VW mode is the largest. The surface energy slightly decreases as the InN fraction increases. The interface energy of the FVM mode is equal to that of the SK mode, and that of the VW mode is the smallest. The interface energy increases as the InN fraction increases. The strain energy becomes much larger than the surface energy as the InN fraction increases. In comparison, the strain energy of the SK and VW modes also increases by introducing the AlN buffer layer, especially when the InN fraction is small. The surface energy of the VW mode also increases using the AlN buffer layer. As can be seen in the phase diagram of the GaInN/AlN/GaN structures (Fig. 7.18), the SK mode appears most in the phase diagram. The FVM mode appears when the InN fraction is small and/or the layer is thin, and the VW mode when the InN fraction is increased. In contrast to the growth on top of GaN, the regions of the FVM and SK modes become larger, and that of the VW decreases by introducing the AlN buffer layer.

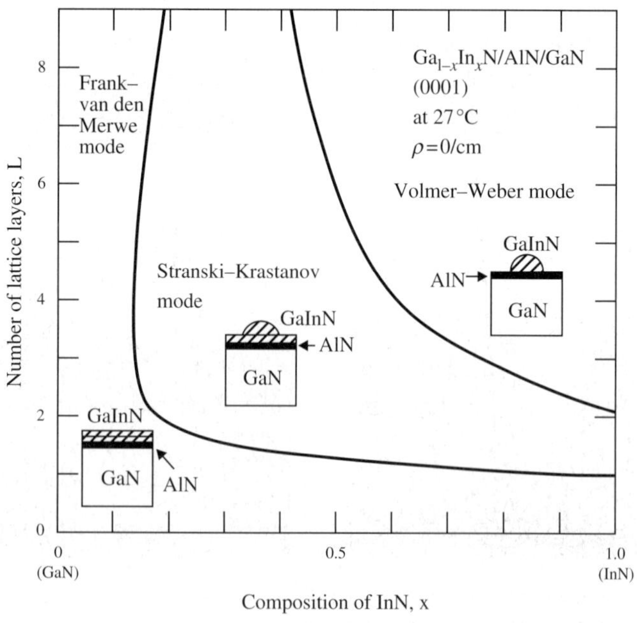

Fig. 7.18 Phase diagram of the strain release during InGaN growth on top of GaN due to the deposition of one monolayer of AlGaN, the FVM mode is extended and the SK mode dominates the phase diagram [40].

7.5 Chemical ordering

Observations carried out on InGaN grown on top of GaN show that the layers have a uniform contrast, no obvious phase separation can be seen [Fig. 7.19(a)]. Along the $[10\bar{1}0]$ zone axis, diffration patterns were recorded inside the ternary layer [Fig. 7.19(b)] and GaN buffer [Fig. 7.19(c)]. As can be seen in Fig 7.19(b), superlattice spots are present in the diffraction pattern of the ternary alloy pointing to ordering on the Ga–In hcp sublattice. Their intensity was found to increase with the In concentration up to 50%, and above this concentration, the growth was found to be disrupted in the analyzed specimens [4].

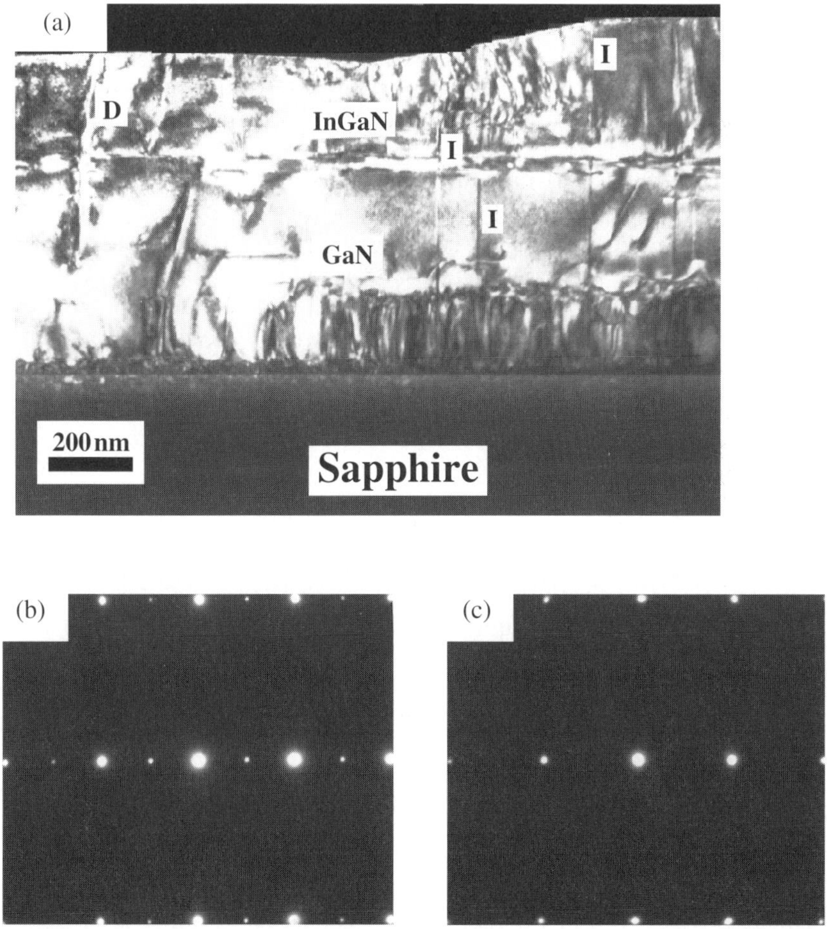

Fig. 7.19 The 1:1 ordering in InGaN layers grown by MOCVD; (a) 0002 dark field image showing a high quality-ordered ternary layer with a nominal composition of In = 0.2 (b) a diffraction pattern along the $[10\bar{1}0]$ zone axis recorded inside the ternary layer, (c) comparable pattern inside the GaN underlying layer with the absence of the 0001 spots.

The wurtzite structure is made of two hcp lattices, and if the origin is taken at 6_3 (3m) as in the International Tables of Crystallography, the Ga and N atoms occupy two, respectively, equivalent positions 1/3, 2/3, 0 and 1/3, 2/3, 0.377, this gives rise to the four atoms of the unit cell on two hcp sublattices related by the 0.377 translation along [0001]. In the In_xGa_{1-x} N phase, if ordering takes place, along the (0001) stacking direction in conformity with the observed diffraction patterns, the two positions of the metallic hcp sublattice are no longer equivalent; they are occupied by two different atoms: In and Ga, respectively. In the hcp lattice, these positions are related by the *c* mirrors parallel to the $\{10\bar{1}0\}$ lattice planes and the screw axes (6_3 and 2_1) along [0001]. Therefore, distinguishing the two atomic sites suppresses the six corresponding symmetry operators(6^+_3 ,6^-_3, 3*c, 2_1), and we are left with a lower symmetry group which no longer belongs to the hexagonal space groups. The remaining operations (3*m, 3_1, 3_2, and 1) give rise to a symmetry group of order 6. As the (0001) diffraction pattern exhibits a pseudohexagonal symmetry, the first choice is to search for a trigonal space group and the most adequate one is the P3m1 group which has the above operations. Moreover, one notices that it contains similar positions as the wurtzite $P6_3mc$, but, in this instance, they are no longer doubly occupied and thus can be used to generate separately the A and B sites of the hcp system. So it is possible to form the N hcp sublattice and two simple hexagonal sublattices for In and Ga, respectively.

Subsequently, additional types of ordering have been shown to take place in InGaN along the [0001] direction [47]. In this case, ordering, planar faults, and phase separation were observed to coexist. Samples with $x = 0.21$ exhibited extra spots of the 0001 reflections due to ordering. Both 1:1 ($In_{0.50}Ga_{0.50}N$) and 1:3 ($In_{0.25}Ga_{0.75}N$) ordering took place in the $x = 0.21$ and 0.49 $In_xGa_{1-x}N$ films [47].

In the 1:1 ordered areas, the unit cell has one Ga atom, one In atom, and two N atoms per unit cell ($a = 3.36$ nm and $c = 5.54$ nm). The Ga, In, and N atoms will occupy the 1b and 1c positions of P3m1, respectively (here, $z = 0$ and 1/2) with the Ga in 1b position at (1/3, 2/3, 0) and In at (2/3, 1/3, 1/2). In this unit cell, the N atoms will occupy the 1b and 1c locations with $z = 0.377$ and 0.877. The unit cell in this structure will have a *c*-axis value close to that of the disordered $In_xGa_{1-x}N$ ($x = 0.5$). This primitive cell will result in the appearance of the forbidden reflections in the electron diffraction patterns such as 0001, 0003, etc.

Inside the 1:3 ordered areas (Ga–Ga–Ga–In–Ga–Ga–Ga–In–Ga–...) the sixfold symmetry of $P6_3mc$ is also broken. And we end with the same P3m1 as the highest symmetry space group. In this type of ordering, there will be three Ga atoms and one In atom per unit cell. These will have the following atomic positions:

1 Ga at *1b* ~ 1/3, 2/3, 0
2 Ga at *1c* ~ 2/3, 1/3,1/4; 2/3, 1/3, 3/4
1 In at *1b* ~ 1/3, 2/3, 1/2

With this stacking, the unit cell will be double that of the disordered $In_{0.25}Ga_{0.75}N$ along the *c*-axis.

It was in AlGaN layers that the (1:1) ordering was pointed out for the first time in MBE layers [6]. Subsequently observations were carried out in MOCVD layers which exhibited homogeneous contrast, as can be seen in Fig. 7.20. Inside some of the samples, we have noticed additional spots particularly along the $[1\bar{1}00]$ and $[11\bar{2}0]$ zone axis. As can be seen in Fig. 7.21, electron diffraction patterns recorded from one of the samples show three types of additional spots. In Fig 7.21(a), the additional spots are located in the 000l ($l=2n+1$) positions, this corresponds to the usual GaN/AlN ordering which was reported by Korakakis *et al.* [6]. Figure 7.21(b) shows a pattern recorded in another area of the same sample where three additional spots can be seen between 0000 and 0002 along the 0001 systematic raw. This corresponds to the 3:1 ordering as in the case of InGaN layers [48].

In Fig 7.21(c), a diffraction pattern recorded along the $[11\bar{2}0]$ zone axis from a different area of the same layer exhibits just two high intensity spots around the GaN ones. These spots indicate a periodicity close to 3.1 nm along the GaN *c*-axis which correspond to nearly 6 lattice parameters. These superlattice spots may originate from the introduction of Al rich lattice planes in a 3.1 nm GaN supercell. The number of Al lattice planes has to be small due to the aluminium concentration (=14% in this sample). Therefore we have built 3.1 nm supercells containing 1, 2, 3 Al monolayers. That is, 11 : 1, 10 : 2, and 9 : 3 ordering schemes. We then carried out diffraction simulations along the $[1\bar{1}00]$ and $[11\bar{2}0]$ zone axes. In fact, the addition of an even number of Al lattice planes brings about the visibility of $2n+1$ spots along the 0001 systematic. However, experimental patterns along the $[1\bar{1}00]$ zone axis do not contain such reflections. Moreover, the simulated patterns along the $[11\bar{2}0]$ zone axis exhibit non-symmetric intensity distributions out of the systematic 0001 raw which is not seen in a straight-forward observation of the experimental

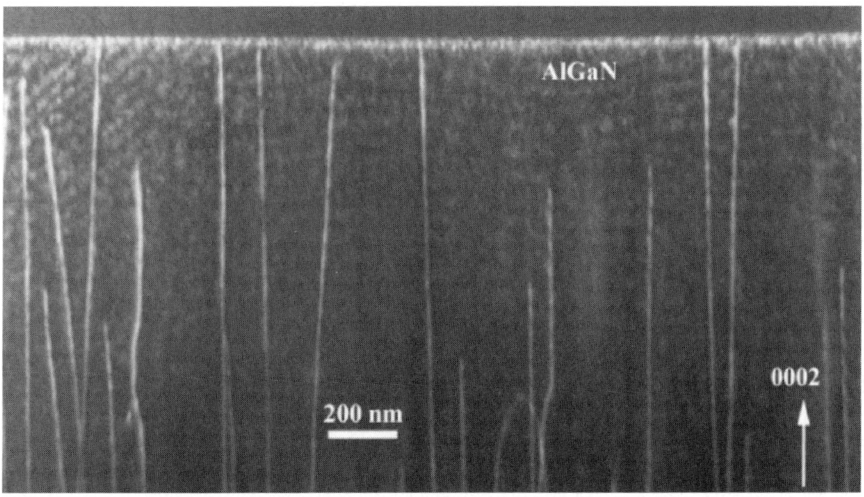

Fig. 7.20 Dark field micrograph (0002) of a high quality AlGaN layer with 3:1 and 10:2 ordering.

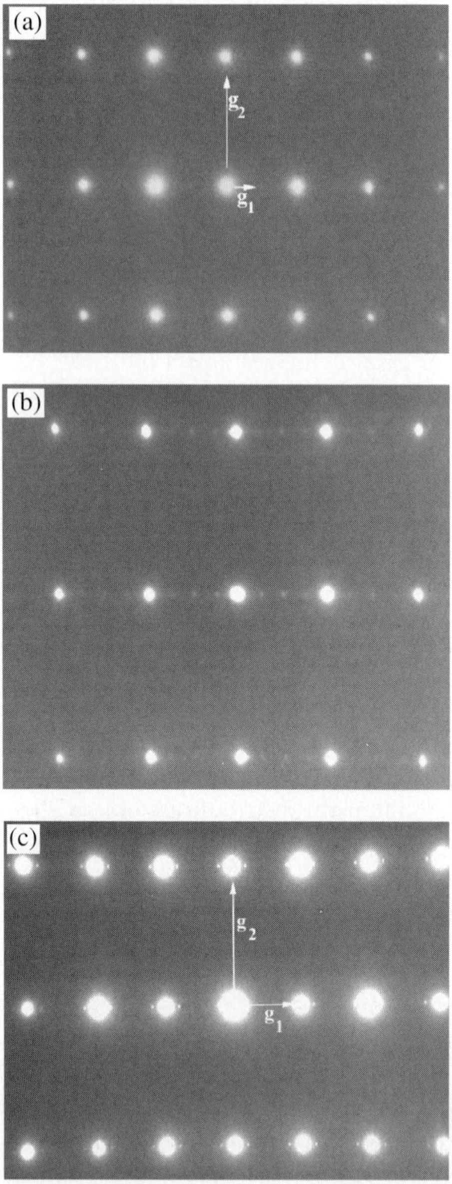

Fig. 7.21 Three diffraction patterns recorded in different areas of the sample of Fig. 7.20; (a) 1 : 1 ordering; (b) 3 : 1 ordering; and (c) 10 : 2 ordering.

patterns. X-ray diffraction analysis showed the same types of superlattice spots [49]. Their position was consistent with the *c*-cell parameter being multiplied by 6. Satellites corresponding to the same criteria were also observed around all the main AlGaN reflections. It is then clear that the additional type of ordering which has taken place in our samples is 10 : 2 which is composed of 5 GaN and 1 AlN unit cells stacks along the *c*-axis. In the investigated samples, such ordered domains were randomly distributed inside the layers and surrounded by a non-ordered matrix. Their extension could be of a few tens of nanometers. The series of AlGaN layers grown by MOCVD exhibited good optoelectronic properties. This seems to agree with their general morphology, size of mosaïc grains ($\approx 100\,\text{nm}$) and dislocation densities in the high $10^9\,\text{cm}^{-2}$. However, these observations mean that the layers do not exhibit a uniform microstructure. One finds the 1 : 1 and the 1 : 3 ordering and the new type in which a series of 5 GaN cells and one AlN form the new supercell (i.e. 10 GaN monolayers and 2 AlN monolayers).

7.6 GaN/AlN quantum dots

The growth of AlN/GaN quantum wells proceeds by the Van der Merwe growth mode when the growth conditions are optimized. As has been shown by many workers [50], there is no interdiffusion and the transitions at interfaces are abrupt, the quantum dots are grown by inducing the Stranski–Krastanov mode at lower temperatures due to reduced surface diffusion. This is better done by MBE between 600–800 °C.

7.6.1 Influence of growth conditions

The effect of growth temperature was investigated by depositing GaN at 685, 705, and 725 °C. Although the density of dots is very high in each sample, it decreases unambiguously with increasing temperature. Furthermore, at 685 °C, numerous dots have coalesced whereas they are isolated when grown at 725 °C. The dot nucleation is partly driven by surface diffusion of Ga and/or N atoms although additional processes cannot be discarded. RHEED studies indicate that in the temperature range of interest for GaN dot growth, the Ga surface mobility is high [51] and this optimized GaN growth by MBE corresponds to a ratio of metal to active nitrogen slightly larger than unity [52]. Departure from stoichiometry produces either surface accumulation of the metal and droplet formation for metal-rich conditions or surface roughening for nitrogen-rich conditions [53]. These results are consistent with a high gallium mobility on a metal-rich surface whereas in the case of a N-rich surface a low Ga mean-free path could be assigned to high Ga–N reactivity. Then, the dependence of dot density on temperature should be related mainly to Ga surface diffusion.

The dot morphology was studied as a function of time under vacuum and under nitrogen plasma flux [52]. For this purpose, GaN dots were grown at 700 °C by depositing the equivalent of 3 MLs of GaN on an AlN surface. The dots were either left under vacuum during three minutes at 700 °C before cooling under vacuum or they were exposed to nitrogen plasma after growth and during cooling. Under vacuum, a ripening effect was observed leading to less dense, less coalesced, and bigger dots than when the surface is exposed to the nitrogen flux. The size changed as a function of time, depending on whether dots were exposed to nitrogen plasma or not, was also studied as a function of growth temperature by recording the intensity of the Bragg spot assigned to dots in the RHEED pattern. In this experiment, growth was interrupted after deposition of about 2.5 MLs of GaN. At high temperature (~700 °C), the variation of the Bragg spot intensity after shuttering off the Ga flux is comparable under vacuum and under nitrogen plasma. In contrast, at low temperature (620–640 °C), two different cases are observed: (i) under vacuum, the Bragg spot intensity increases after Ga flux shuttering and reaches a plateau, as an indication of dot ripening, while (ii) under nitrogen plasma the Bragg spot intensity remains nearly constant after Ga flux shuttering, as an indication of unchanged dot morphology as a function of time. So the mean-free path of Ga is very small when the GaN surface is N-saturated, this is suggested by the influence of the nitrogen flux and the temperature range for the stability of the (2×2) surface reconstruction assigned to nitrogen [53]. In the present growth conditions, the (2×2) reconstruction was observed under active nitrogen flux for temperatures lower than 700 °C. Above 700 °C, the reconstruction is faint and eventually disappears as an indication that the surface is not N-saturated. In this high temperature range, dot ripening was observed both under vacuum and under active nitrogen. In the low temperature range, unchanged morphology of dots under nitrogen is unambiguously correlated with observation of the (2×2) reconstruction. From these experiments, we conclude that Ga surface diffusion is inhibited in the presence of nitrogen. This drastic reduction in Ga mean-free path may result either from occupation of Ga surface diffusion sites by nitrogen atoms and/or from trapping of Ga atoms due to high gallium-nitrogen reactivity [54].

7.6.2 Vertical correlation effects

Taking advantage of the two-dimensional growth mode exhibited by AlN deposited on GaN [54], periodic, stacked layers of GaN dots separated by an AlN spacer were fabricated. Depending on the thickness of the AlN spacer separating the island layers, vertical correlation between the GaN dots was observed. While no correlation was found for an AlN spacer thickness of 20 nm, such a correlation is present for an AlN spacer thickness of 8 nm (Fig. 7.22). The GaN two-dimensional wetting layer, about 2 MLs thick, is clearly seen. An interesting point is the vertical dislocation line going through the vertical line of dots on the left-hand side (D, arrow). This dislocation originates from the AlN buffer layer and acted as a nucleation center for dots. HREM studies were performed to characterize the strains in GaN dots embedded in AlN matrix, in the general case dots are perfectly lattice matched to the AlN matrix. Although dislocations may act as nucleation centers, their density is about ten times lower than that of dots, and most dots are dislocation-free. It is worth noting that the lateral spacing between adjacent dots is about 3.5 times larger than the thickness of the AlN spacer, in agreement with

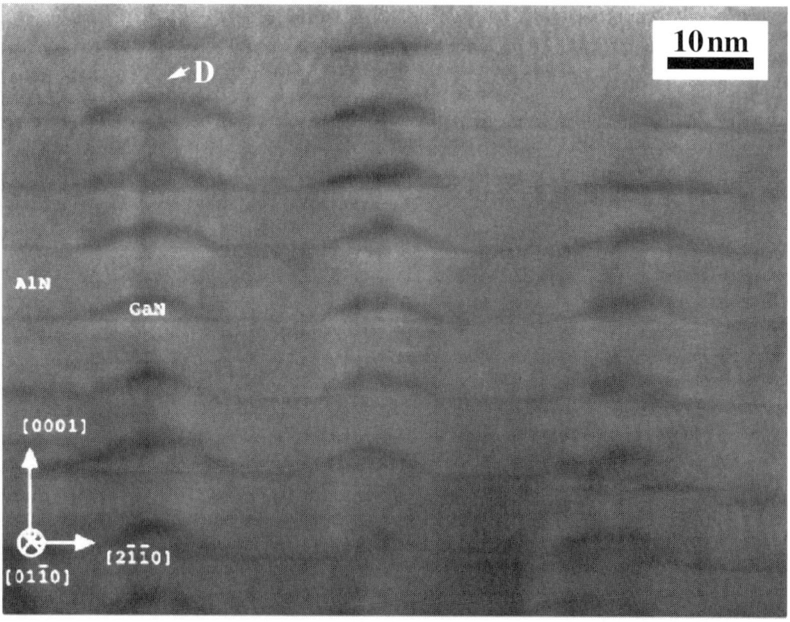

Fig. 7.22 Vertical correlation of GaN/AlN Quantum dots, a threading dislocation crosses one set (D and arrow) [51].

calculations by Tersoff *et al.* [54]. This is assigned to the merging of the strain fields resulting from close spaced islands, leading to the nucleation of a single island above them when depositing the next layer. As can be seen on Fig. 7.22, the GaN wetting layer/AlN interface is relatively abrupt, with weak interdiffusion. Although a quantitative HREM analysis of the chemical composition of dots cannot be performed, due to their three-dimensional character resulting in superimposed Ga and Al atomic rows, it is reasonable to extrapolate the result obtained for the wetting layer and to assume that no interdiffusion occurs at the dot/AlN capping layer interface. In an AFM study a 20 stack of GaN dots separated by 5-nm-thick AlN layers with the top one left uncapped, it was shown that the dots had base diameter of about 20 nm standard deviation of 3 nm and a height of 4 nm standard. This result is similar to what is observed in the case of arsenide materials, where the size homogeneity is known to result from the vertical correlation [55].

7.6.3 Interaction with threading dislocations

In AlN layers with $1.8{*}10^{11}/\text{cm}^2$ threading dislocations, the growth of GaN quantum dots was shown to be in correlation with the *a* edge threading dislocations [56]. This was done by analyzing the layers in planar view and revealed clearly the presence of at least one **a** dislocation at the edge of every QD (Fig. 7.23). This was possible because the density of edge dislocations was of the same order of magnitude as the density of GaN dots (about $1.1\times 10^{11}\,\mathrm{nm}^{-2}$). The quantum dots were systematically located at one end of the two additional $\{10\bar{1}0\}$ lattice planes characterizing the edge threading dislocation (see Fig. 7.5). Moreover, (i) one of the half-planes was aligned along one of the $\{10\bar{1}0\}$ edges of the QD, and (ii) the neighborhood of the dislocation that contains these additional half-planes is in compression. Indeed, this region, which is marked C in Fig. 7.23, has a lattice parameter which is locally smaller than that of bulk AlN. Then the nucleating GaN, whose lattice parameter is larger than that of AlN by about 2.5%, tends to grow at the opposite side of the dislocation where the AlN lattice is stretched. In other words, the strain field of the edge dislocations favors the nucleation of QDs at the dislocation edge. The preferential nucleation of the dot next to an *a* dislocation implies that the location of the islands is not random. As the threading *a* dislocations have a dislocation line along the growth axis, this nucleation mechanism implies that there is a high correlation between the different dot layers. Systematic HRTEM observations and RHEED patterns were used to deduce the shape and characteristic sizes of the dot. They form a hexagonal truncated pyramid with $\alpha=30°$, $h=3.3\,\mathrm{nm}$ and $D=15.3\,\mathrm{nm}$ (Fig. 7.24). The dot itself is dislocation-free, implying that GaN dots are totally strained with respect to the AlN matrix. Analysis of the (0002) *d* spacings across the dots was made

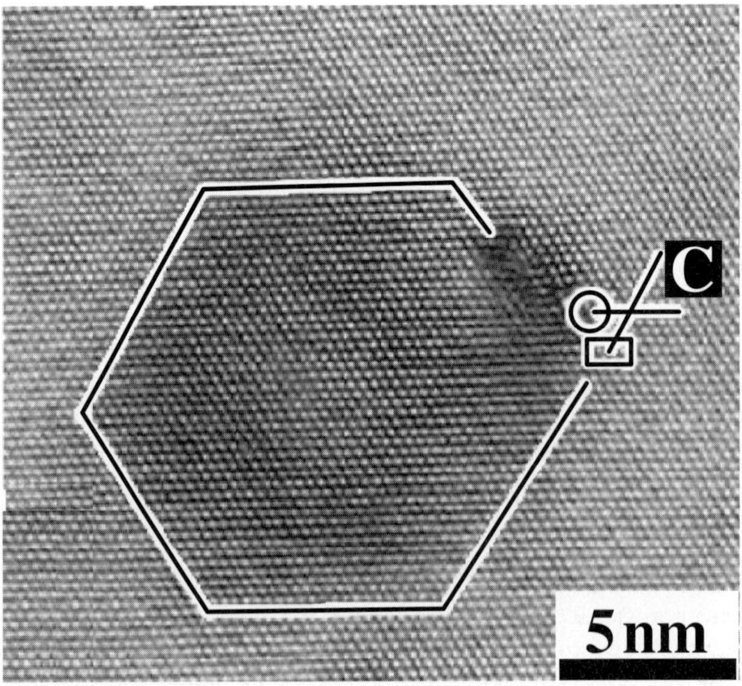

Fig. 7.23 Formation of a GaN quantum dot in the strain field of a pure edge threading dislocation [53].

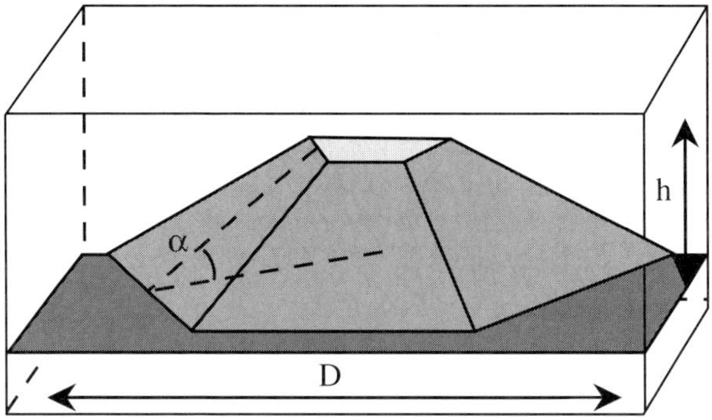

Fig. 7.24 Geometry of a GaN quantum dot as determined by HREM analysis [53].

assuming: (i) a uniform biaxial strain, which is imposed by the coherent growth on AlN, in the whole dot, (ii) a dot composition of pure GaN, and (iii) an averaging effect along the observation axis due to the thickness of the HRTEM specimen as compared to the volume of the dot. The averaging effect geometric was taken into account by introducing the pyramidal shape and fitting the position of the dot within the TEM cross-section. The good reproduction of the experimental decrease within the dot was a strong indication that (i) there is no segregation effect in GaN dots by contrast to InAs dots, and (ii) the strain in the whole dot was uniform, with a value corresponding to the biaxial strain imposed by the AlN layer. These results show that the Stranski–Krastanov growth mode of wurtzite GaN on AlN can be dislocation-free and coherent [40].

7.7 Large misfits and strain effects on the growth of InGaN

For InGaN, results have been published concerning the phase separation which was mainly supposed to give rise to the formation of layers made of InN and GaN patches [57]. This was explained by the fact that the two compounds have a large mismatch of lattice parameters [45]. In practice, the growth of InGaN layers is more demanding and if special care is not taken, one can end with the formation of the ternary alloy $In_xGa_{1-x}N$, InN and GaN, In, and to the same ternary ordered alloys as for AlGaN alloys [6, 49].

7.7.1 Critical thickness

In $In_xGa_{1-x}N$, the increase in film thickness and In composition can be correlated with the surface morphology as shown by El-Masry *et al.* [58]. They measured values of the critical thickness in $In_xGa_{1-x}N$ more than an order of magnitude higher than those reported for arsenide and phosphide compound semiconductors as predicted from the Matthews and Blakeslee model [59]. This was attributed to several reasons. First, the relaxation process in the nitrides occurs in a material system with about 10^{10} dislocations/cm^2, which is not the case in the other III–V compound materials. Second, in the nitride wurtzite structure the formation of misfit or inclined dislocations could be more difficult than in the zinc-blende III–V compounds due to a smaller number of glide planes. In a subsequent work on the InGaN/GaN double heterostructures, Reed *et al.* pointed out that the thickness range over which relaxation occurs in the thermally stressed double heterostructure is reduced to a width of 5 nm as compared to 90 nm for the single heterostructure [60]. This is the situation which probably prevails in ELDs and lasers as shown in Fig. 7.25 where an $a+c$ threading dislocation is shown to bend into the interface when it crosses the GaN/InGaN/AlGaN/GaN first interface in a green Nichia diode with In composition of 0.25% [61].

Finally, energetic processes, such as dislocation formation or a transformation from two-dimensional to three-dimensional growth modes, which are temperature dependent, can control the value of critical thickness [58]. However, the surface morphology degradation of the InGaN ternary layers has been investigated before and does not correspond to a simple VW growth mode. In the temperature range used for the growth of InGaN, the V-shape defects that form have now been investigated experimentally [17, 27] as well as theoretically [62]. Such defects are related to the surface mobility of atoms and they are mainly connected to dislocations with a c component. In the

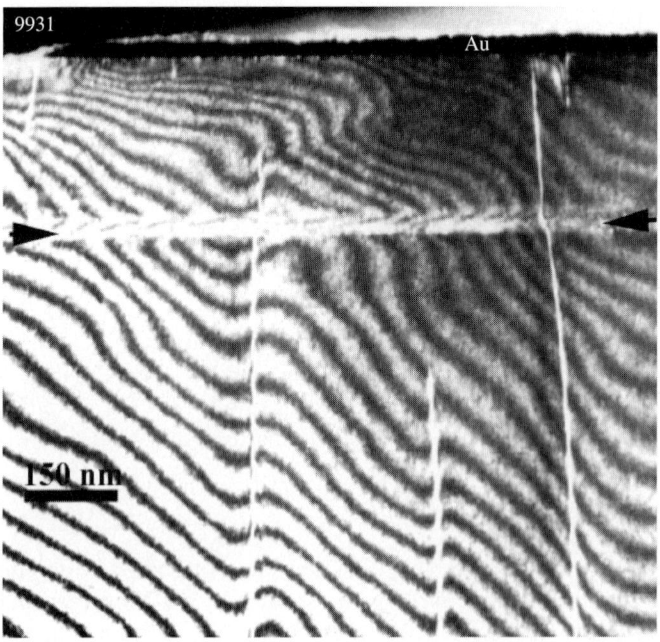

Fig. 7.25 Bending of an $a+c$ threading dislocation at the GaN/InGaN interface of a green Nichia ELD in order to form a misfit dislocation. The nominal In composition is more than 0.2.

10–20% indium composition range they can be deep enough to reach the interface and act as sources of misfit dislocations, they may also release the strain and contribute to the increase of the critical thickness [17].

7.7.2 Phase separation

Theoretical calculations predict that the critical temperature above which InN and GaN are miscible is above 1250 °C, therefore the typical growth temperatures around 700–900 °C fall into the miscibilty gap of the two compounds. As reported by Doppalapudi *et al.* [3], thick (~0.3 μm) InGaN layers grown by molecular beam epitaxy (MBE) contain regions of pure InN for large indium concentrations, in agreement with the theoretical results which showed that at the growth temperatures, the solid solubility of In inside GaN is limited to less than 5% [46]. Then, for compositions between 5% and 20% indium, the phase separation occurs through nucleation and growth. At composition higher than 25% In, the spinodal decomposition is expected. However, the same MBE technique was shown to give rise to GaN/ InGaN /GaN double heterostructures without any X-rays detectable phase separation [63]. This was explained by assuming that phase separation in alloys should require long range diffusion correlated with the length of time for the growth of the layer, but also stabilization of the alloys by the strain inside the quantum wells.

Phase separation has also been investigated in MOCVD quantum wells. In a post-annealing investigation, it was pointed out that at about 30% In composition, long time (8 h) at 900 °C would be necessary for significant phase separation [64]. This is in contrast with a recent report which suggests that spinodal decomposition can take place in quantum wells with indium composition lower than 25% at the growth temperature of 740 °C [18]. In this investigation disk-shaped indium-rich precipitates were observed with a diameter ranging from 5 to 8 nm and a thickness ranging from 2 to 4 nm. With increasing In content, the interfaces between InGaN and GaN became more diffusive and indium-rich agregates extended more into GaN barriers as the nominal indium content increased. In these samples, InN clusters were detected using XRD, and TEM observations showed that the indium-rich InGaN precipitates with diameter range from 5 to 12 nm prefered to agregate near V-shape defects connected to threading dislocations. Therefore, the stress field created by the dislocations might provide the driving force for the migration of indium atoms toward dislocations. This stress-induced migration was supposed to lead to the segregation of indium atoms near dislocations and to the formation of indium-rich clusters [18].

7.7.3 Indium clustering inside quantum wells

Inside the best layers, with lowest densities of V-shaped defects, it is suspected that high brightness of the ternary InGaN layers which constitute the emitting part of the lasers and electroluminescent diodes comes from localization of the carriers at In atoms organized in clusters. Early studies have used EDS analysis in the STEM and pointed out that the indium-rich areas may be as large as 20 nm inside 2–3 nm QWs [65]. Later, QUANTITEM [66] and strain measurements [67] pointed out that segregation of In does take place in the ternary layer, but the nature, size, and geometry of the elemental emitter is still the subject of controversy [68, 69].

We have investigated MOCVD quantum wells with very low V-shape defects density and nominal thickness of 2.5 nm and 17% In composition. We used HREM analysis associated with strain measurement and finite elements modeling of the real sample geometry [70]. This allowed more accurate results to be obtained than when limits between the thin foil aproximation and bulk material approximation were used to calculate the errors bars for composition evaluation [67]. In the QW layers, no precipitates were observed, they appear mainly to have uniform thicknesses, which are rather different to the nominal ones. The average thickness of the deeper QW was 3.6 nm instead of 2.4, whereas the top one measured about 1.7 nm (Fig. 7.26). In such typical two beam images contrast changes inside the wells. These contrast fluctuations correspond to local modification of the strain field. Moreover, the QW layers are elastically strained as no additional dislocations form inside the QWs, or at the interfaces. In the thinnest parts used in HREM (Fig. 7.27), the chemical contrast due to In no longer shows up and the QWs cannot be localized visually. The arrow shows approximately the position of the first thickness fringe corresponding to 5 nm. The uniform contrast in the area is a strong indication that the thickness increase is slow. We therefore can proceed to determine the strain field ε_{xx}. The calculated strain field can be represented as a map of profiles. As shown on Fig. 7.28 and Plate I, for the same area as in Fig. 7.27, its distribution is not homogeneous inside the QWs. It clearly shows that the growth did not give rise to uniform composition layers, but, instead, we see local strain bumps corresponding to In highs. The fluctuations stand at the noise level outside the

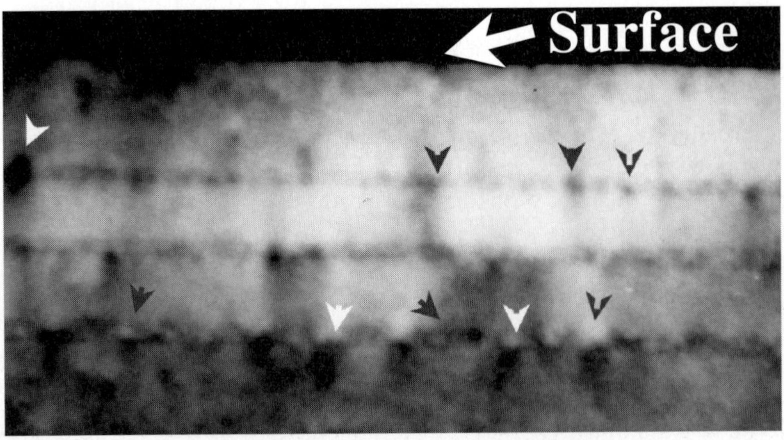

Fig. 7.26 Dark field 0002 image of three InGaN quantum wells of a 17% In nominal composition without V-shaped defects.

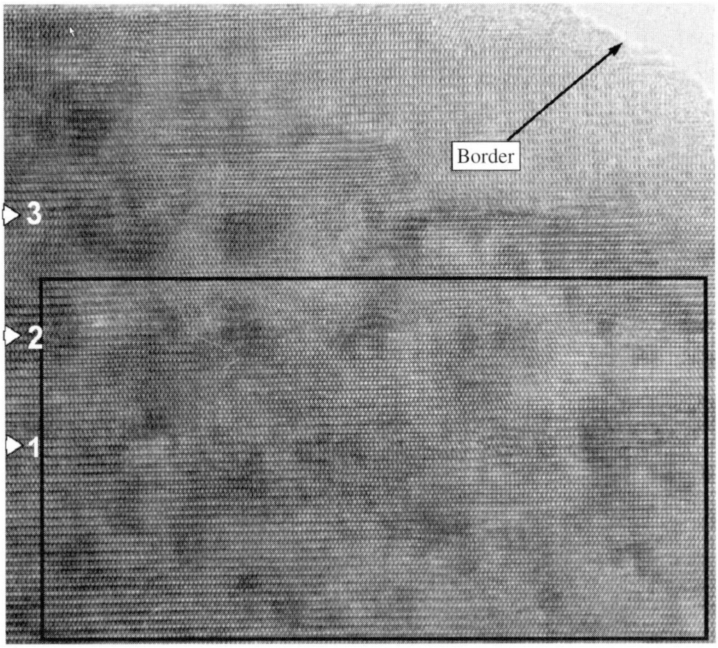

Fig. 7.27 A high-resolution image of the same samples as in Fig. 7.26 recorded along the [1120] zone axis. The marked area was used for quantitative evaluation of the indium composition.

quantum wells. The strain is then used to determine the local indium composition within the Vegard approximation [67, 70]:

$$C_{\mathrm{InN}} = A \frac{c_{\mathrm{GaN}}}{c_{\mathrm{InN}} - c_{\mathrm{GaN}}} \varepsilon_{xx}.$$

From this modeling, the maximum In concentration is found to be as high as 45%. In the dips, the indium concentration is below 10%. Therefore, the growth has resulted in the formation of quantum boxes with a lateral extension of 2 nm in height and 3 nm in width.

7.8 Discussion and conclusion

The low-dimensional structures fabricated in nitrides have better perfomances than similar devices in GaAs and its alloys, despite the presence of huge density of threading dislocations coming from the underlying buffer layers [64]. As can be noticed from Table 7.1, the alloys are made from compounds with large mismatch; it is relatively small between AlN and GaN, but very large between InN and GaN. Therefore these layers will suffer from defects that, as seen above, fall into two categories: (i) The extended crystallographic defects which

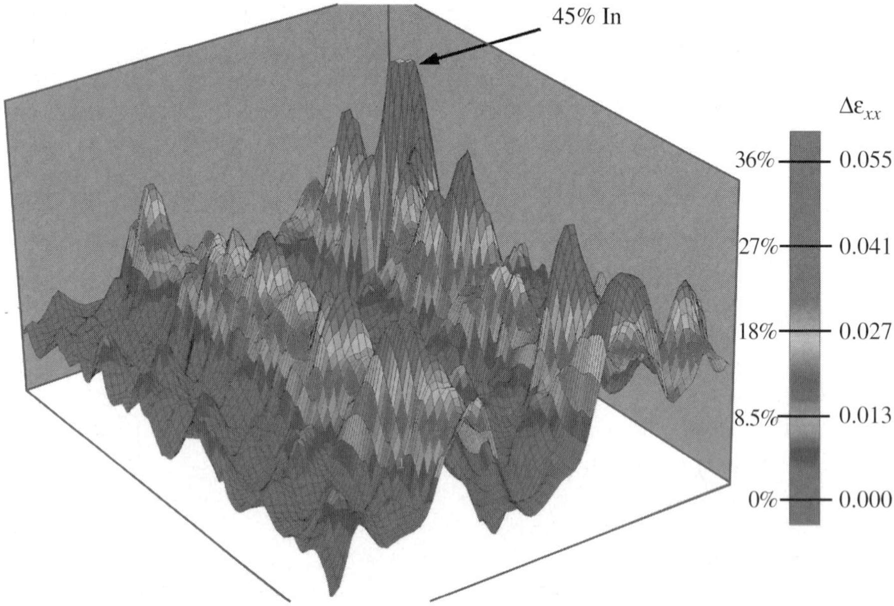

Fig. 7.28 Strain and related composition profiles inside the area of Fig. 7.27, the high indium content areas extend to 2 nm height and 3 nm width. The peak In concentration is 45%. (See also color plate section.)

originate from the underlying layers, but also may form at the various interfaces when the strains are plastically released; (ii) the ordering and phase separation which are probably inherent to the lattice and thermal mismatch of the alloys.

From the thickness–composition phase diagrams of the FVM, SK, and VW modes (GaInN/GaN and GaInN/AlN/GaN), it was shown that the layer-by-layer growth such as the FVM mode will be difficult to obtain for the epitaxial growth of GaInN on GaN. The VW mode is dominant in the phase diagram of the GaInN/GaN structures. The region of the FVM mode is wider and the SK mode becomes dominant by introducing the AlN buffer layer between the GaInN layer and the underlying GaN. This explains nicely the SK mode which is rather straightforward in the growth AlN/GaN QDs (Fig. 7.22). For ordering, the explanations are still not convincing. The same types of ordering have now been observed in AlGaN and GaInN, they are all along the [0001] direction. The 1:1 ordering may be compared to the {111} CuPt ordering in the zinc-blende compounds which has the same close packed symmetry as the (0001) plane of the wurtzite alloys. However, in zinc-blende alloys ordering has not been observed during growth on {111} planes [71].

A theoretical investigation on the surface energetics of InGaN [62] has shown that the ordered bulk compound is not stable with respect to mixing of In and Ga within a given (0001) plane and the 1:1 ordering was explained within a model wherein growth occurs by attachment of atoms at step edges or small $\{10\bar{1}1\}$ facets which have two different positions for the metal atoms.

In practice, a bilayer step of height *c* along the [0001] direction has the same local atomic structure as a $\{10\bar{1}1\}$ microfacet and that In (Ga) atoms preferentially occupy adlayer sites in which they make one(two) bonds to the N atoms. The energy difference between the favorable and unfavorable configurations is 0.5 eV, a value that is large compared to the growth temperature (kT_{growth}=0.09 eV). Consequently as the steps sweep across the surface, Ga incorporation occurs mainly on the Ga planes while In incorporation occurs mainly on the In planes indicated. This mechanism was assumed also to explain the V-shaped defects formation on termination of threading dislocations at the InGaN surface [62], whether it can also explain the similar ordering of AlGaN needs to be further investigated.

The phase separation and growth instabilities which were observed inside InGaN layers can also be explained roughly by the large mismatch between GaN and InN[3, 4]. Recent reports were published on spinodal decomposition in InGaN/GaN multiple quantum wells with indium content in the range of 15%–25% grown by MOCVD. The InGaN/GaN interfaces became more diffusive, and indium-rich precipitates extended into GaN barriers with increasing indium content. Disk-like indium-rich dots were found to be confined near QWs with a relatively lower indium content. The indium-rich precipitates with a diameter range from 5 to 12 nm prefer to agregate near V-shaped defects. This is in contrast with other reports where GaN/$In_{.8}Ga_{.2}N$/GaN layers were grown by MBE [63], and almost $In_{.3}Ga_{.7}N$ from MOCVD [64] was shown to be stable against annealing to less than 900 °C. This means that the growth conditions probably still play a very important part on the quality of the layers and that much work still needs to be done before optimal nitride layers are reproducibly grown.

Acknowledgements

This work was partially supported by the European Commnunity under contract number HPRN-CT-1999-00040 and the French Ministry of Education under 'Action Incitative Nanotechnologies : Project "Boquani". Drs J.D. Ganière and G. Nouet are gratifully acknowledged for critically reading the manuscript. Drs S. Kret, P. Vermaut, and V. Potin are ackowledged for their valuable contribution.

References

1. Nakamura, S. (1998). In *Group III nitride semiconductor compounds*, p. 391 (Ed. B. Gil). Oxford Science Publications.
2. Ruterana, P., Nouet, G., Van der Stricht, W., Moerman, I., and Considine, L. (1998). *Appl. Phys. Lett.* **72**, 1742.
3. Doppalapudi, D., Basu, S. N., Ludwig, K. F., and Moustakas, T. D. (1998). *J. Appl. Phys.* **84**, 1395.
4. Ruterana, P., Aguinet, R., Poisson, M. A. (1999). *Phys. Stat. Solidi* (b) **216,** 663.

5. Widmann, F., Daudin, B., Feuillet, G., Samson, Y., Rouvière, J. L., and Pelekanos, N. (1998) *J. Appl. Phys.* **83**, 7618.
6. Korakakis, D., Ludwig, K. F., and Moutakas, T. D. (1997). *Appl. Phys. Lett.* **71**, 72.
7. Osamura, K., Naka, S., and Murakami, Y. (1975). *J. Appl. Phys*. **46**, 3432.
8. Ponce, F. A., Bour, D. P., Götz, W., Johnson, N. M., Helava, H. I., Grzegory, I. *et al.* (1995). *Appl. Phys. Lett.* **68**, 917.
9. Vermaut, P., Ruterana, P., Nouet, G., Salvador, A., and Morkoç, H. (1995). *Inst. Phys. Conf. Ser.* **146**, 289.
10. Cherns, D., Young, W. T., and Ponce, F. A. (1997). *Mat. Sci. Eng. B* **50**, 76.
11. Rouvière, J. L., Arlery, M., Bourret, A., Niebuhr, R., and Bachem, K. (1995). *Inst. Phys. Conf. Ser.* **146**, 285.
12. Northrup, J. E., Neugebauer, J., and Romano, L. T. (1996). *Phys. Rev. Lett.* **77**, 103.
13. Tanaka, S., Kern, R. S., and Davis, R. F. (1995). *Appl. Phys. Lett.* **66**, 37.
14. Xin, Y., Brown, P.D., Humphreys, C. J., Cheng, T. S., and Foxon, C. T. (1997). *Appl. Phys. Lett.* **70**, 1308.
15. Ning, X. J., Chien, F.R., Pirouz, P., Wang, J. W., and Khan, M. A. (1996). *J. Mater. Res.* **3**, 580.
16. Ponce, F. A. (1997). *MRS Bulletin*, **22**, 51.
17. Jansen, B., Albrecht, M., Dorsch, W., Christiansen, S., Strunk, H. P., Hanser, D., and Davis, R. F. (1998). *MRS Internet J. Nitride Semicond. Res*. **3**, 39.
18. Lin, Y. S., Ma, K. J., Hsu, C., Feng, S. W., Cheng, Y. C., Liao, C. C. *et al.* (2000). *Appl. Phys. Lett.* **77**, 2988.
19. Kaiser, S., Preis, H., Gebhardt, W., Ambacher, O., Angerer, H., Stutzmann, M., Rosenauer, A., and Gerthsen, D. (1998). *Jpn. J. Appl. Phys*. **37**, 84.
20. Pirouz, P. and Yang, J. W. (1993). *Ultramicroscopy* **51**, 189.
21. Ning, X.J., Chien, F.R., Pirouz, P., Wang, J.W., and Khan, M.A. (1996). *J. Mater. Res.* **3**, 580.
22. Ponce, F. A. (1997). *MRS Bulletin* **22**, 51.
23. Elsner, J., Jones, R., Sitch, P. K., Porezag, V. D., Elstner, M., Frauenheim, T., Heggie, M. I., Oberg, S., and Briddon, P. R. (1998). *Phys. Rev.Lett.* **79**, 3672.
24. Vermaut, P., Ruterana, P., Nouet, G., and Morkoç, H. (1998). *MRS Internet J. Nitride Semicond. Res*. **1**, 42.
25. Ruterana, P., Vermaut, P., Potin, V., Nouet, G., Botchkarev, A., Salvador, A. and Morkoç, H. (1997). *Mater. Sci. Eng. B* **50**, 72.
26. Cherns, D., Young,W. T., and Ponce, F. A. E (1997). *Mat. Sci. Eng. B* **50**, 76.
27. Wu, X. H., Elsass, C. R., Abare, A., Mack, M., Keller, S., Petroff, P. M., *et al.* (1998). *Appl. Phys. Lett.* **72**, 693.
28. Stampfl, C., and Van de Walle, C.G. (1998). *Phys. Rev. B* **57**, R15052.
29. Salviatti, G., Albrecht, M., Zanotti-Fregonara, C., Armani, N., Mayer, M., Shreter, Y., Guzzi, M., Melnik, Y. V., Vassilevski, K., Dmitriev, V. A., and Strunk, H. P. (1999). *Phys. Stat. Sol.* (a) **171**, 325.
30. Lee, N. E., Powell, R. C., Kim, Y. W., and Greene, J. E. (1995). *J. Vac. Sci. Technol. A* **13,** 2293.
31. Wu, X. H., Brown, L. M., Kapolnek, D., Keller, S., Keller, B., DenBaars, S. P., and Speck, J. S. (1996). *J. Appl. Phys*. **80**, 3228.
32. Lorenz, K., Gonzalves, M., Kim, W., Narayanan, V., and Mahajan, S. (2000). *Appl. Phys. Lett.* **77**, 3391.
33. Vermaut, P., Ruterana, P., and Nouet, G. (1997). *Phil. Mag. A* **76**, 1215.
34. Blank, H., Delavignette, P., Gevers, R., Amelinckx, S. (1964). *Phys. Stat. Sol.* **7**, 747.
35. Drum, C. M. (1965). *Phil. Mag. A*, **11**, 313.

36. Vermaut, P., Ruterana, P., Nouet, G., Salvador, A., and Morkoç, H. (1997). *Phil. Mag.* **A 75**, 239.
37. Aminoff, G. and Broome, B. (1931). *Z. Kristallogr.* **80**, 355.
38. Potin, V., Ruterana, P., and Nouet, G. (1999). *Appl. Phys. Lett.* **74**, 947.
39. Molina, S. I., Sanchez, A. M., Pacheco, F. J., Garcia, R., Sanchez-Garcia, M. A., Sanchez, F. J., and Calleja, E. (1999). *Appl. Phys. Lett.* **74**, 3362.
40. Nakajima, K., Ujihara, T., Miyashita, S., and Sazaki, G. (2001). *J. Appl. Phys.* **89**, 146.
41. Amano, H., Sawaki, N., Akasaki, I., and Toyoda, Y. (1986). *Appl. Phys. Lett.* **48**, 353.
42. Nakajima, K., Kitahara, K., and Ochiai. S. (1996). *Jpn. J. Appl. Phys.* **35**, 2605.
43. Ochiai, S., Schulte, K., and Peters, P.W.M. (1991). *Compos. Sci. Technol.* **41**, 237.
44. Stringfellow, G. B. (1973) *J. Phys. Chem. Solids* **34**, 1749.
45. Hillert, M. (1961). *Acta Metall.* **9**, 525.
46. Ho, I. and Stringfellow, G.B. (1996). *Appl. Phys. Lett.* **69**, 2701.
47. Behbehani, M. K., Piner, E. L., Liu, S. X., El-Masry, N. A., and Bedair, S. M. (1999). *Appl. Phys. Lett.* **75**, 2202.
48. Ruterana, P., de Saint Jores, G., Omnes, F. (2001). *Mater. Sci. Eng. B* **82**, 203.
49. Ruterana, P., De Saint Jores, G., Laügt, M., Omnes, F., and Bellet-Amalric, E. (2001). *Appl. Phys. Lett.* **78**, 345.
50. Arlery, M., Rouvière, J. L., Widmann, F., Daudin, B., Feuillet, G., and Mariette, H. (1999). *Appl. Phys. Lett.* **74**, 3287.
51. Widmann, F., Daudin, B., Feuillet, G., Samson, Y., Rouvière, J. L., and Pelekanos, N. (1998). *J. Appl. Phys.* **83**, 7619.
52. Daudin, B. and Widmann, F. (1997). *J. Cryst. Growth* **182**, 1.
53. Hacke, P., Feuillet, G., Okumura, H., and Yoshida, S. (1996) *Appl. Phys. Lett.* **69**, 2507.
54. Tersoff, J., Teichert, C., and Lagally, M. G. (1996) *Phys. Rev. Lett.* **76**, 1675.
55. Madhukar, A. (1996). *J. Cryst. Growth* **163**, 149.
56. Rouvière, J. L., Simon, J., Pelekanos, N., Daudin, B., and Feuillet, G. (1999). *Appl. Phys. Lett.* **75**, 2633.
57. Hiramatsu, K., Kawagushi, Y., Shimizu, M., Sawaki, N., Zheleva, T., Davis, R. F., *et al.* (1997). *MRS Internet J. Nitride Semicond. Res.* **2**, 6.
58. El-Masry, N. A., Piner, E. L., Liu, S. X., Bedair, S. M. (1998). *Appl. Phys. Lett.* **72**, 40.
59. Matthews, J. W. and Blakeslee, A. E. (1974). *J. Cryst. Growth* **27**, 118.
60. Reed, M. J., El-Masry, N.A., Parker, C. A., Roberts, J. C., and Bedair, S. M. (2000). *Appl. Phys. Lett.* **77**, 4121.
61. Nakamura, S. (1997). In *GaN and related materials.* p. 41 (Ed. S. J. Peatton). Gordon and Breach Science Publishers.
62. Northrup, J. E., Romano, L. T., and Neugebauer, J. (1999). *Appl. Phys. Lett.* **74**, 2319.
63. Singh, R., Doppalapudi, D., Moustakas, T. D., and Romano, L. T. (1997). *Appl. Phys. Lett.* **70**, 1089.
64. McCluskey, M. D., Romano, L. T., Krusor, B. S., Bour, D. P., Johnson, N. M., Brennan, S. (1998). *Appl. Phys. Lett.* **72**, 1730.
65. Narukawa, Y., Kawakami, Y., Funato, M., Fujita, S., Fujita, S., and Nakamura, S. (1997). *Appl. Phys. Lett.* **70**, 981.
66. Kisielowski, C., Liliental-Weber, Z., and Nakamura, S. (1997). *Jpn. J. Appl. Phys.* **36**, 6932.
67. Gerthsen, D., Hahn, E., Neubauer, B., Rosenauer, A., Schön, O., Heuken, M., and Rizzi, A. (2000). *Phys. Stat. Sol.* (a) **177**, 145.

68. Bellaiche, L., Mattila, T., Wang, L.W., Wei, S.H., and Zunger, A. (1999). *Appl. Phys. Lett.* **74**, 1842.
69. O'Donnell, K. P., Martin, R. W., and Middleton, G. P. (1999). *Phys. Rev. Lett.* **82**, 237.
70. Ruterana, P., Kret, S., Poisson, M. A. (2001). *Mater Sci Eng. B*, (in press).
71. Gomyo, A., Suzuki, T., Iijima, S., Hotta, H., Fujii, H., Kawata, S. *et al.* (1998). *Jpn. J. Appl. Phys.* **27**, L2370.

8 Excitons in GaN-based low-dimensional systems

Bernard Gil

A review is given about the status of the physics of GaN-AlGaN quantum wells and superlattices

8.1 CW spectroscopy in GaN-AlGaN quantum wells

Bulk GaN is characterized by a strong exciton binding energy (25–27 meV) [1, 2] and a large oscillator strength [3]. The coupling with the electromagnetic field is also large. A Rabbi oscillation splitting of some 45 meV has been predicted in GaN-based microcavities [4]. Cubic GaN-AlGaN quantum wells can theoretically lead to an improvement of the optical response compared to the case of bulk c-GaN due to the result of carrier confinement in type I quantum wells [5, 6]. Promising results were published recently and we refer the reader to ref. [6] for detailed optical spectroscopy on such system. Quantum wells based on wurtzite GaN and AlGaN behave differently, due to difference of the spontaneous polarization of the two materials which together with piezoelectric field effect produce an efficient Quantum Confined Stark Effect [7, 8]. Depending on well width, the optical transitions can be boosted either toward shorter wavelengths than the one of bulk GaN or can be boosted in the realm of long wavelength radiation, as shown in Figs 8.1 and 8.2. The magnitude of the internal electric field depends on the aluminium composition [9–19], and of the ratio between the width of the well layers relatively to the width of the barriers, as shown in Figs 8.3, 8.4, and 8.5. In particular, for aluminium compositions ranging up to 27%, for well width larger than a critical value of some 12–13 monolayers (~3 nm), the photoluminescence energy is lower than the bandgap of bulk GaN [15]. Comparing the low temperature photoluminescence and reflectance spectra on Fig. 8.1, one observes a Stokes shift of the photoluminescence peak compared to the reflectance feature as evidence of a localized nature of the excitons that are observed in photoluminescence, at these temperatures. Increasing the lattice temperature gives an excess of kinetic energy that can facilitate the exciton detrapping. The relative ratio of free to bound excitons promotes the free exciton population in such a way that, at high temperature, the photoluminescence signal is dominated by free exciton recombination. Then, the temperature dependence of the

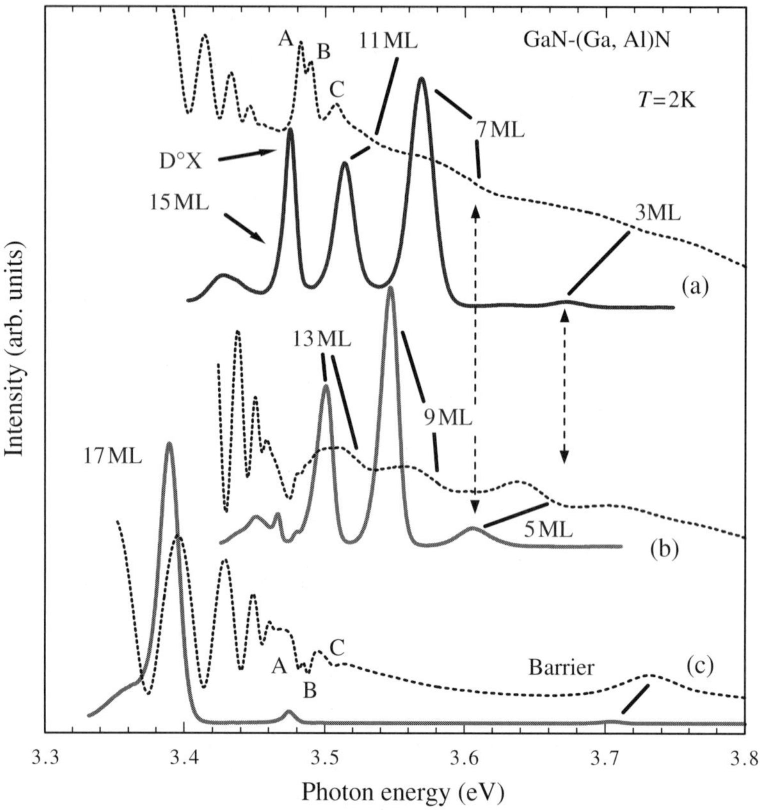

Fig. 8.1 CW PL and reflectivity spectra of a sample, which contains 4 GaN/$Ga_{0.89}Al_{0.11}$N QWs with respective thicknesses: 3, 7, 11, and 15 monolayers (MLs), nominally, from the surface towards the GaN buffer layer. The Stokes shifts are in the range of 40–50 meV between PL lines and reflectance features, indicating that the PL corresponds to the recombination of excitons localized on areas where the QWs are wider by 2 MLs (1 ML=2.59 Å) than their nominal thickness. These Stokes shifts also correspond very well with the energy difference between PL peaks from QWs which differ by 2 MLs in width. This can be checked in Fig. 8.1(b) which is the equivalent but for a sample with three GaN/$Ga_{0.89}Al_{0.11}$N QWs of respective nominal thicknesses 5, 9, and 13 MLs. These lines nicely intercalate between those of the preceding sample. The PL line from the 15 ML-wide QW in Sample 1 overlaps the line of D°X complexes in the GaN buffer. The third sample gives photoluminescence clearly below the one of the GaN template.

photoluminescence peak follows the bandgap of GaN as shown in Figs 8.6 and 8.7. In this particular case of a multiple quantum well for which the nominal thickness of the well layers is 16 monolayers, the low temperature photoluminescence peaks at the energy expected for 18 monolayer wide wells. Increasing the lattice temperature first blue-shifts the photoluminescence maximum to the energy expected if having 17 monolayer well-widths, and then, if further increasing the temperature to the energy expected for 16 monolayer wide wells.

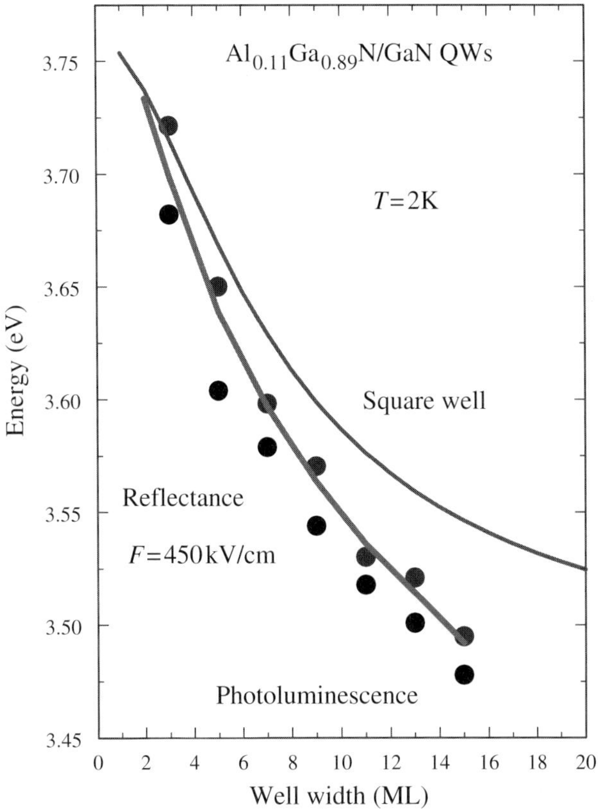

Fig. 8.2 Experimental transition energies in the MQWs with 11% aluminum in the 5-nm-thick barrier layers (high energy data is reflectance, low energy ones are photoluminescence data) together with the result of a square well calculation (thin line), and with the inclusion of an internal electric field in the well layers (here 750 kV/cm). The agreement between theory and experiment together with the crossing of the confined excitonic transitions with the transition in the GaN template indicate the existence of Quantum Confined Stark Effect in these samples.

The internal electric field localizes electrons and holes at different interfaces as is indicated by the results of effective mass calculations shown in Fig. 8.8 [20]. There is also a drastic influence on the oscillator strengths shown in Fig. 8.9 (top) where are plotted both the oscillator strengths measured for several quantum wells from a Fourier transform treatment of some reflectance spectra similar to those plotted in Fig. 8.1. The mathematical treatment being fairly tedious, we refer the reader to ref. [21] for more detailed information. From the drastic reduction of the overlap between the electron and hole wave functions, we measure the oscillator strength to rapidly scale more than two decades when the well width increases. Note that the spectral disribution of the exciton resonances scales a few meV as shown bottom of Fig. 8.9. We will address the origin of this below.

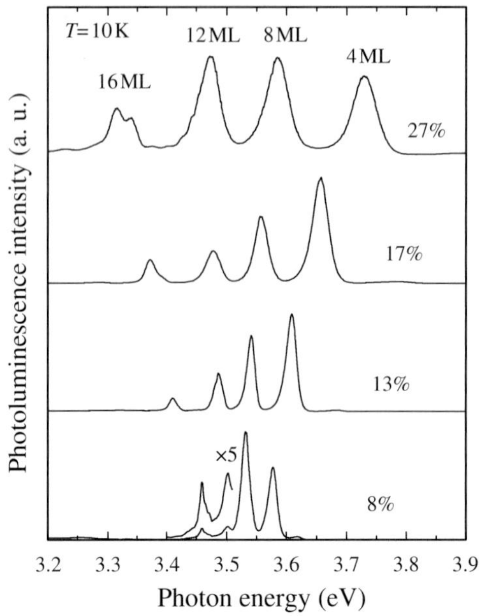

Fig. 8.3 PL spectra of MQW structures with 8, 13, 17, and 27% aluminum in the barrier layers and with constant thickness of the barrier layers (10 nm).

The Quantum Confined Stark Effect has has little influence on the scale of the exciton binding energy due to the long-range nature of this interaction [20]. This is shown in Fig. 8.10 where we plot the variational calculations of the exciton binding energies using 2D and 3D trial functions. Calculations are made in the context of square and triangular wells to quantify both of the effects of confinements, electric fields and refinement of the trial function.The variation of the exciton binding energy has been also computed at different temperatures under photo-injection conditions. In that case electron and hole Fermi seas are created, and the exciton trial function has to obey Pauli's exclusion principle. The mathematical treatment of this problem and the relevant results can be found in refs [22–24].

We now address the localization phenomenon, using an approach derived from the variational calculation of the exciton binding energies that lead to the results plotted in Fig. 8.10. The total exciton wave function is written as (Bloch Waves are ignored for the sake of the simplicity of the mathematical equations):

$$\Psi(Z_e, Z_h, \rho) = N\chi_e(Z_e)\chi_h(Z_h)\exp\left(\frac{-|\rho|}{\lambda}\right)$$

where the χs are the z-dependent envelope functions, the exponential factor is a trial function with one variational parameter λ and N is a normalization constant. As written, the trial function varies like the difference ρ between the in-plane coordinates of the electron and hole that are measured with respect

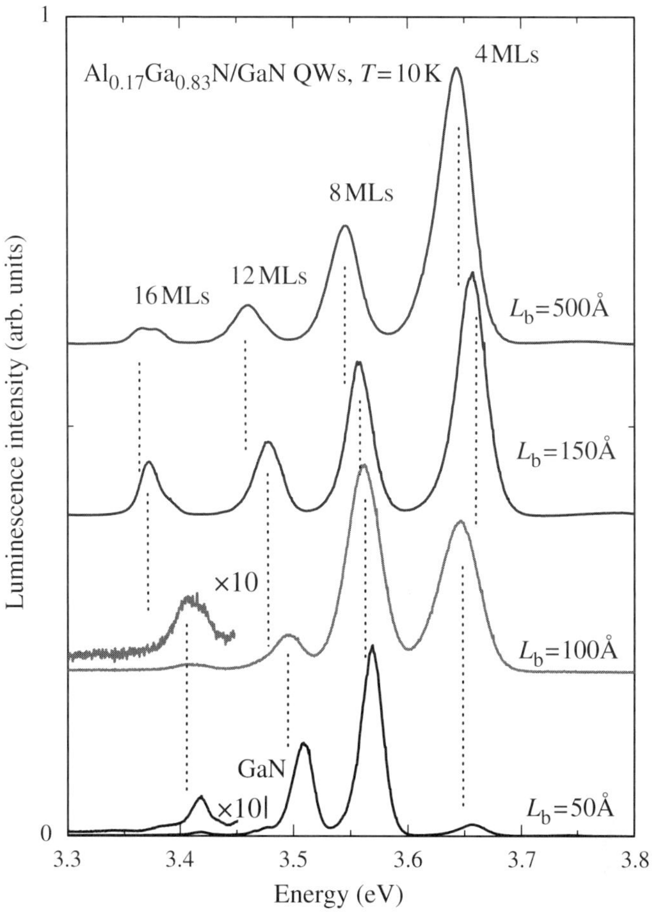

Fig. 8.4 PL spectra of MQW structures with varying the thickness of the barrier layers.

to an arbitrary origin. If we choose for such an origin the center of gravity G of the electron and hole in the plane of the quantum wells, one can define, with respect to G, the in-plane position of the electron (ρ_e) and hole (ρ_h) as

$$\rho_e = \rho \frac{m_h}{m_h + m_e} \quad \text{and} \quad \rho_h = \rho \frac{m_e}{m_h + m_e}$$

The electron and hole masses m_e and m_h are important input parameters. We note that, in that novel writing, $\rho = \rho_e + \rho_h$.

Averaging ψ over all the hole (resp. electron) positions, we get the electron (or the hole) pseudo-envelope functions

$$F_\xi(Z_\xi, \rho_\xi) = N_\xi f(z_\xi) \exp\left(-\frac{\rho_\xi}{\lambda_\xi}\right).$$

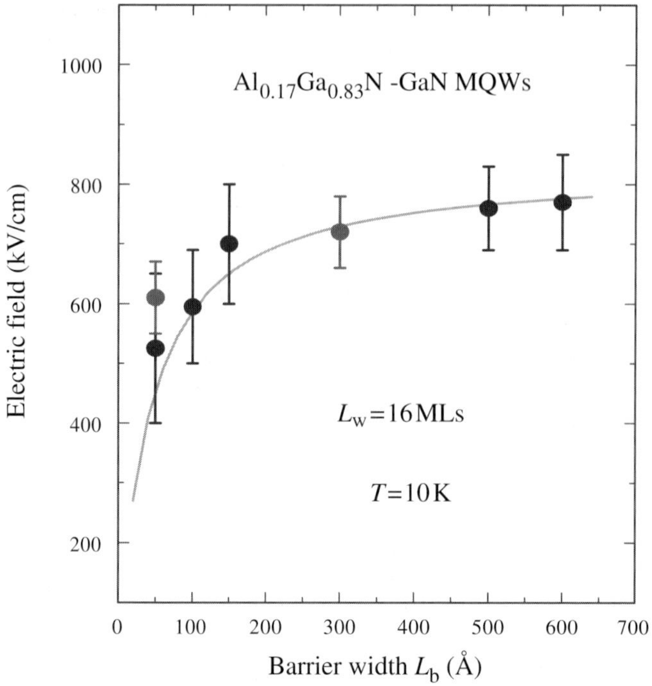

Fig. 8.5 Electric field in the well layer versus width of the barrier layer.

The in-plane extension of the hole and of the electron are

$$\lambda_h = \lambda \frac{m_e}{m_e + m_h} \quad \text{and} \quad \lambda_e = \lambda \frac{m_h}{m_e + m_h}$$

Simple envelope function calculation [25] also indicates that the in-plane extension of the hole and electron wave functions is comparable in size with the average aluminium–aluminium separation for alloy compositions ranging up to at least 30% (~2 nm) as shown in Fig. 8.11. Typically, one gets $\lambda_h = 0.7$ nm for a 30 monolayer wide well and an aluminum composition of 17%. A similar calculation performed for GaAs-GaAlAs quantum wells gives $\lambda_e = 10$–12 nm!. *In GaN-GaAlN quantum wells, the interfacial exciton localization results of the comparable values of the Bohr radii and of the characteristic length of the alloy disorder.*

8.2 Time-resolved spectroscopy

From all that was said above, it is obvious that, given an aluminum composition in the barrier layers, the excitonic oscillator strength decreases when

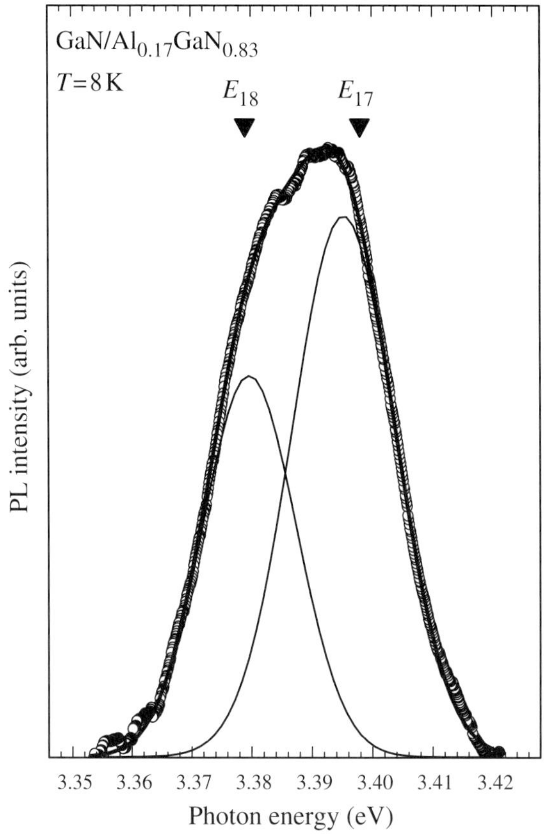

Fig. 8.6 CW photoluminescence spectrum for a nominally 16 ML wide GaN/$Al_{0.17}Ga_{0.83}$ N QW. Exciton energies for the nominal 16-ML-wide QW and for 17 and 18-ML-wide QWs are indicated.

increasing the well width. This statement has been experimentally verified by time-resolved spectroscopy [9, 10]. What was found is both the red shift due to QCSE and a dramatic increase of the recombination time with increasing well width.

These experiments have been reproduced [26]. First, it was extremely important to identify the nature of the recombination in terms of free exciton and localized excitons. From the previous studies made on GaAs-AlGaAs QWs it is now well established that the recombination dynamics of free-excitons is very different from the recombination dynamics of bound or localized ones [27]. The decay time of bound excitons is almost temperature independent at moderate temperatures. Free excitons have a wave function which is translation-invariant in the growth plane; in other words, they are allowed to move in the growth plane of the QW. Quantum mechanics indicates that the radiative lifetime of free excitons increases with the temperature. Mathematics predict a linear

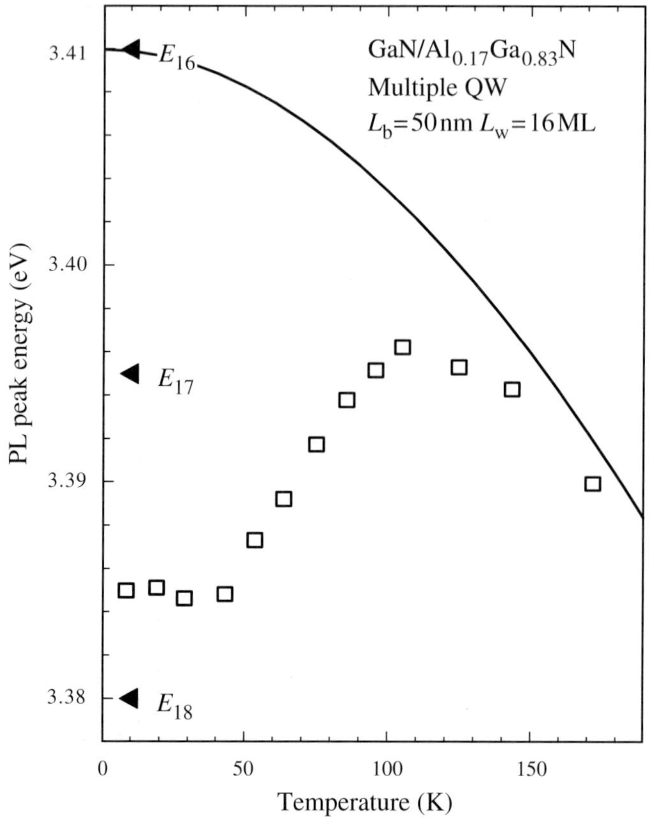

Fig. 8.7 Temperature dependence of the photoluminescence peak energy for the sample of Fig. 8.6. Exciton energies for the nominal 16-ML-wide QW and for 17 and 18-ML-wide QWs are indicated.

dependence with temperature in 2D systems, and a superlinear one in 3D crystals ($\tau_{rad} \sim T^{3/2}$)[28]. The recombination time τ is a function of the radiative recombination time τ_{rad} and of the non-radiative one τ_{nr}. This can be written:

$$\frac{1}{\tau} = \frac{1}{\tau_{rad}} + \frac{1}{\tau_{nr}}.$$

The photoluminescence intensity is inversely proportional to the radiative recombination time and varies as η, where η is given by:

$$\eta = \frac{\tau_{nr}}{\tau_{nr} + \tau_{rad}}.$$

In high-quality quantum wells, it is now well established that non-radiative recombination channels dominate the PL decay at high temperature while

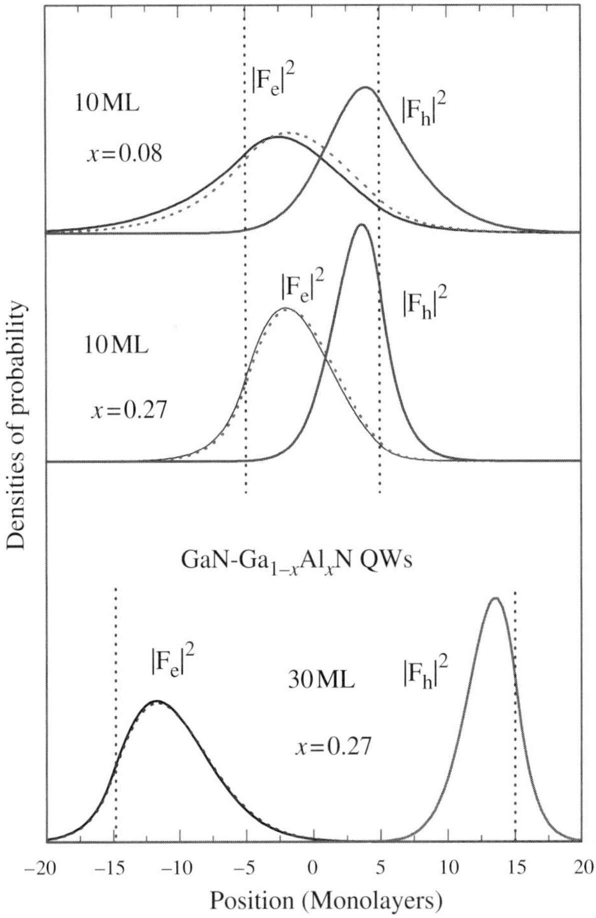

Fig. 8.8 Some typical densities of probability for electron and heavy holes. The correction brought by the self-consistent calculation at the scale of the electron wave function is given using dashed lines. Clearly, in case of wide and deep wells the spatial separation of the carriers is so large that the oscillator strength is extremely small.

radiative recombination rules the photoluminescence dynamics at liquid helium temperature. From the measurement of the decay time and photoluminescence intensity versus temperature one gets the variation of the radiative decay time with temperature as:

$$\tau_{\text{rad}}(T) = \frac{\tau(T)}{\eta(T)}.$$

Quantum mechanics also relates the radiative lifetimes to the excitonic oscillator strengths in the quantum well. The reference quantity is the radiative lifetime at the in-plane exciton–polariton bottleneck: τ_0 at $k = \sqrt{\varepsilon} k_0$. If taking

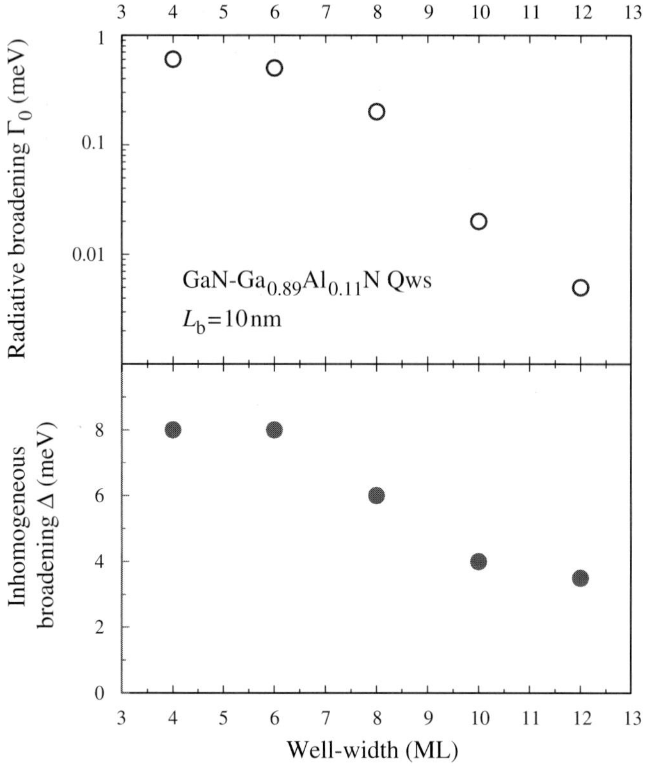

Fig. 8.9 Top: Experimental values of oscillator strength for $GaN\text{-}Ga_{0.89}Al_{0.11}N$ quantum wells. Below: inhomogeneous broadening of the resonance energies in $GaN\text{-}Ga_{0.89}Al_{0.11}N$ quantum wells.

for the 2D exciton envelope function a trial function $\exp(-\rho/\lambda)$, from Fermi's Golden rule, the lifetime τ_0 is *analytically* connected to the oscillator strength as:

$$\tau_0 = \frac{1}{2}\left[\frac{\sqrt{\varepsilon}k_0\omega_{LT}a_B^3}{\lambda^2} I_{eh}^2\right]^{-1}$$

where ω_{LT} is the longitudinal transverse splitting of the exciton–polariton in the bulk, I_{eh} the overlap integral of the envelope function, k_0 is the *in vacuo* wave number of the photon, and a_B is the 3D Bohr radius.

Concerning its temperature dependence, this is written:

$$\frac{\partial\tau}{\partial T} = \frac{\xi M k_B}{\hbar^2 k_0^2}\tau_0 \quad \text{with} \quad \xi = 3$$

if heavy-hole and light-hole excitonic transitions are quite fairly energy split; ξ=2 when they are close in energy. The experiment we show here has been

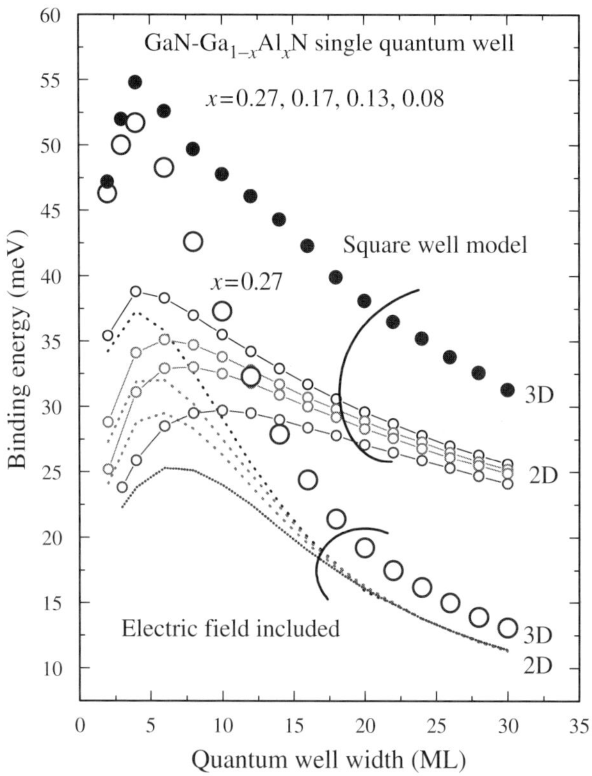

Fig. 8.10 Width dependence of the exciton binding energies computed for GaN-AlGaN quantum wells with square potential line-ups and with the real ones which include the internal electric fields. The results of the 2D calculation are given for all the four aluminum compositions. The results of a 3D calculation are restricted to an aluminum composition of 27% in order to maintain the readability of the figure.

performed on the multiple quantum well structure with 7% aluminum in the barrier layer. The 2 K PL spectrum of this samples exhibit well-marked free and localized excitons. The evolutions of the photoluminescence intensity and of the total decay time with temperature are shown on Fig. 8.12. This offers us the possibility to extract the temperature dependence of the radiative recombination time as shown on Fig. 8.13. From the experimental value $\partial\tau/\partial T = 20.5 \pm 0.7\,\mathrm{ps\,K^{-1}}$ and assuming $k_0 = 1.80\,10^7\,\mathrm{m}^{-1}$, $n = 3$, and $M = 1.2\,m_0$, we obtain $\tau_0 = 2.4\,\mathrm{ps}$. This gives $\hbar\omega_{\mathrm{LT}} = 0.8\,\mathrm{meV}$ in fairly nice agreement with data obtained from line-shape fitting of reflectance on high quality epilayers [29]. This indicates that the time-resolved spectroscopy in nitride has reached that of GaAs-AlGaAs quantum wells. Now, it is obvious that if we deduce the radiative lifetime at the bottleneck from its variation with temperature, one could argue that the absolute value of the decay time results from an overlap of free exciton and bound exciton contributions. At low temperature

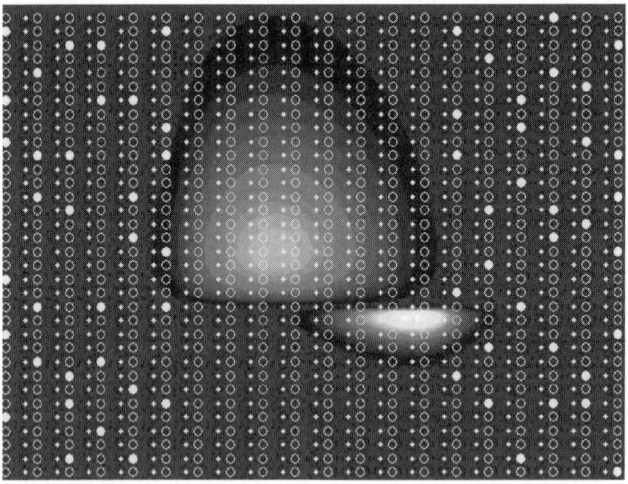

Fig. 8.11 In-plane extension of the densities of probability for electron and hole in case of a 3 nm wide GaN-$Ga_{0.83}Al_{0.17}$ N QW. Aluminium atoms in GaAlN layers are represented using large yellow spots to illustrate the alloy disorder. Note that the in-plane extension of the hole wave function is smaller than the average Al-Al distancy. The alloy disorder is responsible of the interface roughness experienced by the exciton. (See also color Plate II.)

we have in fact a spectral dependence of the decay times as shown in Fig. 8.14 and the asymptotic value at low temperatures in Fig. 8.13 corresponds to the decay time on the high energy wing of the photoluminescence band.

Figure 8.14 shows that increasing the localization energy, the decay time increases. In fact, a thermal detrapping of the localized excitons is thermally activated, and when the temperature increases the population of free exciton increases and the population of localized exciton decreases. This is readily shown in Fig. 8.15 for several QWs for which the localization energies are different. When increasing the temperature, thermal detrapping occurs and the energy of the photoluminescence band follows the trend expected for the reflectance feature. Now from the equation above it is clear that the slope of the radiative lifetime should vary with the thickness of the quantum well. This appears via the evolution of the overlap integral $I_{\theta h}$ and of the 2D variational Bohr radius λ. It is interesting to check this effect. Temperature-dependent time-resolved photoluminescence measurements have been performed on the three samples of Fig. 8.5. The low-temperature decay times are reported in Fig. 8.16 for the different well widths. Although the photoluminescence energies intercalate in Fig. 8.5, the PL decay time does not. We have straight-line dependence but the slopes are different as shown in Fig. 8.17. Lefebvre *et al.* [30] have computed the radiative decay times in these samples (see Fig. 8.18). Obviously there is a discrepancy between the calculation in Fig. 8.18 and the plot in Fig. 8.17. This is due to the escape time of the carrier out of the well in these MQW samples. This escape time is favored by the existence of an

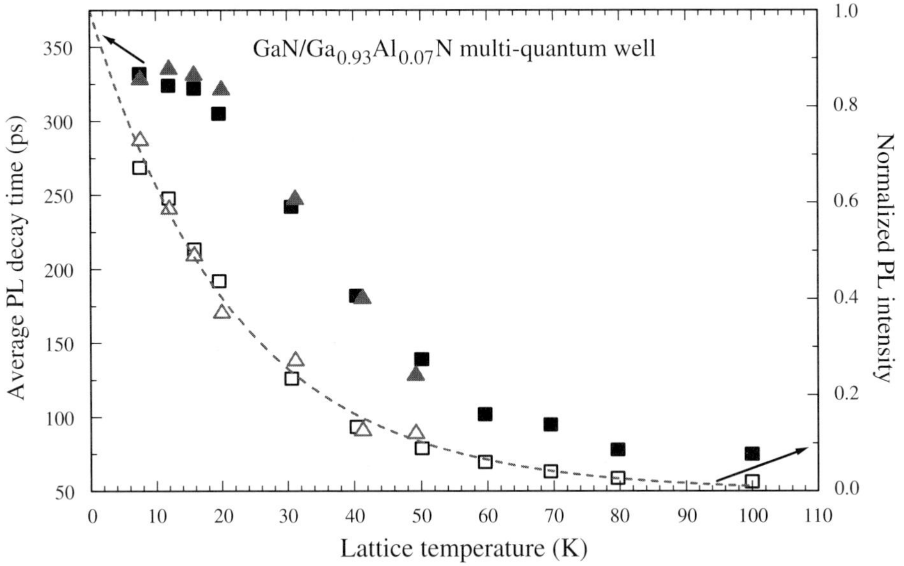

Fig. 8.12 Variation of the PL decay time of excitons in the QWs (solid symbols) and of their spectrally and temporally integrated PL intensity (open symbols) versus temperature. Squares and triangles correspond to the probing of two different spots on the sample surface.

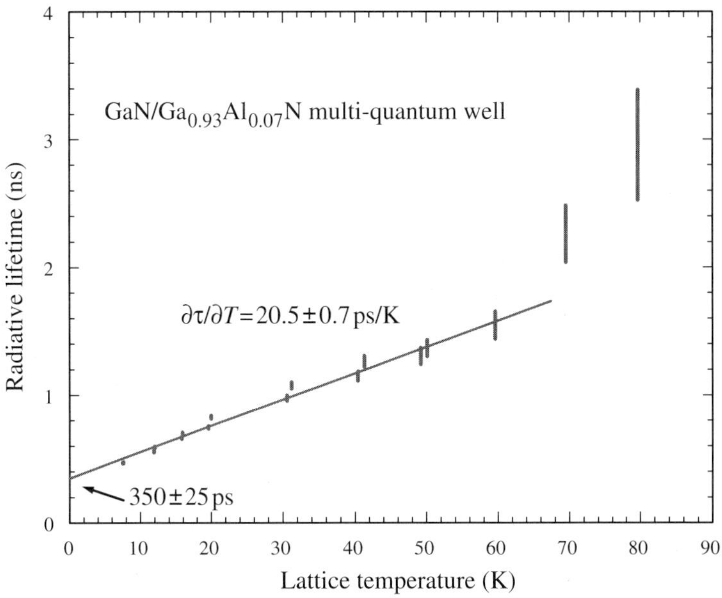

Fig. 8.13 Radiative lifetime of excitons in the GaN/$Ga_{0.93}Al_{0.07}N$ multi-quantum well, deduced from the results of Fig. 8.12. The dashed line shows the linear fitting to the data. Error bars show the total inaccuracy due to both measurements of the decay time and of the time-integrated PL intensity.

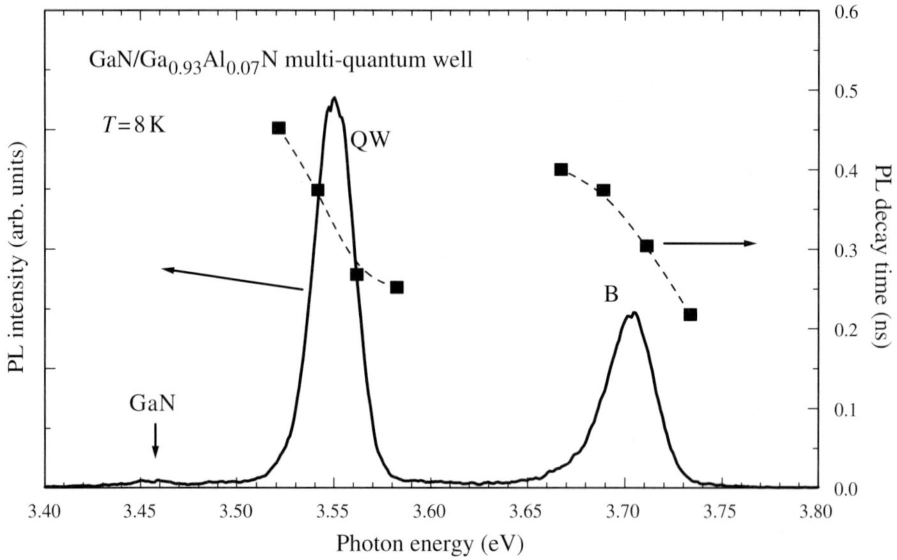

Fig. 8.14 Spectral dependence of the decay-times in the quantum well and barrier layers energy regions.

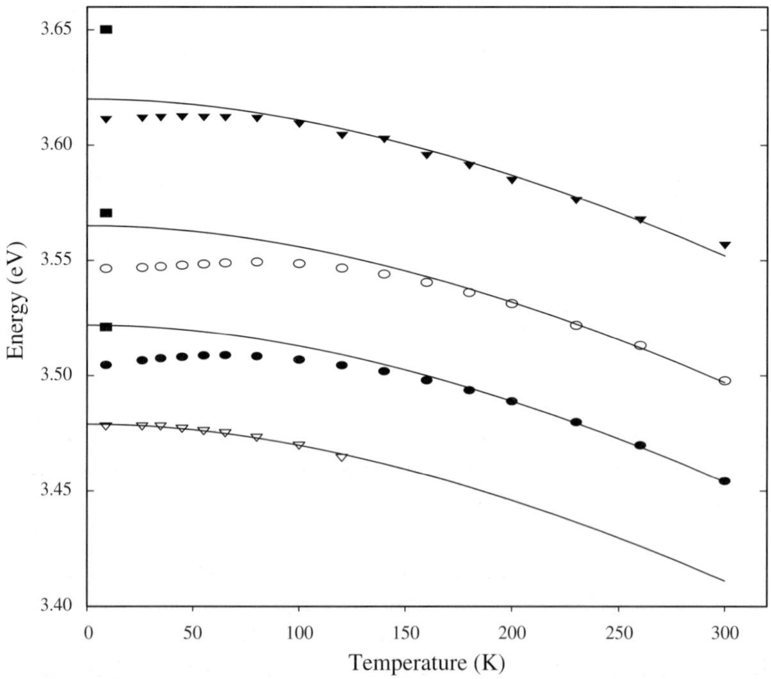

Fig. 8.15 Temperature dependence of the PL energies of the multiple quantum well corresponding to sample middle in Fig. 8.1. The closed squares are free exciton energies deduced from reflectivity.

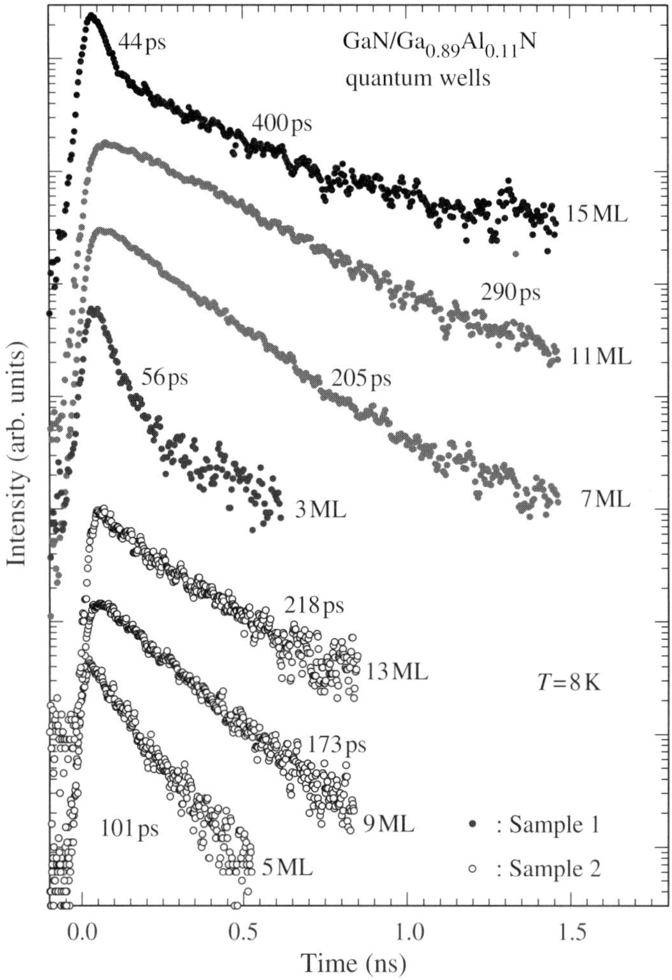

Fig. 8.16 PL intensity decays for the two MQWs top and middle in Fig. 8.1.

electric field in the barrier layers in the one hand and by the existence of a adjacent wider well on the other hand. Then the non-radiative recombination time has two distinct contributions:

$$\frac{1}{\tau_{\text{nr}}} = \frac{1}{\tau_{\text{esc}}} + \frac{1}{\tau_{\text{others}}}.$$

To demonstrate the existence of this effect, we have measured the recombination dynamics in a series of (4, 8, 12, 16) samples with 17% aluminum composition in the barrier layers. The thickness of the barrier layer was changed between 5 nm and 50 nm. The decay time measured at 10 K for the 4 MLs thin quantum

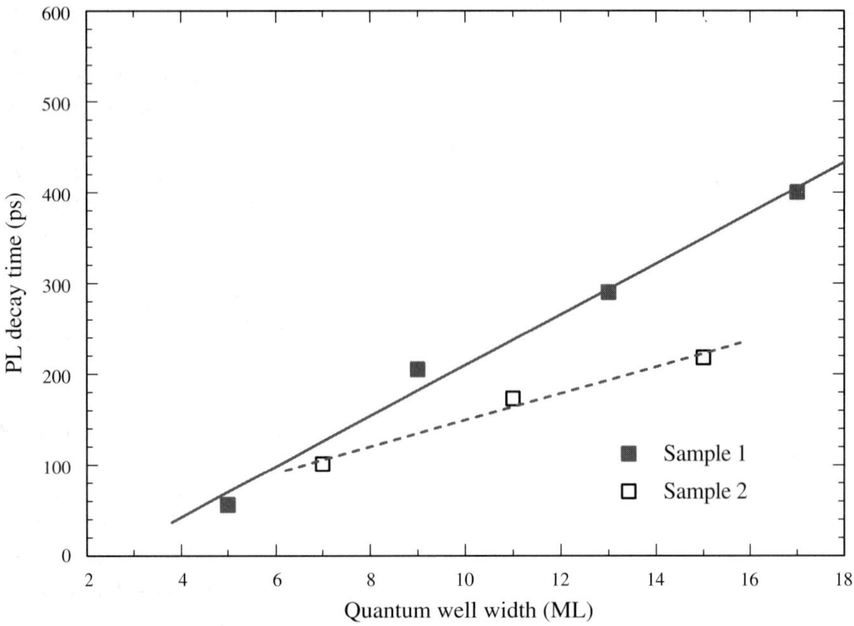

Fig. 8.17 Plot of the PL decay times versus real well widths, i.e. the nominal widths plus 2 monolayers, for samples top and middle in Fig. 8.1.

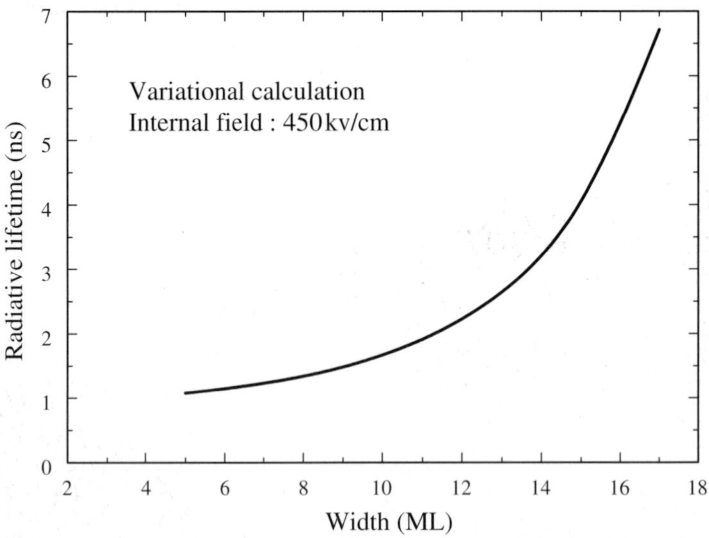

Fig. 8.18 Theoretical calculation of radiative lifetimes for localized excitons in GaN QWs, with longitudinal electric field of 750 kV/cm in the GaN layer.

Plates

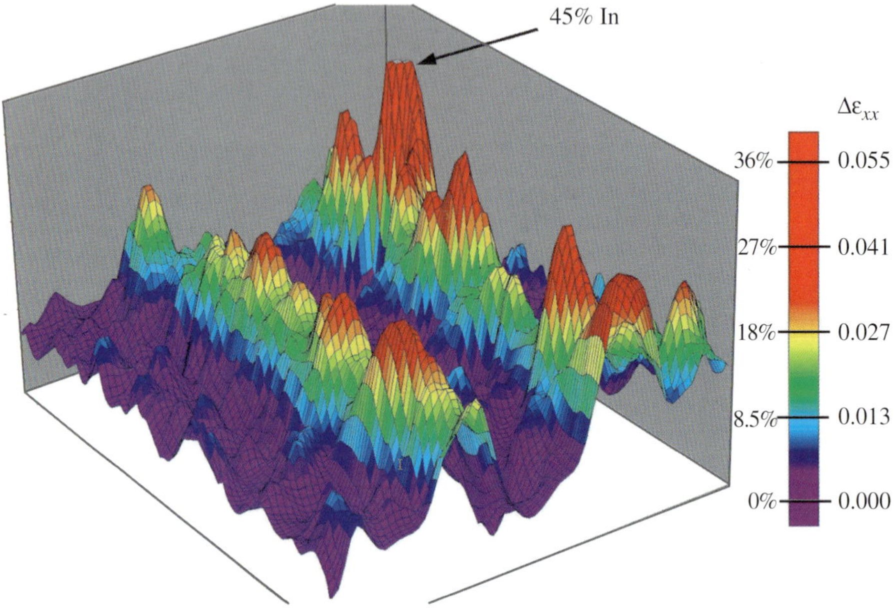

Plate I (Chapter 7) Strain and related composition profiles inside the area of Fig. 7.28, the high indium content areas extend to 2 nm height and 3 nm width. The peak In concentration is 45%.

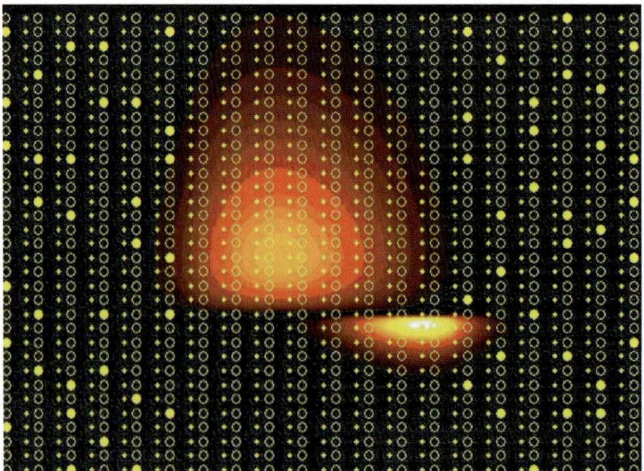

Plate II (Chapter 8) In-plane extension of the densities of probability for electron and hole in the case of a 3 nm wide GaN-$Ga_{0.83}Al_{0.17}N$ QW. Aluminium atoms in GaAlN layers are represented using large yellow spots to illustrate the alloy disorder. Note that the in-plane extension of the hole wave function is smaller than the average Al-Al distancy. The alloy disorder is responsible of the interface roughness experienced by the exciton.

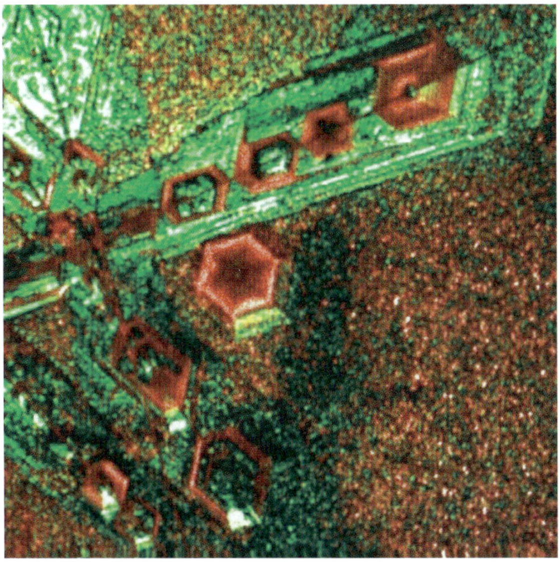

Plate III (Chapter 9) A colourized micrograph of an InGaN epilayer, obtained using the confocal microscope SYCLOPS at the Daresbury Laboratories. The area represented is 500 micron square. Two images are superposed: the green image was obtained using reflected laser light, while the red image shows photoluminescence intensity variations. Note the 'large' hexagonal crystallites and the speckled background.

Plate IV (Chapter 13) Photograph of an ELOG-based blue LED under current injection. There is only a slight modulation of intensity between almost defect-free and highly defective stripes [73].

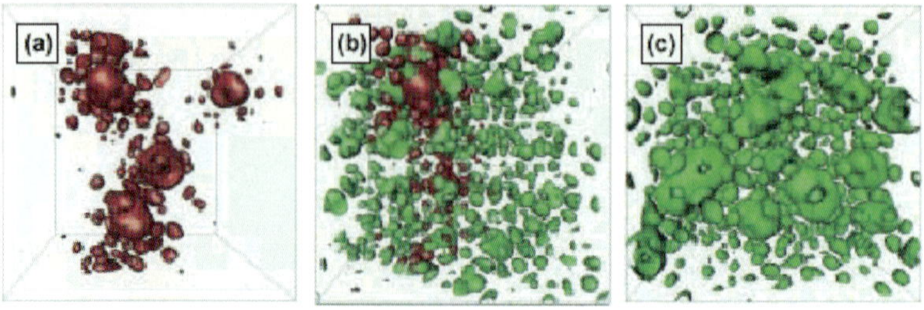

Plate V (Chapter 15) Isosurfaces of the conduction wave functions for (a) $x << x_c$ GaAsN showing nitrogen-localized cluster states within the gap (red), (b) $x \sim x_c$ showing the 'amalgamated state' of localized (red) and the delocalized (Bloch-like) states (green), and (c) $x >> x_c$ extended states (green).

wells increase with barrier width as shown in [31]. For a MLs quantum well, the calculation of the excitonic oscillator strength reveals only some 15% variations for this series of sample. Therefore the radiative decay time is almost constant. Second, the samples have been grown in the same MBE chamber during a campaign of experiments. There are no reasons to deal with different structural qualities, and numbers of photoluminescence killer defects. This indicates that the variations of the decay time which increases when the barrier width increases, i.e. when the barrier layer become less transparent to the carriers is in relation with the variations of the escape time, as expected. Computation of the escape time is difficult, since many mechanisms can contribute to it, impurity-assisted, inter-well resonance via LO phonon emission, which are not easy to estimate numerically at the time this review article is written [31–36].

8.3 Conclusion

In conclusion, wurtzite nitride based low-dimensional semiconductors offer a fascinating field of research for both fundamental physics and for devices designers. In contrast with c-GaN quantum wells, the breaking of the macroscopic parity has a strong impact on the strength and on the mechanisms that rule the light-matter interaction in these materials. The carrier localization at different sides of the well may be favored to specific interface regions, due to the intrinsic chemical barrier disorder. The low-temperature recombination dynamics is ruled by localized exciton, while free excitons may dominate the excitonic recombination dynamics in the high temperature range. Most of the optical properties depend on electric fields. Their value is connected to growth details, doping, growth methods, strain fields, etc. Therefore the determination of their values is a crucial issue for understanding the full set of physical properties of nitride-based heterostructures. In fact, contrary to what one might believe, little has been done so far.

Acknowledgements

Most of this chapter is the result of work performed at the Groupe d'Etude des Semiconducteurs in Montpellier, and at the Centre de Recherche sur l'Heteroepitaxie in Valbonne. I wish to thank (in alphabetical order), Jacques Allegre, Pierre Bigenwald, Mathieu Gallart, Nicolas Grandjean, Andreas Hangleiter, Alexey Kavokin, Pierre Lefebvre, Mathieu Leroux, Guillaume Malpuech, Jean Massies, Aurelien Morel, Hadis Morkoc, Tomasz Ochalski, Thierry Taliercio, Marian Zamfirescu, and many others that I hope will not feel upset if they do not see their name in the list above. This work has been supported by the EU RTN 'CLERMONT' program, contract No. HPRN-CT-1999-00132 and by the French Ministry of Education, Research and Technology within the 'BOQUANI' and 'NANILUB' research programs.

References

1. Leroux, M. and Gil, B. (1999). In *Properties, processing and applications of GaN and related semiconductors*, p. 117. EMIS Data review series, (eds J.H. Edgar, S. Strite, I. Akasaki, H. Amano, and C. Wetzel. Published by INSPEC, London, UK.
2. Alemu, A., Gil, Julier, M., and Nakamura, S. (1998). *Phys. Rev. B*, **57**, 3761.
3. Kavokin, A. and Gil, B. (1999). *Mater. Sci. Eng. B* **59**, 261.
4. Gil, B. (1999). *Semiconductors and semi-metals*, (eds J.I. Pankove, and T.D. Moutsakas), Wiley, New York, vol. 57, Chapter 6, p. 209.
5. Martinez-Guerrero, E., Adelmann, C., Chabuel, F., Simon, J., Pelekanos, N. T., Mula, G., *et al.* (2000). *Appl. Phys. Lett.* **77** (16), 809.
6. Simon, J., Martinez-Guerrero, E., Adelmann, C., Mula, G., Daudin, B., Feuillet, G., *et al.* (2001). *Physica. Status Solidi* (b) **224,** 13.
7. Bernardini, F., Fiorentini, V., and Vanderbilt, D. (1997). *Phys. Rev. B* **56**, R10026.
8. Bechstedt, F., Grossner, U., and Furthmülller, J. (2000). *Phys. Rev. B* **62**, 8003.
9. Hangleiter, A., Im, J. S., Kollmer, H., Heppel, S., Off, J., and Scholz, F. (1998). *MRS Internet J. Nitride Semicond. Res.* **3**, 15.
10. Im, J. S., Kollmer, H., Off, J., Sohmer, A., Scholz, F., and Hangleiter, A. (1998). *Phys. Rev. B* **57**, R9435.
11. Leroux, M., Grandjean, N., Laügt, M., Massies, J., Gil, B., Lefebvre, P., and Bigenwald, P. (1998). *Phys. Rev. B* **58**, R13371.
12. Leroux, M., Grandjean, N., Massies, J., Gil, B., Lefebvre, P., and Bigenwald, P. (1999). *Phys. Rev. B* **60**, 1496.
13. Widmann, F., Simon, J., Daudin, B., Feuillet, G., Rouvière, J. L., Pelekanos, N. T., and Fishman, G. (1998). *Phys. Rev. B.* **58**, R15989.
14. Grandjean, N., Damilano, B., Dalmasso, S., Leroux, M., Laügt, M., and Massies, J. (1999). *J. Appl. Phys.* **86**, 3714.
15. Grandjean, N., Massies, J., and Leroux, M. (1999). *Appl. Phys. Lett.* **74**, 2361.
16. Takeuchi, T., Sota, S., Katsuragawa, M., Komori, M., Takeuchi, H., Amano, H., and Akasaki, I. (1997). *Jpn. J. Appl. Phys.*, Part 2, No. 4A, **36**, L382.
17. Filippetti, A., Fiorentini, V., Cappellini, G., Bosin, A. (1999). *Phys. Rev. B* **59**, 8026.
18. Gandjean, N., Damilano, B., Massies, J., Neu, G., Teissere, M., Grzegory, I., *et al.* (2000). *J. Appl. Phys.* **88**, 183.
19. Gil, B., Lefebvre, P., Allegre, J., Mathieu, H., Grandjean, N., Leroux, M., *et al.* (1999). *Phys. Rev. B* **59**, 10246.
20. Bigenwald, P., Lefebvre, T., Bretagnon, and Gil, B. (1999). *Phys. Stat. Sol.* **216**, 371.
21. Zamfirescu, M., Gil, B., Grandjean, N., Malpuech, G., Kavokin, A., Bigenwald, P., and Massies, J. (2001). *Phys. Rev. B* **64**, 121304R.
22. Fiorentini, V., Bernardini, F., Della Sala, F., Di Carlo, A., and Lugli, P. (1999). *Phys. Rev. B* **60**, 8849.
23. Bigenwald, P., Kavokin, A., Gil, B., and Lefebvre, P. (2000). *Phys. Rev. B* **61**, 15621.
24. Bigenwald, P., Kavokin, A., Gil, B., and Lefebvre, P. (2001). *Phys. Rev. B* **63**, 035315.
25. Gallart, M., Morel, A., Taliercio, T., Lefebvre, P., Gil, B., Allègre, J., *et al.* (2000). *Phys. Stat. Sol.* (a) **180**, 127.
26. Lefebvre, P., Allègre, J., Gil, B., Mathieu, H., Bigenwald, P., Grandjean, N., *et al.* (1999). *Phys. Rev. B* **59**, 15363, and references cited therein.
27. Colocci, M., Gurioli, M., and Martinez-Pastor, J. (1993). *J. Physique* IV (France). Colloque C5, suppl. *J. Phys.* (France) II, vol. 3, p.3.
28. Andreani, L. C., Tassone, F., and Bassani, F. (1991). *Solid State Commun.* **77**, 641.
29. Hoffmann, A. (1997). *Mater. Sci. Eng. B* **43**, 185.

30. Lefebvre, P., Allegre, J., Gil, B., Kavokin, A., Mathieu, H., Morkoc, H., *et al.* (1998). *Phys. Rev. B* **57**, R9447.
31. Lefebvre, P., Gallart, M., Taliercio, T., Gil, B., Allègre, J., Mathieu, H., *et al.* (1999). *Phys. Stat. Sol.* **216**, 361.
32. Reale, A., di Carlo, A., Lugli, P., and Kavokin, A. (2001). *Phys. Stat. Sol.* (a), **183,** 127.
33. Sanchez-Rojas, J. L., Garrido, J. A., and Muñoz, E. (2000). *Phys. Rev. B* **61**, 2773.
34. Ambacher, O., Foutzm, B., Smart, J., Shealy, J. R., Weimann, N. G., Chu, K., *et al.* (2000). *J. Appl. Phys.* **87**, 334.
35. Cingolani, R., Botchkarev, A., Tang, H., Morkoç, H., Traetta, G., Coli, G., Lomascolo, M., DiCarlo, A., Della Sella, P., Lugli, F. (2000). *Phys. Rev. B* **61**, 2711.
36. DiCarlo, A., Reale, A., Lugli, P., Traetta, G., Lomascolo, M., Passaseo, A., Cingolani, R., Bonfiglio, A., Berti, M., Napolitani, E., Natali, M., Sinha, S.K., Drigo, A.V., Vinatierri, A., and Colocci, M. (2001). *Phys. Rev. B* **63**, 235305.

9 Indium gallium nitride (InGaN)—optical properties and technological implications

Kevin P. O'Donnell

The successful introduction of indium into gallium nitride using epitaxial growth techniques has lead to the recent introduction of the InGaN-based blue laser diode and of light-emitting diodes that span the visible spectrum. However, it is believed that indium nitride is *insoluble* in gallium nitride at the growth temperatures used in fabricating such devices, with the result that a vigorous debate has started on the state of aggregation of the light-emitting material on the nanoscale. While most research has focused upon materials with modest indium nitride fractions, it is at high and low concentrations that we observe the most interesting and puzzling optical effects.

This contribution reviews optical properties of InN–GaN solid solutions with a view to relating these properties to the local atomic structure of the material. At very low levels of indium, the luminescence efficiency increases dramatically. Internal scaling laws connect the optical absorption and emission spectra of InGaN epilayers, quantum wells and quantum boxes. In epilayers, where atomic concentrations can be reliably measured, the optical energies are found to depend upon the ratio of InN to GaN fractions. The mean atomic spacings have also been determined as a function of composition.

Difficulties of interpretation begin when we try to account for the observations: residual lattice strain, alloy decomposition, spontaneous quantum dot formation and internal electric fields: all of these seem to play some role in determining the optical properties of any given sample. A unified view of the material's essential nature remains elusive.

9.1 Introduction

Indium gallium nitride has recently become one of the most important materials in the world of semiconductors. Although, at the moment, it falls well behind silicon and gallium arsenide in the commercial value of its products, it easily outstrips the established materials in terms of technological promise. The challenge of fully utilizing nitride semiconductors has only recently presented itself; the potential rewards of our succeeding in this endeavour are very great.

InGaN layers, about ten atomic monolayers (ML) thick, form the active regions of a range of commercial light-emitting devices: blue, green, or amber light-emitting diodes (LEDs); 'white' diodes; and violet lasers. These devices have potential or actual application to display technologies, reprographics and digital data storage and read-out. With the addition of As, InGaN becomes GaInNAs (sometimes nick-named 'guinness', but not the real thing), which introduces applicability of nitrides to fibre-optic telecommunications in the near infrared spectral region. No other material promises such a wide range of optoelectronic functionality.

Despite successful industrial exploitation of nitrides (see Chapter 1, J.-Y. Duboz and J.-C. Garcia), and a considerable amount of attention from a large and growing army of researchers, InGaN retains a certain enigmatic quality: no one is quite sure 'how it works'. Partly, this attitude reflects the shock of success: In the late 1980s, it was apparent to only a handful of people, perhaps only one or two, that the very serious problems attending the production of efficient nitride LEDs could be solved. The story of the early years of struggle has been told very affectingly by one of the major figures in the development of nitride technology, Shuji Nakamura, in his book, *The blue laser diode* [1]. A description of the advances made in our understanding of nitrides during the first decade of commercial device development can be found in ref. [2]. But more to the point, the great surprise of InGaN has been that it works so well: the devices have outstandingly high efficiencies and are robust. They work well in spite of the relatively poor structural quality of the material (which some investigators regard as a sort of richness) and, at least at the moment, comfortably outperform other semiconductors, such as their faded rivals, the II–VI family, and organics.

It is the purpose of this chapter to examine the optical properties of InGaN and to comment upon their influence on the technological implications of the material. The bridge between these two realms of discourse could be founded on a proper understanding of the origin of the optical effects, but, in the opinion of the author, this understanding is absent. However, we do have some clues. I propose to lay out the evidence, in as clear a manner as possible, in order to help future investigators to address the problem.

In the history of technologies, there is sometimes found to be a point at which the scientific impulse appears to be overtaken by the technological. At the point where it becomes easier to buy than to build, scientists (hopefully seeking problems) become outnumbered by engineers (desperately seeking solutions). This is an aspect of the law of diminishing returns. For instance, we all know that photocathodes and phosphors work, without really understanding the chemical processes involved in their manufacture, many of which may be empirically determined and refined on a day-to-day basis. There is some evidence that InGaN has already reached this stage of development. It is nowadays considered naïve to suggest that a scientific investigation, of defect morphology, for example, will lead to technical improvements in laser operation. While we can not guarantee that an increased understanding of InGaN will result in a monetary pay-off, we may be sure of the intellectual benefits of such research. When technology takes over, science can begin to pay attention to its own, more abstract but equally valid, concerns.

9.2 InGaN basics

The III-Nitrides, AlN, GaN and InN, are a family of semiconductors that crystallizes predominantly in the wurtzite (hexagonal) system. This fact distinguishes them from better known semiconductors, like silicon, which are cubic materials, and leads to some distinguishing physical properties, such as an increased tendency to exhibit piezo- and pyroelectricity. Nitrides are partly ionic in bonding and have high melting points, which allows the growth only of relatively small crystals (see Chapter 3, I. Grzegory and S. Porowski). Most commercial production is done through epitaxial growth on lattice-mismatched substrates, such as silicon carbide and sapphire (Al_2O_3) (see Chapter 1, J.-Y. Duboz and J.-C. Garcia).

The wurtzite III-Nitrides have direct optical bandgaps of 6.2 eV (AlN), 3.5 eV (GaN) and 1.9 eV (InN), respectively [3]. The longitudinal-optical phonon energies and the exciton binding energies are both relatively high, typically greater than the thermal equipartition energy at room temperature (k_B.300 K ~ 0.026 eV, where k_B represents Boltzmann's constant). This means that electron–hole (e–h) pairs, produced by electrical or optical excitation across the band gap, tend to form stable excitons, which can survive the effects of thermal dissociation at relatively high temperatures. In any case we expect excitons to dominate the optical spectra at low temperatures and this is found to be the case in the photoluminescence (PL) spectrum of high-quality GaN, as shown in Fig. 9.1. On the other hand, the exciton spectrum of InN has probably never been seen. Whether this is due to the low structural quality of available samples, or a high impurity content, is not known. PL spectra of InN are more than scarce in the literature: the luminescence band reported in ref. [4] probably originates in the substrate material [5].

In the spirit of Vegard's law and the virtual crystal approximation, we expect the optical properties of InGaN to be intermediate between those of its binary

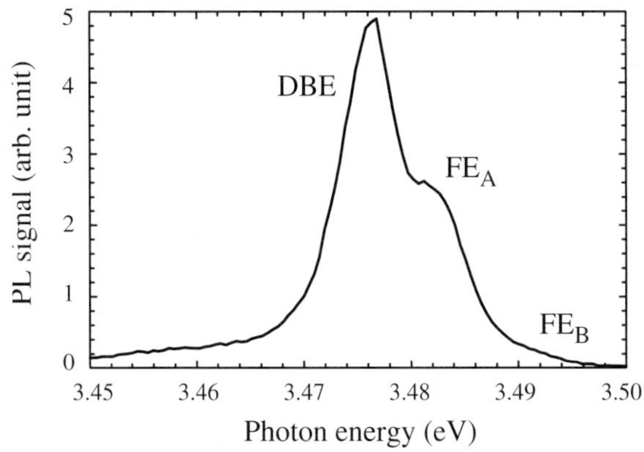

Fig. 9.1 Low-temperature photoluminescence spectrum of a GaN epilayer near the band-edge. Two components of the free exciton, FE_A and FE_B, are dominated by a donor bound exciton, DBE.

constituents, GaN and InN. Formally, InGaN is a solid solution of InN and GaN:

$$In_xGa_{1-x}N = x\text{InN} + (1-x)\text{GaN} \quad \text{with } 0<x<1 \tag{9.1}$$

Using epitaxial techniques, it seems to be difficult to grow InGaN alloys with x larger than about 0.5, because In tends to evaporate from the growing surface, unless the temperature is made very low. However, as x increases in the range $0.25<x<0.5$, and the growth temperature decreases towards 600 K, the crystalline quality decreases markedly.

InGaN is usually not grown directly onto a substrate of sapphire or SiC. But then neither is GaN. First, a thin (~50 nm) amorphous layer of GaN is grown, at about 500 K, to act as an accommodation layer or template for the subsequent growth of a thick GaN layer at about 1400 K. (These figures apply to MOCVD growth. MBE uses lower temperatures.) InGaN is usually grown on top of thick layers of GaN or AlGaN. As mentioned previously, most commercial InGaN layers are not more than a few nanometres thick. They are therefore likely to grow pseudomorphically to the under-layer, which means that they adopt the same in-plane lattice constant, $a = a_{\text{GaN}}$, and are consequently in a state of biaxial compression (since $a_{\text{InGaN}} > a_{\text{GaN}}$ when both materials are in a relaxed condition). Thicker, and at least partially relaxed, layers of InGaN have been grown for research purposes.

In the early days of InGaN research, variations in growth conditions might produce macroscopic inhomogeneities of composition or structure. Some samples would feature a population of distinct hexagonal sub-crystals, of various sizes up to about 100 micron in diameter. Figure 9.2 shows a striking example of this. The indium content might also vary from point to point, leading to spatially inhomogeneous PL spectra. Nowadays, although samples are much more uniform, this certainly does not mean that all samples of a given specification are actually the same, or even equal.

9.3 Photoluminescence of InGaN epilayers

Figure 9.3 shows a typical PL spectrum from a thick InGaN/GaN bilayer. Very similar spectra are recorded from InGaN quantum wells, sandwiched between layers of material of higher bandgap, usually GaN, and from self-organized InGaN quantum dots, whether 'bare' (uncapped) or immersed in GaN or InGaN.

The spectrum shown here is relatively clean. Some samples emit an additional yellow luminescence band, which originates in the GaN buffer layer.

9.3.1 General features of the InGaN PL band

Rather little systematic work has been done on analysing the profile of the InGaN PL band [6,7]. The shape is asymmetric: it rises (on the high-energy side) more rapidly than it falls. The rise and fall are both approximately exponential

Fig. 9.2 A colourized micrograph of an InGaN epilayer, obtained using the confocal microscope SYCLOPS at the Daresbury Laboratories. The area represented is 500 micron square. Two images are superposed: the green image was obtained using reflected laser light, while the red image shows photoluminescence intensity variations. Note the 'large' hexagonal crystallites and the speckled background. (See also color Plate III.)

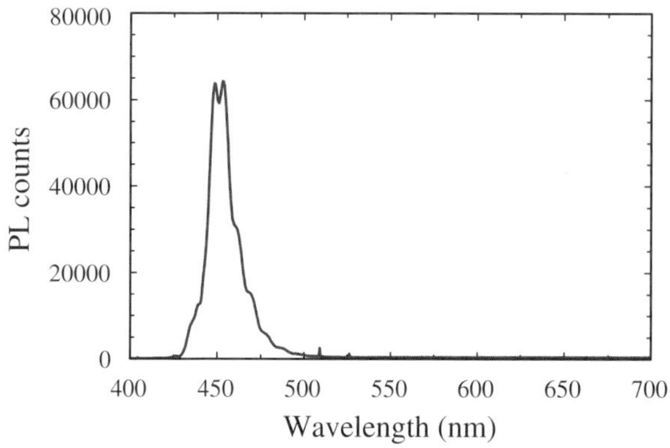

Fig. 9.3 PL spectrum of InGaN/GaN bilayer. The regularly spaced 'fine structure' is due to thin-film interference in the layer. Note the absence of 'yellow-band' emission near 570 nm.

in energy. The high-energy side of the band may be thermally activated [7]. The half-width (full width at half-maximum height, or FWHM) tends to increase with increasing peak wavelength, which scales with the indium content, as we shall see later.

It is generally acknowledged that the efficiency of InGaN PL tends to a maximum around 450 nm. Certainly, a very great deal of the literature concerns samples that emit near this wavelength. It has been reported that the addition of a rather small indium nitride fraction increases the electroluminescence efficiency of ultraviolet nitride LEDs [8]. This effect may have several contributions: an improvement of material quality through the surfactant action of indium [9]; the increasing luminescence efficiency associated with a deeper quantum well [10]; an increasing density of concentration fluctuations, which can localise excitation, and so on.

Figure 9.4 shows, for one inhomogeneous sample [11], with a monotonic variation of composition along its length, the PL spectra obtained at different points on the specimen.

The spectral intensities first increase, with increasing separation from the GaN band-edge (near 363 nm at room temperature) and then rapidly decrease. Figure 9.5 plots the relative decrease of intensity for InGaN PL across the visible spectrum.

We can illustrate such spectral variation, or emphasize its absence in uniform samples, by producing a PL map to show the spectral peak wavelength as a function of position on the sample. Figure 9.6 shows the PL map of an unusually non-uniform specimen.

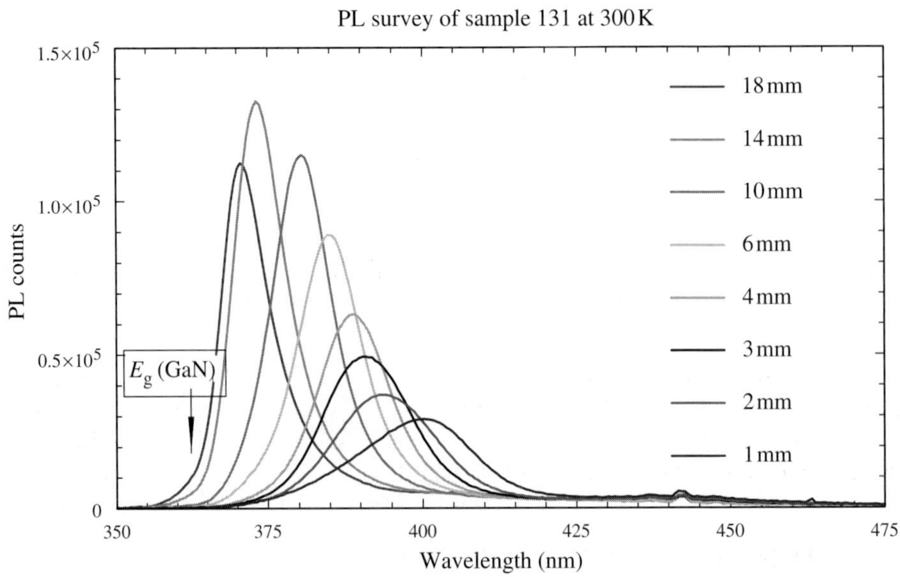

Fig. 9.4 PL survey of an InGaN/GaN bilayer that shows macroscopic spectral inhomogeneity.

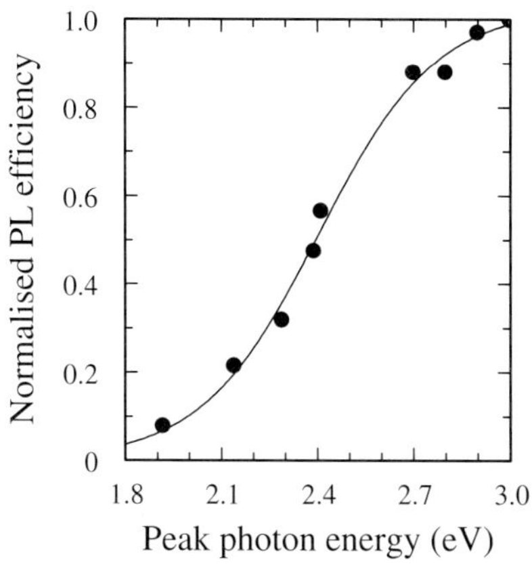

Fig. 9.5 Representative InGaN PL efficiency as a function of spectral peak energy.

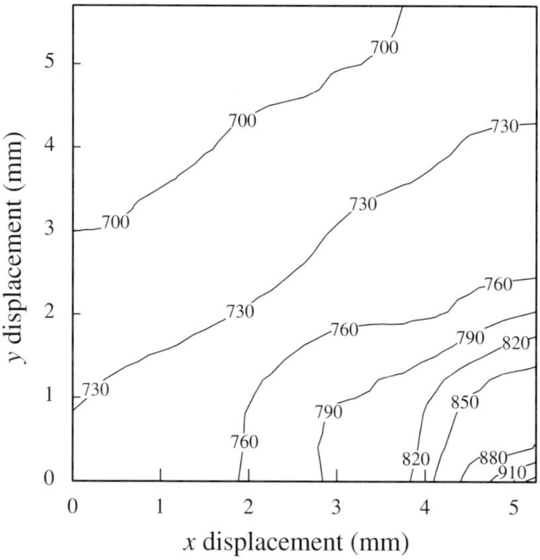

Fig. 9.6 PL map of the most extreme InGaN sample ever grown. The contour lines show peak wavelengths of the InGaN bands.

Somewhat surprisingly, the PL peaks represented on Fig. 9.6 are ALL at lower photon energy than the InN band gap ($E_g \sim 1.9$ eV; $\lambda_g \sim 650$ nm). Representative PL spectra across nearly the entire range of InGaN luminescence, reported to date, are shown in Fig. 9.7.

9.3.2 Dependence of the PL peak energy on *x*, the InN fraction

It is important at this point to say what the PL peak energy of InGaN emission does NOT represent. It is firstly not the same as the alloy bandgap, which in

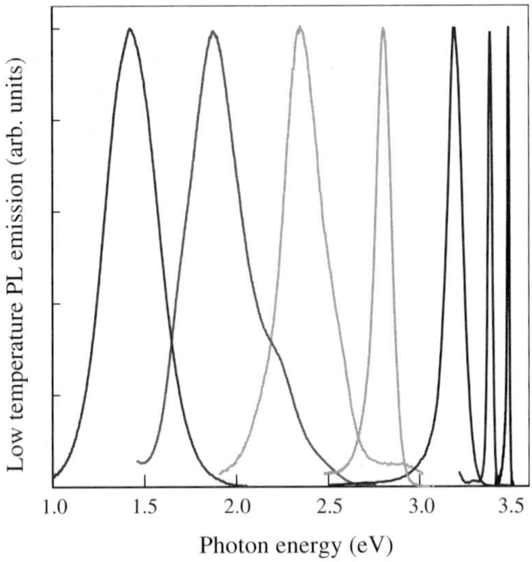

Fig. 9.7 A selection of InGaN PL spectra from a number of samples.

any case is rather more difficult to define than that of a binary compound (see below). Neither is it the 'excitonic gap' of the alloy in any sensible definition.

The clearest definition is a statistical one, which applies to all semiconductors: the PL peak energy is the *modal energy* for electron–hole recombination. The photon energy at which an e–h pair recombines is always likely to be lower than the energy with which the pair is generated. As stated by Stokes, who provided the first comprehensive description of the excitation–fluorescence cycle: 'the tendency is to a lowering' [8]. The energy lowering is usually called the Stokes' shift.

We should be careful not to overemphasize the significance of the modal peak in relation to the rest of the PL band. Detected with sufficiently high sensitivity, the PL of *all* InGaN samples shows *some* intensity of emission as far to the high-energy side as the GaN band-edge, and as far to the low-energy side as our instruments reach.

To summarize, the PL peak energy is an empirical measure of the most likely energy for e–h recombination. Clearly, we might expect this energy to depend in some simple way on the composition of the material. In a binary semiconductor, for example, we expect to see light emission in the vicinity of the band-edge. For alloys, all that is needed for a systematic study is some means of measuring the composition and the PL peak energy of a number of different samples. In the following survey of composition measurement techniques, we shall refer to the InN content, defined in eqn (9.1), simply as x.

9.3.2.1 *X-ray diffraction (XRD) determination of x in* $In_xGa_{1-x}N$

XRD is the most popular technique used by crystal growers to measure the average value of the lattice constants of (possibly strained) layers.

In favourable circumstances, it can also measure the layer thickness. Figure 9.8(a) shows a θ-2θ scan of GaN and InGaN *(00.2)* diffraction peaks from a representative InGaN/GaN bilayer. The lattice constants in the growth direction, c(GaN) and c(InGaN), for any allowed *(00.l)* reflection, are given directly by Bragg's law:

$$c = l^{*}\lambda/(2\sin\theta_{\mathrm{B}}) \tag{9.2}$$

with the Bragg angle, θ_{B}, obtained from the peak of the XRD distribution. For the example shown here, the respective lattice constants obtained by use of eqn (9.2) are c(GaN)=5.1838 Å and c(InGaN)=5.2516 Å. In a similar way, a scan through the asymmetric *(10.5)* reflection, shown in Fig. 9.8(b), provides estimates of the lattice parameters a(GaN)=3.1908 Å and a(InGaN)= 3.207 Å.

In biaxially strained wurtzite structures, such as InGaN/GaN, the hexagonal unit cell distorts. In order to separate the influence of strain and composition on the relative positions of GaN and InGaN diffraction peaks, both parameters of the wurtzite lattice, c(InGaN) and a(InGaN), should be compared to their relaxed values.

We define the ratio of the longitudinal to the transverse strain, $\xi(x)$, as:

$$\xi(x) = -\frac{c(\mathrm{InGaN}) - c_0(x)}{c_0(x)} \cdot \frac{a_0(x)}{a(\mathrm{InGaN}) - a_0(x)}, \tag{9.3}$$

where $c_0(x)$ and $a_0(x)$ are the relaxed lattice constants. In order to determine the composition from the measured lattice constants, we can evaluate both sides of eqn (9.3) separately and seek the value of x at which they coincide. This procedure is equivalent to solving a cubic equation with only one physically meaningful root, in the range $0 < x < 1$. According to Vegard's law, the lattice parameters can be calculated for a given value of x by a linear interpolation between the constants of the binary compounds, GaN and InN. A value for $\xi(x)$ can be interpolated in a similar way from the measured or calculated binary values of Poisson's ratio.

Symmetric XRD reflections measure only the out-of-plane lattice constant, c(InGaN), but are stronger and easier to obtain experimentally than asymmetric reflections. In the absence of further information, it is customary to assume one or other of the following extreme cases:

1. Layers are fully relaxed: hence, $c(\mathrm{InGaN})=c_0$. This assumption is likely to be justified when layers are thick. According to Vegard's law, the *relaxed* lattice constant $c_0(x)$ of the binary mixture of compounds, $\mathrm{In}_x\mathrm{Ga}_{1-x}\mathrm{N}=x\mathrm{InN}:(1-x)\mathrm{GaN}$, scales linearly with x:

$$c_0(x) = c(\mathrm{GaN}) + x(c\mathrm{InN}) - c(\mathrm{GaN}) = (1 - x)c(\mathrm{GaN}) + xc(\mathrm{InN}). \tag{9.4}$$

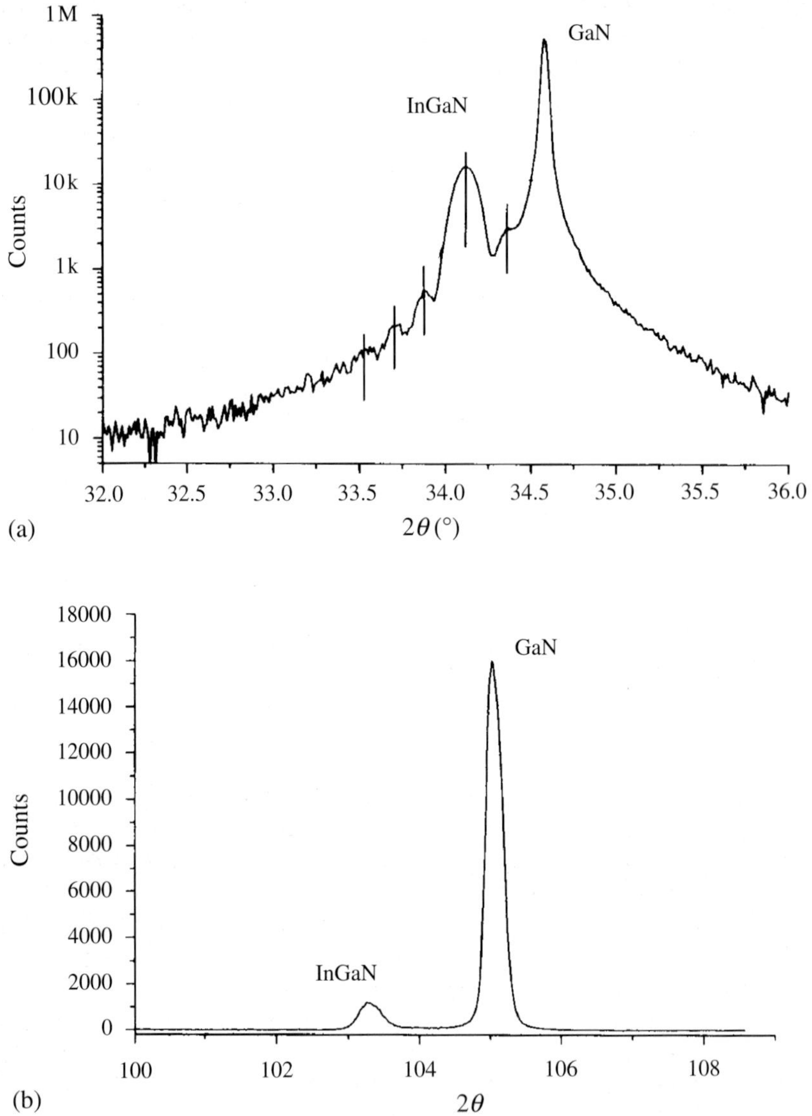

Fig. 9.8 (a) symmetric *(00.2)* and (b) asymmetric *(10.5)* X-ray reflections from an InGaN/GaN bilayer. Note the very different sensitivities. Some authors make use of the stronger, symmetric, reflection only, combined with an assumption about the strain state of the sample, in order to make an estimate of the InGaN composition.

Hence x can be determined immediately from c(InGaN) (assumed to be the same as c_0) if 'relaxed' values of c(GaN) and c(InN) are known. Benchmark values of the relaxed lattice constants are available from measurements on 'bulk' crystals of binary materials.

2. Layers are pseudomorphic: $a(\text{InGaN}) = a(\text{GaN})$. This is often the case for InGaN layers of thickness less than about 75 nm, grown on GaN. With the assumption that the in-plane lattice constants of epilayer and substrate are equal, x can again be estimated, knowing only $c(\text{InGaN})$, by solving eqn (9.3).

The partial relaxation expected in real samples, or in ideal samples of intermediate thickness, is an important issue. If the assumption is made that relaxation is always complete, the indium nitride fraction will be consistently overestimated. Indium nitride fractions tend to be underestimated if layers are assumed to be pseudomorphic.

If $a(\text{InGaN})$ is not measured, we can infer from Fig. 9.8(a) only the fact that the indium nitride fraction falls between 8% and 13%. The range of uncertainty in estimating x in this way obviously increases with increasing InN content. On the other hand, if $a(\text{InGaN})$ is also known, we can readily determine x. The intersection point of the curves,

$$y_1 = \xi(x); \quad y_2 = -\frac{c(\text{InGaN}) - c_0(x)}{c_0(x)} \cdot \frac{a_0(x)}{a(\text{InGaN}) - a_0(x)}$$

gives the required solution. An uncertainty remains in choosing the value of the elastic constants to calculate Poisson's ratio. Widely accepted theoretical estimates are $\xi\,(\text{GaN}) = 0.509$ and $\xi\,(\text{InN}) = 0.821$ [13]. Further purely experimental uncertainties in measuring $c(\text{InGaN})$ and $a(\text{InGaN})$ lead to a surprisingly high estimated error in x. While our best value of $x = 10.0\%$ is obtained by adopting $a(\text{InGaN}) = 3.207\,\text{Å}$, a value of $a(\text{InGaN}) = 3.210\,\text{Å}$, higher by only one part in a thousand, would raise this estimate to 11%. Finally, closely spaced XRD satellites, indicated by vertical bars in Fig. 9.8(a), allow a direct estimate of layer thickness, in this case 45 nm.

From the discussion above, it is evident that the strain state of *each* sample should be determined if a valid estimate of x is sought from XRD measurements. Contrary to expectation, the layer in the example was found to be partially relaxed, although very thin. In principal, the strain state and composition of any layer can be determined from independent measurements of both of its lattice constants, $c(\text{InGaN})$ and $a(\text{InGaN})$, if we know Poisson's ratio for the layer. Uncertainty in the Poisson ratio will be less damaging to accuracy than uncertainties in the lattice constants. However, potentially large uncertainties will arise if the layer is non-uniform, either in terms of its indium content or state of strain. XRD provides a *structural determination of x* on the basis of Vegard's law. A method that provides information of a similar kind, with the added bonus of very high spatial resolution, is Digitally Aided Lattice Imaging (DALI), an extension of high-resolution transmission electron microscopy. Applied to cross-sectional images of InGaN/GaN quantum well samples, DALI compares the positions of atom pairs in the InGaN well with the positions of pairs in the GaN buffer. In other words, it measures lattice displacements that occur due to alloying, on an atomic scale of length.

9.3.2.2 *Chemical determinations of x*

In contrast to techniques which depend upon structural models for their results, chemical analytical methods aim to determine composition by measuring a signal that depends directly on the numbers of atoms of each species that are present in the sample. The chemical techniques most commonly used in the determination of x in $In_xGa_{1-x}N$ are Rutherford backscattering spectrometry (RBS), X-ray fluorescence spectroscopy (either in energy-dispersive (EDX) or wavelength-dispersive (WDX) form) and Extended X-ray Fine Structure (EXAFS) analysis. None of these techniques even looks at a test-tube: the chemical tag refers to the identification of atomic elements in a number of different ways. We will describe the various techniques by use of examples from previously published work.

RBS When a helium ion recoils from a solid target, it does so because there is a finite probability of its meeting a nucleus 'head-on' in the target. This is Rutherford back-scattering. The recoil energy is chemically determined, since the atomic mass of the target atom enters the kinematic equations that describe the collision. A measurement of the number of recoils as a function of energy loss is the basis of an analytical spectrometry of the atoms that are present in a mixed target, such as an alloy.

Figure 9.9 shows a representative RBS spectrum of an InGaN/GaN bilayer, grown on sapphire [14]. It is conventional to compare spectra obtained with an incident beam in the direction of a crystal axis with those obtained when the sample is rotated off-axis by a few degrees. The ratio of the two scattering

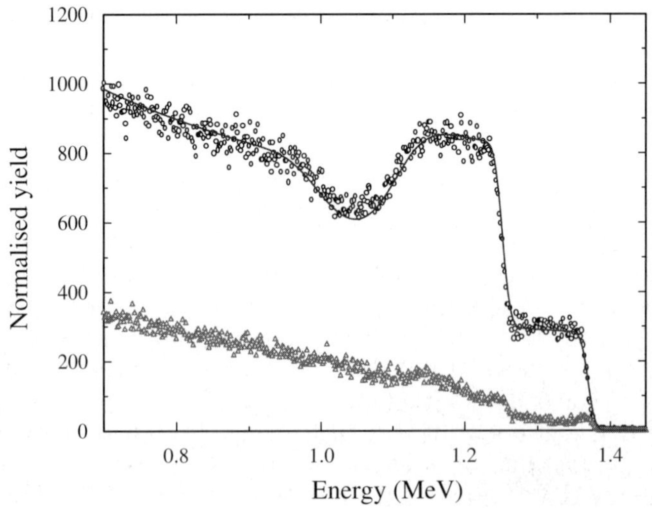

Fig. 9.9 RBS spectrum of an InGaN/GaN bilayer, measured on (o) and off (Δ) axis, with a solid line data fit (—) to a structure model with InGaN composition of $x=0.18$ and a thickness of 250 nm.

intensities provides a measure of crystalline quality, χ_{min}. Analysis of the data is carried out by using a computer code to simulate the energy loss spectrum of a model sample, which is systematically refined until agreement with the experimental spectrum is reached. Accuracy levels of 1% are claimed in good conditions.

EDX and WDX Atoms emit characteristic X-ray fluorescence (XRF) when they are excited by suitably energetic photons or fast particles. In EDX and WDX analysis, the excitation of XRF is carried out by using electrons, with a typical beam energy of 3–30 keV. Such a beam is readily provided by a scanning electron microscope (SEM): elemental images of the sample can therefore be formed to complement the topographical images that are normally available from this instrument. Given suitable calibration standards for different elements, XRF provides a powerful tool to determine the composition of samples, with the advantage of very high spatial resolution.

EDX is achieved by tuning solid-state detectors and is a routine add-on to SEM. It is commonly regarded as at best a semi-quantitative technique. WDX uses X-ray spectrometers and is preferable for quantitative work, because it suffers less interference from background and offers much better discrimination of elements through its superior energy resolution.

Figure 9.10 compares elemental maps of In and Ga from an InGaN sample with an average x of 0.21. These images were obtained by using the WDX facility of a Cameca SX-100 Electron Probe Micro-Analyser (EPMA) [15].

Compositional non-uniformities show up on such a pair of images as *complementary* variations in contrast, with equivalent areas bright in one image and dark in the other. These occur, for this sample, only on a length scale of about one micron.

With careful calibration, WDX measurements of x can match the accuracy attainable by RBS and provide much better spatial resolution.

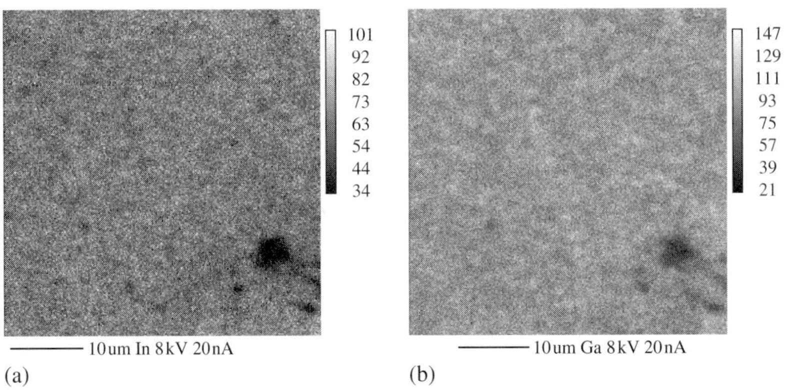

Fig. 9.10 (a) In and (b) Ga elemental maps of InGaN epilayer sample. Note length scale.

EXAFS

High energy X-ray photons are preferentially absorbed by atoms above characteristic energies, the absorption edges, which correspond to ionization of their inner electrons. EXAFS originates in the self-interference of the resultant photoelectrons, as they propagate from a target atom and scatter back from neighbouring atomic shells. As a result of the interference, oscillations appear in the *x*-ray absorption spectrum, at energies above the edge. The chemical identity, the co-ordination number and the separation from the target atom of the surrounding atomic shells determine the period and phase of these oscillations. Usually, EXAFS of a model environment is calculated and refined systematically, shell by shell, in order to match the experimental spectrum or its Fourier Transform. In addition, the data fitting also yields a measure of disorder of the neighbouring atom shells through the well-known Debye–Waller factors. Hence EXAFS determines local structure. In suitable circumstances, it allows us to measure average composition.

Figure 9.11 shows, by way of example, a two-shell data fit to the EXAFS oscillations above the In K-edge of an InGaN sample with $x=0.25$.

EXAFS does not yield its information content easily. There are no clearly characteristic signals that can be directly associated with particular elements. In fact, elements distinguish themselves by characteristic phase shifts of the reflected waves. Such phase shifts are usually determined from theory. However, we shall see later that accurate determinations of x are, in fact, possible with this technique.

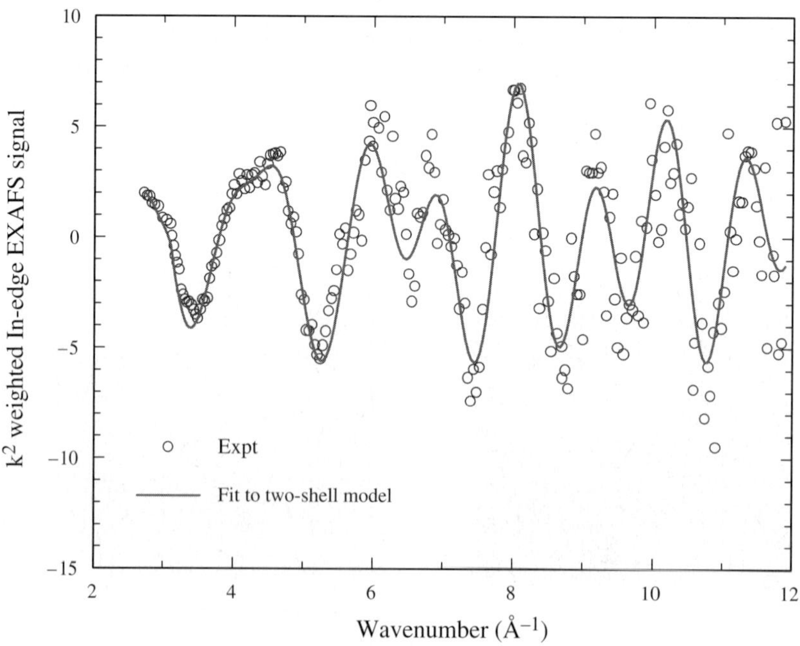

Fig. 9.11 EXAFS dataset and fit.

9.3.2.3 *Summary of experimental results*

When applied to the same sample set, all of the techniques described above are capable of a fairly high level of accuracy. Comparative measurements agree within a fraction of a per cent in favourable circumstances, namely when samples are laterally and vertically uniform and have flat surfaces.

Figure 9.12 shows the dependence of the PL peak energy on the measured InN content for thick InGaN epilayers.

Several features of this plot are worth noting. (1) The intercept on the energy axis lies very close to the band-edge energy of pure GaN (3.42 eV at room temperature). Pure GaN is expected to emit close to its edge, as we saw earlier. (2) The dependence of E_p upon x is linear to a very good approximation in the range $0 < x < 0.4$. Several authors have confirmed this linear behaviour over more restricted ranges of composition, using a variety of techniques [16]. (3) There is no *experimental* evidence for the existence of a bowing parameter in the relation $E_p(x)$.

9.3.3 Relation of PL peak energy to optical absorption spectrum

The optical absorption spectrum of a binary semiconductor ideally measures the joint density of states (JDOS) of the electron and hole states in the conduction and valence bands, respectively. If momentum is conserved, this should produce, in a direct gap semiconductor, a rising parabolic absorption profile, as sketched in Fig. 9.13.

Ideally, there is no absorption at all below the optical bandgap energy, E_g(opt), and there should be a well-defined dependence of the absorption strength on the photon energy in excess of this gap. In reality, a number of processes, related to

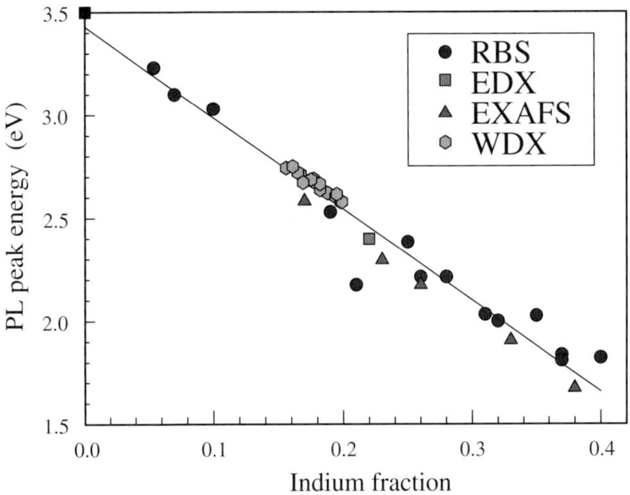

Fig. 9.12 Relationship between indium content (InN fraction) and PL peak energy of epilayers.

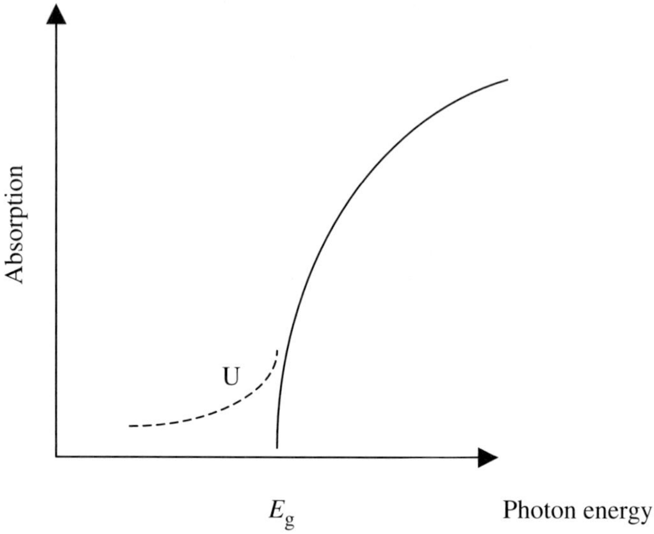

Fig. 9.13 Optical absorption spectrum of a nearly ideal direct-gap semiconductor.

static and thermal disorder, contribute to an exponential tail of absorption, U, usually named after its discoverer, F. Urbach [17], which appears below the gap.

At a given length scale, an alloy can be regarded as a superposition of many different semiconductors, with a statistical distribution of composition and band gap. It is not straightforward to calculate the expected absorption profile in the absence of microscopic statistical information about the distribution. Martin *et al.* [18] avoided this problem by introducing an empirical description of alloy bandgaps, based upon a fitting of the experimental absorption spectrum to the sigmoidal function:

$$A(E) = \frac{A_0}{1 + \exp\left[(E - E_g)/\Delta E\right]}. \tag{9.5}$$

This broadened step function (which is not, however, a thermodynamic function) is quite similar in shape to the Urbach-broadened absorption edge shown in Fig. 9.13. In fact the broadening parameter ΔE probably generalises the idea of the Urbach tail to include a contribution from alloy fluctuations (to the uncertainty of the bandgap position). Of course, we may regard such fluctuations as just another aspect of disorder. The parameter E_g, which emerges from the data fitting, is *formally identified* with the band gap energy of the alloy.

Figure 9.14 shows a data fit to the absorption spectrum of an epilayer, together with the corresponding PL emission spectrum.

Rather good fits are obtained when we apply eqn (9.5) to the absorption profiles of thick InGaN epilayers. These can be measured with quite good

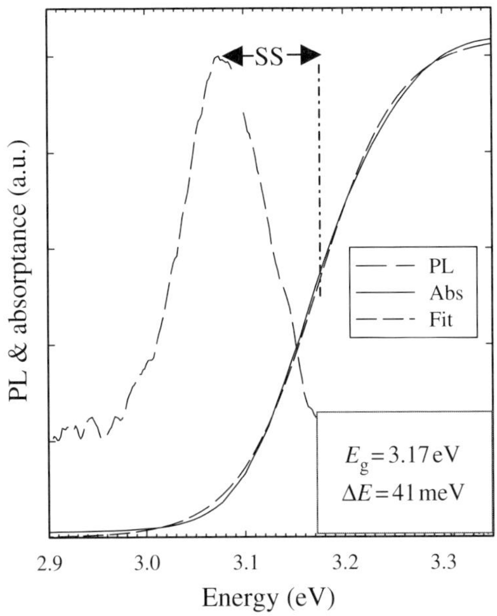

Fig. 9.14 The definition of the Stokes' shift for InGaN epilayers.

spatial resolution on inhomogeneous samples by appropriate masking of the sample surface. Continuing with the formality, we now define the Stokes' shift as the difference in energy between the alloy band gap, obtained by fitting a sigmoidal profile to the absorption spectrum, and the peak of the emission spectrum:

$$\text{ss} = E_g - E_p. \tag{9.6}$$

We are just about to discover something remarkable. If we carry out combined measurements of PL peak energy (E_p) and energy bandgap (E_g) for a number of different samples, with a range of indium content, a plot of E_g against E_p looks like Fig. 9.15.

The Stokes' shift, defined above, is therefore also a linear function of E_p (and E_g), which would appear to vanish, as expected, when these quantities become equal, within experimental uncertainty, *at the band-edge of GaN!*

This remarkable optical behaviour is referred to as universality: The optical spectra of *any* InGaN epilayer can be deduced from a single measurement [18]. In the previous section, we have seen that the peak emission energy depends solely on the indium content of the layer. Now we can go further and say that the complete optical behaviour of $In_xGa_{1-x}N$ depends upon x, in a remarkably simple fashion. A set of linear relations connects the optical

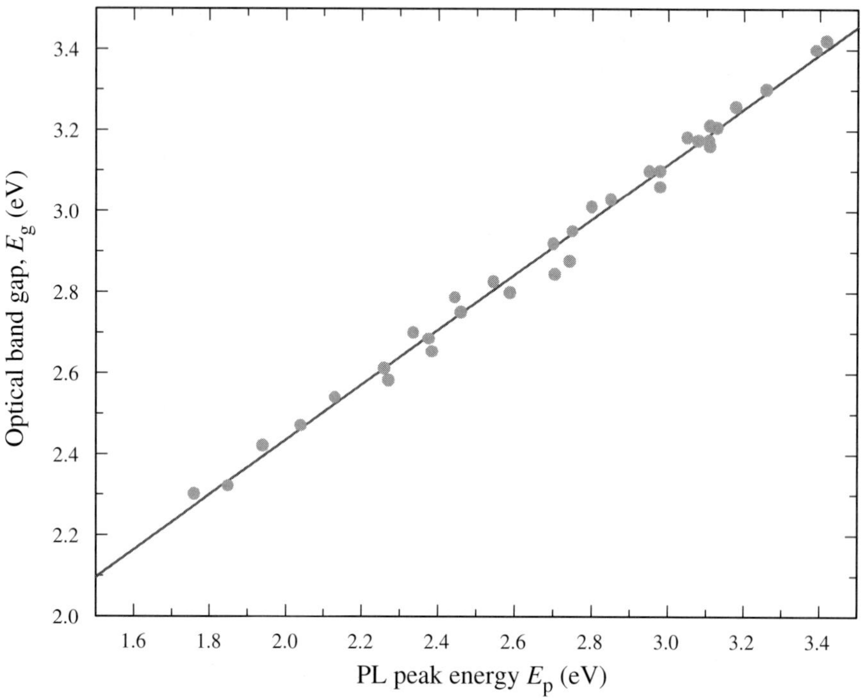

Fig. 9.15 Linear relationship between optical energies of InGaN epilayers.

energies, the Stokes' shift and the composition. At the time of writing, the best estimates of the numerical constants in this set of relations (with error estimates in parentheses) are:

$$\begin{aligned} E_p &= 3.41(3) - 4.36(16)x \\ E_g &= 3.44(12) - 3.0(3)x \\ \mathrm{SS} &= 0.03(13) + 1.36(35)x. \end{aligned} \tag{9.7}$$

9.4 Origin of luminescence in InGaN devices

A lot of the work done on InGaN to date has concerned itself, as described above, with relatively thick layers (>100 nm) of material, grown on an intermediate GaN buffer in order to avoid the deleterious effects of gross lattice mismatch with an underlying sapphire substrate. The active regions in light-emitting devices consist of much thinner layers, typically only 3 nm thick. The equivalence of thick and thin layers of the 'same' material must not be taken for granted.

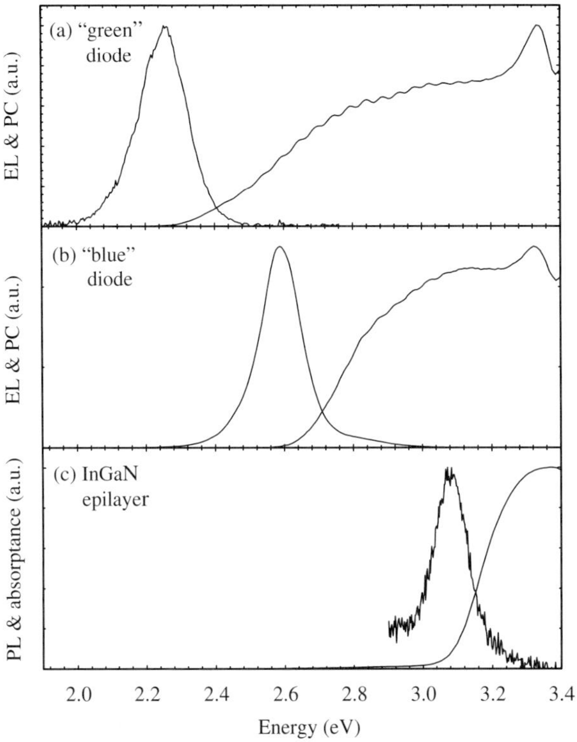

Fig. 9.16 Comparing the optical spectra of LEDs (EL and PC) with those of an epilayer (PL and OA).

There is some evidence, however, that the *optical properties* of InGaN quantum wells are *indistinguishable* from those of thick epilayers. While it is usually not possible to perform a direct optical absorption (OA) experiments on a single quantum well, it is possible to make an equivalent measurement on an active device, which contains a single quantum well, by using the photocurrent (PC) spectrum [18]. Photoluminescence excitation (PLE) spectroscopy provides another means to obtain the same information [20].

We compare the PC and electroluminescence (EL) spectra of green and blue Nichia diodes in Fig. 9.16. It is practically impossible to distinguish such spectra from the optical absorption/photoluminescence spectra obtained for thick epilayers.

In Fig. 9.17, we plot the Stokes' shift against luminescence peak energies, measured by the different respective techniques for epilayer and diode samples.

The similarity of the optical spectra suggests a common origin for optical effects in epilayers and the relatively thick quantum wells used in diodes. When a quantum well is sufficiently thin, however, we expect to observe blue shifts due to quantum confinement effects. Such shifts are indeed observed when the well thickness decreases below about 3 nm [21].

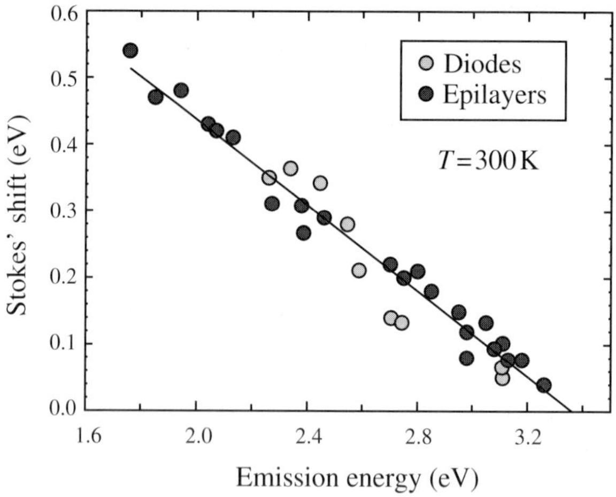

Fig. 9.17 Comparison of optical energies for InGaN epilayers and diodes.

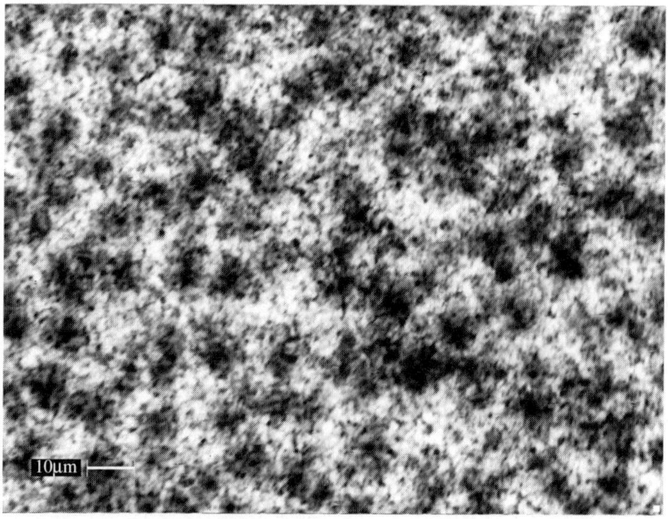

Fig. 9.18 Cathodoluminescence micrograph of a single InGaN QW sample. Note the 10 micron scale bar.

Fluorescent micrographs of epilayers and quantum well material are also similar in showing a mottled or spotty appearance, reproduced in Fig. 9.18. The width of the smallest of the objects that make up the granular structure in the bright areas of this micrograph is less than 500 nm, comparable with the wavelength of the light emitted by the layer. Recently, the spectra of individual grains has been measured in the confocal microscope and found to differ little from the spectrum of a large area of sample [20]. This observation strongly suggests that emission originates in objects that are much smaller than 500 nm wide and, necessarily, less high than the quantum well width. The composition of these elementary lumophores is unknown, but they are very likely to be indium-rich. It has been suggested that they are pure InN quantum dots [19].

9.5 Prospects

At the time of writing (mid-2001) the major commercial players in nitrides technology are battling each other not only in the marketplace but also through the law courts. In the event of a satisfactory outcome of the many legal disputes, the prospects for nitrides are, as stated in the introduction, extremely good. Perhaps the most exciting possibility is the complete replacement of the tungsten lamp by highly efficient, long-lived diode arrays.

One 'failure' of nitrides at the present time is the limited wavelength range of the commercial devices. Although amber (~590 nm) and red (~620 nm) light-emitting diodes have been produced, they are less efficient than the competition, based on phosphides. Again, the range of lasing wavelengths is rather limited: only UV and blue lasers have been produced so far. It may be that the observed loss of efficiency with increasing wavelength is an intrinsic property of the materials, caused by the lowering of recombination probability of carriers separated by internal electric fields. On the other hand, perhaps extrinsic effects play a role, so that the continuing reduction of defect densities will allow further progress to be made. After all, InGaN luminescence bands with peak wavelengths close to 1000 nm have been observed, albeit in somewhat exceptional material [22].

Whatever may happen in future, the last ten years of nitride research has brought so many surprises that we can surely look forward to the prospect of one or two more.

References

1. Nakamura, S. and Fasol, G. (2000). *The blue laser diode*, 2nd edn. Springer-Verlag, Berlin.
2. Gil, B. (ed) (1998). *Group III nitride semiconductor compounds, physics and applications.* Series on Semiconductor Science and Technology 6. Oxford Science Publications, Oxford University Press.
3. Edgar, J. H. (ed.) (2000). EMIS data book. 116, IEE Publishing, London.
4. Inushima, T. *et al.* (1995). *Inst. Phys. Conf. Ser.* **142**, 971.
5. Similar spectra, obtained recently at Strathclyde, have been spatially located at scratches by microspectroscopic analysis.
6. Zolina, K.G., Kudryashov, V.E., Turkin, A.N., Yunovich, A.E., and Nakamura, S. (1996). *MRS Internet J. Nitride Semicond. Res.* **1**, 11.
7. O'Donnell, K. P., Breitkopf, T., Kalt, H., Van der Stricht, W., Moerman, I., Demeester, P. and Middleton, P.G. (1997). *Appl. Phys. Lett.* **70**, 1843.
8. Mukai, T., Morita, D. and Nakamura, S. (1998). *J. Cryst. Growth* **189/190**, 778.

9. Shu, C. K., Ou, J., Lin, H. C., Chen, W. K. and Lee, M. C. (1998). *Appl. Phys. Lett.* **73**, 641.
10. Schneider, H. and Vonklitzing, K. (1998). *Phys. Rev. B* **38**, 6160.
11. Van der Stricht, W. (1999). PhD thesis, University of Ghent, unpublished.
12. Stokes, G.G. (1887). *On light.* Macmillan, London & New York.
13. Wright, A.F. (1997). *J. Appl. Phys.* **82**, 2833.
14. Vantomme, A., Wu, M. F., Hogg, S., Langouche, G., Jacobs, K., Moerman, I., *et al.* (2000). *MRS Internet J. Nitride Semicond. Res*. **5S1**, W11.38.
15. Micrographs by courtesy of M. Outrequin, Cameca S. A., Courbevoie, France.
16. See O'Donnell, K.P. (2001). *J. Phys. Cond. Matt.* (to be published).
17. Pankove, J. I. (1976). *Optical processes in semiconductors*. Dover, New York.
18. Martin, R. W., Middleton, P. G., O'Donnell, K. P. and Van der Stricht, W. (1999). *Appl. Phys. Lett.* **74**, 263.
19. O'Donnell, K.P., Martin, R. W. and Middleton, P.G. (1999). *Phys. Rev. Lett.* **82**, 237.
20. White, M. E., O'Donnell, K. P., Martin, R.W., Pereira, S., Deatcher, C.J., and Watson, I.M. (2001). Proceedings of E-MRS Spring Meeting, Strasbourg.
21. Keller, S., Keller, B.P., Kapolnek, D., Abare, A.C., Masui, H., Coldren, L.A. Mishra, U.K., and DenBarrs, S.P. (1996). *Appl. Phys. Lett.* **68**, 3147.
22. O'Donnell, K. P., Martin, R.W., Pereira, S., Bangura, A., White, M.E., Van der Stricht, W., and Jacobs, K. (1999). *Phys. Stat. Sol.* (b), **216**, 141.

10 The optical properties of InGaN-based quantum wells and quantum dots

Yoichi Kawakami

10.1 Introduction

Recent developments in the growth technology of $In_xGa_{1-x}N/GaN/Al_yGa_{1-y}N$ heterostructures have opened the way to the realization of incandescent violet, blue, green, and amber light-emitting diodes (LEDs) [1, 2], as well as of laser diodes (LDs) [3, 4] operated from ultraviolet (366.4 nm) [5] to blue (460 nm) [6] spectral region. However, further improvement of emission efficiency, or reduction of lasing threshold in wider spectral region is desired in order to expand the application field of such devices. This can be facilitated by an understanding of recombination mechanism, and by achieving a positive feedback for the appropriate growth condition, and for the design of device structures.

Currently, there has been controversy concerning the emission mechanism in $In_xGa_{1-x}N$-based semiconductors [7]. Some groups claim that the large Stokes-like shift between absorption and emission is due to the effect of exciton localization caused by inhomogeneous distribution of In in $In_xGa_{1-x}N$ active layers [8–12]. On the contrary, other groups have reported that this is just as a result of the quantum confined Stark effect [13–15], as well as of the Franz–Keldysh effect [16], both of which are induced by large internal electric fields whose origins are piezoelectric effects. There are probably two reasons for such a discrepancy. First, the macroscopic optical data such as photoluminescence (PL) and absorption do not provide enough information. Second, the degree of localization, as well as the contribution from internal electric fields differs between samples because of the difference in growth conditions, such as growth method, reactor, gas flow, pressure, substrate treatment, growth temperature, mean In mole fraction, background impurities, and so on.

In this paper, material parameters such as the bowing parameter and piezoelectric field and strain-induced effect on transition energies are summarized, and then localization behavior of excitons and its proposed model is generally surveyed. Finally, several approaches are introduced for making $In_xGa_{1-x}N$-based quantum dots.

10.2 Bowing parameter of bandgap energies in $In_xGa_{1-x}N$ alloys

In order to discuss the emission mechanism in $In_xGa_{1-x}N$-based quantum well structures, it is very important to estimate the bandgap energies of $In_xGa_{1-x}N$

ternary alloys, and then to take into account the effects of quantum confinement under an electric field, and/or of localization induced by alloy disorder of well-width fluctuation. The bandgap energies (E_g) of GaN and InN are 3.42 eV and 1.89 eV, respectively. Therefore, E_g of $In_xGa_{1-x}N$ alloys can be calculated using the equation,

$$E_g = 3.42\text{ eV}\,(1 - x) + 1.89\text{ eV}\,x - bx\,(1 - x), \qquad (10.1)$$

where b denotes the bowing parameter. The b value was initially reported to be 1.0 eV [17].

However, it was found that this value is underestimated, although the exact value is still controversial. Wetzel *et al.* [18] derived the effective bandgap bowing parameters for pseudomorphically stressed $In_xGa_{1-x}N$ on unstrained GaN: $b = 2.6$ eV for bandgap from absorption taken with photoreflectance (PR), and $b = 3.2$ eV for bandgap from PL in localized states. They then extrapolated that $b = 3.8$ eV for unstrained $In_xGa_{1-x}N$ materials. On the other hand, Bellaiche *et al.* [19] reported the theoretical values of bandgap bowing

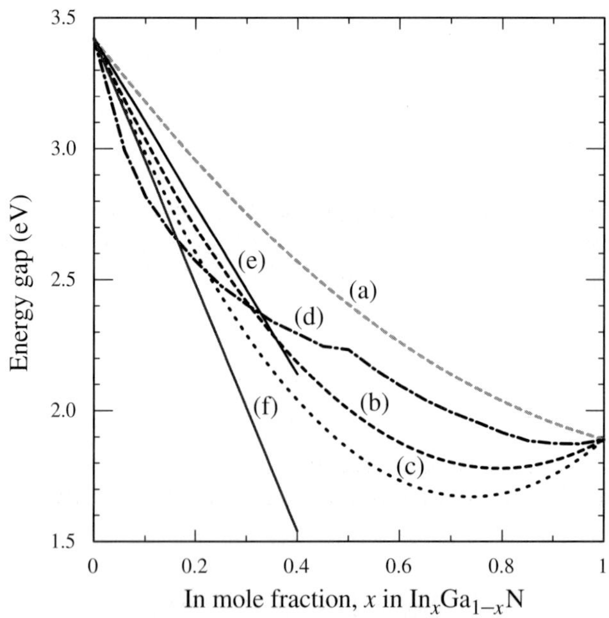

Fig. 10.1 Relationship between In mole fraction x in $In_xGa_{1-x}N$ and energy gap calculated, assuming various bowing parameters: (a) $b = 1.0$ eV (Osamura *et al.*, ref. [17], (b), and (c) $b = 2.6$ eV for bandgap assessed by PR, $b = 3.2$ eV by PL for localized states (Wetzel *et al.*, ref. [18]), (d) variable b values (Bellaiche *et al.*, ref. [19]), (e) and (f) linear relationship between energies and x values for $0 < x < 0.4$, with a slope of –3.2 eV for the bandgap and –4.7 eV for the PL peak.

parameters using large supercell empirical pseudopotential calculations, where b values range from about 1.5 eV for large In compositions to over 5 eV for small ($x<0.1$) compositions. On the other hand, O'Donell *et al.* [20] claimed a linear relationship between PL peak energy (bandgap energy as well) and measured indium fraction for $In_xGa_{1-x}N$ epilayers with $0<x<0.4$, with a slope of –4.7 eV for the PL peak and –3.2 eV for the bandgap. Their estimation is based on the experimental interpretation that the Stokes' shift between PL and absorption peaks is proportional to the mean In mole fraction, x in $In_xGa_{1-x}N$ [21]. Bandgap energies of $In_xGa_{1-x}N$ alloys are plotted as a function of temperature using various bowing parameters.

10.3 Piezoelectric fields in strained $In_xGa_{1-x}N$ layers

Large biaxial compressive strain is generated in $In_xGa_{1-x}N$ layers under pseudomorphic growth condition to unstrained GaN [13]. The strain elements in the case of a c-axis grown hexagonal structure are expressed by

$$\varepsilon_{xx} = \varepsilon_{yy} = \frac{a_{\mathrm{GaN}} - a_{\mathrm{InGaN}}}{a_{\mathrm{GaN}}} \tag{10.2}$$

$$\varepsilon_{zz} = -\frac{2c_{13}}{c_{33}}\varepsilon_{xx} \tag{10.3}$$

$$\varepsilon_{yz} = \varepsilon_{zx} = \varepsilon_{xy} = 0 \tag{10.4}$$

where a_{GaN} and a_{InGaN} are a-axis lattice constants of GaN and $In_xGa_{1-x}N$ under zero-strain [$a_{\mathrm{GaN}}=0.3189$ nm, $a_{\mathrm{InGaN}}=0.3189(1-x)+0.3548x$], respectively and c_{13} and c_{33} denote stiffiness constants. The piezoelectric polarization to the c-axis (P_z) is then given by

$$P_z = e_{31}\varepsilon_{xx} + e_{31}\varepsilon_{yy} + e_{33}\varepsilon_{zz}, \tag{10.5}$$

where e_{31} and e_{33} are piezoelectric constants.

The internal electric field along the c-axis (E_z) due to piezoelectric effect is then given by

$$E_z = -\frac{P_z}{\varepsilon_r\varepsilon_0}, \tag{10.6}$$

where ε_r and ε_0 are the dielectric constants of $In_xGa_{1-x}N$ and the permittivity of free space. Assuming that parameters such as c_{13}, c_{33}, e_{31} and e_{33} in $In_xGa_{1-x}N$ do not change with In mole fraction (x value), E_z is approximated as a linear relationship with $x(0 \leqq x \leqq 1)$, where $E_z=kx$. The k constant varies from

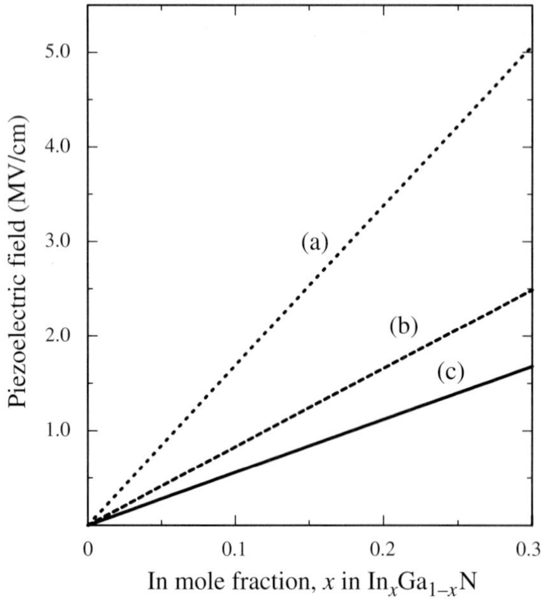

Fig. 10.2 Piezoelectric fields plotted as a function of In mole fraction x in $In_xGa_{1-x}N$, assuming various piezoelectric constants: (a) $e_{31}=-0.49\,C/m^2$ and $e_{33}=0.73\,C/m^2$ (Vernardini *et al.*, ref. [22]), (b) $e_{31}=-0.22\,C/m^2$ and $e_{33}=0.43\,C/m^2$ (Bykhovski *et al.*, ref. [23]), while (c) is obtained by the fitting of PL peak in $In_xGa_{1-x}N$ SQWs as a function of external bias (Chichibu *et al.*, ref. [24]).

5.6 MV/cm to 16.9 MV/cm according to the reported values [22–24] as shown in Fig. 10.2.

10.4 Strain-induced shift of bandgap

Free excitons in hexagonal-$In_xGa_{1-x}N$ (h-$In_xGa_{1-x}N$) are composed of three bands, labeled E_{XA}, E_{XB}, and E_{XC}, that are transitions from the conduction (Γ_{7c}) to the A(Γ_{9v}), B(Γ_{7uv}) and C(Γ_{7l}) valence bands, respectively. Among them, E_{XA} and E_{XB} bands are much more pronounced than the E_{XC} band, considering their oscillator strength. Strain-induced shift of bandgap energy in GaN ($x=0$) is precisely studied both experimentally and theoretically. From refs [25, 26], the energy shifts of E_{XA} and E_{XB} in GaN (ΔE_{XA} and ΔE_{XB}) as a function of the strain are expressed as $\Delta E_{XA}=15.4\varepsilon_{zz}$ and $\Delta E_{XB}=19.1\varepsilon_{zz}$, respectively. However, such characteristics have not been clarified for $In_xGa_{1-x}N$ probably because the inhomogeneous broadening effect, and also the exciton localization, hinders precise analysis of experimental results of strain-induced shift of bandgap energies. Nevertheless, data of GaN may be substituted for those of $In_xGa_{1-x}N$ with small x value for a rough estimation.

10.5 Transition energy in $In_xGa_{1-x}N$/GaN quantum wells

Transition energy in $In_xGa_{1-x}N$/GaN quantum well (QW) is thus highly dependent on both well width and carrier density, as schematically depicted in Fig. 10.3. With increasing well width, transition energy decreases due not only to the down-shift of quantized energy but also to the quantum confined Stark effect (QCSE) [13]. The lifetime of PL also shows large well width dependence where it varies from sub-ns to more than μs if the well width is increased from 2 nm to 7 nm [14]. This is because QCSE leads to the separation of wave function between electron and hole, resulting in a strong reduction of the oscillator strength for larger well width. Another important feature is that transition energy causes blue-shift with increasing injected carrier density because the internal electric field is screened by charged carriers. Therefore, one has to be careful in identifying transition energy and correlating it with the structure, since it also changes with background carrier concentration, as well as with excitation energy density.

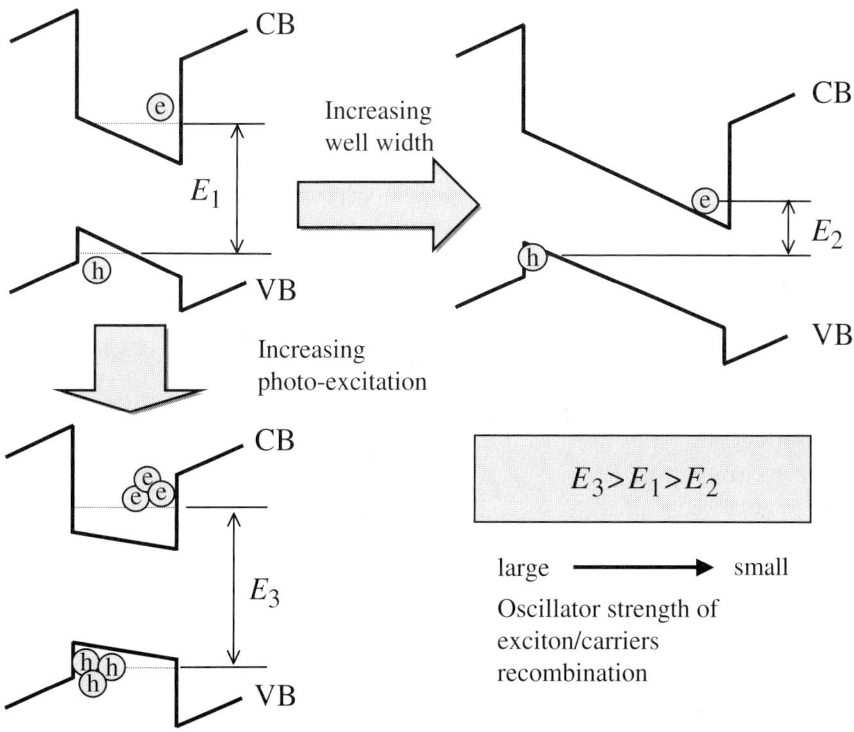

Fig. 10.3 Band line-ups of $In_xGa_{1-x}N$ SQWs under electric fields. Significant red shift in transition energy is observed with increasing well width due to QCSE. Internal electric fields are reduced by the screening effect induced by photo-generated carriers with increasing photo-excitation energy density.

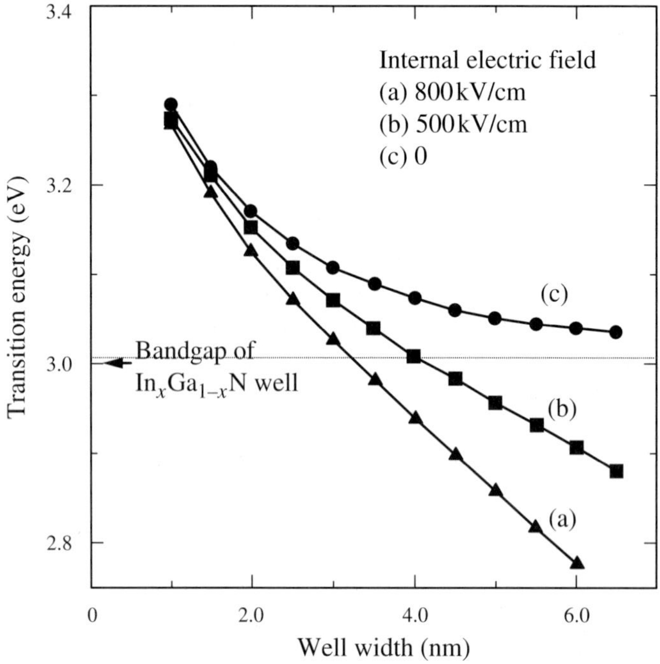

Fig. 10.4 Well width dependence of the ground state transition energies in an $In_xGa_{1-x}N$/GaN SQW, whose bandgap energy of $In_xGa_{1-x}N$ well is 3.0 eV, under electric fields of (a) 800 kV/cm; (b) 500 kV/cm; and (c) 0, assuming that the band offset ratio ($\Delta E_C/\Delta E_V$) is 4 between conduction and valence bands (Sota, ref. [27]).

Sota has calculated the well width dependence of the ground state transition energies in an $In_xGa_{1-x}N$/GaN single quantum well (SQW) whose bandgap energy of $In_xGa_{1-x}N$ well is 3.0 eV [27]. The results are plotted in Fig. 10.4 for the case of various internal electric fields. It should be noted that the typical well width is as small as 2 to 3 nm for $In_xGa_{1-x}N$/GaN-based light-emitting devices in order to avoid the QCSE effect as small as possible.

10.6 Miscibility gap and compositional fluctuation in $In_xGa_{1-x}N$ alloys

The large difference in interatomic spacing between GaN and InN leads to an unstable mixing region depending on the mean In mole fraction, as well as the growth temperature. Such characteristics of the miscibility gap can be calculated from the free energy of mixing [28, 29]. Figure 10.5 shows spinodal isotherms for $In_xGa_{1-x-y}Al_yN$ at 0–4000 °C obtained assuming that the interaction parameters are given by the model of delta lattice parameter. An unstable mixing region appears from $x = 0.07$ to 0.93 for the ternary $In_xGa_{1-x}N$ alloys in the case of a typical growth temperature of 800 °C.

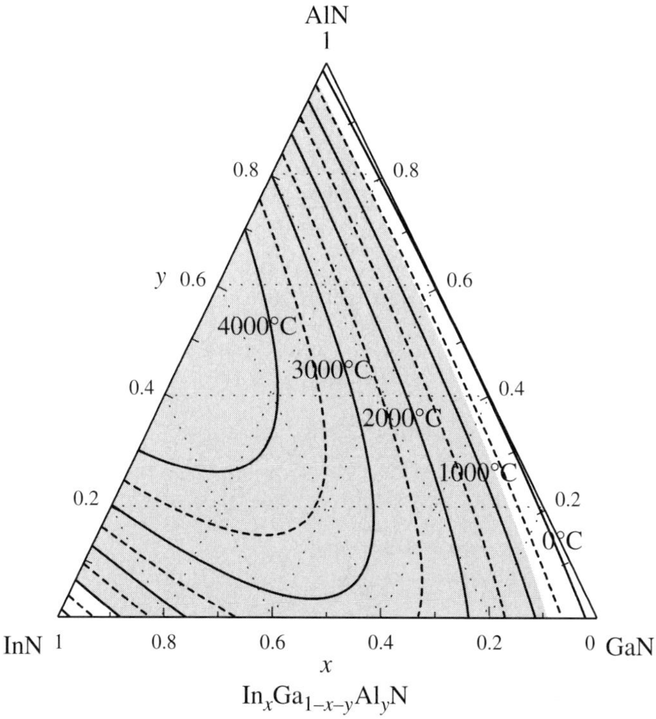

Fig. 10.5 Calculated spinodal isotherms for $In_xGa_{1-x-y}Al_yN$ alloy system at 0–4000 °C by the model of delta lattice parameter. The gray area is an unstable mixing region at growth temperature of 800 °C.

Although the metalorganic chemical vapor phase epitaxy (MOVPE) commonly used for the device fabrication is not a thermal equilibrium growth method, it is evident that the fluctuation of In composition is energetically favored [30]. Okumura *et.al.* performed a numerical study of the pattern formations of $In_xGa_{1-x}N$ caused by phase separation by the approach of cell dynamical system [31]. The results showed that In-rich dot-like structures appear in a wide range of order parameters, pattern formation (density or diameter of dots) of which depends on the diffusion constants of substituting atoms at the surface of $In_xGa_{1-x}N$ during the growth, on the lattice-mismatch between $In_xGa_{1-x}N$ and GaN, on the density of defects, as well as on the growth temperature. Actually, phase separation or formation of In-rich region has been reported by several groups [9, 32–36].

10.7 Localization-induced effect

If In compositions within the $In_xGa_{1-x}N$ layer are fluctuated, excitons and/or carriers are transferred localized to potential minima in real space and recombine radiatively [8–12]. The Bohr radius of excitons is much smaller (approximately 3 nm for $In_{0.1}Ga_{0.9}N$) than other semiconductors like GaAs. Moreover, as described in Section 10.2, a bandgap bowing parameter of $In_xGa_{1-x}N$ is anomalously large. These features lead to preferential exciton localization even with a small clustering of In composition as schematically illustrated in Fig. 10.6. Therefore, such localization may account for the large Stokes' shift reported because the density of tail states (DOS) is very small, as shown in Fig. 10.7. The fluctuation may not be randomly distributed but may take place as an In-rich clustering with a specific size of less than a several nanometers. In such a case, localization centers can act as quantum dots (QDs).

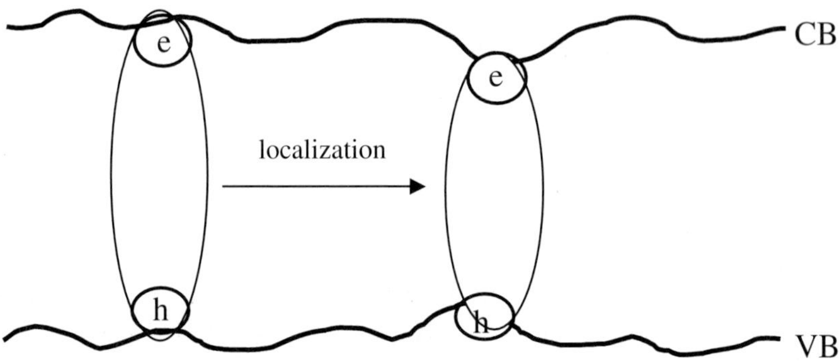

Fig 10.6 The model of exciton localization due to potential fluctuation in $In_xGa_{1-x}N$ layer.

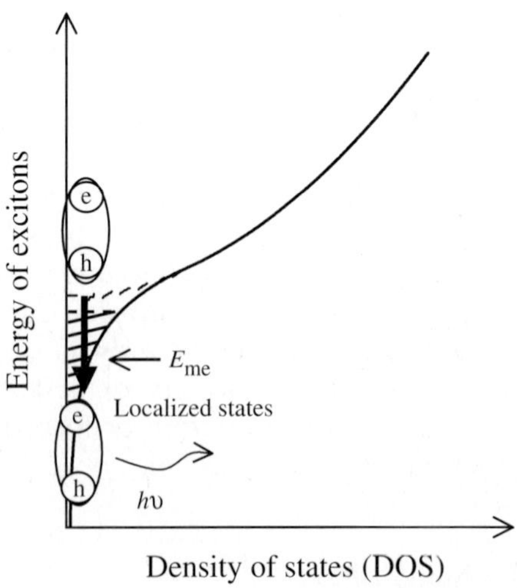

Fig. 10.7 Density of states (DOS) of exciton energy distribution induced by the random potential fluctuation. Excitons are transferred to lower energy lying states, and then recombine radiatively.

Here, recombination dynamics of localized exciton is theoretically treated for the system of random fluctuation. Suppose that pulsed photo-exciation or carrier-injection was made at energy levels of mobile states (continuum states). In this case, luminescence decay time depends on monitored photon energy with a function of $\tau_L(E)$. This is because exciton transition is due not only to radiative and/or nonradiative recombination processes but also transfer process to the lower-lying energy levels. If the density of tail states is approximated as an exponential function of $g(E)=A\exp(-E/E_0)$, where E_0 represents the depth of localization, and if radiative lifetime is constant within the emission band neglecting the non-radiative process, it is possible to estimate both E_0 and radiative recombination time (τ_{rad}) by fitting the experimental data with theoretical equation as shown below [37].

$$\frac{dn(E,t)}{dt} = -\frac{n(E,t)}{\tau_{rad}} - \sum_{E'<E} W(E \to E')n(E,t) + \sum_{E'>E} W(E' \to E)n(E',t) \quad (10.7)$$

$$W_{tr}(E) \propto \int_{E}^{\infty} g(E')dE' \propto \exp(-E/E_0) \quad (10.8)$$

where $n(E, t)$ is the density of excitons with energy of E at time t after pulsed excitation, $W(E' \to E)$ denotes the transfer probability from E to E' and $W_{tr}(E)$ shows the net transfer probability of exciton having energy E to lower-lying energy levels. By solving the above equations, $n(E, t)$ and $\tau_L(E)$ are given by the following equation,

$$n(E,t) = n_0 \exp\{-t/\tau_L(E)\} \quad (10.9)$$

$$\tau_L(E) = \frac{\tau_{rad}}{1 + \exp\{(E - E_{me})/E_0\}}, \quad (10.10)$$

where E_{me} represents the characteristics energy analogous to the mobility edge.

Figure 10.8 shows a typical example of time-resolved PL spectra under various times after pulsed excitation taken at a 0.1-μm-thick $In_{0.08}Ga_{0.92}N$ single epilayer [38]. The PL peak is located at 3.227 eV just after photo-excitation, and it makes red shift with increasing time reaching 3.208 eV at 1500 ps. This clearly shows the localization dynamics of excitons, the origin of which is a generated potential fluctuation due to the alloy disorder. The spectra of $\tau_L(E)$ and time-integrated PL are shown in Fig. 10.9. The best fitting of $\tau_L(E)$ was made using values.

$$\tau_{rad} = 440\,\text{ps}, \quad E_0 = 21\text{ meV} \quad \text{and} \quad E_{me} = 3.242\,\text{eV}.$$

However, one has to be very careful in assessing $In_xGa_{1-x}N$ QW structures by means of time-resolved luminescence spectroscopy because the

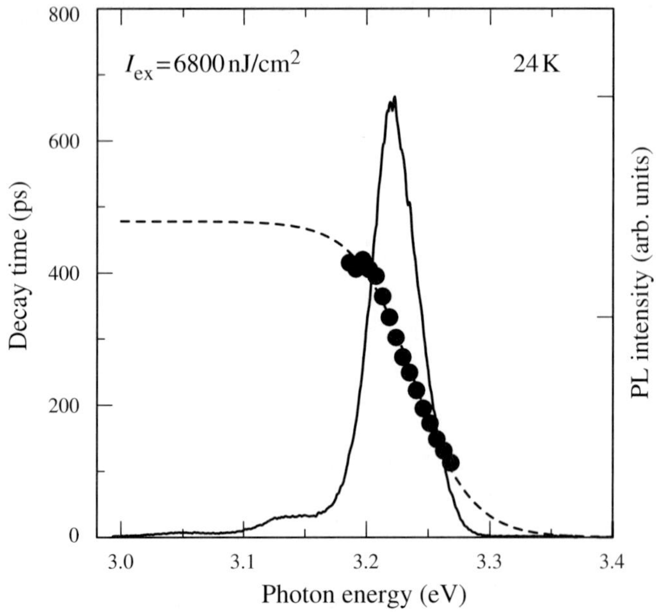

Fig. 10.8 Time-integrated PL spectrum and the PL decay time as a function of monitored photon energy at 24 K taken at a 0.1-μm-thick $In_{0.08}Ga_{0.92}N$ single epilayer (Omae *et al.* ref. [38]).

localization-like behavior can also be contributed from the QCSE. As mentioned in the previous section, photo-generated carriers or excitons reduce the QCSE because of the screening of internal electric field, where PL energy makes red-shift and PL lifetime increases according with the reduction of carriers [13–15]. It has been reported that the Si doping in the GaN barriers results in the reduction of Stokes-like shift and in the decrease of radiative recombination lifetime [39]. This is probably due, at least in part, to the screening of the internal electric field induced by the redistribution of doped electrons, besides the increase in structural and interface quality with increasing Si-doping. However, there exist phenomena even in macroscopic optical spectroscopy that cannot be interpreted by such QCFK and/or QCSE effects. They are, for example, (1) anomalous temperature dependence of emission peak energies, (2) near temperature independence of radiative recombination lifetimes, and (3) mobility edge behavior.

10.8 Anomalous temperature dependence of emission peak energies

Generally, bandgap energies of semiconductors decreases with increasing temperature, namely, $E_0(T)=E_0-\alpha T^2/(T+\beta)$, where α and β are known as Varshini's fitting parameters. For wurzite GaN, it has been reported

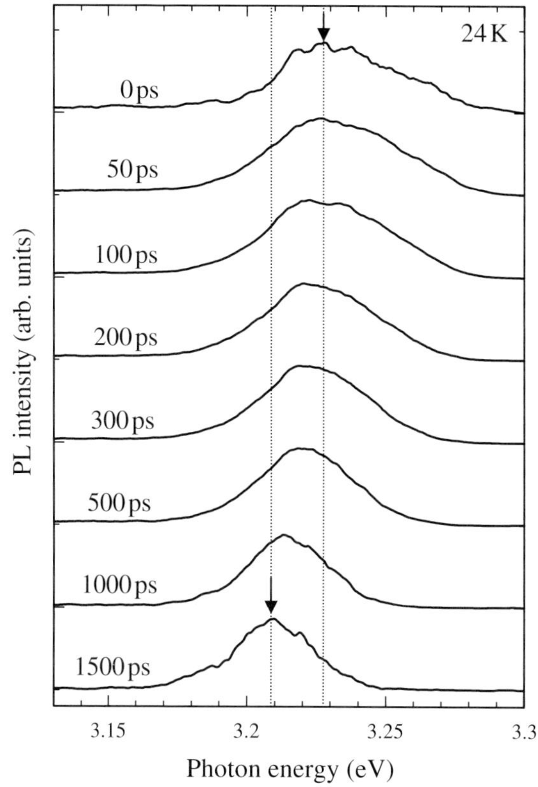

Fig. 10.9 Time-resolved PL spectra of a 0.1-μm-thick $In_{0.08}Ga_{0.92}N$ single epilayer at 24 K monitored at various time delays after pulsed photo-excitation (Omae *et al.* ref. [38]).

that α=0.375 meV and β=270 K for the temperature variation of A-exciton transition [40]. This shows that the temperature-induced red shift of transition energies is about 60 meV between 0 K and 300 K. However, anomalous temperature dependence has been observed in $In_xGa_{1-x}N$ active layers, where luminescence peak energies make blue shift with respect to the values predicted by the Varshini's formula. Eliseev *et al.* [41] reported that such blue-shifting behavior can be interpreted by the effect of localized tail states, assuming that the density of state. (DOS) of excitons induced by potential fluctuation is in Gaussian form having dispersions of σ^2. In such a case, the luminescence peak $E_L(T)$ has an additional positive term as shown below.

$$E_L(T) = E_{L0} - \frac{\alpha T^2}{T + \beta} + \frac{\sigma^2}{k_B T^2} \tag{10.11}$$

Here σ shows the energy representing the degree of localization, and α and β are fitting parameters. In Fig. 10.10, PL peak energy of a green $In_xGa_{1-x}N$

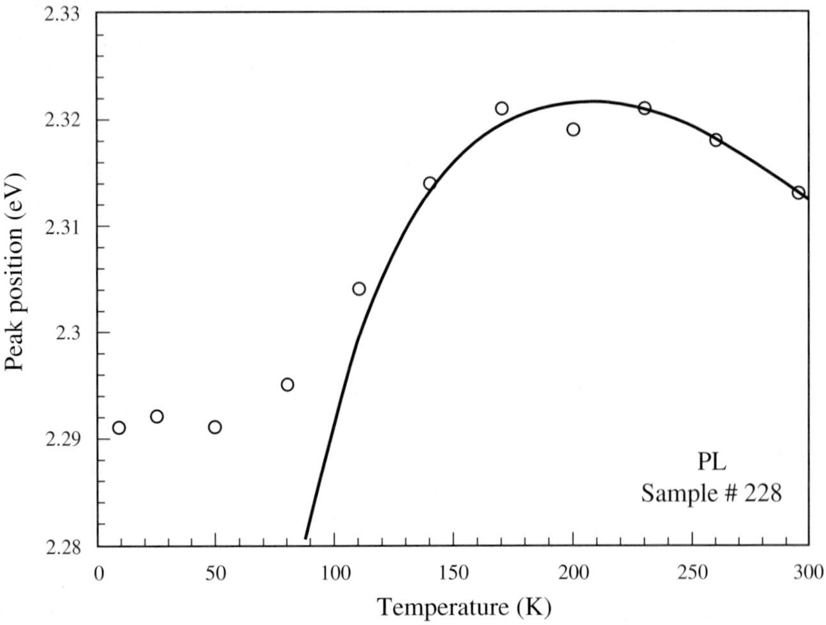

Fig. 10.10 PL peak position (open circles) plotted as a function of temperature for a green $Al_yGa_{1-y}N/In_xGa_{1-x}N/GaN$ SQW LED. The solid curve is calculated using the band-tail model with parameters described in the text (Eliseev *et al.* ref. [41]).

SQW LED structure is plotted as a function of temperature, where the fitting was made using the band-tail model with parameters: $E_{L0}=2.394$ eV, $\alpha=0.5$ meV/K, $\beta=600$ K, $\sigma=28.6$ meV. The value of σ scatters in a wide range according to the literature [36, 42, 43]. This suggests that the localization strongly depends on the growth conditions of $In_xGa_{1-x}N$ active layers.

10.9 Temperature dependence of radiative recombination lifetimes

Besides the potential fluctuation, it is very important to know the spatial extent of localization centers in the plane of quantum wells (dimensionality is another name for it) that is an important factor in naming the system, such as compositional fluctuation with micron-scale, quantum disk, mesoscopic dots, or QDs. Temperature dependence of radiative lifetimes is an effective tool for the evaluation of dimensionality despite the macroscopic characterization. This is because the DOS of excitons in K(momentum)-space is determined by the dimensionality [44–46], and that the temperature dependence of exciton-oscillator-strength is estimated by counting the number of excitons which satisfy the momentum conservation law. In the actual QWs, the depth and the spatial dimension of localization centers differ according to the structures. In the case of weak localization, excitons are easily thermalized to higher sub-bands within localization centers due to the small energy difference between the lowest quantized-level and upper sub-bands. In this case, radiative lifetime increases with temperature because the oscillator strength is averaged by the contribution from other upper-transitions. If temperature is raised further, excitons may be delocalized to extended states with 2D-DOS. In such a case, radiative lifetime increases in proportion to temperature. Therefore, temperature dependence of radiative lifetime becomes less dominant with increasing the potential depth of exciton localization.

Such behavior has been theoretically derived assuming that the degree of localization center formed in the $In_{0.2}Ga_{0.8}N$ well ($L_z=3$ nm) is expressed with quantum boxes having a plain size of $L_x \times L_y$ and a potential depth of ΔV as depicted in Fig. 10.11 [47]. Calculations have been made for the case of (a), (b), and (c) with In compositions of 40% and 50% ($\Delta V=0.83$ eV and 1.00 eV, respectively). The results described in Fig. 10.12 show that the radiative lifetimes are almost independent of temperature up to 300 K if the excitons are confined in localization centers whose size is less than 3 nm. Such temperature independence behavior has been observed in $In_xGa_{1-x}N$-SQW blue LED structure [47], as well as in $In_xGa_{1-x}N$-based low-dimensional structures grown by molecular beam epitaxy (MBE) [43].

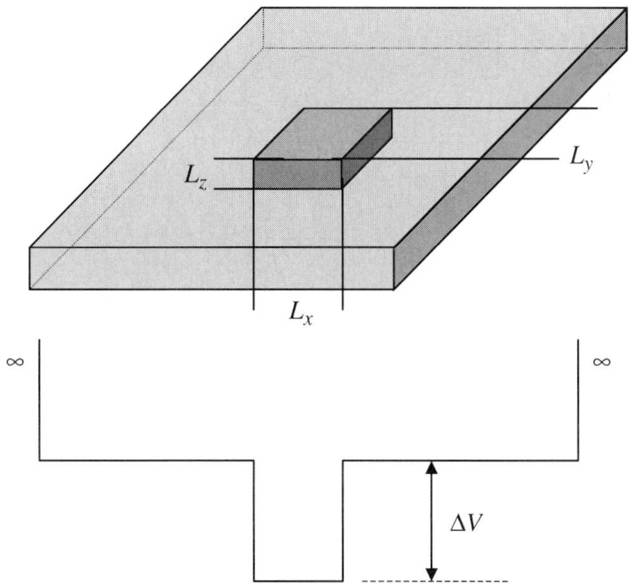

Fig. 10.11 Schematic of localization model in an $In_xGa_{1-x}N$ SQW with well width of L_z. The localization center is represented as a quantum box having a plain size of $L_x \times L_y$ and potential depth of ΔV (Kawakami *et al.* ref. [47]).

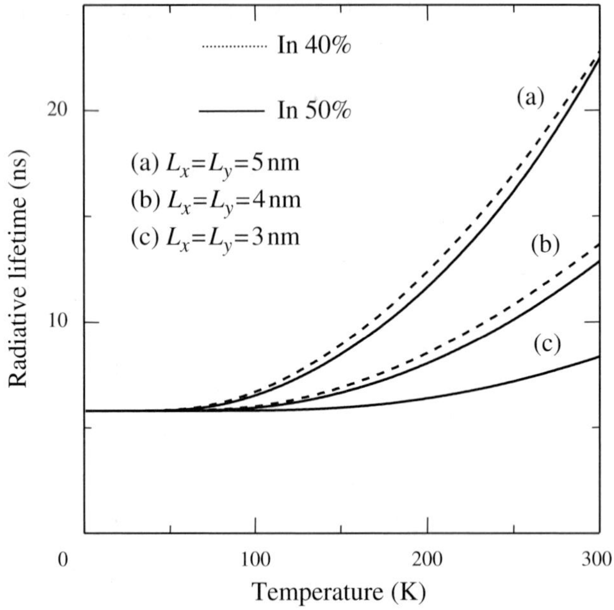

Fig. 10.12 Temperature dependence of radiative lifetime theoretically calculated in an $In_{0.2}Ga_{0.8}N$ SQW (L_z=3 nm) whose localization centers are composed of quantum boxes with (a) L_x=L_y=5 nm, (b) L_x=L_y=4 nm and (c) L_x=L_y=3 nm, and with In mole fraction x of 40% or 50% (Kawakami *et al.* ref. [47]).

10.10 Mobility edge behavior

'Mobility edge'-type behavior has been observed in $In_xGa_{1-x}N$-QWs in both spontaneous emission and stimulated ones, where the luminescence peak is insensitive to photo-excitation energy (E_{exc}) higher than a certain energy (E_m), but as E_{exc} is tuned below E_m, the peak position makes redshift quickly [11, 12]. Such phenomena can be interpreted by an energy relaxation for localized excitons. Once excitons are photo-generated, they relax rapidly into localized states where further relaxation and spatial migration are possible only through the tunneling between localized states. The relaxation rate for such a process strongly depends on the density of available final states with lower energy, and can be enhanced if the photo-excitation is made below the mobility edge. In Fig. 10.13, this kind of behavior reported by Satake *et al.* [11] is shown for the PL spectra taken under various excitation laser energies in an undoped 0.36-μm-thick $In_{0.08}Ga_{0.92}N$ single-layer.

10.11 Spatial inhomogeneity of luminescence

Chichibu *et al.* [48] performed the cathodoluminescence (CL) spectra mapping of $In_{0.1}Ga_{0.9}N$ single quantum wells (SQWs), and found that the broad CL

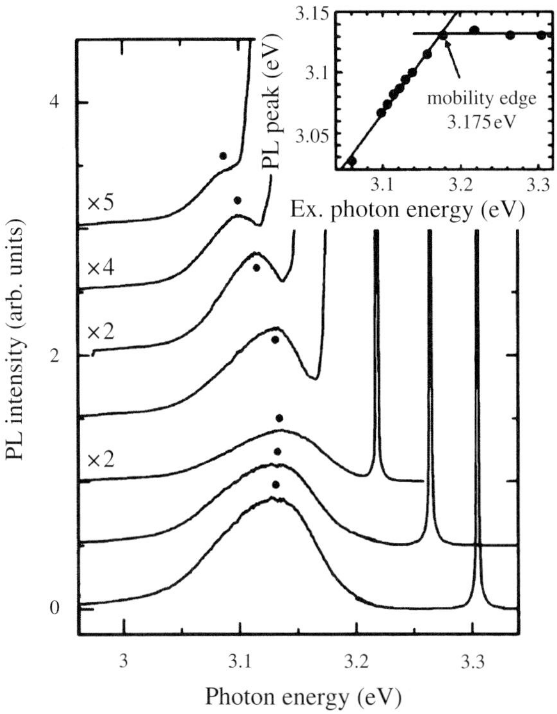

Fig. 10.13 PL spectra of an undoped 0.36-μm-thick $In_{0.08}Ga_{0.92}N$ single-layer under the site-selective excitation in the range of 3.122–3.305 eV. Closed circles show the PL peaks. The inset shows the PL peak energy as a function of the excitation photon energy (Satake *et al.* ref. [11]).

band taken from wide-area consists of several sharp emission bands having different peak energies, as shown in Fig. 10.14(1). Moreover, monochromatic CL images in Fig. 10.14(2) showed that the emission area is spatially distributed with size less than 60 nm whose resolution is limited by the diffusion length of the carriers (or excitons).

Zhang *et al.* [49] reported that a time-resolved CL study shows that carriers generated by electron-beam excitation will diffuse toward and recombine at neighboring InN-rich centers, resulting in a strong lateral excitonic localization prior to radiative recombination.

Scanning near-field optical microscopy (SNOM) is another effective tool to investigate the PL mapping. Vertikov *et al.* [50] performed the SNOM-PL mapping in thin InGaN/GaN epilayers and QWs with the spatial resolution of 100 nm. Most of SNOM results were obtained using a home-built illumination-mode system, where the laser light was focused into a few hundred nm^2 spot through the SNOM tip composed of tapered optical fiber, and the PL spectra were probed by objective lens. A drawback of such a system is that it is very difficult to know whether excitons and/or carriers producing PL

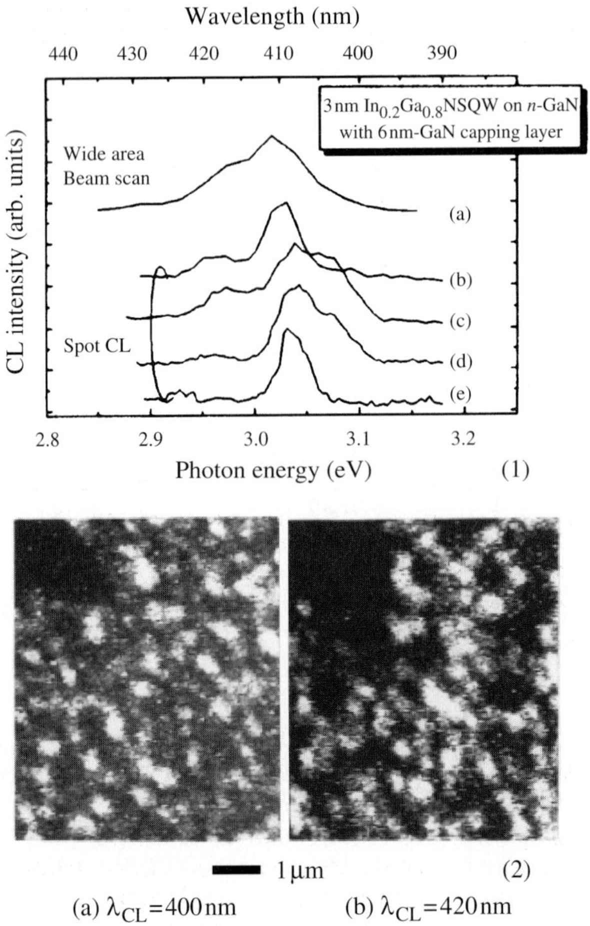

Fig. 10.14 (1) Wide area scanning (a) and spot excitation CL spectra (c)–(e) of 3-nm-thick $In_{0.1}Ga_{0.9}N$ SQW. (2) Monochromatic CL images taken at 400 and 420 nm showing primarily the complimentary relation between dark and bright region (Chichibu *et al.* ref. [48]).

are directly photo-generated at the probe region or are diffused from outside. This problem may be overcome by means of illumination–collection mode developed recently for the assessment of widegap semiconductors, where photo-excitation and the PL probing are performed using the same fiber tip [51].

10.12 Origin of localization center

It has been reported that the self-formation of an In-rich region may be the intrinsic nature of $In_xGa_{1-x}N$ ternary alloys, since the large difference in interatomic spacing between GaN and InN was found to give rise to a solid phase miscibility gap [28, 29]. Actually, cross-sectional transmission electron microscopy (TEM) and energy-dispersive X-ray microanalysis (EDX) revealed that the contrast of light and shade in the well layers observed in TEM images corresponds to the difference in In compositions as described in Fig. 10.15 [9]. Since the In-rich region has an almost isotropic dot structure whose diameter is 3 nm in average, it acts as a QD at which excitons and/or carriers are deeply localized. It has been pointed out that the high-quantum efficiency of $In_xGa_{1-x}N$-based LEDs is mostly due to the large exciton localization because the pathways of non-radiative recombination are hindered once the excitons are captured at small volume [9].

Kisielowski *et al.* [32] observed local compositional fluctuation with estimated diameters of 1–2 nm in the lattice images of a $GaN/In_{0.43}Ga_{0.57}N/Al_{0.1}Ga_{0.9}N$ QW structure obtained by high-resolution TEM. After that, they made a systematic assessment for $In_xGa_{1-x}N$ samples with wider indium concentration and well-thickness [52]. They reported that a piezoelectric-field based mechanism is shown to dominate in samples with thick wells ($L > 3$ nm) of low indium concentration ($x < 0.15$–0.20), while spatial indium-concentration fluctuations dominate luminescence properties in samples of higher indium concentration in thinner wells. It should be noted that O'Donnell *et al.* claim that the origin of localization is not In-rich $In_xGa_{1-x}N$ alloys but nearly pure InN QDs [34, 53].

Some groups claim that the localization behavior is not due to In-rich clustering but may be interpreted by the peculiar band structure of $In_xGa_{1-x}N$ ternary alloys. Yamaguchi *et al.* [54] reported that an antibonding *s*-like state produced by nitrogen vacancy (V_N) is predicted to appear at 0.3 eV below the conduction-band (CB) edge for GaN, which gets shallower and then resonant with the CB for $In_xGa_{1-x}N$. Therefore, the V_N was proposed as a candidate for the strong localization of excitons in $In_xGa_{1-x}N$. On the other hand, (Bellaiche *et al.* [19]) alloying of GaN with In induces localization in the hole wave function, resonating with the valence band even with perfectly homogeneous In distribution by means of large supercell empirical pseudopotential calculations [55]. Another model for the localization-like behavior is that a polaron state of electrons contributes to the recombination process [56].

10.13 Artificially grown $In_xGa_{1-x}N$ quantum dots

Self-assembled formation of $In_xGa_{1-x}N$ QDs (diameter: approx. 10 nm, height: approx. 5 nm) with $x = 0.22$ to 0.52 has been reported by means of MOVPE, where three-dimensional nanoscale island growth of $In_xGa_{1-x}N$ epilayers was achieved by using silicon as an antisurfactant on $Al_yGa_{1-y}N$ surfaces [57]. Intense PL was observed even at room temperature (RT) from $In_xGa_{1-x}N$ QDs embedded with the GaN capping layers. More recently, Tachibana *et al.* [58] fabricated $In_xGa_{1-x}N$ QDs, the minimum average diameter of which are as small as 8.4 nm, without the use of an antisurfactant, by modifying the growth condition in atmospheric-pressure two-flow MOVPE. Narrow peaks (170 μeV at 3.5 K)

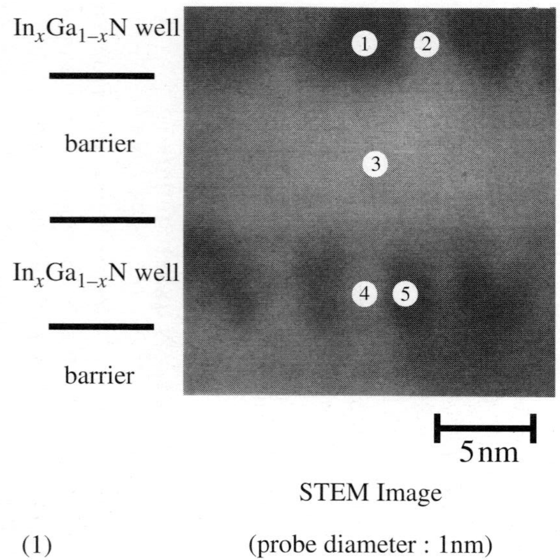

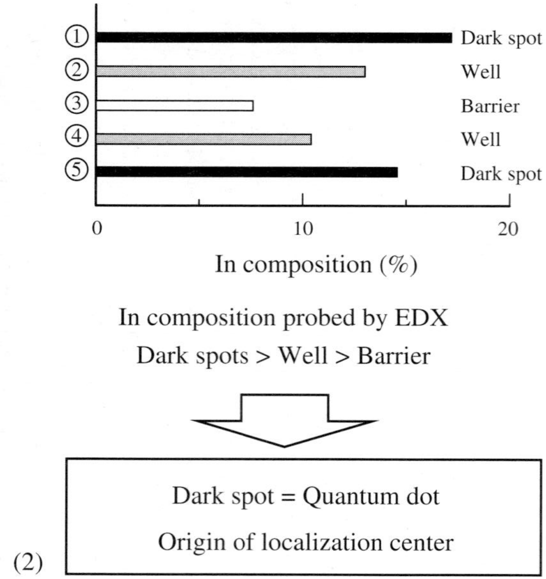

Fig. 10.15 (1) Cross sectional TEM image of an $In_xGa_{1-x}N$ multiple quantum well (MQW) structure. Positions probed by EDX are marked 1–5. (2) Indium composition at positions 1–5 determined by EDX data (Narukawa *et al.* ref. [9]).

were observed in microscopic PL spectra from the $In_xGa_{1-x}N$ QDs, indicating that excitons are strongly confined in such structures [59]. They reported that it remained unclear whether these QDs arose by strain-induced island formations like Stranski–Krastanow (SK) growth modes, or rather by phase separation. This was because the nucleation behavior differs from that of the SK growth mode, where $In_xGa_{1-x}N$ QDs are formed even when the deposition amount was 19.1 monolayer (ML) [58].

On the other hand, the clear SK growth modes have been performed by MBE for the self-formation of $In_xGa_{1-x}N$ QDs [60, 61], where the epitaxial growth of $In_xGa_{1-x}N$ exhibits a 2D–3D growth transition above a critical In composition of 12% as observed *in situ* by reflection high-energy electron diffraction (RHEED). Lefebvre *et al.* [43] reported that the size of $In_xGa_{1-x}N$ QDs thus obtained is 5–10 nm in diameter and 2 nm in height, while almost no temperature dependence of radiative lifetime suggests that carriers localize to potential fluctuations with spatial extension much smaller than the size of QDs.

The fabrication methods described above still have problems, such as the controllability of positioning, size and density of each QD. Such drawbacks can be overcome to some extent by the selective growth technique. Wang *et al.* [62] grew $In_xGa_{1-x}N$ QDs on a Si/GaN/sapphire substrate patterned by focused ion beam etching. They achieved the growth of GaN plinths and that of $In_xGa_{1-x}N$ QDs on them, by the selective nucleation on nanoscale windows opened in the Si-mask using MOVPE. More recently, $In_xGa_{1-x}N$ QDs was fabricated on a SiO_2/GaN sapphire substrate patterned even with conventional photolithography [63]. Micro-PL mapping revealed that intense PL at RT arose from the $In_xGa_{1-x}N$ QDs area grown on top of hexagonal GaN pyramids.

10.14 Conclusion

Material parameters, such as bowing parameter, piezoelectric constants of $In_xGa_{1-x}N$, have been summarized in this chapter. However, their reported values fluctuated, making precise analysis of transition energies difficult. Moreover, degrees of compositional fluctuation, interfacial disorder, background impurity density, defect density, and so on, differ in each sample depending on growth conditions. Therefore, it will still be important to accumulate a lot of reliable data reported by various groups in order to reach a general understanding of emission mechanism in $In_xGa_{1-x}N$ QWs. The assessment of optical properties of artificially grown $In_xGa_{1-x}N$ QDs and careful comparison of the data with those of $In_xGa_{1-x}N$ QWs may lead to the advanced understanding on the dimensionality of excitons.

References

1. Nakamura, S., Senoh, M., Iwasa, N., Nagahama, S., Yamada, T., and Mukai, T. (1995). Superbright green InGaN single-quantum-well-structure light-emitting diodes. *Jpn. J. Appl. Phys.* **34**, L1332–5.

2. Mukai, T., Narimatsu, H., and Nakamura, S. (1998). Amber InGaN-based light-emitting diodes operable at high ambient temperatures. *Jpn. J. Appl. Phys.* **37**, L479-81.
3. Nakamura, S., Senoh, M., Nagahama, S., Iwasa, N., Yamada, T., Matsushita, T., Kiyoku, H., and Sugimoto, Y. (1996). InGaN-based multi-quantum-well-structure laser diodes. *Jpn. J. Appl. Phys.* **35**, L74–6.
4. Nagahama, S., Iwasa, N., Senoh, M., Matsushita, T., Sugimoto, Y., Kiyoku, H., *et al.* (2000). High-power and long-lifetime InGaN multi-quantum-well laser diodes grown on low-dislocation-density GaN substrates. *Jpn. J. Appl. Phys.* **39**, L647–50.
5. Nagahama, S., Yanamoto, T., Sano, M., and Mukai, T. (2001). Characteristics of ultraviolet laser diodes composed of quaternary $Al_xIn_yGa_{1-x-y}N$. *Jpn. J. Appl. Phys.* **40**, L788–L91.
6. Nagahama, S., Yanamoto, T., Sano, M., and Mukai, T. (2001). Characteristics of InGaN laser diodes in the pure blue region. *Appl. Phys. Lett.* **79**, 1948–50.
7. For review, see for example, Monemar, B., Bergman, J.P., Dalfors, J., Pozina, G., Sernelius, B. E., Holts, P. O., *et al.* (1999). Radiative recombination in $In_{0.15}Ga_{0.85}N$/GaN multiple quantum well structures. *MRS Internet J. Nitride Semicond. Res.* **4**, 16 (http://nsr.mij.mrs.org/4/16/).
8. Chichibu, S., Azuhata, T., Sota, T., and Nakamura. S. (1996). Spontaneous emission of localized excitons in InGaN single and multiquantum well structures. *Appl. Phys. Lett.* **69**, 4188–90.
9. Narukawa, Y., Kawakami, Y., Funato, M., Fujita Sz., Fujita, Sg., and Nakamura, S. (1997). Role of self-formed InGaN quantum dots for exciton localization in the purple laser diode emitting at 420 nm. *Appl. Phys. Lett.* **70**, 981–3.
10. Narukawa, Y., Kawakami, Y., Fujita, Sz., Fujita, Sg., and Nakamura, S. (1997). Recombination dynamics of localized excitons in $In_{0.20}Ga_{0.80}N$-$In_{0.05}Ga_{0.95}N$ multiple quantum wells. *Phys. Rev. B* **55**, R1938–41.
11. Satake, A., Masumoto, Y., Miyajima, T., Asatsuma, T., Nakamura, F., and Ikeda, M. (1998). Localized exciton and its stimulated emission in surface mode from single-layer $In_xGa_{1-x}N$. *Phys. Rev. B* **57**, R2041–4.
12. Cho, Y. H., Schmidt, T. J., Bidnyk, S., Gainer, G. H., Song, J. J., Keller, S., *et al.* (2000). Linear and nonlinear optical properties of $In_xGa_{1-x}N$/GaN heterostructures. *Phys. Rev. B* **61**, 7571–88.
13. Takeuchi, T., Sota, S., Katsuragawa, M., Komori, M., Takeuchi, H., Amano, H., and Akasaki, I. (1997). Quantum-confined Stark effect due to piezoelectric fields in GaInN strained quantum wells. *Jpn. J. Appl. Phys.* **36**, L382–5.
14. Hangleiter, A., Im, J. S., Kollmer, H., Heppel, S., Off, J., and Scholz (1988). The role of piezoelectric fields in GaN-based quantum wells. *MRS Internet J. Nitride Semicond. Res.* **3**, 15 (http://nsr.mij.mrs.org/3/15/).
15. Riblet, P., Hirayama, H., Kinoshita, A., Hirata, A., Sugano, T., and Aoyagi, Y. (1999). Determination of photoluminescence mechanism in InGaN quantum wells. *Appl. Phys. Lett.* **75**, 2241–3.
16. Wetzel, C., Takeuchi, T., Amano, H., and Akasaki, I. (1999). Piezoelectric Franz–Keldysh effect in strained GaInN/GaN heterostructures. *J. Appl. Phys.* **85**, 3786–91.
17. Osamura, K., Naka, S., and Murakami, Y. (1975). Preparation and optical properties of $Ga_{1-x}In_xN$ thin films. *J. Appl. Phys.* **46**, 3432–7.
18. Wetzel, C., Takeuchi, T., Yamaguchi, S., Katoh, H., Amano, H., and Akasaki, I. (1998). Optical band gap in $Ga_{1-x}In_xN$ ($0<x<0.2$) on GaN by photoreflection spectroscopy. *Appl. Phys. Lett.* **73**, 1994–6.
19. Bellaiche, L., Mattila, T., Wang, L. W., Wei, S. H., and Zunger, A. (1999). Resonant hole localization and anomalous optical bowing in InGaN alloys. *Appl. Phys. Lett.* **74**, 1842–4.

20. O'Donnell, K. P. (2001). A mystery wrapped in enigma: optical properties of InGaN alloys. *Phys. Stat. Sol.* (a) **183**, 117–20.
21. Martin, R. W., Middleton, P. G., O'Donnell, K. P., and Van der Stricht, W. (1999). Exciton localization and the Stokes' shift in InGaN epilayers. *Appl. Phys. Lett.* **74**, 263–5.
22. Bernardini, F., Fiorentini, V., and Vanderbilt, D. (1997). Spontaneous polarization and piezoelectric constants of III–V nitrides. *Phys. Rev. B* **56**, R10024–7.
23. Bykhovski, A. D., Kaminski, V. V., Shur, M. S., Chen, Q. C., and Khan, M.A. (1996). Piezoresistive effect in wurtzite *n*-type GaN. *Appl. Phys. Lett.* **68**, 818–19.
24. Chichibu, S., Sota, T., Wada, K., Brandt, O., Ploog, K. H., DenBaars, S.P., and Nakamura, S. (2001). Impact of Internal electric field and localization effect on quantum well excitons in AlGaN/GaN/InGaN light emitting diodes. *Phys. Stat. Sol.* (a) **183**, 91–8.
25. Gil, B., Briot, O., and Aulombard, R.L. (1995). Valence-band physics and the optical properties of GaN epilayers grown onto sapphire with wurtzite symmetry. *Phys. Rev. B* **52**, R17028–31.
26. Shikanai, A., Azuhata, T., Sota, T., Chichibu, S., and Nakamura, S. (1997). Biaxial strain dependence of exciton resonance energies in wurtzite GaN. *J. Appl. Phys.* **81**, 417–424.
27. Sota, T. (2000). Emission Mechanisms and Excitons in GaN and InGaN Bulk and QWs. *Introduction to nitride semiconductor blue lasers and light emitting diodes,* Chapter 5, pp. 192–7. Taylor & Francis, London and New York.
28. Ho, I., and Stringfellow, G.B. (1996). Solid phase immiscibility in GaInN. *Appl. Phys. Lett.* **69**, 2701–3.
29. Matsuoka, T. (1997). Calculation of unstable mixing region in wurtzite $In_{1-x-y}Ga_xAl_yN$. *Appl. Phys. Lett.* **71**, 105–6.
30. Koukitu, A., Takahashi, N., Taki, T., and Seki, H. (1996). Thermodynamic analysis of $In_xGa_{1-x}N$ alloy composition grown by metalorganic vapor phase epitaxy. *Jpn. J. Appl. Phys.* **35**, L673–5.
31. Okumura, T., Ishida, M., and Kamikawa, T. (2000). Numerical study of InGaN pattern formation caused by phase separation. *Jpn. J. Appl. Phys.* **39**, 1044–50.
32. Kisielowski, C., Weber, Z.L., and Nakamura, S. (1997). Atomic scale indium distribution in a $GaN/In_{0.43}Ga_{0.57}N/Al_{0.1}Ga_{0.9}N$ quantum well structure. *Jpn. J. Appl. Phys.* **36**, 6932–6.
33. El-Masry, N. A., Piner, E. L., Liu, S. X., and Bedair, S. M. (1998). Phase separation in InGaN grown by metalorganic chemical vapor deposition. *Appl. Phys. Lett.* **72**, 40–2.
34. Nistor, L., Bender, H., Vantomme, A., Wu, M. F., Van Landuyt, J., O'Donnell, K. P., *et al.* (2000). Direct evidence of spontaneous quantum dot formation in a thick InGaN epilayer. *Appl. Phys. Lett.* **77**, 507–9.
35. Strassburg, M., Hoffman, A., Krestnikov, and Ledentsov (2001). Optical and structural properties of quantum dots in wide-bandgap semiconductors. *Phys. Stat. Sol (a)* **183**, 99–104.
36. Moon, Y. T., Kim, D. J., Park, J. S., Oh, J. T., Lee, J. M., Ok, Y. W., *et al.* (2001). Temperature dependence of photoluminescence of InGaN films containing In-rich quantum dots. *Appl. Phys. Lett.* **79**, 599–601.
37. Gourdon, C., and Lavallard, P. (1989). Exciton transfer between localized states in $CdS_{1-x}Se_x$ alloys. *Phys. Stat. Sol* (b) **153**, 641–52.
38. Omae, K., Kawakami, Y., Fujita, Sg., Yamada, M., Narukawa, Y., and Mukai, T. (2001). Unpublished data.
39. Cho, Y. H., Song, J. J., Keller, S., Minsky, M. S., Hu, E., Mishra, U. K., and DenBaars, S. P. (1998). Influence of Si doping on characteristics of InGaN/GaN multiple quantum wells. *Appl. Phys. Lett.* **73**, 1128–30.

40. Chichibu, S., Azuhata, T., Sota, T., and Nakamura, S. (1996). Excitonic emissions from hexagonal GaN epitaxial layers. *J. Appl. Phys.* **79**, 2784–6.
41. Eliseev, P. G., Perlin, P., Lee, J., and Osinski, M. (1997). 'Blue' temperature-induced shift and band-tail emission in InGaN-based light sources. *Appl. Phys. Lett.* **71**, 569–71.
42. Wang, T., Saeki, H., Bai, J., Shirahama, T., Lachab, M., Sakai, M., and Eliseev, P. (2000). Effect of silicon doping on the optical and transport properties of InGaN/GaN multiple-quantum-well structures. *Appl. Phys. Lett.* **76**, 1737–9.
43. Lefebvre, P., Taliercio, T., Morel, A., Allègre, J., Gallart, M., Gil, B., *et al.* (2001). Effects of GaAlN barriers and of dimensionality on optical recombination processes in InGaN quantum wells and quantum boxes. *Appl. Phys. Lett.* **78**, 1538–40.
44. Gotoh, H., Ando, H., Takagahara, T., Kamada, H., Chavez-Pirson, A., and Temmyo, J. (1997). Effects of dimensionality on radiative recombination lifetime of excitons in thin quantum boxes of intermediate regime between zero and two dimensions. *Jpn. J. Appl. Phys.* **36**, 4204–8.
45. Lefebvre, P., Allègre, J., Gil, B., Kavokine, A., Mathieu, H., Kim, W., *et al.* (1998). Recombination dynamics of free and localized excitons in GaN/$Ga_{0.93}Al_{0.07}N$ quantum wells. *Phys. Rev. B* **57**, R9447–50.
46. Narukawa, Y., Kawakami, W., Fujita, S., and Nakamura, S. (1999). Dimensionality of excitons in laser-diode structures composed of $In_xGa_{1-x}N$ multiple quantum wells. *Phys. Rev. B* **59**, 10283–8.
47. Kawakami, Y., Narukawa, Y., Omae, K., Fujita, Sg., and Nakamura, S. (2000). Dimensionality of excitons in InGaN-based light emitting devices. *Phys. Stat. Sol.* (a) **178**, 331–6.
48. Chichibu, S., Wada, K., and Nakamura, S. (1997). Spatially resolved cathodoluminescence spectra of InGaN quantum wells. *Appl. Phys. Lett.* **71**, 2346–8.
49. Zhang, X., Rich, D. H., Kobayashi, J. T., Kobayashi, N. T., and Dapkus, P. D. (1998). Carrier relaxation and recombination in an InGaN/GaN quantum well probed with time-resolved cathodoluminescence. *Appl. Phys. Lett.* **73**, 1430–2.
50. Vertikov, A., Kuball, M., Nurmikko, A. V., Chen, Y., and Wang, S. Y. (1998). Near-field optical study of InGaN/GaN epitaxial layers and quantum wells. *Appl. Phys. Lett.* **72**, 2645–7.
51. Kaneta, A., Izumi, T., Okamoto, K., Kawakami, Y., Fujita, Sg., Narita, Y., *et al.* (2001). Spatial inhomogeneity of photoluminescence in an InGaN-based light-emitting diode structure probed by near-field optical microscopy under illumination-collection mode. *Jpn. J. Appl. Phys.* **40**, 110–11.
52. Shapiro, N. A., Perlin, P., Kisielowski, C., Mattos, L. S., and Weber, E. R. (2000). The effects of indium concentration and well-thickness on the mechanism of radiative recombination in $In_xGa_{1-x}N$ quantum wells. *MRS Internet J. Nitride Semicond. Res.* **5**, 1 (http://nsr.mij.mrs.org/5/1/).
53. O'Donnell, K. P., Martin, R. W., and Middleton, P.G. (1999). Origin of luminescence from InGaN diodes. *Phys Rev. Lett.* **82**, 237–40.
54. Yamaguchi, E. and Junnakar, M. R. (1998). Effects of nitrogen vacancy on optical properties of nitride semiconductors. *J. Cryst. Growth* **189/190**, 570–4.
55. Kent, P. R. C. and Zunger, A. (2001). Carrier localization and the origin of luminescence in cubic InGaN alloys. *Appl. Phys. Lett.* **79**, 1977–9.
56. Kudo, H., Tanabe, T., Ishibashi, H., Zheng, R., Yamada, Y., and Taguchi, T. (2000). Radiative recombination dynamics of carriers in $In_xGa_{1-x}N$ epitaxial layers revealed by temperature dependence of time-resolved photoluminescence spectra. *Phys. Stat. Sol.* (a) **180**, 27–31.

57. Hirayama, H., Tanaka, S., Ramvall, P., and Aoyagi, Y. (1998). Intense photoluminescence from self-assembling InGaN quantum dots artificially fabricated on AlGaN surfaces. *Appl. Phys. Lett.* **72**, 1736–8.
58. Tachibana, K., Someya, T., and Arakawa, Y. (1999). Nanometer-scale InGaN self-assembled quantum dots grown by metalorganic chemical vapor deposition. *Appl. Phys. Lett.* **74**, 383–5.
59. Moriwaki, O., Someya, T., Tachibana, K., Ishida, S., and Arakawa, Y. (2000). Narrow photoluminescence peaks from localized states in InGaN quantum dot structures. *Appl. Phys. Lett.* **76**, 2361–3.
60. Damilano, B., Grandjean, N., Dalmasso, S., and Massies, J. (1999). Room-temperature blue-green emission from InGaN/GaN quantum dots made by strain-induced islanding growth. *Appl. Phys. Lett.* **75**, 3751–3.
61. Adelmann, C., Simon, J., Feuillet, G., Pelekanos, N. T., Daudin, B., and Fishman, G. (2000). Self-assembled InGaN quantum dots grown by molecular-beam epitaxy. *Appl. Phys. Lett.* **76**, 1570–2.
62. Wang, J., Nozaki, M., Lachab, M., Ishikawa, Y., Qhalid Fareed, R. S., *et al.* (1999). Metalorganic chemical vapor deposition selective growth and characterization of InGaN quantum dots. *Appl. Phys. Lett.* **75**, 950–2.
63. Tachibana, K., Someya, T., Ishida, S., and Arakawa, Y. (2000). Selective growth of InGaN quantum dot structures and their microphotoluminescence at room temperature. *Appl. Phys. Lett.* **76**, 3212–14.

11 Electron–phonon interaction in GaN and its low-dimensional structures

Xue Bing Zhang and Bernard Gil

11.1 Introduction

GaN and its low-dimensional structures have been extensively investigated in the past years because of their excellent optical and electrical properties. Progress in this field is so fast that the nitrides are expected to revolutionize the optoelectronics and electronics in the new century. The nature of high thermal conductivity, high luminous efficiency, and mechanical robustness have made the GaN and its alloys the superior material system for light-emitting devices operation from the UV to the entire visible spectra region [1]. In addition, new applications in high power, high temperature, and high frequency electronic devices based on GaN field effect transistors and heterobipolar transistor are under extensive exploration.

Despite their viable technological applications and remarkable achievements, the fundamental intrinsic properties, such as electron–phonon interaction, and physical parameters of the III-nitride materials, are less known. While the extrinsic effects such as growth-generated defects, fabrication-induced defects, which degrade the device operations, can be controlled through the improved methods of growth and fabrications, the intrinsic properties, however, cannot be controlled, but must be known since they play an essential role in the device operation.

The interaction between electrons and phonons has a great influence on the optical and electrical properties of semiconductors [2]. For example, phonons can participate in both the near-bandgap emission and the absorption processes, which leads to an enhanced absorption near the fundamental band-edge and broadens the emission peak. The cooling of hot carriers is also strongly governed by the electron–phonon interactions. For the electrical properties, the interaction with phonons will scatter the electron from one state to another, which results in a reduction of the carrier mobility and in an increase in resistivity. The basic knowledge of electron–phonon interaction is very important for improving conditions of device operation.

In this chapter, we review the influence of lattice vibrations on the optical properties of nitride semiconductors and we present recent studies of the electron–phonon interaction in this system. The organization of this chapter is as follows: in Section 11.2, we introduce the basic physics about the electron–phonon interaction. After that, we deal with several phenomena in GaN, which are results of electron–phonon interaction. They are: phonon replica in the emission spectrum (Section 11.3), broadening of excitonic line widths (Section 11.4), cooling of hot carriers in GaN (Section 11.5). Finally, in Section 14.6, the electron–phonon interaction in nitride low-dimensional structures is discussed.

11.2 Basic theory of electron–phonon interaction in GaN

The intrinsic properties of semiconductors are mainly governed by the motion of the valence electrons in the crystal, which is a many body effect. There is no realistic way to solve the Hamiltonian involving a large number of electrons and nuclei. Approximations have to be introduced, such as the well-known adiabatic approximation, in which the electrons are assumed to follow instantaneously the motions of the ion core of the lattice atom. This approximation is based on the fact that the ion core of the lattice atom is three to five orders of magnitude heavier than the electron. In the context of this approximation, the Hamiltonian for the total crystal system can be expressed as the sum of an electronic part H_e, of a lattice vibrational part H_{ph}, and of an electron–lattice interaction part $H_{e\text{-}ph}$. The electronic and lattice vibrational parts can be separated from the total Hamiltonian and treated individually, while the electron–lattice interaction is treated as a perturbation. As a result, the wave function of the total system can be written as a direct product of the electronic state and of the lattice vibrational state. This treatment has been found to be successful for many semiconductor systems.

11.2.1 Electronic band structure

The electronic term in the Hamiltonian is usually treated within the one-electron approximation. Various methods have been developed in order to calculate the electronic band structure. A review on this topic in nitride semiconductors has been given in the earlier book in this series [3]. Generally, for III–V nitride semiconductors with a wurtzite structure belonging to the space group C_{6v}^4, the conduction band minimum is located at the center of the Brillouin zone and its vicinity is almost composed by the *s* states of nitrogen and cation atoms. The valence band maximum is also located at the center of the Brillouin zone and its vicinity is almost composed by the *p* states of nitrogen. Due to the reduced symmetry and to the spin–obit interaction, the valence band splits into the well-known A, B, and C bands with Γ_9, Γ_7, and Γ_7 symmetries, respectively. The conduction band has the Γ_7 symmetry. The valence band dispersion relations are anisotropic and strongly k dependent. The effects of the stress on the band structure and optical properties of wurtzite III nitrides have been reviewed in ref. [4].

11.2.2 Phonons

When solving the lattice vibrational part of the Hamiltonian, both *ab initio* calculations [5–10] and the phenomenological model have been used to obtain the phonon dispersion curves [11–15]. In *ab initio* calculations, the phonon dispersions and frequencies of phonons at the Brillouin center depend strongly on the method used [6, 9, 10]. While in the phenomenological model, the experimental phonon frequencies are used as parameters in fitting the model. A group theory analysis of phonons at the Γ point of the Brillouin zone predicts the following set of eight modes: ($2A_1 + 2B_1 + 2E_1 + 2E_2$), which belong to the irreducible representations of the C_{6v}^4 space group. Among them, the acoustic modes Γ_{ac} have A_1+E_1 symmetries; the optical modes Γ_{opt} have $A_1 + 2B_1 + E_1 + 2E_2$ symmetries. The A_1 and E_1 optical modes are both Raman and infrared active, the E_2 modes are only Raman active, and the B_1 modes are silent. The six optical modes A_1(TO) + A_1(LO) + E_1(TO) + E_1(LO) + $2E_2$ have been widely investigated by the first-order Raman scattering [5, 12, 13, 16–18]. Recently, the frequencies of phonons in GaN, including those of silent modes, at other points of the Brillouin zone were determined by inelastic X-ray scattering [10]. The results were found in very good agreement with the *ab initio* calculations when the calculated results were scaled by a factor of 0.97. The frequencies of the zone center modes in GaN, AlN, and InN bulk materials, obtained from the first-order Raman scattering, the inelastic X-ray scattering, and the calculations, are summarized in Table 11.1. In AlGaN and GaInN alloys, the phonons were also calculated and investigated by Raman spectroscopy. It was found that the phonons in GaInN alloys display a one-mode behavior, while in AlGaN alloys the mechanism is more complicated. The LO phonon displays a one mode behavior, but both one-mode and two-mode behaviors have been reported for the TO phonons [19–22].

11.2.3 Exciton–phonon interaction

The calculation of electronic band structures within the adiabatic approximation gives essential information about the optical and electrical properties of materials. However, atoms of the lattice are not at rest. Their displacement away from the equilibrium positions perturbs the periodic potential acting on the electrons in the crystal, leading to an electron–phonon interaction energy. Due to the different ways that the lattice vibrations perturb the motion of electrons, there are various types of electron–phonon interaction [2], such as the Fröhlich interaction with longitudinal optical (LO) phonons, the deformation–potential interaction with both optical and acoustic phonons, and the piezoelectric interaction with acoustic phonons. The Fröhlich interaction is a Coulomb interaction between the electrons and the longitudinal electric field produced by the LO phonons. The deformation–potential interaction is caused by the displacement of atoms from equilibrium positions, which affects the electronic band structure similarly to the effect of applying a strain. In noncentrosymmetric crystals, the strain field produces the electric field through the piezoelectric effect. Acoustic phonons can be piezoelectric active in certain directions. The piezoelectric interaction of electrons with acoustic phonons is similar to the Fröhlich interaction with LO phonons.

Table 11.1 Zone center phonon frequencies (in cm^{-1}) of hexagonal GaN, AlN, and InN.

Mode	GaN	AlN	InN
E_2 (low)	153[a] 140.4[b*] 144[a]	252[a] 248.6[a] 237[d]	87[e,f] 88[g] 93[g]
B_1 (low)	319[a] 328.5[b*]	571[a]	200[e] 202[g]
A_1 (TO)	546[a] 528.8[b*] 531.8[a] 533[c]	615[a] 611.0[a] 619[d]	480[e] 447[f] 440[g] 443[g]
E_1 (TO)	554[a] 554.5[b*] 558.8[a] 561[c]	673[a] 670.8[a] 677[d]	476[e,f] 470[g]
E_2 (high)	565[a] 563.6[b*] 567.6[a] 569[c]	667[a] 657.4[a] 667[d]	488[e,f] 490[g] 492[g]
B_1 (high)	728[a] 697.3[b*]	778[a]	540[e] 568[g]
A_1 (LO)	734[a] 728.7[b*] 734.0[a] 735[c]	881[a] 890.0[a] 893[d]	580[e] 586[f] 590[g] 589[g]
E_1 (LO)	739[a] 729.6[b*] 741.0[a] 743[c]	922[a] 912[a] 919[d]	570[e] 593[f] 605[g]

[a]Davydov, V. Yu., Kitaev, Yu. E., Goncharuk, I. N., Smirnov, A. N., Graul, J., Semchinova, O., *et al.* (1998). *Phys. Rev. B* **58**, 12899.

[b]Ruf, T., Serrano, J., Cardona, M., Pavone, P. Pabst, M., Krisch, M., *et al.* (2001). *Phys. Rev. Lett.* **86**, 906.

[c]Azuhata, T., Sota, T., Suzuki, K., and Nakamura, S., (1995). *J. Phys. Condens. Mater*. **7**, L129.

[d]Filippidis, L. Siegle, H., Hoffman, A., Thomsen, C., Karch, K., and Bechstedt, F. (1996). *Phys. Stat. Sol.* (b) **198**, 621.

[e]Inushima, T., Shiraishi, T., and Davydov, V. Yu., (1999). *Solid Stat. Commun*. **110**, 491.

[f]Davydov, V. Yu., Emtsev, V. V., Goncharuk, I. N., Smirnov, A. N., Petrikov, V. D., Mamutin, V. V., *et al.* (1999). *Appl. Phys. Lett.* **75**, 3297.

[g]Kaczmarczyk, G., Kaschner, A., Reich, S., Hoffmann, A., Thomsen, C., As, D. J., *et al.* (2000). *Appl. Phys. Lett.* **76**, 2122.

*According to ref. b, the values have been scaled by a factor of 0.97, which is in excellent agreement with the experiment.

In most cases, the electrons in the conduction bands and holes in the valence bands are pair to pairs created and annihilated, we can write the electron–phonon interaction Hamiltonian in terms of the exciton–phonon interaction $H_{\text{ex-ph}}$ [23]:

$$H_{\text{ex-ph}} = \sum_{n,n',k,q} V^{nn'}(q) B^{+}_{n,k+q} B_{n',k}(a_q + a^{+}_{-q}), \tag{11.1}$$

where (B^+, B) and (a^+, a) are exciton and phonon (creation, annihilation) operators, respectively. q is the phonon wave vector. $V^{nn'}$ is the matrix element for the exciton–phonon interaction. n' and n specify the initial and the final electronic states, respectively. The exciton–phonon interaction is usually weak and can be written as the sum of electron–phonon and hole–phonon interactions acting separately on the electron and hole bound in the exciton complex. In this case, $V^{nn'}(q)$ can be written as [23, 24]:

$$V^{nn'}(q) = \int d^3r \phi^*_n(r)\phi_{n'}(r)[F^{\mathrm{e}}_q \exp(-iq \cdot rm_{\mathrm{e}}/M - F^{\mathrm{h}}_q \exp(iq \cdot rm_{\mathrm{h}}/M)], \tag{11.2}$$

where m_e and m_h are the electron and hole masses, respectively. $M = m_e + m_h$ is the total mass. $\phi_n(r)$ is the exciton wave function. F_q^e and F_q^h are electron– and hole–phonon coupling constants, respectively.

For the interaction of exciton with acoustic phonon, two types of interactions have been considered. The first one is the deformation potential interaction, for which we have:

$$F_q^{e(h)} = \sqrt{\frac{\hbar}{2\rho v V q}}\mathrm{D}_{e(h)} \tag{11.3}$$

where v is the velocity of sound, ρ the density of the crystal, V the volume of the crystal, and D_e, D_h are the deformation potential constants of conduction band and valence band, respectively. Since the transverse acoustic (TA) phonons produce a shear strain and hence have a smaller deformation potential than the longitudinal acoustic (LA) phonons, the dominant contribution is from the LA phonons. The second type is the piezoelectric interaction, which in wurtzite structures writes [25]:

$$\begin{aligned} F^{e}_{q,\mathrm{TA(LA)}} &= F^{h}_{q,\mathrm{TA(LA)}} \\ &= \frac{4\pi e}{\varepsilon_0}\sqrt{\frac{\hbar}{2\rho v_{\mathrm{TA(LA)}} V q}}\left[e_{15}\sin^2\theta(U_z\sin\theta + U_y\cos\theta)\right. \\ &\quad \left. + e_{13}U_y\sin\theta\cos\theta + e_{33}U_z\cos^2\theta\right] \end{aligned} \tag{11.4}$$

where v_{TA}, v_{LA} are the transverse and the longitudinal sound velocities, respectively. $(i=x, y, z)$ is the displacement of atoms in the unit cell. θ is the angle between $\vec{q}$ and z axis, and e_{15}, e_{33}, e_{13} the nonzero components of the piezoelectric tensor. The angular factors in the above equation can be spherically averaged to give effective piezoelectric constants for LA and TA phonons, respectively.

$$< e_{\mathrm{LA}}^2 >_\theta = \frac{1}{7}e_{33}^2 + \frac{4}{35}e_{33}(e_{31} + 2e_{15}) + \frac{8}{105}(e_{31} + 2e_{15})^2, \tag{11.5}$$

$$< e_{\mathrm{LA}}^2 >_\theta = \frac{2}{35}(e_{33} - e_{31} - e_{15})^2 + \frac{16}{105}e_{15}(e_{33} - e_{31} - e_{15}) + \frac{16}{35}e_{15}^2. \tag{11.6}$$

The interaction of the exciton with non-polar transverse optical (TO) phonons can be simply described as a deformation potential. This mechanism of interaction can be written as [26]:

$$F_q^{e(h)} = \sqrt{\frac{\hbar}{2\mu N\omega}}\frac{D_{e(h)}}{a}, \tag{11.7}$$

where μ is the reduced mass of the atoms in the unit cell, N is the number of unit cells in the crystal, a is the lattice constant and $\hbar\omega$ the optical phonon energy.

In the case of LO phonons, besides the non-polar deformation potential interaction, there is an additional Fröhlich interaction. The F_q for the Fröhlich interaction is:

$$F_q^{\mathrm{e}} = F_q^{\mathrm{h}} = \left[\frac{2\pi e^2\hbar\omega_{\mathrm{LO}}}{Vq^2}(\varepsilon_\infty^{-1} - \varepsilon_0^{-1})\right]^{1/2}, \tag{11.8}$$

where $\hbar\omega_{\mathrm{LO}}$ is the LO phonon energy. ε_∞, ε_0 are the high- and low-frequency dielectric constants, respectively. It can be seen that the Fröhlich interaction is strongly wave vector q dependent. In wurtzite III-nitride semiconductors, the A_1 (TO) and E_1 (TO) phonons are not purely transverse modes, the Fröhlich interaction with these TO-like modes can be strongly enhanced over a range of angles with respect to the c-axis [27]. The Fröhlich interaction with the LO-like modes, i.e. A_1 (LO) and E_1 (LO), is similar to eqn (11.8) [27]. Due to the strong ionic nature of nitride semiconductors, the Fröhlich interaction is the strongest one among other interactions. It plays a dominant role in determining the cooling of hot carriers and the band-edge optical transitions, which will be addressed in the sections below.

11.3 LO phonon replica

A most important result of exciton–phonon interactions in semiconductors is the appearance of phonon-assisted emissions of free and bound excitons in the luminescence spectra. In the past decades, phonon replicas have been observed in the photoluminescence (PL) spectra in most II–VI semiconductors [28–32] and in some ionic crystals [33]. Their origin has been thoroughly investigated. The distribution of the emission intensities between the main peak and the phonon replicas has been found to depend strongly on the exciton–phonon coupling strength. Within the Frank–Condon approximation, this coupling is expressed by the Huang–Rhys factor S [34]. At low temperature, the relationship between the intensity of the nth phonon replica I_n and the main emission I_0 is given by:

$$I_n = I_0\frac{S^n}{n}, \quad n = 0, 1, 2, \ldots \tag{11.9}$$

where the S factor is defined as [34–36]:

$$S = \sum_q \frac{|V(q)|^2}{(\hbar\omega_{\mathrm{LO}})^2}, \tag{11.10}$$

$V(q) = \int d^3r\phi(r)^2[F_q^{\mathrm{e}}\exp(-iq \cdot rm_{\mathrm{e}}/M) - F_q^{\mathrm{h}}\exp(iq \cdot rm_{\mathrm{h}}/M)]$. The parameters in $V(q)$ have been discussed in Section 11.2.3. Since the LO phonon Fröhlich interaction is the strongest, usually only the LO phonon replicas are $V(q)$. observed. Thus, usually, the Fröhlich interaction is taken into account in calculating $V(q)$. The S value for LO phonons therefore depends on the ionic nature of the semiconductor, on the wave vector q, and on the separation of Fourier transformed charge distributions of electron and hole.

For bound excitons, S depends on the spreading of wave function of the bound exciton in k-space. In the case of shallow donors or acceptors, the extension of the wave function in k-space is restricted to the vicinity of the Brillouin zone center. Only those phonons with wave numbers close to the Brillouin zone center can participate to the emission process. The S factor for a bound exciton can be written as [35, 37]:

$$S = \frac{e^2}{\hbar\omega_{\mathrm{LO}}}\left(\frac{1}{\varepsilon_\infty} - \frac{1}{\varepsilon_0}\right)\int |\rho(q)|^2 d^3q, \qquad (11.11)$$

where $\rho(q)$ is the Fourier transform of the charge density. In the limiting case of the hydrogen-like impurity, the S factor can be simplified as [35]:

$$S = \frac{\mathrm{e}^2}{\hbar\omega_{\mathrm{LO}}}\left(\frac{1}{\varepsilon_\infty} - \frac{1}{\varepsilon_0}\right)\frac{5}{16a^*}, \qquad (11.12)$$

where a^* is the effective Bohr radius of the impurity.

For free Wannier exciton in the lowest-lying 1S state, the $|V(q)|^2$ for Fröhlich interaction is strongly q dependent [2, 26, 32], its maximum value is at $q \approx 1/a_{\mathrm{B}}$, where a_{B} is the Bohr radius of the 1S exciton. In this case, the S value for free exciton is similar to those for hydrogen-like impurities and is proportional to $1/a_{\mathrm{B}}$. However, there is a difference between the free exciton and bound excitons, the free exciton can migrate in the crystal. It has a kinetic energy and thus a center-of-mass wave vector K. Those excitons with $K \neq 0$ are not allowed to recombine radiatively due to the momentum conservation law. The interaction with LO phonon transfers the momentum to the phonon and allows the exciton to recombine. At higher temperatures, this momentum transfer by LO phonons is more likely to take place. Detailed investigations of the line-shape of LO phonon replicas showed that the shape of phonon replicas represents a good accuracy to the equilibrium Maxwell distribution of the exciton kinetic energies and can be expressed as the following power law equation [28, 29, 32]:

$$I_n(E) \propto E^{1/2}\exp(-E/kT)W_n(E), \qquad (11.13)$$

where $E^{1/2}$ reflects the density of states in the parabolic exciton band (valid for a three-dimensional, bulk crystal). The exponential term reflects the exciton statistics, which can be taken as a Maxwell–Boltzmann distribution at low excitation intensities. T is the lattice temperature. $W_n(E)$ is the probability that the free exciton of a given kinetic energy E recombines via nth phonon-assisted process. For 1 LO phonon-assisted process, experimentally, it is found that the probability $W_1(E)$ is proportional to the exciton kinetic energy E. Based on a simple argument, when q is small, it is theoretically found that $W_1(E) \propto |V(q)|^2 \propto q^2$. Due to the conservation of momentum in the phonon assisted recombination process, $q \approx K$. Thus, $W_1(E)$ is also proportional to the exciton kinetic energy.

For the second LO phonon-assisted processes, it is found experimentally that the probability $W_2(E)$ is independent of the exciton kinetic energy [29]. Theoretically, a perfect calculation of $W_2(E)$ should include the scattering of exciton from the 1S state through the intermediate states, involving discrete and continuum exciton states, to the final radiative state by emission of 2 LO phonons [28, 32]. It is impractical to calculate this sum of intermediate states exactly. After several simplifications [28], it can be shown that the lineshape of the second LO phonon replica around the threshold is also independent on the exciton kinetic energy. The independence of the lineshape of $I_2(E)$ on the exciton kinetic energy is important experimentally. It gives a direct information about the concentration and about the distribution of free excitons in the semiconductor.

The above power law lineshape of the LO phonon replicas results in the following important predictions [32]:

1. The lineshape of phonon replica is asymmetric. Its peak energy depends on the temperature. The shift Δ of the line maximum from its low-energy threshold increases linearly with temperature, i.e., $\Delta = (3/2)\,kT$ for 1 LO and $\Delta = (1/2)\,kT$ for 2 LO phonon replica.
2. The line width also increases linearly with temperature. The line width is $3kT$ for 1 LO and $2kT$ for 2 LO replica.
3. The ratio of the integrated intensity of 1 LO replica to that of 2 LO replica, i.e., $I_{1\mathrm{LO}}/I_{2\mathrm{LO}}$, also depends linearly on temperature.

These predictions show that the lattice temperature has a great influence on the distribution of luminescence among the LO phonon replicas and the zero-phonon line.

The LO phonon replicas of free and bound excitons in GaN were observed 30 years ago [38], but it was only recently, with the breakthrough in the growth methods and the improvement in the quality of GaN material, that the LO phonon replicas in GaN could be systematically studied. The study of the LO phonon replica in GaN showed that the physics of the phonon replica, obtained mostly in II–VI semiconductors, works well when applied to nitride semiconductors [39–47]. Figure 11.1 shows a typical temperature dependent PL study of an undoped GaN sample [39, 40]. At low temperatures, the spectra are dominated by neutral donor bound exciton emission (I_2 line). Free exciton A and B emissions are also seen in the spectra. At the low energy side of these main peaks, two groups of peaks can be well resolved in the spectra. The peaks marked as A-LO (I_2-LO) and A-2LO (I_2-2LO) correspond to the

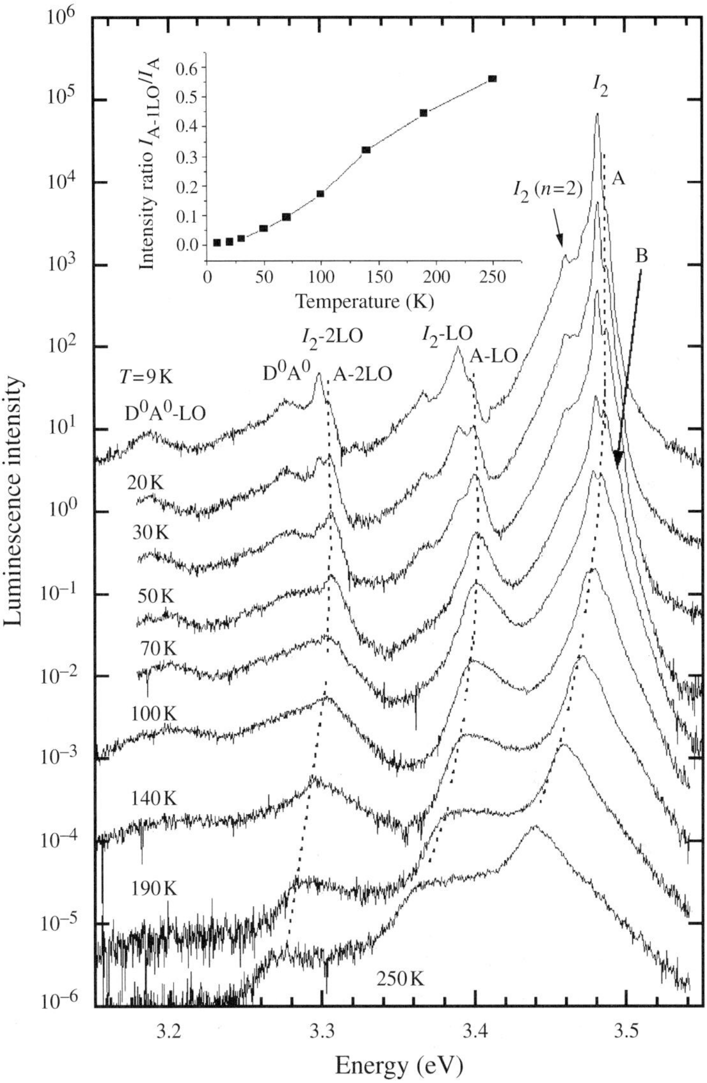

Fig. 11.1 Temperature-dependent PL spectra of a high-quality undoped GaN sample grown on sapphire. The temperature for each spectrum is indicated in the figure. The inset of the figure shows the intensity ratio of $I_{A\text{-}1LO}/I_A$ as a function of temperature. (After ref. [40].)

recombinations of A exciton (neutral donor bound exciton) with simultaneous emission of 1 LO and 2 LO phonons, respectively. The emission band near 3.27 eV comes from donor–acceptor pairs (DAP) recombinations, their 1 LO phonon replica around 3.18 eV is also seen in the spectra. A close investigation of spectrum at 9 K shows that the Huang–Rhys factor S for free exciton A, neutral donor bound exciton I_2, and DAP recombinations are around 0.0077, 0.0013, and 0.7, respectively. The values of the S factor for the free exciton and for the neutral donor bound exciton are close to those reported in ref. [46]. The small value of S for the neutral donor bound exciton indicates that the coupling between the LO phonon and the neutral donor bound exciton is weak. This weak coupling is caused by the less spreading of the wave function of donor bound exciton in k-space. Thus, fewer phonons are involved in the coupling [41]. The large S value for DAP is due to the fact that the S value of DAP depends on both the distance between and the localization of charges at the donor and the acceptor centers [35, 37]. The more distant the donor and acceptor pair, the larger the S value observed. A slightly smaller S value of 0.6 for DAP was also reported in GaN [42]. It is possibly due to the difference in excitation intensity in these PL measurements. Since the value of S for DAP is very sensitive to the distance between the donor and acceptor pair, increasing the excitation intensity would make the closer DAP to have more chance to recombine. Thus, a smaller value of S is expected [48].

With an increase in temperature, the I_2 line progressively quenches to the benefit of the free exciton line A. The free exciton line A becomes the dominant one in the spectrum taken at 100 K. Relative to the free exciton A, the A-1LO replica is getting remarkably strong at higher temperatures. The inset of the figure shows the intensity ratio of $I_{A-1\mathrm{LO}}/I_A$ as a function of the temperature. The ratio is about 0.6 at temperature of 250 K, significantly larger than that at temperature of 9 K. This indicates that the coupling strength between the free exciton A and the LO phonon increases with the temperature. This increase in the coupling strength is also reported by other groups [43, 46]. Within Segall and Mahan's theory [28], the phonon-assisted excitonic transition probability depends on the kinetic energy of exciton [eqn (11.13)]. Since the kinetic energy of exciton increases with the temperature, an increase in intensity ratio $I_{n\mathrm{LO}}/I_0$ with temperature would be expected. Various values of S factor reported by different authors might be due to the temperature effect [44–47]. With an increase in temperature, the linewidths of A-LO replicas becomes broader. The influence of temperature on the linewidth of A-LO replicas agrees well with the theoretical predictions [32]. The dotted lines in the figure indicate the temperature variation of the peak energies of the free exciton A and its phonon replicas. The peak energy of free exciton A tracks the bandgap of GaN in the whole temperature range, while for the 1LO and 2LO phonon replicas, the temperature shifts of the peak energies are small, compared with the shift of free exciton A peak. The smaller shifts also agree well with the theoretical predictions. The line-shape of phonon replicas shows some departure from eqn (11.13), possibly due to the scattering by point defects [49]. The dependences of line-shape and linewidth of LO phonon replicas, and the intensity ratio of 1 LO to 2 LO replicas on temperature in high-quality GaN sample agree well with the three predictions [44–46].

Nitride semiconductors are known for their high efficiency in luminescence despite the huge densities of defects. Study of the defect properties may provide additional knowledge on the device performances. It has been found that defects not only affect the distribution of luminescence between the phonon–replicas and the main peak, but also modify the shape of the phonon replicas [49]. When GaN contains high density of residual donors, the efficiency of phonon-assisted free excitonic recombination is low and a deviation of the lineshape of 1 LO phonon replica from eqn (11.13) is observed [49]. It was speculated that these unusual observations were caused by the enhanced scattering of exciton–polariton into the radiative branch of the zero phonon-free exciton A by point or extended defects, which assist the momentum transfer. In another report [50], similarly low efficiency of phonon-assisted free excitonic recombination was observed in GaN with p-type doping. The origin was also considered to be the defect enhancement of the scattering. Later in a PL study of GaN films grown with different polarity, it was found that the polarity of GaN film also has a great influence on the observation of LO phonon replicas [51]. When growing samples on Ga-terminated surface, relative to the zero phonon peak, the LO phonon replicas are strong; whereas when growing on N-terminated surface, the replica is hardly seen. Since the polarity affects the growth mode and thus affects the incorporation of defects in GaN epilayers, the incorporated defects was also thought to rule the observation of phonon replicas. It is interesting to compare the effect of polarity in GaN [51] with recent observation in ZnO [52], where the phonon replicas are much stronger in the PL spectra measured from the O-terminated surface than from Zn-terminated surface. Compared with the GaN, the effect of polarity is reversed in ZnO. It was argued that the Zn surface is easier to be contaminated. As such a defect mechanism was also thought to be responsible for the observations.

11.4 Band-edge excitonic linewidth

The band-edge optical properties of GaN are dominated by excitonic effects even at room temperature. Such excitonic effects strongly affect the linear and nonlinear optical properties of the material and hence have an interest in device applications. The interactions between excitons and phonons, however, will ionize the exciton into free electron and hole in the continuum of states or scatter the exciton into discrete exciton states. At low temperatures, since the population of LO phonon is vanishingly small, the scattering is mainly ruled by acoustic phonons. At high temperatures, the LO phonon Fröhlich scattering dominates. The lifetime of exciton, as a result of the scattering by phonons, will be significantly reduced at high temperatures. An increase in homogeneous linewidth of exciton can be observed in the temperature-dependent absorption or emission or reflectance spectra. Conversely, study of the temperature-dependent excitonic linewidth will provide important information about the exciton–phonon interactions.

According to Segall's expression [23], the total linewidth of exciton contains two contributions: the inhomogeneous broadening and the homogeneous broadening. Within the adiabatic approximation, the linewidth of the lowest-lying $1S$ exciton state can be written as:

$$\Gamma(T) = \Gamma_0 + \gamma_{\text{ph}} T + \Gamma_{\text{LO}} N_{\text{LO}}, \tag{11.14}$$

where Γ_0 is the inhomogeneous linewidth due to the exciton–exciton, exciton–carrier interaction [53], and the scattering by defects, impurities, and alloy fluctuations. In low-dimensional heterostructures, the size fluctuations also contribute to Γ_0. It is independent of temperature. The second term $\gamma_{\text{ph}} T$ is due to acoustic phonon scattering. The quantity γ_{ph} represents the acoustic phonon coupling strength, mainly caused by the deformation potential mechanism. The T in $\gamma_{\text{ph}} T$ is a result of linear dependence of the acoustic phonon occupation number on the temperature T. In most semiconductors, due to the small sound velocity, the scattering by acoustic phonon occurs within the same excitonic band state [23]. The third term is the linewidth due to the LO phonon scattering. Γ_{LO} is the exciton-LO phonon coupling strength, and $N_{\text{LO}} = 1/[\exp(\hbar\omega_{\text{LO}}/kT) - 1]$, is the LO phonon occupation number. For those excitons, which are not in the lowest-lying states, the phonon occupation number N_{LO} is replaced with $2N_{\text{LO}} + 1$ [54], since both phonon absorption and emission processes can now take place.

Various models, slightly different from eqn (11.14), appear in the literature. For example, a Bose–Einstein type expression, in which both acoustic and optical phonons are taken into account as an average effect on the excitonic line width, is used to describe the temperature dependent excitonic linewidth. The linewidth is then written as [54]:

$$\Gamma = \Gamma_0 + \Gamma_{\text{ep}}/[\exp(\theta/T) - 1], \tag{11.15}$$

where θ is an average of phonon temperature, Γ_{eq} is an average exciton–phonon coupling strength. This model gives the information about an average strength of the exciton–phonon scattering by all phonon branches. In another model [55], scattering by ionized impurities is included in fitting the experimental line width. Care should be taken when using this model since more parameters are involved in the fitting process [56].

Temperature-dependent linewidths of excitons in GaN have been measured by a variety of spectroscopic methods. These methods include absorption [57], PL [58, 59, 46], reflectance [60, 61], electroreflectance [62], spectroscopic ellipsometry [63] and femto second four-wave-mixing studies [64]. However, the mechanism responsible for the linewidth broadening is under debate. Scattering of exciton by different phonon branches and different models have been considered by different authors. In refs [57, 58, 60, 61, 64], both acoustic phonon and LO phonon scattering are considered, while in refs [59, 61], instead of acoustic phonon, the authors showed that scattering by the low frequency branch of E_2 phonon gives a better fit to the experiment linewidth. In refs [62, 63], the contribution by acoustic phonon is completely neglected, an exciton average LO phonon interaction model was used to fit the experimental results. In ref. [46], a complicated model involving contributions of the acoustic phonon and the average LO phonons was used to fit the experimental linewidth. In these reports, some authors used the full width at half maximum (FWHM) of exciton peak to obtain the exciton–phonon coupling parameters while others used the half-linewidth of exciton peak. It is worthwhile to mention that the values obtained by fitting the FWHM of exciton peak are two times larger than those obtained by fitting the half-linewidth. Table 11.2 summarizes the exciton–phonon coupling parameters, obtained by fitting the experimental linewidth, and measured by various spectroscopic methods in these reports. Note that some values of parameters in the table are divided by 2, since they were obtained by fitting the FWHM of exciton peak.

The parameter Δ in the table is a half-value of the homogeneous linewidth at 300 K, calculated by the model used in the reference. From the table, it is clearly seen that the values for exciton–phonon coupling strength, even when obtained by considering the same model, are not consistent. The values cover a wide range. We note that the calculated Δ, which represents the observed homogeneous linewidth at 300 K very well, is much smaller when the spectra were measured by electroreflectance spectroscopy, and is much larger by PL. The other methods give moderate values of Δ. This difference suggests that the mechanism for linewidth broadening or errors generated during the fitting of the experimental results is different from measurement to measurement. In refs [62] and [63], the inhomogeneous linewidths of free excitons A and B are larger than the homogeneous ones at 300 K. Moreover, the A and B excitons are not well resolved even at very low temperature. This may plague the analysis of experimental linewidth, and errors would be generated during the fitting of the experimental results. In PL measurements, when at high temperatures, the linewidths of A and B excitons are much broader than the separation between the A and B peaks [65], and the two transitions will overlap significantly. This overlap, together with an increase in the relative intensity of B exciton at high temperature, would also make the experimental results difficult to be fitted. Errors would be generated during the fitting. It has been argued that four-wave-mixing studies might give the most accurate parameters presently for exciton–phonon coupling strength, since the excitonic features are several orders of magnitude stronger and the background, coming from the contribution of free carriers, is significantly suppressed [64]. Improving the quality of GaN may give more accurate values of exciton–phonon coupling strengths in the measurements.

It is interesting to compare Γ_{LO} in GaN with those in other III–V semiconductors and II–VI semiconductors. The Γ_{LO} values for GaAs [66], ZnSe [67], CdTe [23], CdS [23], and ZnO [68] are 5, 81, 17, 41, and 876 meV, respectively (note that the large value of Γ_{LO} obtained in ZnO is by fitting the FWHM of the excitonic peak in the absorption spectra and should be divided by 2). We see that the values of Γ_{LO} in GaN, obtained by various methods, are much larger than those in GaAs and in most II–VI semiconductors. The large value of Γ_{LO} suggests that the exciton-LO phonon Fröhlich interaction significantly affects the room temperature and high temperature performances of GaN-based devices.

Table 11.2 Reported line width parameters for free exciton transitions in GaN

Method	Γ_0 (meV)	γ_{ph} (μeV/K)	Γ_{LO} (meV)	Γ_{ep} (meV)	θ (K)	Γ_{E1} (meV)	Δ (meV)
[a] Absorption (A exciton)*	10/2	15/2	375/2				7.8
[b] PL (A exciton)	2.8	21	525				21.9
[b] PL (B exciton)	1.4	22	495				21.3
[c] PL (A exciton)*	5.69/2	41.8	165/2			10/2	7.5
[d] PL (A exciton)	3.15	15.3		140	748		25.2
[e] Reflectance (A exciton)	~1.5	9.5	208				10.8
[f] Reflectance (A exciton)	2.4		295				11.6
[f] Reflectance (A exciton)	2.4		220			3.7	10.3
[g] Electroreflectance (A exciton)	15			60	731		5.7
[g] Electroreflectance (B exciton)	13			74	773		6.1
[h] Spectroscopic ellipsometry	34			104	800		7.8
[i] TI Four-wave-mixing (B exciton)*	2.4/2	16/2	390/2				8.2
[i] SR Four-wave-mixing (B exciton)*	2.5/2	13/2	470/2				8.9

[a]Fischer, A.J., Shan, W., Song, J.J., Chang, Y.C., Horning, R. and Goldenberg, B. (1997). *Appl. Phys. Lett.* **71**, 1981.

[b]Viswanath, A.K., Lee, J.I., Kim, D. Lee, C.R., and Leem, J.Y., (1998). *Phys. Rev.* B **58**, 16333.

[c]Chichibu, S., Okumura, H., Nakamura, S., Feuillet, G., Azuiiata, T., Sota, T., and Yoshida, S., (1997). *Jpn. J. Appl. Phys.* **36**, 1976.

[d]Liu, W., Li, M.F., Xu, S.J., Uchida, K., and Matsumoto, K., (1998). *Semicond. Sci. Technol.* **13**, 769.

[e]Korona, K.P., Wysmolek, A., Pakula, K. Stepniewski, R. Baranowski, J.M., Grzegory, I. *et al.* (1996). *Appl. Phys. Lett.* **69**, 788.

[f]Siozade, L., Colard, S., Mihailovic, M., Leymarie, J., Vasson, A., Grandjean, N., *et al.* (1999). *Phys. Stat. Sol.* (b) **216**, 73.

[g]Li, C.F. Huang, Y.S., Malikova, L., and Pollak, F.H., (1997). *Phys. Rev.* B **55**, 9251.

[h]Petalas, J., Logothetidis, S., Boultadakis, S., Alouani, M., and Wills, J.M., (1995). *Phys. Rev.* B **52**, 8082.

[i]Fischer, A.J., Shan, W., Park, G.H., Song, J.J., Kim, D.S., Yee, D.S., *et al.* (1997). *Phys. Rev.* B **56**, 1077.

*The parameters are obtained by fitting the experimental full width at half maximum of exciton peak. Note that the parameters obtained by fitting the full width are two times larger than those obtained by fitting the half-width of exciton peak. Thus, the values of these parameters are divided by 2, as shown in the table.

11.5 Cooling of hot carriers in GaN

Nitride semiconductors are known for their applications in high power and high temperature devices. Therefore, the knowledge about the fundamental relaxation process of hot carriers and their distributions is very important for device engineers. Optical excitation of semiconductors at energies well above the bandgap is followed by a rapid relaxation in both momentum and energy of the photo-generated carriers. A variety of scattering processes contribute to the relaxation of hot electrons and holes and affect their distributions. There may be electron–electron scattering, hole–hole scattering, electron–hole scattering, electron–phonon interactions, hole–phonon interactions, and phonon–phonon interactions. Their relative importance in the relaxation of hot carriers is determined by the actual experimental conditions, in particular by the photon energy and the intensity of exciting laser light, and by the lattice temperature. Carrier–carrier scattering leads to the relaxation of both momentum and energy in the carrier system and drives the monochromatic or quasi-monochromatic energy distribution of carriers into a quasi-equilibrium distribution. The scattering rate depends strongly on the carrier density [69, 70]. Interactions with phonons transfer the excess energy of the carrier system to the lattice by creation of various phonons. The phonon–phonon interaction will affect the efficiency of this energy transfer.

The excitation intensity, and thus the density of carriers, has a great influence on the carrier–carrier scattering rates [71]. With an increase in excitation intensity, the carrier–carrier scattering rate increases. A simple model predicts that the carrier–carrier scattering rate increases linearly with the carrier density [70]. There is a critical carrier density, below which the carrier–LO phonon interaction dominates, and above which the carrier–carrier scattering dominates. This critical density is about $10^{19}\,\text{cm}^{-3}$ in GaN [72]. In the following, we discuss the relaxation of hot carriers in the low and high excitation intensity regions.

At low excitation intensities, carrier LO–phonon interaction dominates. The carrier system loses its kinetic energy through a cascade emission of LO phonons before it is thermalized with a well-defined temperature. In GaAs, the photo-excited electrons and holes are relaxed individually toward their respective band extrema by successive emission of LO phonons, until they do not have sufficient energy to emit a further LO phonon. The cooling of the remaining energy is followed by the acoustic phonon deformation potential interaction. In GaN, however, due to the large exciton binding energy, photo-generated electrons and holes directly form excitons assisted by emission of an LO phonon. The subsequent relaxation of the hot exciton occurs via a cascade emission of LO phonons. Evidence of this relaxation of electron–hole pairs comes from the PL excitation studies [45, 73]. The PL excitation spectra show periodic structure with an energy separation exactly equal to $\hbar\omega_{\text{LO}}$. Increasing the carrier density reduces the oscillation strength of the periodic structure.

At high excitation intensities, carrier–carrier scattering becomes more important. Moreover, the strengths of carrier–phonon interactions are greatly influenced by the high density of carriers [71]. The strength of carrier–phonon interactions is controlled by the hot phonon effect and the screening effect [71]. In GaN, the relative magnitudes of the energy loss rates, through the interactions with different branch of phonons, have been calculated in ref. [74]. Figure 11.2 shows the energy loss rates of electrons and holes, through the polar LO phonon Fröhlich interaction, TO, LA phonon deformation potential interactions, and the LA phonon piezoelectric interaction. In the case of electron–phonon interactions [Fig. 11.2(a)], the polar LO phonon interaction is the dominant one. At low carrier densities, it is about 2 orders of magnitude stronger than that of TO phonon deformation potential interaction, and 4 orders of magnitude stronger than the LA phonon piezoelectric interaction. Compared with the energy loss rates of electrons, a slight difference for holes can be seen in Fig. 11.2(b). The non-polar interaction with TO phonon becomes as important as the polar interaction with LO phonons in relaxing the energy of holes, especially when the carrier density is high. Interactions with LA phonons through both the deformation potential and the piezoelectric mechanisms are greatly enhanced for hole. The energy loss rates, through the piezoelectric interaction with the TA phonons, which is not shown in Fig. 11.2, should be comparable to those through the piezoelectric interaction with the LA phonons [75]. A characteristic feature in the figure is that, at high carrier densities, due to the screening effect and the hot phonon effect, the energy loss rate through the LO phonon emission reduces significantly, especially for electrons. A comparison between the screening and hot phonon effects shows that the hot phonon effect is the dominant one in reducing the energy loss rate. The hot phonons, generated by the relaxation of hot carriers, enable the hot carriers to have an increased probability of reabsorbing phonons. Thus, a reduction in energy loss rate of hot carriers is expected.

The lifetimes of phonons determine the hot phonon effect. At high excitation intensities, phonons, especially the LO phonons, which are created by the relaxation of hot carriers, do not have enough time to decay. A build-up of non-equilibrium phonon populations is expected. This build-up of non-equilibrium LO phonon populations is evidenced from a low temperature Raman study [76]. Figure 11.3 shows the polarized Raman spectra of GaN sample taken at 25 K. The sharp peak at 757 cm^{-1} in (a) is the scattering by E_g phonon of the substrate sapphire, the shoulder around 741 cm^{-1} comes from the A_1 (LO) phonon of GaN. The peaks at 574 cm^{-1} and 538 cm^{-1} in spectrum (b) correspond to E_2 and A_1 (TO) phonons, respectively. The anti-Stokes Raman spectra in (c) and (d) show that the A_1 (LO) phonon is clearly seen but E_2 and A_1 (TO) phonons are not. Similar results for E_1 (LO) and E_1 (TO) phonons in GaN are also observed. Under thermal equilibrium conditions, the ratio of anti-Stokes to Stokes intensities for a phonon mode in Raman spectra is proportional to $N_{LO}/(N_{LO} + 1)$. For LO phonon with $\hbar\omega_{LO} \sim 92$ meV, the ratio is below 10^{-10} at 25 K. Thus, the anti-Stokes peaks should be too weak

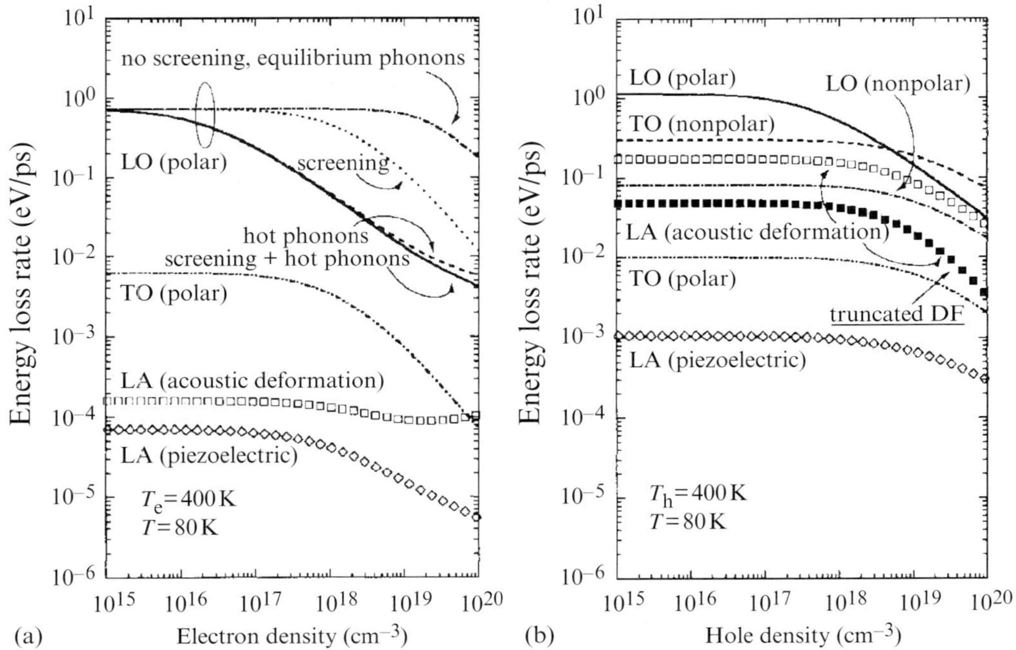

Fig. 11.2 The calculated energy loss rates of electrons (a) and holes (b), as a function of the carrier density in GaN. The energy loss rates due to the scattering by various phonon modes are indicated in the figure. (After ref. [74].)

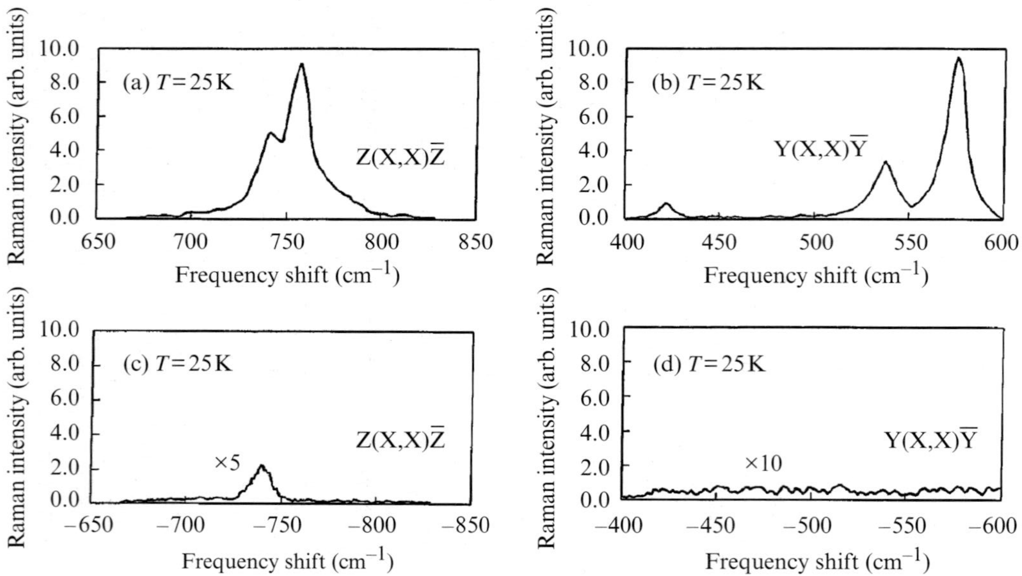

Fig. 11.3 Polarized–Stokes Raman spectra (a) and (b) and anti-Stokes Raman spectra (c) and (d). The spectra were measured from an undoped GaN sample at 25 K. The scattering geometry for each spectrum is indicated in the figure. Notice that the anti-Stokes Raman spectrum is multiplied by a factor of 5 in (c) and 10 in (d). (After ref. [76].)

to be observable at this temperature. The observation of anti-Stokes A_1 (LO) and E_1 (LO) peaks indicates that a significant number of non-equilibrium LO phonons have been created in the material. The non-equilibrium LO phonons are created by the relaxation of energetic electron–hole pairs, excited by incident laser light, through the strong Fröhlich interaction.

In GaN, the lifetime of A_1 (LO) phonons, obtained from time-resolved Raman studies, is about 5 ps [77, 78], while in AlN, the lifetime of A_1(LO) phonons, obtained from the phonon linewidth in Raman spectra, has been reported in the range from 0.45 to 0.75 ps [79, 80]. The lifetimes of phonons are determined by the anharmonic decay of a phonon into other low energy Brillouin zone phonons such that the energy and momentum are conserved [81, 82]. Crystal imperfections such as impurities and defects are also shown to affect the lifetimes of phonons, since the crystal imperfections can destroy the phonon propagation [80].

Besides the theoretical calculations, the rates of carrier-LO phonon interactions are also determined experimentally. Femtosecond pump and probe technique has been used to probe the relaxation rate of hot electrons in n-type doped GaN [83] and the relaxation rate of hot holes in p-type doped GaN [84]. In these reports, the measured LO phonon emission time is $\tau_e \sim 0.2$ ps for electron and $\tau_h \sim 0.6$ ps for hole. These values are about 20 times larger than those calculated theoretically, possibly due to the hot phonon effect. The time for electron–A_1(LO) phonon scattering, determined from subpicosecond time-resolved Raman scattering, is ~0.05 ps [85]. The scattering time for E_1 (LO) phonons was also around 0.05 ps. Considering the interaction with the two branchs of LO phonons, it showed that the electron–LO phonon scattering rate in GaN is about one order of magnitude larger than that obtained in GaAs [85].

The hot phonons have a great influence on the exciton stability. When the occupation number of hot phonon is large, due to the scattering with hot phonons, exciton will be ionized even at very low temperature. As a result, electron–hole pairs lose their correlation immediately after their generation and rapidly form a hot plasma, and then thermalize among themselves with a characteristic temperature much higher than the lattice temperature. This is known as the heating effect. The heating of an electron and hole plasma has a great influence on the performance of GaN-related devices, which operate in the high injection regime. This heating effect is directly reflected by an enhancement of the high-energy wing of the near band emission in PL spectra under high photo excitations. Figure 11.4 shows the near bandgap PL spectra of a GaN sample at temperature of 80 K [86]. The excitation photon energy is 4.66 eV, thus the excess energy of a photo-created electron–hole pair is 1.18 eV, much larger than the energy of LO phonon. Relaxation of hot carriers through a cascade emission of LO phonons is possible. At high excitation intensities, hot phonon effects, which reduce the energy loss rate of the hot carriers, would be expected. If the characteristic temperature of the thermalized carriers has not reached the lattice temperature before the recombination has taken place, the temperature of carriers will be reflected in the PL spectra. In this case, the

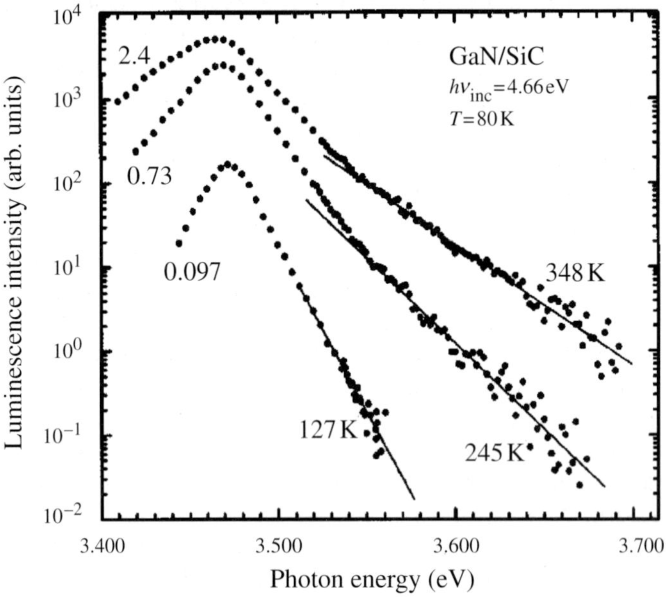

Fig. 11.4 PL spectra of a GaN epilayer grown on SiC with a lattice background doping below 10^{16} cm^{-3}. The exciting photon energy is 4.66 eV and the temperature is 80 K. The excitation power density in MW/cm^2 for each spectrum is indicated on the left-hand side of the curve. Electron temperature is indicated on the right-hand side of the curves. (After ref. [86].)

high-energy tail of the band-edge PL can be well fitted by:

$$I(\hbar\omega) \propto \exp[-(\hbar\omega - E_g)/kT_e], \tag{11.16}$$

where T_e is the carrier temperature. The fitted value of T_e for each spectrum, taken under different excitation intensity, is indicated in the figure. An increase in T_e with an increase in excitation intensity is clearly seen. A detailed analysis of the relationship between T_e and the excitation intensity suggests that hot holes are unable to establish a smooth distribution above the threshold of optical phonon emission before they recombine with electrons [86].

The dynamic evolution of the exciton temperature can be observed in time-resolved PL studies. In the case of low excitation conditions, the near band emission is dominated by excitonic transitions. The photo-excited hot exciton relaxes down by a cascade emission of LO phonons until its kinetic energy is not sufficient enough to emit a further LO phonon. The residual energy is dissipated primarily through the acoustic phonon interactions. Since the interaction with acoustic phonons is about 3–4 orders of magnitude weaker than that with LO phonons, the interaction of excitons among themselves will establish a quasi-equilibrium distribution with a characteristic temperature. However, the temperature of excitons cannot be extracted directly from the free exciton recombination line, since those excitons with momentum $K \neq 0$ are not allowed to recombine as a result of the momentum conservation law. As discussed in Section 11.3, the momentum selection rule can be relaxed by the participation of phonons in the recombination. The line-shape of LO phonon replicas represents the Maxwell distribution of exciton kinetic energies with a characteristic temperature. The dynamic evolution of the shape of LO phonon replica thus directly reflects the cooling process of excitons through the emission of acoustic phonons. By fitting the 1LO phonon replica of free exciton in GaN with eqn (11.13), the exciton temperature at each time, delayed after the excitation pulse, can be obtained. Figure 11.5 (a) and (b) show the exciton temperatures as a function of the time delay after the excitation, at lattice temperatures of 20 K and 55 K, respectively [87]. The exciton temperatures rapidly cool down toward the lattice temperature within 150 picoseconds. The cooling is slightly faster in GaN than in GaAs [88] and in ZnSe [89]. Since the deformation potential interaction with the acoustic phonons is the dominant mechanism responsible for the cooling of excitons in this regime, the decrease in exciton temperature can be modeled with a standard deformation potential theory. The energy loss rate due to the deformation potential scattering is [90, 91]:

$$\left\langle \frac{dE}{dt} \right\rangle = \frac{8\sqrt{2}}{\pi^{3/2}} \frac{(m_x)^{5/2} D^2}{\rho \hbar^4} (k_B T)^{3/2} \left(1 - \frac{T_{\mathrm{L}}}{T}\right), \tag{11.17}$$

where m_x is the average mass of exciton, D is the acoustic deformation potential for exciton, ρ is the mass density, and T_{L} is the lattice temperature. The dashed lines in Fig. 11.4 are fitted with the above energy loss rate equation. The value of the deformation potential deduced from the fitting is 13 ± 1 eV, close to those obtained by other methods [92].

11.6 Exciton–phonon interaction in nitride low-dimensional structures

Exciton–phonon interactions in low-dimensional structures are more complicated due to the effects of quantum confinements on the electronic states and to the modifications of phonon modes [2]. In II–VI and GaAs-based quantum wells (QWs) or superlattices, when the temperature dependent excitonic linewidth is studied, a reduction in exciton-LO phonon coupling strength [93–96] with a decrease in QW thickness has been reported. This reduced coupling strength is caused by (1) the wave vector q dependent nature of Fröhlich interaction [2, 26, 32] and (2) the fewer density of states available to accommodate the scattered exciton [94]. When the thickness of QW is of certain values, where the separation of the confined states matches the LO phonon energy, an enhancement in the exciton–LO phonon coupling strength is

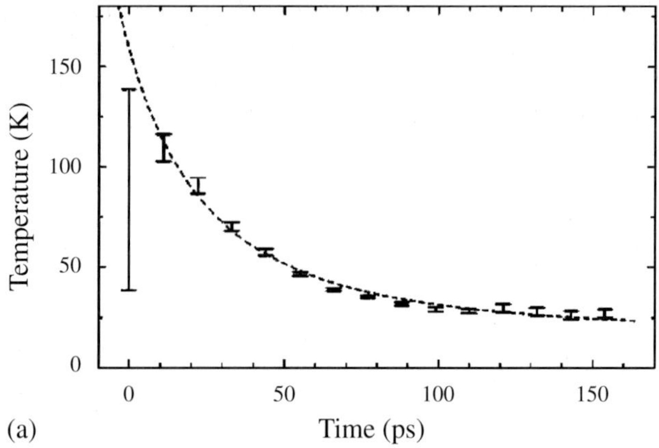

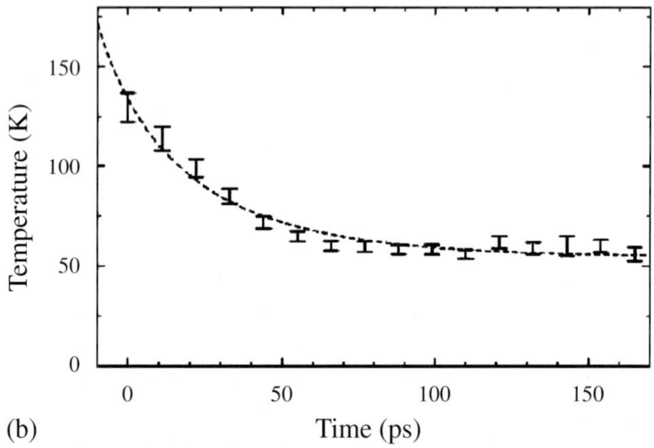

Fig. 11.5 The excitonic temperature in a GaN sample as a function of the time delayed after the excitation pulse. The temperatures were obtained by fitting the high-energy tail of the first LO phonon replica with eqn (11.13). The measurements were performed at lattice temperatures of 20 K (a) and 55 K (b). The dashed lines are obtained from fit to the data by using eqn (11.17). (After ref. [87].)

observed [95, 97]. In InGaAs/InP QWs, different values of the Huang–Rhys factor observed were attributed to the difference in the degree of exciton localization in the plane of the QWs [98, 99]. Since the hole has a larger effective mass than the electron, it localizes easier. As a result, charge separation between the localized hole and the delocalized electron is expected. The difference in charge separation, caused by the difference in localization, is responsible for the observation of different values of the exciton–phonon coupling strength in these QWs.

In the case of quantum dots (QDs), the situation is more complicated, and conflicting results have been reported. Based on the nature of wave vector q dependent Fröhlich interaction, it was first argued that due to the strong quantum confinement, which enhances the overlap of electron and hole wave functions, the exciton–LO phonon Fröhlich interaction should be vanished in small QDs [100]. This reduced interaction strength was supported experimentally by the Raman study of CdSe QDs [101]. By using a donor-like exciton model and an uncorrelated electron–hole wave function, a size-independent coupling strength was predicted theoretically and was also observed in the Raman study of CdSe QDs [102]. The charge neutrality makes the exciton–phonon coupling strength to decrease in small QDs. However, this decreasing polar coupling is accompanied by an increasing coupling to short wavelength phonons. These two effects may exactly compensate each other and a size-independent coupling strength would be expected [102]. In another report, by taking into account the valence band degeneracy and the conduction band non-parabolicity, the LO phonon coupling strength was shown to decrease first, then to increase below a certain value of the QD size [103]. However, the calculated absolute value of Huang–Rhys factor by this model is significantly smaller than the experimental ones. To explain the experimental observations of much larger Huang–Rhys factor in QDs, a non-adiabatic treatment of the exciton–phonon system has been considered [104, 105]. The calculation showed that the effects of non-adiabaticity are responsible for the observed enhancement of phonon-assisted excitonic transition probabilities. Alternatively, the localization of hole at the surface of QDs [106], and the spatial separation of electron and hole charge density distributions, as a result of particular quantum confinement and piezoelectric potential in the QDs [107], were used to explain the enhanced phonon assisted processes.

Compared with II–VI and other III–V semiconductors, the exciton–phonon interactions in GaN-based QWs and QDs are less studied. Moreover, the reported results are controversial, both reduced and enhanced exciton–phonon interactions have been reported. In the temperature dependent PL study of GaN/AlGaN QDs [108], a slow red shift of the PL peak energy (compared with the shift of GaN band gap) was found when the temperature was increased. This slow red shift was explained within the picture of a reduced exciton–phonon coupling strength in the QDs. In fact, the observed slow shift of PL peak energy has several origins. Since the inhomogeneous linewidth of the PL of QDs is very large [108], exciton localization and thermally activated exciton transfer between QDs may occur when the temperature is changed. These processes will lead to a complicated temperature evolution of the PL energy of QDs. Moreover, screening of the built-in electric field, induced by spontaneous and piezoelectric polarization, may be more important at high temperature due to the thermal activation of carriers [109]. To get a reliable strength of the exciton–phonon interaction from the temperature dependent PL study, these possibilities have to be considered. In another temperature dependent PL study of GaN/AlGaN superlattices [110], a fast red shift of PL peak energy and a large broadening of PL linewidth (about 30 meV from 150 K to 300 K) were observed when the temperature exceeds 150 K.

In addition to the temperature dependent PL studies, LO phonon assisted excitonic recombinations have been observed in GaN-based low-dimensional structures. Strong LO phonon replicas have been observed in the low-temperature PL spectra of GaInN/GaN QWs [111, 112], GaN/AlGaN QWs [113, 114], and GaInN/GaN QDs [115]. The observed Huang-Rhys factor S at low temperature is in a range from 0.07 to 0.8, much larger than those observed in bulk GaN. This indicates that the exciton–phonon interaction is greatly enhanced in these low-dimensional structures. The experimental results [111, 114] suggested that the symmetric properties of QWs, together with the alloy disorder, are responsible for the observed large S factor. Further PL studies of GaInN/GaN QWs and QDs show that the value of S factor depends strongly on the thickness of QWs or on the height of QDs. A continuous increase in S factor with an increase in thickness of QWs or height of QDs has been observed. A typical size-dependent PL study of $Ga_{1-x}In_xN$/GaN QD samples is shown in Fig. 11.6. The indium composition in all samples is the same and is in a range of 0.15 to 0.2. Microscopic structural characterizations showed that the height of QDs is in a range of 2–5 nm, whereas the full-width at half height is from 5 to 10 nm [115]. The PL spectra in the figure are normalized at the same peak intensity. Obviously, an increase in QD size from sample (a) to (c) is manifested by a decrease in PL peak energy. The 1 LO and 2 LO phonon

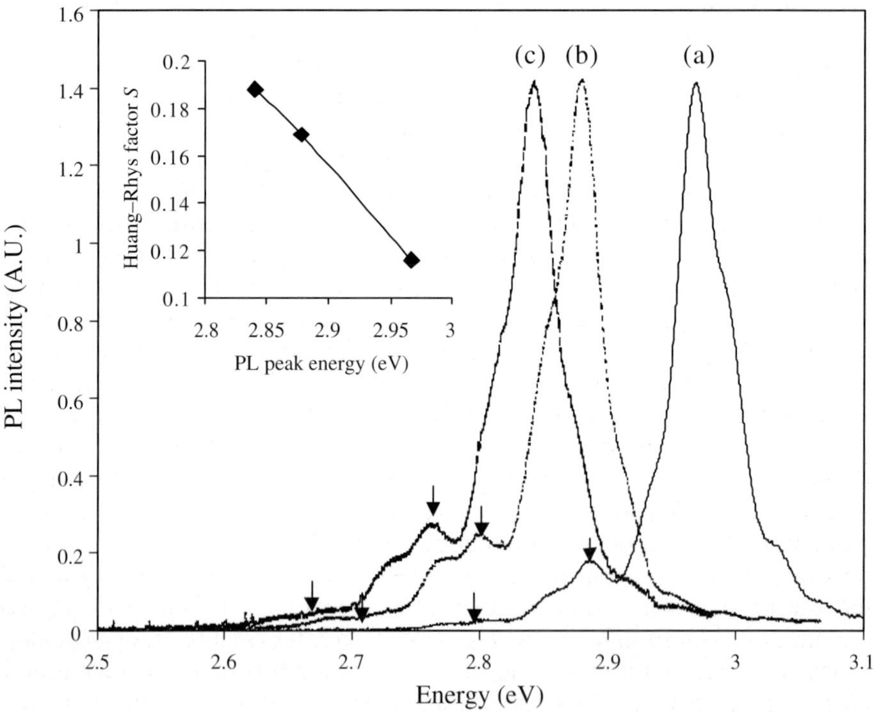

Fig. 11.6 (a), (b), and (c) represent the PL spectra measured at 8 K for three $Ga_{1-x}In_xN$/GaN QD samples. The LO phonon replicas are indicated as arrows in the figure. The inset of the figure shows the S factor as a function of the zero phonon peak energy.

replicas, as indicated by arrows in the figure, are clearly seen in each spectrum. The energy separation is 91 meV, which is very close to the energy of the LO phonon in bulk wurtzite GaN. By fitting the spectrum with Gaussian functions, the Huang-Rhys factor S is obtained. The value of S factor is plotted versus the zero phonon peak energy in the inset of the figure. An increase in S factor with a decrease in PL energy shows that the exciton– phonon interaction is stronger in larger QDs.

Since the Huang–Rhys factor depends strongly on the distributions of electron and hole charge densities, we speculate that the charge distributions have been changed in these QDs. A characteristic feature in GaInN/GaN low-dimensional structures is the strong internal electric field caused by piezoelectric and spontaneous polarizations. Such an electric field can be as large as 2.45 MV/cm. A direct effect of such electric field on the photo-generated electron–hole pairs is the progressive separation of electron and hole charge density with an increase in QD height. The reduced overlap of electron and hole charge densities is responsible for the larger S factor observed in larger QDs. By using the same Hamiltonian of exciton LO phonon interaction as in bulk GaN, and by using the electron and hole wave functions described in ref. [116], the calculated values of S factor agree fairly well with the experimental observation [117]. The calculation predicts that the in-plane localization of exciton in QWs also enhances the phonon-assisted recombinations.

It is known that temperature has a great influence on the phonon-assisted recombination processes. Within Segall and Mahan's theory [28], where the phonon assisted excitonic transition probability depends on the kinetic energy of exciton, a continuous increase in intensity ratio $I_{n\mathrm{LO}}/I_0$ with the temperature would be expected. This increase has been shown in the inset of Fig. 11.1. The temperature dependent LO phonon-assisted recombinations in GaN-based low-dimensional structures was studied recently. Figure 11.7 shows the intensity ratio of the first LO phonon replica to that of the zero phonon peak, as well as the PL peak energy of the zero phonon peak, measured from a GaN/$Al_{0.17}Ga_{0.83}N$ multiple QWs sample, as a function of the temperature. The well width is 16 monolayers (ML), i.e., 4.14 nm, and the barrier width is 30 nm. The expected electric field in the QWs is about 814 kV/cm [118]. An example of PL spectrum, taken at 8 K, is shown in the inset of the figure. Note the logarithmic scale of the PL intensity. Two replicas and a main zero phonon peak are clearly seen. The separation between adjacent peaks is 92 meV, which is in excellent agreement with the LO phonon energy in GaN. With an increase in temperature, the PL peak energy approximately tracks the band gap of bulk GaN until 50 K, above which a blue shift is observed. This blue shift is known as thermal delocalization of excitons from '+2 ML' to '+1 ML' regions, where '+2 ML' ('+1 ML') represents places in QWs where the well thickness is 2 ML (1 ML) thicker than the average thickness [118]. It is interesting to note that the intensity ratio of $I_{1\mathrm{LO}}/I_0$ is nearly constant in the region (i.e., from 80 K to 125 K) where the blue shift of PL peak energy is observed, while in other temperature ranges it increases with the temperature. The increase in the intensity ratio $I_{1\mathrm{LO}}/I_0$ with the temperature can be understood, similarly to the case of

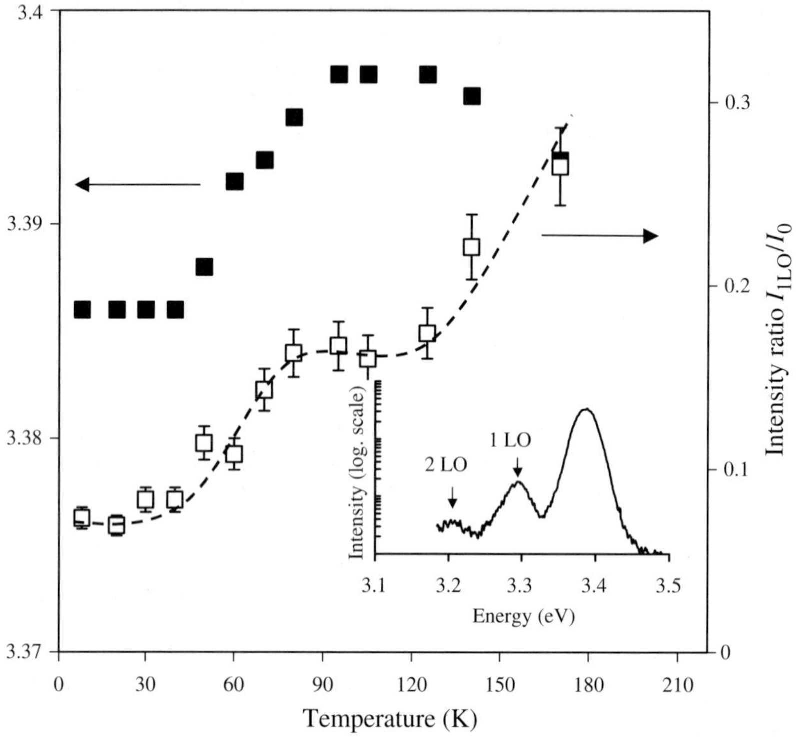

Fig. 11.7 The intensity ratio of the first LO phonon replica relative to that of the zero phonon peak, and the peak energy of the zero phonon peak, measured from a $GaN/Al_{0.17}Ga_{0.83}N$ multiple QWs sample, as a function of temperature. The inset of the figure shows a PL spectrum taken at 8 K. Note the logarithmic scale of the PL intensity.

bulk GaN, as a result of the thermal enhancement of the phonon-assisted excitonic transition probability. The small variation of the ratio in the temperature range of 80K to 125 K can be understood as a result of the change in Huang Rhys factor S. As discussed above, the *S* factor depends on the thickness of QW. The *S* factor, and thus the intensity ratio I_{1LO}/I_0, is larger in the +2 ML region than in the +1 ML region. The constant value of I_{1LO}/I_0 in the temperature range of 80 K to 125 K is a result of the balance between two competitive processes: (1) a thermally enhanced phonon assisted transition probability and (2) a thermally activated transfer of excitons from +2 ML to +1 ML region, which leads to a reduction of *S* factor and concomitantly a reduction in I_{1LO}/I_0.

Acknowledgements

We acknowledge fruitful discussions with Dr P. Lefebvre, Dr T. Taliercio, and Mr S. Kalliakos. We are particularly grateful to Prof. P. Y. Yu for illuminating discussions about exciton–phonon interactions and valuable advice, to Dr M. Leroux for valuable suggestions and providing the electronic data of Fig. 11.1, and to Dr S. K. Hark for valuable suggestions. Results presented in Fig. 11.6 and 11.7 were taken from samples prepared for us by Dr Nicolas Grandjean. Xue Bing Zhang acknowledges the hospitality of the University of Montpellier II and the French Ministry of Education for financial support.

References

1. Nakamura, S. (1999). *Semicond. Sci. Technol.* **14**, R27.
2. Yu, P. Y. and Cardona, M. (1996). In *Fundamentals of semiconductors, physics and materials properties*. Springer-Verlag, Berlin.
3. Suzuki, M. and Uenoyama, T. (1998). In *Group III nitride semiconductor compound, physics and applications* (ed. *B*. Gil), pp. 307–42. Clarendon Press, Oxford.
4. Gil, B. (1999). In *Semiconductors and semimetals*. Vol. 57, (eds J. I. Pankove, and T. D. Moustakas), pp. 209–274, Academic Press, San Diego.
5. Karch, K., Wagner, J.-M., and Bechstedt, F. (1998). *Phys. Rev. B* **57**, 7043.
6. Miwa, K. and Fukumoto, A. (1993). *Phys. Rev. B* 48, 7897.
7. Wagner, J.-M. and Bechstedt, F. (2000). *Phys. Rev. B* **62**, 4526.
8. Zhang, J. M., Ruf, T., Cardona, M., Ambacher, O., Stutzmann, M., Wagner, J.-M., and Bechstedt, F. (1997). *Phys. Rev. B* **56**, 14399.
9. Parlinskim, K., and Kawazoe, Y., (1999). *Phys. Rev. B* **60**, 15511.
10. Ruf, T., Serrano, J., Cardona, M., Pavone, P., Pabst, M., Krisch, M., *et al.* (2001). *Phys. Rev. Lett.* **86**, 906.
11. Siegle, H., Kaczmarczyk, G., Filippidis, L. Litvinchuk, A. P. Hoffmann, A. and Thomsen, C. (1997) *Phys. Rev. B* **55**, 7000.
12. Davydov, V., Yu. Kitaev, Yu. E., Goncharuk, I. N., Smirnov, A. N. Graul, J., Semchinova, O., *et al.* (1998). *Phys. Rev. B* **58**, 12899.
13. Kaczmarczyk, G., Kaschner, A., Reich, S., Hoffmann, A., Thomsen, C., As, D. J., *et al.* (2000). *Appl. Phys. Lett.* **76**, 2122.
14. Coffey, D. and Bock, N. (1999). *Phys. Rev. B* **59**, 5799.
15. Davydov, V., Yu. Klochikhin, A. A., Smirnov, M. B., Emtsev, V. V., Petrikov, V. D. Abroyan, I. A. *et al.* (1999). *Phys. Stat. Sol.* (b) **216**, 779.
16. Azuhata, T. Sota, T. Suzuki, K. and Nakamura, S. (1995). *J. Phys. Condens. Mater.* **7**, L129.
17. Filippidis, L., Siegle, H., Hoffman, A., Thomsen, C., Karch, K., and Bechstedt, F. (1996). *Phys. Stat. Sol.* (b) **198**, 621.
18. Demangeot, F. Frandon, J. Renucci, M. A. Briot, O., Gil, B. and Aulombard, R. L. (1996). *Solid Stat. Commun.* **100**, 207.
19. Klochikhin, A. A., Davydov, V., Yu. Goncharuk, I. N., Smirnov, A. N., Nikolaev, A. E., Baidakova, M. V., *et al.* (2000). *Phys. Rev. B* **62**, 2522.
20. Bungaro, C. and Gironcoli, S. de. (2000). *Appl. Phys. Lett.* **76**, 2101.
21. Bechstedt, F. and Grille, H. (1999). *Phys. Stat. Sol.* (b) **216**, 761.
22. Yu SeGi, Kim, K. W. Bergman, L. Dutta, M. Stroscio, M. A. and Zavada, J. M. (1998). *Phys. Rev. B* **58**, 15283.
23. Rudin, S., Reinecke, T. L., and Segall, B. (1990). *Phys. Rev. B* **42**, 11218.
24. Weisbuch, C. and Ulbrich, R. G. (1982). In *Light scattering in solids III* (eds M. Cardona and G. Güntherodt), pp. 210. Springer-Verlag, Berlin, Heidelberg, New York.
25. Cohen, E. and Sturge, M. D. (1982). *Phys. Rev. B* **25**, 3828.

26. Yu, P. Y. (1979). In *Excitons* (ed. K. Cho), pp. 211. Springer-Verlag, Berlin, Heidelberg, New York.
27. Lee, B. C., Kim, K. W., Dutta, M., and Stroscio, M. A. (1997). *Phys. Rev. B* **56**, 997.
28. Segall, B. and Mahan, G. D. (1968). *Phys. Rev*. **171**, 935.
29. Gross, E., Permogorov, S., and Razbirin, B., (1966). *J. Phys. Chem. Solids* **27**, 1647.
30. Klyuchikhin, A. A., Permogorov, S. A., and Reznitski, A. N. (1976). *Sov. Phys. JETP*, **44**, 1176.
31. Weiher, R. L., and Tait, W. C., (1968). *Phys. Rev*. **166**, 791.
32. Permogorov, S. (1982). In *Excitons* (eds E. I. Rashba and M. D. Sturge), pp. 179. North-Holland, Amsterdam.
33. O'Connel-Bronin, A. A. and Plekhanov, V. G. (1979). *Phys. Stat. Sol.* (b) **95**, 75.
34. Huang, K. and Rhys, A. (1950). *Proc. Roy. Soc. (London)*, A **204**, 406.
35. Gurskii, A. L. and Viotikov, S. V. (1999). *Solid State Commun*. **112**, 339.
36. Duke, C. B. and Mahan, G. D. (1965). *Phys. Rev*. **139**, A1965.
37. Soltani, M. Certier, M. Evrard, R. Kartheuser, E. (1995). *J. Appl. Phys.* **78**, 5626.
38. Dingle, R., Sell, D. D., Stokowski, S. E., and Ilegems, M., (1971). *Phys. Rev. B*. **4**, 1211.
39. Leroux, M. and Gil, B. (1999). In *Properties, processing and applications of gallium nitride and related semiconductors* (eds J. H. Edgar, S. Strite, I. Akasaki, H. Amano, and C. Wetzel), pp. 58. Inspec, IEE, London.
40. Leroux, M., Grandjean, N., Beaumont, B., Nataf, G., Semond, F., Massies, J., and Gibart, P. (1999). *J. Appl. Phys*. **86**, 3721.
41. Korona, K. P., Wysmolek, A., Baranowski, J. M., Pakula, K., Bergman, J. P., Monemar, B., *et al.* (1998). *Mater. Res. Soc.* **482**, 501.
42. LeVaillant, Y.-M. (1999). Ph.D thesis, pp. 130. Université de Montpellier II, France.
43. Oh, E., Lee, S. K., Park, S. S., Lee, K.Y., Song, I. J., and Han, J. Y. (2001). *Appl. Phys. Lett.* **78**, 273.
44. Wojdak, M., Wysmolek, A., Pakula, K., and Baranowski, J. M. (1999). *Phys. Stat. Sol.* (b) **216**, 95.
45. Kovalev, D., Averboukh, B., Volm, D., Meyer, B. K., Amano, H., and, Akasaki, I., (1996) *Phys. Rev. B* **54**, 2518.
46. Liu, W., Li, M. F., Xu, S.J., Uchida, K., and Matsumoto, K. (1998). *Semicond. Sci. Technol.* **13**, 769.
47. Xu, S. J., Liu, W., and Li, M. F. (2000). *Appl. Phys. Lett*. **77**, 3376.
48. Gurskii, A. L., Voitikov, S. V., Hamadeh, H., Kalisch, H., Heuken, M., and Heime, K. (2000). *J. Cryst. Growth* **214/215**, 567.
49. Buyanova, I. A., Bergman, J. P., Monemar, B., Amano, H., and Akasaki, I. (1997). *Mater. Sci. Eng. B* **50**, 130.
50. An, H. Y., Cha, O. H., Kim, J. H., Yang, G. M. Lim, K. Y. Suh, E.-K. and Lee, H. J. (1999). *J. Appl. Phys*. **85**, 2888.
51. Sumiya, M., Ohnishi, T., Tanaka, M., Ohtomo, A., Kawasaki, M., Yoshimoto, M., *et al.* (1999). *Mater. Res. Soc*. **537**, G 6.23.
52. Sherriff, R. E., Reynolds, D. C., Look, D. C., Jogai, B., Hoelscher, J. E., Collins, T. C., *et al.* (2000). *J. Appl. Phys*. **88**, 3454.
53. Litvinenko, K., Birkedal, D., Lyssenko, V. G., and Hvam, J. M. (1999). *Phys. Rev. B* **59**, 10255.
54. Logothetidis, S., Cardona, M., Lautenschlager, P., and Garriga, M. (1986). *Phys. Rev. B* **34**, 2458.
55. Lee, J., Koteles, E. S., and Vassell, M. O. (1986). *Phys. Rev. B* **33**, 5512.
56. Benzaquen, R., Leonelli, R., and Charbonneau, S. (1999). *Phys. Rev. B* **59**, 1973.
57. Fischer, A. J., Shan, W., Song, J. J., Chang, Y. C., Horning, R., and Goldenberg, B. (1997). *Appl. Phys. Lett.* **71**, 1981.

58. Viswanath, A. K., Lee, J. I., Kim, D., Lee, C. R., and Leem, J. Y. (1998). *Phys. Rev B* **58**, 16333.
59. Chichibu, S., Okumura, H., Nakamura, S., Feuillet, G., Azuiiata, T., Sota, T., and Yoshida, S. (1997). *Jpn. J. Appl. Phys.* **36**, 1976.
60. Korona, K. P., Wysmolek, A., Pakula, K., Stepniewski, R., Baranowski, J. M. Grzegory, I., *et al.* (1996). *Appl. Phys. Lett.* **69**, 788.
61. Siozade, L. Colard, S. Mihailovic, M. Leymarie, J. Vasson, A. Grandjean, N., *et al.* (1999). *Phys. Stat. Sol.* (b) **216**, 73.
62. Li, C. F., Huang, Y. S. Malikova, L. and Pollak, F. H., (1997). *Phys. Rev. B* **55**, 9251.
63. Petalas, J., Logothetidis, S. Boultadakis, S. Alouani, M. and Wills, J. M. (1995). *Phys. Rev. B* **52**, 8082.
64. Fischer, A. J., Shan, W. Park, G. H., Song, J. J., Kim, D. S., Yee, D. S., *et al.* (1997). *Phys. Rev. B* **56**, 1077.
65. Tchounkeu, M. Briot, O. Gil, B., Alexis, J. P., and Aulombard, R. L. (1996). *J. Appl. Phys.* **80**, 5352.
66. Lee, D., Johnson, A. M., Zucker, J. E., Feldman, R. D., and Austin, R. F., (1991). *J. Appl. Phys.* **69**, 6722.
67. Fischer, A. J., Kim, D. S., Hays, J., Shan, W., Song, J. J., Eason, D.,B., *et al.* (1994). *Phys. Rev. Lett.* **73**, 2368.
68. Makino, T. Chia, C.H. Tuan, N. T. Segawa, Y. Kawasaki, M. Ohtomo, A., *et al.* (2000). *Appl. Phys. Lett.* **76**, 3549.
69. Young, J. F., Gong, T., Fauchet P. M., and Kelly, P. J. (1994). *Phys. Rev. B* **50**, 2208.
70. Shah, J. (1978). *Solid State Electron.* **21**, 43.
71. Shah, J. (1998). In *Hot electrons in semiconductor physics and devices* (ed. N. Balkan), pp. 55. Clarendon Press, Oxford.
72. Binet, F. and Duboz, J. Y. (1999). *Phys. Rev. B* **60**, 4715.
73. Shan, W. Fischer, A. J., Hwang, S. J., Little, B. D., Hauenstein, R. J., Xie. X. C., *et al.* (1998). *J. Appl. Phys.* **83**, 455.
74. Zukauskas, A. Z., Tamulaitis, G., Gaska, R., Shur, M. S., Khan, M. A., and Yang, J. W. (1999). *Phys. Stat. Sol.* (b) **216**, 495.
75. Yu, P. Y., private communication.
76. Sheih, S. J., Tsen, K. T., Ferry, D. K., Botchkarev, A., Sverdlov, B., Salvador, A., and Morkoc, H. (1995). *Appl. Phys. Lett.* **67**, 1757.
77. Tsen, K. T., Ferry, D. K., Botchkarev, A., Sverdlov, B., Salvador, A., and Morkoc, H. (1998). *Appl. Phys. Lett.* **72**, 2132.
78. Tsen, K. T., Joshi, R. P., Ferry, D. K., Botchkarev, A., Sverdlov, B., Salvador, A., and Morkoc, H., (1996). *Appl. Phys. Lett.* **68**, 2990.
79. Kuball, M., Hayes, J. M., Shi, Y., and Edgar, J. H. (2000). *Appl. Phys. Lett.* 77, 1958.
80. Bergman, L., Alexson, D., Murphy, P. L., Nemanich, R. J., Dutta, M., Stroscio, M. A., *et al.* (1999). *Phys. Rev. B* **59**, 12977.
81. Klemens, P. G. (1966). *Phys. Rev.* **148**, 845.
82. Ridley, B. K. (1996). *J. Phys.: Condens. Mater.*, **8**, L511.
83. Ye, H., Wicks, G. W., and Fauchet, P. M. (1999). *Appl. Phys. Lett.* **74**, 711.
84. Ye, H., Wicks, G. W., and Fauchet, P. M. (2000). *Appl. Phys. Lett.* **77**, 1185.
85. Tsen, K. T., Ferry, D. K., Botchkarev, A., Sverdlov, B., Salvador, A., and Morkoc, H. (1997). *Appl. Phys. Lett.* **71**, 1852.
86. Tamulaitis, G. Zukauskas, A. Z., Yang, J. W., Khan, M. A., Shur, M. S., and Gaska, R. (1999). *Appl. Phys. Lett.* **75**, 2277.
87. Hagele, D., Zimmermann, R., Oestreich, M., Hofmann, M. R., Ruhle, W. W., Meyer, B. K., *et al.* (1999). *Phys. Rev. B* **59**, R7797.
88. Leo, K., Ruhle, W. W., and Ploog, K. (1988). *Phys. Rev. B* **38**, 1947.

89. Poweleit, C. D., Smith, L. M., and Jonker, B. T. (1997). *Phys. Rev. B* **55**, 5062.
90. Ulbrich, R. (1973). *Phys. Rev. B* **8**, 5719.
91. Snoke D. W. and Wolfe J. P. (1990). *Phys. Rev. B*, **42** 7876
92. Morkoc, H. (1999). In *Nitride semiconductors and devices*, pp. 245. Springer-Verlag, Berlin.
93. Pollak, F. H. (1993). In *Phonons in semiconductor nanostructures*, vol 236 of *nato advanced study institute, series E: Applied Sciences* (ed. J. P. Leburton, J. Pascual and C. Sotomayor Torres), pp. 341. Plenum Press, New York.
94. Pelekanos, N. T., Ding, J., Hagerott, M., Nurmikko, A. V., Luo, H., Samarth, N., and Furdyna, J. K. (1992). *Phys. Rev. B* **45**, 6037.
95. Young, P. M., Runge, E., Ziegler, M., and Ehrenreich, H., (1994). *Phys. Rev. B* **49**, 7424.
96. Nashiki, H., Suemune, I., Kumano, H., Suzuki, H., Obinata, T., Vesugi, K., and Nakahara, J. (1997). *Appl. Phys. Lett.* **70**, 2350.
97. Venu Gopal, A., Kumar, R., Vengurlekar, A. S., Melin, T., Laruelle, F., and Etienne, B. (1999). *Appl. Phys. Lett.* **74**, 2474.
98. Skolnick, M. S., Nash, K. J., Tapster, P. R., Mowbray, D. J., Bass, S. J., and Pitt, A. D. (1987). *Phys. Rev. B* **35**, 5925.
99. Nash, K. J., Skolnick, M. S., Claxton, P. A., and Roberts, J. S., (1989). *Phys. Rev. B* **39**, 5558.
100. Schmitt-Rink, S. Miller, D. A. B., and Chemla, D. S., (1987). *Phys. Rev. B* **35**, 8113.
101. Shiang, J. J., Risbud, S. H., and Alivisatos, A. P. (1993). *J. Chem. Phys.* **98**, 8432.
102. Klein, M. C., Hache, F., Ricard, D., and Flytzanis, C. (1990). *Phys. Rev. B* **42**, 11123.
103. Nomura, S. and Kobayashi, T. (1992). *Phys. Rev. B* **45**, 1305.
104. Fomin, V. M., Gladilin, V. N., Devreese, J. T., Pokatilov, E. P., Balaban, S. N., and Klimin, S. N. (1998). *Phys. Rev. B* **57**, 2415.
105. Fomin, V. M., Gladilin, V. N., Klimin, S. N., Devreese, J. T., Koenraad, P. M., and Wolter, J. H. (2000). *Phys. Rev. B* **61**, R2436.
106. Nirmal, M., Murray, C. B., and Bawendi, M. G., (1994). *Phys. Rev. B* **50**, 2293.
107. Heitz, R., Mukhametzhanov, I., Stier, O., Madhukar, A., and Bimberg, D., (1999). *Phys. Rev. Lett.* **83**, 4654.
108. Ramvall, P., Tanaka, S., Nomura, S., Riblet, P., and Aoyagi, Y., (1999). *Appl. Phys. Lett.* **75**, 1935.
109. Riblet, P., Hirayama, H., Kinodhita, A., Hirata, A., Sugano, T., and Aoyagi, Y. (1999). *Appl. Phys. Lett.* **75**, 2241.
110. Bergman, L., Dutta, M., Stroscio, M. A., Komirenko, S. M., Nemanich, R. J., Eiting, C.J. *et al.* (2000). *Appl. Phys. Lett.* **76**, 1969.
111. Smith, M., Lin, J. Y., and Jiang, H. X., Khan, A., Chen, Q., Salvador, A., *et al.* (1997). *Appl. Phys. Lett.* **70**, 2882.
112. Davidson, J. A., Dawson, P., Wang, T., Sugahara, T., Orton, J. W., and Sakai, S. (2000). *Semicond. Sci. Technol.* **15**, 497.
113. Alderighi, D., Vinattieri, A., Bogani, F., Colocci, M., Gottardo, S., Grandjean, N., and Massies, J. (2001). *Phys. Stat. Sol.* (a) **183**, 129.
114. Smith, M., Lin, J. Y., and Jiang, H. X., Salvador, A., Botchkarev, A., Kim, W., and Morkoc, H. (1996). *Appl. Phys. Lett.* **69**, 2453.
115. Morel, A., Gallart, M., Taliercio, T., Lefebvre, P., Gil, B., Allegre, J., *et al.* (2000). *Phys. Stat. Sol.* (a) **180**, 375.
116. Gallart, M., Morel, A., Taliercio, T., Lefebvre, P., Gil, B., Allegre, J., *et al.* (2000). *Phys. Stat. Sol.* (a) **180**, 127.
117. Kalliakos, S., Zhang, X.B., Taliercio, T., Lefebvre, P., Gil, B., Grandjean, N., Damilano, B., and Massies, J. (2002) *Appl. Phys. Lett.* **80**, 428.
118. Gallart, M., Lefebvre, P., Morel, A., Taliercio, T., Gil, B., Allegre, J., *et al.* (2001). *Phys. Stat. Sol.* (a) **183**, 61.

12 Low-dimensional nitrides: a laboratory for ultrafast physics

Alexey Kavokin and Guillaume Malpuech

Exciton–polariton will be the hero of this chapter. Light-matter coupling phenomena in nitrides will be considered from the point of view of both quantum mechanics and classical optics. We will show that extremely large exciton oscillator strength and peculiar band structure make nitrides excellent candidates for study of the most striking phenomena of solid state optics. Exciton–polaritons, i.e. quasi-particles appearing as a result of photon coupling with an elementary crystal excitation, govern the coherent optical response of low-dimensional semiconductor structures on a femtosecond to picosecond time-scale. Ultra-rapid transport of a neutral electronic excitation by a blue light is a close reality in nitrides. The challenge of the year 2001 is achievement of the strong-coupling regime in GaN-based microcavities that would allow studying of the stimulated scattering of bi-dimensional blue exciton–polaritons that could give way to a new generation of optoelectronic devices (*polariton lasers*). The principal difficulty for observation of a wide class of striking phenomena predicted theoretically, is a huge inhomogeneous broadening of excitons caused by a potential disorder in presently available nitride-based structures. The problem of exciton–polariton propagation in disordered media is one of the most essential for the correct description of ultrafast optical phenomena in nitrides.

This chapter proposes an original approach to this problem, that allows all kinds of coherent time-resolved optical experiments (reflection, transmission and resonant Rayleigh scattering) performed on thin films of nitrides as well as on quantum wells and microcavities to be described. In first paragraphs the basic aspects of the formalism will be presented (scattering state technique, non-local dielectric response theory, quantum dot scattering model for description of the exciton inhomogeneous broadening). In the next paragraph the resonant Rayleigh scattering (RRS) of light in GaN/AlGaN quantum wells (QWs) will be addressed. The RRS is extremely sensitive to the interplay of structural disorder and polariton coupling, it is expected to exhibit a quite peculiar dynamics in nitrides. Comparing the RRS with time-resolved reflection data, one can obtain a complex information on the statistics of potential fluctuations in nitride-based QWs and excitonic characteristics. We shall demonstrate a dramatic decrease of the exciton oscillator strength as a function of the quantum well width in GaN/AlGaN quantum wells caused by the polarization fields. In the final part of this chapter we shall analyze the possible ways to achieve the exciton–polariton Bose condensation in GaN-based microcavities, that would open a way to a new generation of semiconductor lasers and polariton lasers.

12.1 Introduction

The exciton binding energy in GaN is about 25 meV [1], which corresponds exactly to the thermal activation energy $k_b T$ at room temperature. Morover, the exciton stability is considerably enhanced in quantum structures which generally form the elementary cells of the new generations of light-emitting devices. That is why the excitons are expected to play a major role in optical properties of nitride-based quantum structures even at room temperature. The excitons are only active in optics if coupled to light. This coupling results in exciton–polaritons [2, 3], quasi-particles combining properties of photons and excitons. The ultrafast optical response of quantum structures is governed by the dynamics of the exciton–polaritons. Their group velocity, ranging from almost zero to the light velocity in the media, will be a characteristic of the information transfer rate in future optical processors, thus it has a key importance for quite intriguing applications.

Though the theory of exciton–polaritons in bulk semiconductors was built up almost 40 years ago by Hopfield and Agranovich [2], to calculate the polariton propagation in nitrides one should revise all existing theories. The huge additional complication comes from the potential disorder that is extremely strong even in high-quality nitride-based heterostructures. The disorder comes from the random polarization fields, alloy and interface fluctuations, dislocations, and impurities. Due to a potential disorder the number of discrete localized exciton states coupled to light amounts to a few millions in a standard optical experiment. To describe the polariton propagation in such a complicated system important approximations have to be adapted. The first of these is to *ignore* the microscopic parameters of each individual localized exciton state and to operate with macroscopic characteristics. These characteristics are: the exciton radiative decay rate Γ_r, the value of the polarization field, the average distance d between the exciton localization centers, the average size r_{QD} of these localization centers, and the broadening Δ of the frequency distribution of the exciton resonance. Any model operating with these macroscopic parameters is a phenomenological model. We would like to formulate and apply here a phenomenological model that takes into account two most important effects that dominate the light-matter interaction in the system, namely, the exciton–polariton effect and the disorder scattering effect.

In this chapter, we present experimental and theoretical tools for the study of exciton properties versus disorder in nitride-based quantum structures. We show that resonant Rayleigh scattering (RRS) [4–5] combined with conventional reflectivity measurements can easily allow determination of the most essential parameters of quantum well (QW) excitons if they are interpreted in the framework of an advanced Fourier spectroscopy analysis [6].

The RRS of light by quantum wells excitons is due to the breakdown by the disorder of the translation invariance in the plane of the QWs. In other words, the disorder potential induces the localization of excitons in certain 'quantum dot-like' states, and these localized excitons coherently scatter the light in all directions. The time-resolved RRS spectra are a direct probe of the statistical properties of the disorder [5]. The low intensity of the light scattered in a given direction makes the RRS experiment quite non-trivial to perform. In nitrides this problem is less important than in arsenides because of the strong disorder. Basically, scattering experiments are certainly the most adapted tool to study the exciton properties in nitride-based QWs. Time-resolved reflection measurements have equally been used extensively in conventional III–V semiconductors in order to determine the exciton properties [7]. In general, coherent time-resolved techniques have been found to be much more precise than the conventional *cw* reflectivity or transmission. However, the intensity of the time-resolved signal is proportional to $(\Gamma_r/\Delta)^2$ whereas the decay is roughly given by $1/\Delta$. The two last quantities are extremely small in the most of GaN/AlGaN QWs and the direct measurement of the time-resolved reflection remains a challenge from the experimental point of view. An indirect alternative way can be to perform the numerical Fourier transform of the *cw* experimental signals. Such kind of treatment, currently performed in other fields of physics [8], has already been applied successfully in InGaAs QWs [9, 10]. This allows signals to get extremely close to the real time-resolved reflectivity and is potentially promising for the extraction of the exciton oscillator strength and inhomogeneous broadening.

In the next paragraph we present a new theory describing all types of coherent optical experiments (reflection, transmission, scattering) of confined localized excitons in the framework of the unified formalism. The model properly takes into account the renormalization effects associated with the exciton–light coupling as well as the structural disorder effects. In the third part we analyze theoretically RRS spectra of QWs and show how to extract from these spectra the information on excitonic statistics. We demonstrate that all the key parameters of the QW excitons can be obtained by performing angle- and time-resolved RRS experiments. The fourth part is devoted to application of our formalism to the analysis of the experimental reflection spectra of GaN/AlGaN QWs. The numerical Fourier transform of the steady-state reflection spectra of MBE grown GaN/AlGaN QWs is performed. This has allowed us to obtain the exciton oscillator strength and inhomogeneous broadening as a function of the QW width. A rapid decrease of the oscillator strength with the QW width increase has been found to be a consequence of the exciton distortion by the polarization fields. This tendency is confirmed by a self-consistent variational calculation. Finally, in the fifth part, we address the perspectives of observation of the polariton–laser effect in nitride-based microcavities.

12.2 Formalism

Our formalism (see also [6]) is based on the Green-function approach to the problem of light–exciton coupling in low-dimensional structures [11] and the generalized scattering state technique for the light–pulse propagation in multilayer systems [4–12]. We consider each QW as an irregular array of quantum dots (QDs). The QDs represent shallow potential islands that weakly localize exciton wave functions, so that the exciton wave functions attributed to the neighboring QDs may overlap. Each QD is characterized by a given exciton resonance frequency ω_n, and radius-vector $\vec{R}_n$ that denotes the center of the localized exciton wave-function in the dot. We assume ω_n to be distributed with a function

$$f(\omega_1) = \frac{1}{\sqrt{\pi\Delta}} \exp\left[-\left(\frac{\omega_1 - \omega_0}{\Delta}\right)^2\right], \tag{12.1}$$

where Δ describes the exciton inhomogeneous broadening.

Neglecting the deviations in the shape of exciton wave function localized in different dots, we solve the Maxwell equation:

$$\nabla \times \nabla \times \vec{E} = k_0^2 \vec{D}, \tag{12.2}$$

where

$$\vec{D} = \varepsilon_b \vec{E} + 4\pi \vec{P}_{\text{exc}} \tag{12.3}$$

and

$$4\pi \vec{P}_{\text{exc}} = \sum_{n=1}^{N} T_n \Phi(\vec{r} - \vec{R}_n) \vec{E}(\vec{r}\,')\Phi(\vec{r}\,' - \vec{R}_n) d\vec{r}\,', \tag{12.4}$$

with

$$T_n = \frac{\varepsilon_b \omega_{\text{LT}} \pi^{3}{}_{\text{B}}}{\omega_n - \omega - i\gamma},$$

k_0 being the wave vector of the incident light in the media, ω_{LT} being the longitudinal-transverse splitting, ε_b being the background dielectric constant, a_{B} being the Bohr radius of the exciton in the bulk, γ being the homogeneous broadening, $\Phi(\vec{r})$ being the localized exciton wave function taken with equal electron and hole coordinates. Further, we assume equal shapes of $\Phi(\vec{r})$ for all QDs. Solving eqn (12.2) we represent an electric field as

$$\vec{E}(\omega, \vec{r}) = \vec{E}_0 \exp(i\vec{k}\vec{r}) + k_0^2 \sum_{n=1}^{N} T_n \int d\vec{r}\,' \Phi(\vec{r}\,' - \vec{R}_n) G_0(\vec{r} - \vec{r}\,') \int \vec{E}(\vec{r}\,'')\Phi(\vec{r}\,'' - \vec{R}_n) d\vec{r}\,'' \tag{12.5}$$

where $G_0(\vec{r}-\vec{r}\,')=e^{ik|\vec{r}-\vec{r}\,'|}/4\pi|\vec{r}-\vec{r}\,'|$ is the Green function for a 0-dimensional system, $\vec{k}=\vec{k}_0\sqrt{\varepsilon_b}$. is the wave vector of the incident light in the media. It is convenient to multiply left and right parts of eqn (12.5) by $\Phi(\vec{r}-\vec{R}_m)$ and integrate over $\vec{r}$. This procedure yields an expression for

$$\Lambda_m \equiv \int \vec{E}(\vec{r})\Phi(\vec{r}-\vec{R}_m)d\vec{r} = \frac{\int \vec{E}_0 \exp(i\vec{k}\cdot\vec{r})\Phi_{\rm QD}(\vec{r}-\vec{R}_m)d\vec{r} + \sum_{n\neq m}^{N}\Theta_{nm}\Lambda_n}{1-\Theta_{mm}} \tag{12.6}$$

where

$$\Theta_{nm} = k_0^2 T_n \iint G_0(\vec{r}-\vec{r}\,')\Phi_{\rm QD}(\vec{r}\,'-\vec{R}_n)\Phi_{\rm QD}(\vec{r}-\vec{R}_m)d\vec{r}\,'d\vec{r}. \tag{12.7}$$

In the normal incidence case, the solution of eqn (12.6) can be approximated by

$$\Lambda_m = \frac{\int \vec{E}_0 e^{i\vec{k}\vec{r}}\Phi(\vec{r})d\vec{r}}{1-\sum_{n=1}^{N}\Theta_{nm}}, \tag{12.8}$$

that allows expansion of the electric field as:

$$\vec{E}(\omega,\vec{r}) = \vec{E}_0\exp(i\vec{k}\vec{r}) + k_0^2\sum_{n=1}^{N}T_n\int d\vec{r}\,'\Phi(\vec{r}\,'-\vec{R}_n)G_0(\vec{r}-\vec{r}\,')\frac{\int \vec{E}_0 e^{i\vec{k}\vec{r}}\Phi(\vec{r})d\vec{r}}{1-\sum_{n=1}^{N}\Theta_{nm}}. \tag{12.9}$$

The Fourier transform of the electric field (12.9) yields its directional dependence which can be represented in form

$$\vec{E}_d(\omega,\vec{k}_s) = \vec{E}_0\delta_{\vec{k},\vec{k}_s} + \vec{E}_0\sum_{m=1}^{N}\frac{i\Gamma_0^{\rm QD}}{\omega_m-\omega-i\gamma}\frac{1}{1-\sum_{n=1}^{N}\Theta_{nm}}\exp(i\vec{k}_s.\vec{R}_m), \tag{12.10}$$

where

$$\Gamma_0^{\rm QD} = \frac{1}{6}\omega_{\rm LT}k_0^3a_{\rm B}^3\left(\int d\vec{r}\cos(\vec{k}\cdot\vec{r})\Phi(\vec{r})\right)^2, \tag{12.11}$$

$\delta_{\vec{k},\vec{k}_s} = 1$ if $\vec{k}_s = \vec{k}$, $\delta_{\vec{k},\vec{k}} = 0$ if $\vec{k}_s \neq \vec{k}$. To obtain eqn (12.11) we used the $\vec{k}$-space representation of the Green-function G_0 and separated variables of in-plane and normal-to-plane motion.

Note, that in the oblique incidence case, eqn (12.10) should be slightly modified. In particular, Θ_{nm} should be multiplied by $\cos\alpha$ in the TE-polarization and divided by $\cos\alpha$ in the TM-polarization, where α is the light propagation angle in the media. Moreover, in general, at oblique incidence, the exciton resonance is split into longitudinal, transverse, and Z-polariton modes, that is neglected in our consideration since in realistic nitride-based structures these effects are too fine to be observed.

We can further transform eqn (12.10), bearing in mind that:

$$\sum_{n=1}^{N} \Theta_{nm} \cong \beta = i\Gamma_r \int dv \frac{f(v)}{v - \omega - i\gamma} \text{ with}$$

$$\Gamma_r = \frac{k}{2d^2}\omega_{\mathrm{LT}}\pi a_B^3 \left[\int \Phi(\vec{r}) \cos(\vec{k}\vec{r}) d\vec{r} \right]^2, \tag{12.12}$$

where d is the average distance between two dots. Taking the exciton localization radius equal to the average radius r_{QD} of a QD, one can approximate the integral in formula (12) by $r_{\mathrm{QD}}/a_{\mathrm{B}}^{2\mathrm{D}}$ [16] where $a_B^{2\mathrm{D}}$ is the two-dimensional exciton Bohr radius. The radiative broadening of the QW becomes

$$\Gamma_{\mathrm{r}} = \frac{k}{2}\omega_{\mathrm{LT}}\pi a_{\mathrm{B}}^3 \left(\frac{r_{\mathrm{QD}}}{a_{\mathrm{B}}^{2\mathrm{D}} d} \right)^2. \tag{12.13}$$

We then obtain:

$$\vec{E}_d(\omega, \vec{k}_s) = \vec{E}_0 \delta_{\vec{k},\vec{k}_s} + \frac{\vec{E}_0 \sum_{m=1}^{N} \frac{i\Gamma_0^{\mathrm{QD}}}{\omega_m - \omega - i\gamma} \exp(i\vec{k}_s . \vec{R}_m),}{1 - \beta}. \tag{12.14}$$

Equation (12.14) is quite important for the following discussion, since it describes all kinds of coherent optical experiments (reflection, transmission, and Rayleigh scattering) within the same semi-classical formalism, properly taking into account the polariton effect. Γ_{r} is the radiative broadening of the exciton within the QW. It is proportional the oscillator strength per unit area:

$$f = \frac{\sqrt{\varepsilon_b}}{\pi} \frac{mc}{e^2} \Gamma_{\mathrm{r}}. \tag{12.15}$$

In the specular reflection direction, the formula (12.16) yields

$$r_{QW} = \frac{\beta}{1-\beta}, \; t_{QW}\frac{1}{1-\beta}. \tag{12.16}$$

Equation (12.16) is formally equivalent to the expressions for the reflection and transmission coefficients of a QW with an inhomogeneously broadened exciton resonance by Andreani *et al.* [13] if one takes d and the in-plane size of a QD equal to the in-plane Bohr radius of a free exciton in a QW. We recover in that case $\Gamma_r = \Gamma_0$, the radiative broadening of the free exciton. The present formalism proves the validity of the phenomenological model [13] using a micro-model of a disordered array of QDs. Additionally, it offers a way to calculation of the amplitude of scattered light. The passage from the frequency-resolved to the time-resolved reflection and RRS spectra is simply achieved by the Fourier transform of eqn (12.16).

In conclusion of this part, we have presented a unified semi-classical theory of propagation and scattering of exciton–polaritons in heterostructures that has allowed to describe time-resolved reflection and RRS spectra of QWs within the same formalism.

12.3 RRS measurements as a tool to characterize the QW excitons

Here we discuss the optical ways to characterize the spatial distribution of localized excitons in a QW, their energies, and oscillator strengths. As before, attention will mostly be paid to the RRS experiments.

Consider that a laser pulse illuminates a single quantum well (SQW) under the normal incidence at $t = 0$, and creates a coherent population of N excitons having a zero in-plane wave vector. These excitons in their turn coherently re-emit light in the reflected or transmitted directions within the life-time $1/\Gamma_r$. They can also be coherently scattered within the time-constant $1/\Delta$ and incoherently scattered within the time-constant $1/\gamma$. Within this description, the number $N_0\,(t)$ of excitons having a zero in-plane wave vector at the time t is:

$$N_{k=0}(t) = N \exp\left(-\frac{\Delta^2 t^2}{2} - \Gamma_r t - \gamma t\right). \tag{12.17}$$

The integrated number of reflected photons N_{ref} is given by:

$$N_{\text{ref}} = \int_0^\infty \Gamma_r N_{k=0}(t)\,dt. \tag{12.18}$$

The integrated intensity of the reflected light is proportional to N^2_{ref}. In the nitride-based structures that are currently available, the condition $\Delta \gg \Gamma_0$ is fulfilled, that yields:

$$I_{\text{ref}} = \frac{\pi}{2} N^2 \frac{\Gamma_r^2}{\Delta^2}. \tag{12.19}$$

At the same time the total number of the scattered excitons is given by:

$$N_{k\neq 0} = \int_0^{\infty} \Delta N_{k=0}(t)dt. \tag{12.20}$$

The intensity emitted in a single speckle is given by:

$$I_{\text{scat}} = \frac{N_{k\neq 0}^2}{N} = \frac{\pi}{2} N. \tag{12.21}$$

The ratio of the reflected intensity to the average scattered intensity is thus:

$$I_{\text{ref}}/I_{\text{scat}} \approx N\left(\frac{\Gamma_r}{\Delta}\right)^2 \tag{12.22}$$

Figure 12.1 shows the integrated intensity as a function of the scattering angle calculated with the formula (12.14). The intensity for a scattering angle equal to zero corresponds to the reflected intensity. The average intensity obtained for the other angles will be further referred to as scattered intensity. The calculation has been performed for $N=10^6$, $\Gamma_r=0.5\,\text{meV}$ and $\Delta=5\,\text{meV}$

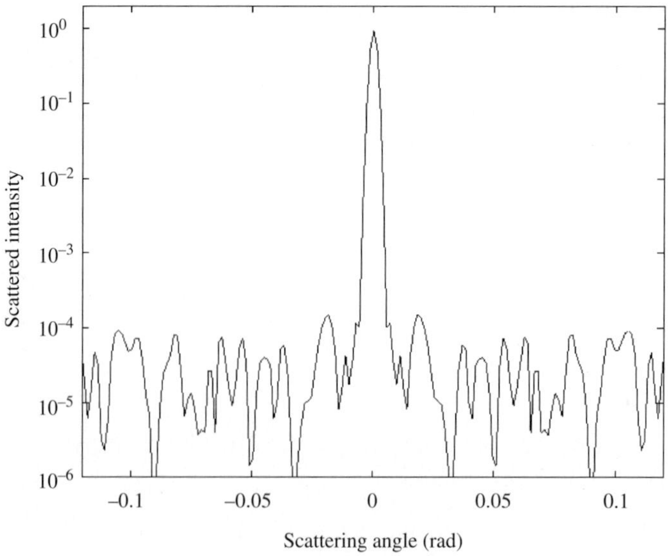

Fig. 12.1 Integrated scattered intensity as a function of the scattering angle of a GaN/AlGaN single quantum well calculated with the formula (12.14). The parameters used are $N=10^6$, $\Gamma_r=0.5\,\text{meV}$ and $\Delta=5\,\text{meV}$ and $n_b = \sqrt{\varepsilon_b} = 2.6$.

and for the background optical index $n_b = \sqrt{\varepsilon_b} = 2.6$. The ratio between reflection and scattering is 10^4 is perfect agreement with the estimation obtained from the formula (12.22).

As we will see in the third part of this article, one can extract the values of Γ_r and Δ from the *cw* reflectance spectra by a Fourier analysis. Once these two quantities are known, the number N of localized states within the light spot can be obtained using the formula (12.22), as well as the average distance between the dots.

It should be noted, however, that to obtain precise information on the statistics of excitons in a QW, the time-resolved RRS measurements are needed.

Typically a time-resolved RRS spectrum of an SQW exhiits a fast rise time $1/\Delta$ proportional to the scattering time of the exciton population created by the incident pulse followed by a decay time proportional at low temperature to $2/(\Gamma_r + \gamma)$. This means that the contributions of the inhomogeneous and of the radiative and homogeneous broadenings can easily be separated if one analyses the shape of the time-resolved RRS. Note also that γ is extremely small at low temperatures and can easily be obtained from independent photoluminescence measurements.

Figure 12.2 shows on the same scale the time resolved reflection and the time resolved RRS calculated with the formula (12.14). The parameters are $\Delta = 5\,\text{meV}$, $\Gamma_r = 0.5$ meV, $\gamma = 0.001$ meV, $N = 10^6$. The diameter of the area exited by the incident laser pulse is taken to be 50 μm. From eqns (12.22, 12.14), the average ratio between the reflection speckle and a scattered speckle is 10^4. The

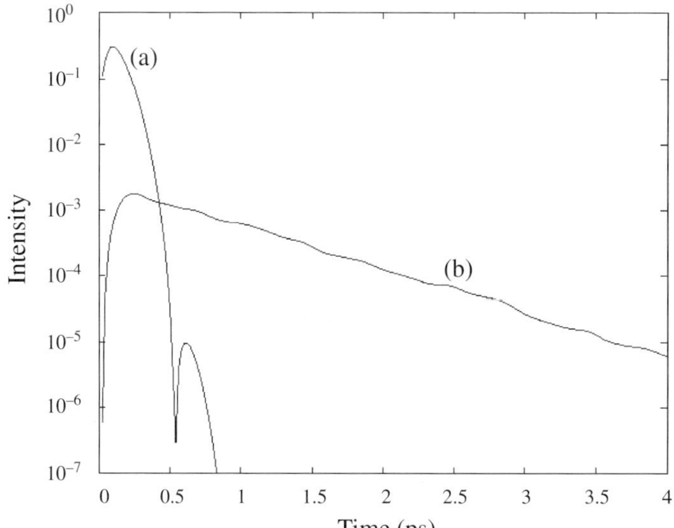

Fig. 12.2 (a) Time-resolved reflection of a GaN/AlGaN single quantum well. (b) Time-resolved RRS of a GaN/AlGaN single quantum well. Both curves are obtained with the formula (12.14). The parameters used are $\Delta = 5\,\text{meV}$, $\Gamma_r = 0.5\,\text{meV}$, $\gamma = 0.001\,\text{meV}$, $N = 10^6$. The diameter of the area exited by the incident laser pulse is taken to be 50 μm.

reflection signal in Fig. 12.2 is obtained from a single speckle, whereas the scattering signal is obtained from the averaging over 100 speckles. The ratio between the reflected and the scattered intensities is thus given by $10^4/100 = 10^2$ as one can see in Fig. 12.2. Note also the fast decay of the reflection that corresponds to the fast rise of the scattered signal. This latter becomes stronger than the reflection after only 500 fs, and then slowly decays with a characteristic time $2/\Gamma_r = 1.3$ ps.

Figure 12.3 (a) shows the RRS of a single QW calculated for different value of Γ_r. We used the following set of parameters: $\Delta = 5$ meV, $\gamma = 0.001$ meV, $N = 10^6$. The values of Γ_r are 1 meV, 0.1 meV, and 0.01 meV for the curves (a), (b), and (c), respectively. The advantage of this kind of experiment is here clearly seen. Indeed, it makes it possible to measure with a good accuracy the radiative broadening even if it is strongly dominated by the inhomogeneous broadening, that would be impossible in reflection. Figure 3(b) shows the RRS of a single QW calculated for different value of Δ. We always used $\Gamma_r = 0.5$ meV, $\gamma = 0.001$ meV, $N = 10^6$. The values of Δ are 20 meV, 2 meV and 0.2 meV for the curves (a), (b), and (c), respectively. Once again, the values of Δ and Γ_r can easily be obtained with the use of such technique, the only limitation in case of large values of Δ comes from the time-resolution of the experimental set-up.

12.4 Analysis of the reflection spectra of GaN/AlGaN QWs

In our earlier papers [14, 15] we have presented the variational approach that allowed calculation of the main excitonic parameters in GaN/AlGaN QWs, taking into account the polarization fields and eventual screening of them by

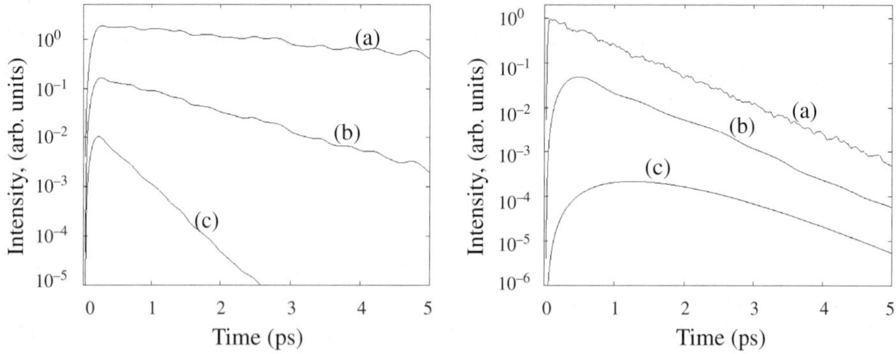

Fig. 12.3 On the left: time-resolved RRS of a single GaN/AlGaN QW calculated for different value of Γ_r. We used the following set of parameters: $\Delta = 5$ meV, $\gamma = 0.001$ meV, $N = 10^6$. The values of Γ_r are 1 meV, 0.3 meV and 0.1 meV for the curves (a), (b), and (c) respectively. On the right: time-resolved RRS of a single GaN/AlGaN QW calculated for different value of Δ. We used the following set of parameters $\Gamma_r = 0.5$ meV, $\gamma = 0.001$ meV, $N = 10^6$. The values of Δ are 20 meV, 2 meV and 0.2 meV for the curves (a), (b), and (c), respectively.

a photoinduced free electron–hole plasma. To compare the results of these calculations with the experiment one has to find a way to extract the exciton oscillator strength from the available optical spectra.

This problem, that seems to be trivial at first glance, represents in fact a major difficulty for the optical characterization of nitrides. Actually, the exciton resonances in typical reflection or photoluminescence spectra of GaN-based QWs have a broadening of a few tens of meV. The origin of this broadening is in the potential disorder in-plane of the QWs that affects the excitons. This disorder is caused by the QW width fluctuations, alloy fluctuations and polarization field fluctuations. The excitonic spectrum is composed of contributions from millions of localized exciton states having different energies and oscillator strengths. In this situation the frequently used single-free-exciton-resonance model [3] fails to describe the real optical spectra.

Here we present an alternative model, allowing analysis of the exciton reflection spectra, taking into account the exciton inhomogeneous broadening, and separating it numerically from the radiative broadening. This model has been applied to a series of reflection spectra taken by us from a series of high-quality MBE-grown $GaN/Al_{0.07}Ga_{0.93}N$ QW samples containing the QWs of different thicknesses. The most essential conclusion of the performed analysis is that the oscillator strength in GaN/AlGaN QWs decreases with the QW width *much faster* than in conventional QWs. The collapse of the exciton oscillator strength in 10–12 monolayer-thick QWs is caused by distortion of excitons by the polarization fields, as our variational calculation confirms [14, 15]. Thus, only excitons in very thin QWs have the oscillator strength exceeding that in the bulk GaN.

Our method consists in the numerical Fourier transform of the experimental reflection spectrum of our sample and its comparison with the calculated time-resolved reflection coefficient, as is illustrated by Fig. 12.4, that shows experimental and calculated frequency-resolved and time-resolved spectra of a 1-μm-thick layer of GaN. A- and B-exciton resonances induce broad peaks of the magnitude of about 2%, while the background reflectivity is about 20%. In this case, the reflection coefficient of the structure can be written in form

$$r(\omega) = r_0 + \Delta r(\omega), \qquad (12.23)$$

where r_0 is the background coefficient, and the excitonic contribution can be expressed as $\Delta r(\omega) = \Delta a(\omega) + i\Delta\mathscr{C}(\omega)$, with $\Delta a(\omega)$ being an odd real function, and $\Delta b(\omega)$ being an even real function [9].

We convolute the measured reflectivity spectrum $|r(\omega)|^2$ with a broad Gaussian function $g(\omega)$ (that would correspond to a femtosecond incident light pulse in the case of a time-resolved measurement). For times longer than the pulse duration the Fourier transform of this quantity allows to obtain:

$$\int_{-\infty}^{+\infty} |g(\omega)r(\omega)|^2 e^{-i\omega t} d\omega \cong 2r_0 \int_{-\infty}^{+\infty} g^2(\omega)\Delta a(\omega) e^{-i\omega t} d\omega. \qquad (12.24)$$

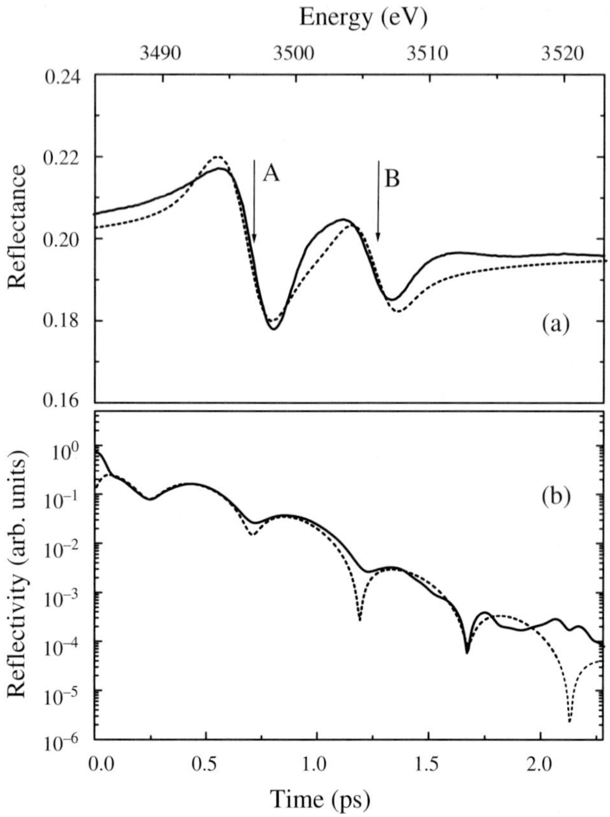

Fig. 12.4 Frequency spectra (a) and time-resolved spectra (b) of GaN layer grown on sapphire. Time-resolved reflectivity is determined by Fourier transform of the of experimental reflectance (solid lines), calculated reflectivity is shown by dashed lines.

The approximate equality (12.24) is correct if $\Delta a, \Delta b \ll r_0$. The time-resolved reflection of a structure illuminated by a pulse $g^2(\omega)$ is

$$r_{g^2}(t) = \int_{-\infty}^{+\infty} g^2(\omega) r(\omega) e^{-i\omega t} d\omega \cong 2 \int_{-\infty}^{+\infty} g^2(\omega) \Delta a(\omega) e^{-i\omega t} d\omega,$$

that is proportional to the quantity given by eqn (12.24). Thus, we have shown that at our experimental conditions, the numerical Fourier transform of the experimental reflectivity convoluted with a large spectral function yields formally the same spectrum (with a very high accuracy) as the direct measurement of the time-resolved reflection by the femtosecond optical spectroscopy. The solid line in Fig. 12.4(b) represents the time-resolved reflection of the layer of bulk GaN obtained from the experimental spectrum shown in Fig. 12.4 (a) as described above. It exhibits quite a rapid Gaussian-like decay typical for the systems with a strong exciton inhomogeneous broadening [16], and the pronounced oscillations that are nothing but beats between A- and B-exciton resonances.

The dashed curves in Fig. 12.4 are calculated using the generalized scattering state technique as described in ref. [16]. We used the longitudinal transverse splittings for A and B excitons $\hbar\omega_{LT}^{A} = 1.7$ meV, $\hbar\omega_{LT}^{B} = 0.8$ meV, respectively, and their inhomogeneous broadenings of $\Delta_A = 2.2$ meV, $\Delta_B = 1.9$ meV. The difference between the two curves for extremely short times (< 100 fs) comes from the finite pulse width we use (at the times comparable with the inverse pulse width, the approximate formula (12.25) is no more valid). Otherwise, one can observe a fairly good agreement between theory and experiment, both in the *cw* and time-resolved spectra. The excitonic parameters are obtained with a quite good accuracy.

Once the technique allowing extraction of the excitonic characteristics from extremely broad spectral lines is established, we apply it to the series of GaN/$Al_{0.07}Ga_{0.93}N$ single QWs having different widths.

Figure 12.5 (a) shows the measured (solid) and calculated (dashed) reflectivity of the GaN/$Al_{0.07}Ga_{0.93}N$ QW of 12 monolayer (ML) width. In

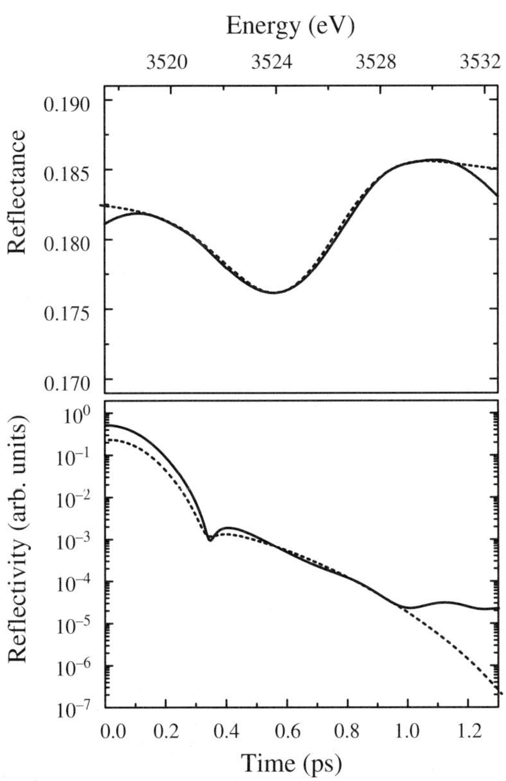

Fig. 12.5 Frequency-resolved spectra (a) and time-resolved spectra (b) of 8 monolayer-thick GaN-AlGaN QW.

Fig. 12.5(b) the solid line shows the time-resolved reflection obtained from the experimental reflectivity according to the formula (12.25) compared to the calculation with use of the semi-classical formalism [16] (dashed line). As well as for the bulk GaN sample, the fast decay of the time-resolved spectra is due the high value of the inhomogeneous broadening. The inhomogeneous broadening has been assumed to be Gaussian function characterized by a parameter Δ. The two others parameters of this model are the homogeneous broadening γ and the radiative broadening Γ_0 of the exciton line. At low temperature the value of the homogeneous broadening is negligible in comparison with Δ and Γ_0, thus only two parameters are to be determined. Δ characterizes the disorder within the QW and should decrease with the increase of the well thickness. Γ_0 characterizes the strength of exciton-light interaction and is proportional to the exciton oscillator strength [13]. The agreement between experiment and theory is quite good in Fig. 12.2 Once again, the difference at extremely short times comes from the limited accuracy of the transformation (12.24) in this time-range. For times longer than 1.2 ps the experimental signal is dominated by noise and should not be taken into account. The comparison between experimental and theoretical curves allows us to extract Γ_0 and Δ [see Fig. 12.6 (a,b)].

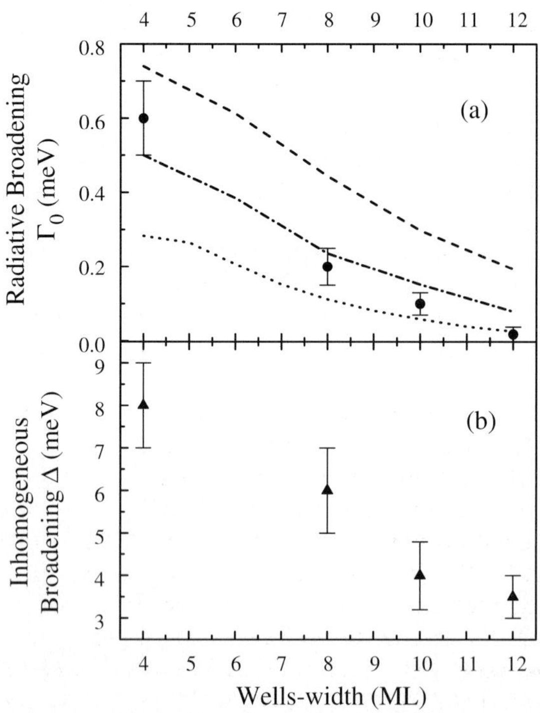

Fig. 12.6 Well-width dependence of the oscillator strength Γ_0 (a) and inhomogeneous broadening Δ (b) of GaN-AlGaN QWs obtained from the spectra (solid circles) and calculated variationally with different values of the piezoelectric field F (lines): F=600 kV/cm (dashed line), F=700 kV/cm (dashed-dotted line), F=800 kV/cm (dotted line).

In order to observe the evolution of the radiative and inhomogeneous broadening as a function of the wells thickness we have applied our procedure to a series of four high-quality $GaN/Al_{0.07}Ga_{0.93}N$ QWs of 4, 8, 10 and 12 MLs. The results are summarized in Fig. 12.6 that shows the extracted points in comparison with theoretical curves calculated by the variational method [14, 15] for different values of the piezoelectric field inside the QW. One can observe an extremely strong dependence of the radiative broadening on the thickness of the QW. The exciton oscillator strength decreases by a factor of 30 between 4 and 12 ML thick QWs, while the same variation of thickness of a GaAs/AlGaAs QW would cause the oscillator strength decrease by only a factor of less than 2 [3]. This drastic dependence of the oscillator strength on the QW width is an important peculiarity of GaN/AlGaN QWs, that should be carefully taken into account when using these wells in light-emitting optical devices.

This dependence is mostly due to the strong polarization fields that are present in the nitride-based QWs. The fields spatially separate the electron and hole mass-centers by a distance close to the well thickness. This considerably suppresses the overlap of electron and hole wave functions leading to a collapse of the oscillator strength. This tendency has been predicted theoretically by variational calculations of the oscillator strength [14, 15]. We present here the first direct experimental evidence of this dependence.

The comparison of our data with the result of a variational calculation shows that the value of the polarization field in our QWs was about 700 kV/cm. Note that this value is sensitive to the aluminum concentration in the barriers and the barrier thickness. The value of 700 kV/cm seems in average be the closest of the real value of the field, even if it is different in all the wells because of the different values of the strain and of the ratio of the well thickness over barrier thickness.

We would like to stress that despite the decrease of the exciton oscillator strength with the QW width, its absolute value is quite large in thin GaN/GaAlN QWs. So, the well of 8 ML width, exhibits the radiative broadening that is still one order of magnitude larger than in conventional GaAs or InGaAs QWs.

Figure 12.6 (b) shows the evolution of the exciton inhomogeneous broadening Δ with the QW width. It is quite large compared to the typical GaAs-based structures and it strongly increases with the decrease of the well thickness. However, it is worth noting that the ratio Γ_0/Δ is in the thin GaN/AlGaN wells higher that in the most of the conventional III–V QWs (0.1 for the 8 MLs well against 0.03 in a 10-nm-thick GaAs/AlGaAs QW [3]). The high value of this ratio gives a great deal of hope for observation of the exciton–polariton propagation effects in the nitride-based heterostructures [4, 7, 17, 18]. Let us note at this point that a multiple QW structure containing N wells has the ratio of the radiative broadening of the exciton–polariton mode to the inhomogeneous broadening about $[N(\Gamma_0/\Delta)]$. The coherent coupling of an exciton with light starts to dominate the disorder when this ratio becomes equal or superior to one, that we would expect to occur at $N > 10$. This proves, in particular, that the observation of the strong coupling regime [19] in nitride-based microcavities is a quite realistic task. This realization would open the way to the achievement of the stimulated scattering of microcavity polaritons at room temperature which would represent a huge step forward in the realization of a new class of opto-electronic devices. In the next paragraph we address the dynamics of polaritons in microcavities, and perspectives of fabrication of the *polariton laser* based on GaN microcavities.

12.5 Polariton laser on GaN: utopia or realistic perspective?

Dynamics of exciton–polaritons in semiconductor microcavities has become a subject of numerous publications and debates in recent few years. Microcavities represent low-dimensional structures providing the confinement of optical modes and exhibiting excitonic transitions resonant with these optical modes [20]. As a result of coupling between an exciton and a cavity photon, two exciton–polariton eigen-modes appear. In the strong-coupling regime that will be discussed here, two polariton branches experience an anti-crossing characterized by so-called Rabi-splitting [21]. The shape of the in-plane dispersion curves of two polariton modes is strongly dependent on the detuning between the bare cavity mode and the exciton resonance at $k=0$, where k is the in-plane wave vector. At zero or negative detuning (cavity mode has the lower energy than the exciton in the latter case) the characteristic bottleneck appears in the dispersion curve of the lower polariton mode. At small values of k the dispersion is governed by a very light 2D photon mass, while at large k it is dominated by the heavy exciton mass. This bottleneck prevents the polaritons to relax at $k=0$ state, because the acoustic phonon assisted relaxation is suppressed in the 'photonic' part of the dispersion curve [22]. This is one of the main obstacles on the way towards realization of the polariton laser based on the Bose condensation of exciton–polaritons at $k=0$ [23].

Very recently, a number of experimental papers appeared demonstrating a new nonlinear relaxation mechanism in a microcavity, namely, the resonant polariton–polariton scattering [17, 18, 24]. Simple and straightforward modeling of this scattering has been performed by Ciuti *et al.* [25]. Despite quite spectacular evidence of the stimulation of the polariton relaxation to $k=0$, the polariton–polariton scattering could not resolve the bottleneck problem, since the 'magic angle' [17] of the resonant excitation of the system required for the polariton–polariton scattering to be efficient, lies far below the bottleneck. Attempts to recover the bottleneck by organizing other kinds of exciton–exciton, exciton–polariton, polariton–polariton scattering processes [26] do not seem to be efficient up until today, because the high excitonic concentration and very specific conditions are needed to allow for these effects.

Here we propose a new mechanism of the cavity–polariton relaxation from the bottleneck region to the $k=0$ state in the linear optical regime. We propose to introduce in the quantum well (QW) embedded in a microcavity the small concentration of free conductivity electrons, preferably accelerated by the applied in-plane bias. Such microcavity structures based on GaAs are currently available [27]. We consider the resonant polariton–electron scattering. In this case, the polariton gives excess of its energy and wave vector to a free electron, as is described by the following equations:

$$e^-(\mathbf{k}_e) + P(\mathbf{k}) \rightarrow e^-(\mathbf{k}_e + \mathbf{k}) + P(0), \tag{12.26}$$

$$E(0) + \frac{\hbar(k + k_e)^2}{2m_e} - \frac{\hbar^2 k_e^2}{2m_e} = E(\mathbf{k}), \tag{12.27}$$

where e^- and P denote an electron and a polariton, respectively, E is the polariton energy, m_e is the electron effective mass, the free electron wave vector is dependent on the applied bias. In the following we shall consider the case of non-zero in-plane bias in order to scatter the polaritons from the 'bottleneck' region, while the discussed scattering process equally applies to the no-bias case, when the polaritons having large in-plane wave vectors can be scattered. In our case, only a very low concentration of free electrons is needed, so that the Fermi energy (in the absence of the electric field) should only slightly exceed the exciton line-width. The condition (12.27) can be satisfied because of the peculiar form of the exciton–polariton dispersion curve in a microcavity (see Fig. 12.7). Actually, the free conductivity electron is lighter than the exction and heavier than the cavity photon. The value of initial electron wave vector can be estimated as

$$k_e = \frac{1}{\hbar}\sqrt{2m_e \lambda e F}, \tag{12.28}$$

where F is the external electric field, and λ is the mean free path length in the plane of the QW. Tuning k_e by varying the external bias, one can have the condition (12.27) fulfilled for any point of the bottleneck region of the polariton dispersion curve.

In Fig. 12.7 the cavity polariton dispersion calculated for a model structure with the parameters taken from ref. [27] is compared to the free electron dispersion in order to reveal the resonance condition (12.27). One can check the efficiency of the scattering process we propose by performing the pump-probe experiment similar to one described in ref. [6]. The pump pulse must be sent at the oblique incidence in order to create the polariton population at the point **A** of the dispersion curve. The probe pulse sent at normal incidence will stimulate the scattering of polaritons from **A** to **B** with simultaneous acceleration of free electrons. The value of **k** (or of the required incidence angle) for which the resonance is achieved decreases with increase of the external bias value, as is shown in the inset to Fig. 12.7. ($\lambda = 10\,\mu$m was taken).

Note, that the coexistence of excitons and free electrons in a QW is possible at intermediate free charge densities (see e.g. [26] and references therein). An additional effect that can modify the cavity dispersion in this regime is the appearance of a trion state coupled to the cavity mode [27]. In this case the stimulated scattering of exciton–polaritons can be accompanied by dissociation

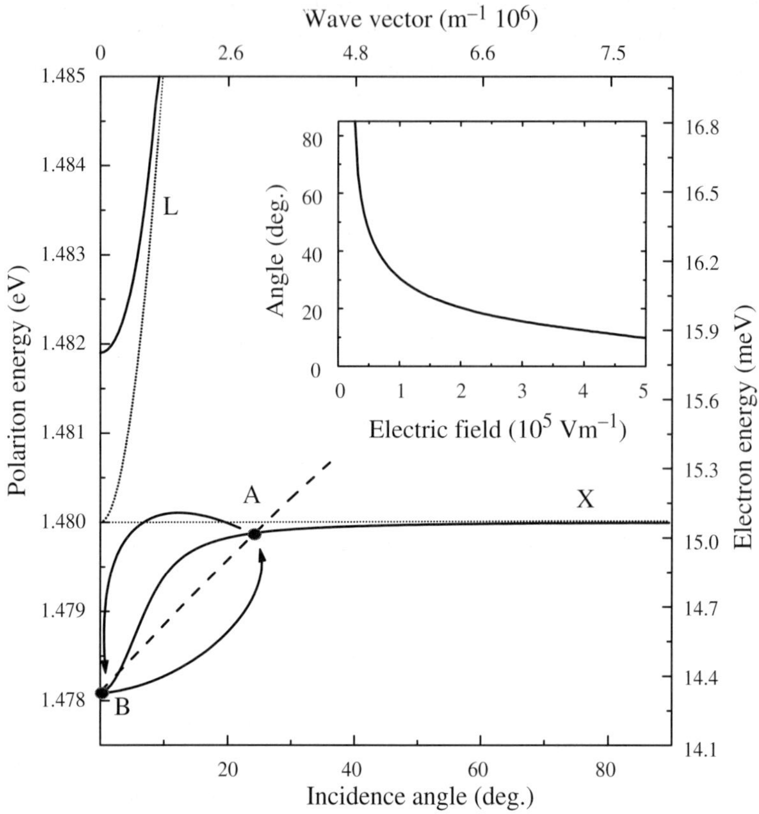

Fig. 12.7 Bare exciton (X) and photon (L) modes (dotted) and exciton–polariton modes (solid) in a model microcavity structure containing a two-dimensional electron gas. Dashed line shows the part of the dispersion curve of free electrons in GaAs. The inset shows the incidence angle for which the resonance condition (12.27) is satisfied as a function of the applied electric field. The electric field is applied in a QW plane and in the incidence plane.

of trions while the charge acceleration effect remains. Further, we shall neglect the trions for simplicity.

We describe the process of electron acceleration using the polariton rate equations. We consider three different polaritonic states, namely the ground state ($\mathbf{k}=0$) characterized by the population $N_0\ (t)$, the initial state ($\mathbf{k}$) excited by the pump pulse and characterized by a population $N_k(t)$, and the ensemble of polariton states $S_k(t)$ situated on the 'elastic circle' [22, 24, 26], i.e. whose wave vectors have the same absolute value as $\mathbf{k}$. In our notation S_k also includes the population N_k. We consider three kinds of processes, namely, the radiative escape of the polaritons, their resonant scattering to the ground state and backward by interaction with free electrons, and their elastic scattering due to the in-plane potential disorder. We shall neglect the exciton–phonon scattering process that is frozen since the state excited by the pump pulse is situated in the bottleneck region. We shall use C_k and X_k being the photon and exciton fractions of the exciton–polariton wave function having a wave vector k, respectively, τ_c being the radiative life-time of a cavity polariton (supposed by independent of k), Δ being the exciton inhomogeneous broadening, and τ_d being the characteristic time of the stimulated scattering. The equations which describe the temporary evolution of N_0 and S_k are:

$$\frac{dN_0}{dt} = -\frac{C_0}{\tau_c}N_0 + \frac{X_k}{\tau_d}(N_0 + 1)N_{\mathbf{k}}, \tag{12.29}$$

$$\frac{dN_{\mathbf{k}}}{dt} = -\frac{C_k}{\tau_c}N_{\mathbf{k}} - \frac{X_k}{\tau_d}(N_0 + 1)N_{\mathbf{k}} + \frac{X_k\Delta}{n - 1}(S_k - nN_{\mathbf{k}}), \tag{12.30}$$

$$\frac{dS_k}{dt} = -\frac{C_k}{\tau_c}S_k - \frac{X_k}{\tau_d}(N_0 + 1)N_{\mathbf{k}}, \tag{12.31}$$

Here n is the number of states on the elastic circle given by $n = kR$, where R is the size of the light-spot ($R = 100$ μm will be taken in the following calculation). The stimulated scattering time constant is given by

$$\frac{1}{\tau_d} = \frac{2\pi|W|^2X_k^2X_0^2}{\hbar\,(E(k) - E(0))}\,f(\mathbf{k}_e)\,(1 - f\,[\mathbf{k}_e + \mathbf{k}]), \tag{12.32}$$

where W is the matrix element of the exciton–electron scattering in a QW, $f(\mathbf{k})$ is the probability of finding an electron having a wave-vector $\mathbf{k}$. The electron–exciton scattering in bulk has been addressed in a number of theoretical papers [28–30]. In order to find it in a QW we have adapted the formalism developed by Ciuti *et al.* [31] for the exciton–exciton scattering in quantum wells having canceled the second hole, thus reducing their four-particle to our three-particle problem and accounting for the screening of the Coulomb interaction within the Thomas–Fermi approximation [14, 15]. A numerical estimate of W is of 0.2 meV in our model system, that yields $\tau_d = 10$ ps (we supposed $f(\mathbf{k})=1$, $f(\mathbf{k}_e+\mathbf{k})=0$). Using the cavity life-time of 1 ps, we estimate from eqn (12.29) the critical value of polariton pumping in the initial state required for the optical amplification (in a low disorder case) as:

$$N^s_{\text{pump}} = \frac{C_0\tau_d}{X_k\tau_c} = 5, \tag{12.33}$$

that is extremely small.

The electron–polartion scattering induces the current of accelerated electrons $N^e_{\mathbf{k}+\mathbf{k}_e}$ given by:

$$\mathbf{j} = e\frac{dN^e_{\mathbf{k}+\mathbf{k}_e}}{dt} = e\frac{X_k}{\tau_d}(N_0 + 1)N_{\mathbf{k}}. \tag{12.34}$$

The average current generated by a series of pulses is

$$I = \Omega \int j(t)dt, \tag{12.35}$$

where Ω is the pulsation of the incident light (taken in the following calculations to be 82 MHz as in ref. [6]).

Figure 12.8 (a) shows the evolution of S_k, N_0, and N_k as functions of time in the case of $N_{\text{pump}} = 6$ (i.e. close to the amplification threshold) and $k_e = 4.4\ 10^8 \text{m}^{-1}$. One can observe almost no amplification at $k=0$ and a slow decay of S_k. Contrary, the decay of N_k is fast, since it is governed by the inhomogeneous broadening taken to be $\hbar\Delta = 0.4$ meV. The inset shows the shape of the current pulse $j(t)$ generated as a function of time that is quite weak in this case. Figure 12.8(b) shows the evolution of the same quantities for the same pump pulse as in Fig. 12.8(a), but now a probe pulse creating a population $N_0^{\text{probe}} = 3$ is sent to the sample with a delay of 1 ps. One can observe the strong depopulation of the S_k states after the arrival of the probe pulse. The inset shows the shape of the generated current as a function of time. Contrary to the previous case, one can see an intensive short current pulse generated by the stimulated scattering process.

Figure 12.8(c) shows the intensity of the average current $\boldsymbol{I}$ as function of the delay between the pump and the probe pulses. One can observe a fast decay governed by the short elastic scattering time constant. The stimulating scattering process itself is not affected by the disorder effect thus the number of polaritons at the ground state is independent of the delay [dashed line, Fig. 12.8(c).]

Figure 12.9(a) shows the evolution of S_k, N_0, and N_k as a function of time for a pump pulse well above the amplification threshold $N_{\text{pump}} = 50$ and without any probe pulse. One can observe a very fast and efficient transfer of the polariton population from the initial state to the ground state before the depopulation of the initial state by the elastic scattering process. This gives rise to a well-shaped, intense, and extremely short current emission (see the inset). Note that this is also a regime of the expected polariton laser action. Figure 12.9(b) shows the average current I generated by the stimulated scattering process as a function of N_{pump} (no probe pulse). On can see the initial quadratic increase of the current with N_{pump} followed by a linear dependence above the amplification threshold. This is quite natural, since at low pumping the effect is quite weak (without stimulation by a probe pulse), while in the strong pumping case almost all the created polartions experience scattering with electrons.

Note, that the detection of ultra-short pulses of current is a separate problem addressed in a number of publications in the recent years (see e.g. ref. [32] and references therein). The detection of current-induced radiation seems to be the easiest way to detect the effect we have predicted.

Finally, at present the effect described above can only be observed in high-quality doped GaAs microcavities at low tempeatures. However, for application of the Bose-condensation of exciton–polaritons in polariton lasers, GaAs-based microcavities are clearly not adapted. The typical value of

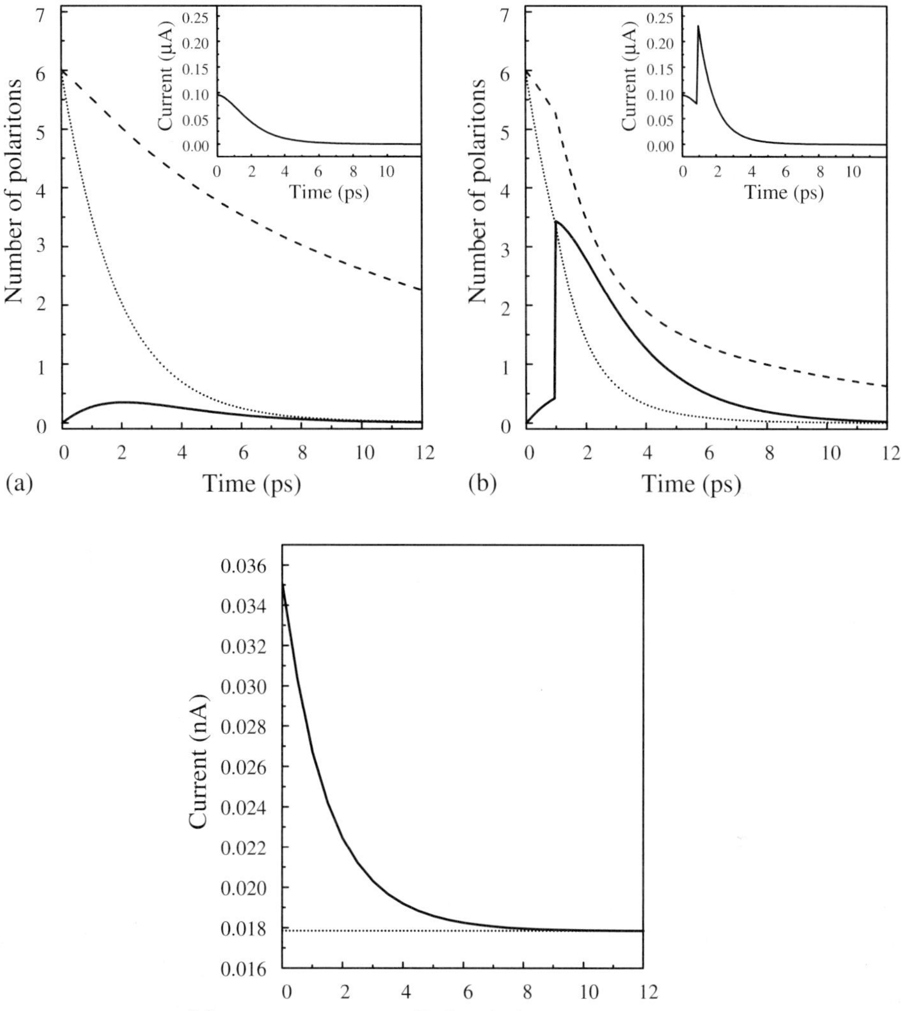

Fig. 12.8 Temporary evolution of the polariton population at $k=0$ (solid) and on the elastic circle (dashed), in the case of $N^{\text{pump}}=6$, without a probe pulse (a) and with a probe pulse delayed by 1 ps with respect to the pump pulse (b). The insets show the pulses of current $j(t)$ induced in these regimes. The solid line (c) shows the average current I as a function of the delay between pump and probe pulses. The dotted line (c) shows the limiting case of no probe pulse.

Rabi-splitting in these cavities is of 5–7 meV [20], that is also the order of magnitude of the exciton binding energy in InGaAs/GaAs QWs. Already at liquid nitrogen temperature the exciton homogeneous broadening exceeds its binding energy, and the strong coupling regime cannot be observed in this kind of structures. On the contrary, in GaN-based microcavities the Rabi-splitting of

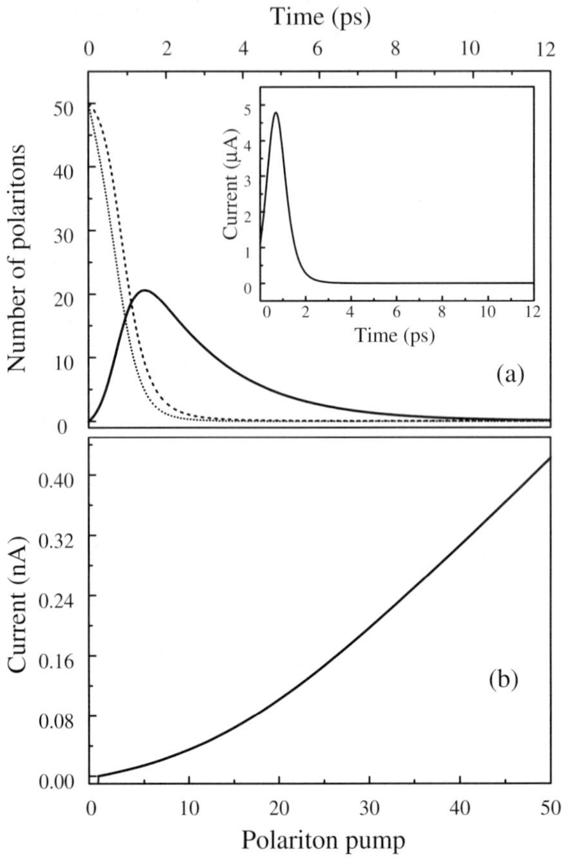

Fig. 12.9 Time evolution of the polaritons number at $k = 0$ (solid), on the elastic circle (dashed), and in the initial state (dotted) in the strong pumping regime (a), and generated average current I as a function of the pumping (b). The inset shows the shape of the generated pulse of current $j(t)$.

about 50 meV is expected [19], and the excitons are still stable at room temperature. This makes nitrides excellent candidates for realization of a polariton laser working at room temperature. Two main technological steps towards the polariton lasers seem to us to be the growth of high-quality GaN-based microcavities exhibiting the strong coupling regime and solving of the problem of contacts in order to be able to suppress the polariton bottleneck effect applying the in-plane bias as we have described above.

12.6 Conclusion

We have developed here a semi-classical theory of light-exciton coupling in strongly disordered GaN-based quantum structures. We have shown that the resonant Rayleigh scattering governs the coherent optical spectra of these structures, and that studying the RRS spectra one can obtain a complex information on the exciton statistics, radiative and non-radiative broadening of exciton resonances, oscillator strength, and localization radius. We have presented the first direct experimental study of the QW width effect on the radiative broadening and inhomogeneous broadening of excitons in GaN/$Al_{0.07}Ga_{0.93}N$ QWs. The major limitation of the accuracy of the experimental determination of these characteristics due to extremely broad exciton lines in the spectra has been eliminated by use of the original technique of numerical analysis of the spectra. The values of the radiative broadening we have obtained are in a good agreement with the results of a variational calculation assuming the polarization field within the wells of 700 kV/cm. The collapse of the radiative broadening with the increase of the well width, and its high value for thin QWs are evidenced. The significant value of the radiative broadening compared to the inhomogeneous broadening in thin GaN/AlGaN QWs makes them extremely promising for the observation of fine exciton–polariton effects. The stability of the excitons in nitrides at room temperature and their strong coupling with light provide a hope for the realization of new classes of optoelectronic devices based on microcavity polaritons. We have given theoretical evidence of free carrier acceleration by stimulated scattering of exciton–polaritons in microcavities and have proposed a principle of the polariton laser based on this effect. We believe that pump-probe experiments on the high quality GaN-microcavities will soon provide an experimental confirmation of all the above effects.

Acknowledgement

This work would have been impossible without numerous enlightening discussions with Bernard Gil, who also provided us with all the optical spectroscopy data used in this paper. The samples were grown by N. Grandjean and J. Massies. We are indebted to M. Zamfirescu and P. Bigenwald for their numerical calculations used in this article. This work has been supported by the EU RTN 'CLERMONT' program, contract No. HPRN-CT-1999-00132.

References

1. Alemu, A., Gil, B., Julier, M., and Nakamura, S. (1998). *Phys. Rev. B* **57**, 3761.
2. Hopfeld, J. J. (1958). *Phys. Rev.* **112**, 1555; Agranovich, V. M. (1960). *Sov. Phys. JETP* **37**, 307. For a review see, e.g., Agranovich, V. M. and Ginzburg, V. L. (1966). *Spatial dispersion in crystal optics and the theory of excitons.* Interscience Publisher, London.
3. Andreani, L. C. (1995). In *Confined electrons and photons* (Eds E. Burstein and C. Weisbuch), Plenum Press, New York, p. 57.
4. Malpuech, G., Kavokin, A., Langbein, W., and Hvam, J. M. (2000). *Phys. Rev. Lett.* **85**, 650.

5. Langbein, W., Hvam, J. M.,, and Zimmermann, R. (1999). *Phys. Rev. Lett.* **82**, 1040.
6. Malpuech, G. and Kavokin, A. (2001). *Phys. Stat. Sol.* (a) **183**, 75.
7. Baumberg, J. J., Heberle, A. P., Kavokin, A. V., Vladimirova, M. R., and Köhler, K. (1998). *Phys. Rev. Lett.* **80**, 3567.
8. EXAFS technique, for example.
9. Malpuech, G., Kavokin, A., Leymarie J., and Vasson, A. (2000). *Solid State Commun.* **113**, 185.
10. Malpuech, G., Kavokin, A., Leymarie, J., Disseix, P., and Vasson, A. (1999). *Phys. Rev. B* **60**, 13298.
11. Ivchenko, E. L. and Kavokin, A. V. (1992). *Sov. Phys. Solid State* **34**, 1815.
12. Malpuech, G. and Kavokin, A. (2001). *Semicond., Sci. and Technol. Topical Review* **16**, 3, R1.
13. Andreani, L. C., Panzarini, G., Kavokin, A. V., and Vladimirova, M. R. (1998). *Phys. Rev. B* **57**, 4670.
14. Bigenwald, P., Kavokin, A. V., Gil, B., and Lefebvre, P. (2000). *Phys. Rev. B* **61**, 15621.
15. Bigenwald, P., Kavokin, A., Gil, B., and Lefebvre, P. (2001). *Phys. Rev. B* **63**, 5315.
16. Kavokin, A., Malpuech, G., and Panzarini, G. (1999). *Phys. Rev. B* **60**, 16788.
17. Huang, R., Tassone, F., and Yamamoto, Y. (2000). *Phys. Rev. B* **61**, R7854; Savvidis, P. G., Baumberg, J. J., Stevenson, R. M., Skolnick, M., Whittaker, D. M., and Roberts, J. S. (2000). *Phys. Rev. Lett.* **84**, 1547.
18. Stevenson, R. M., Astratov, V. N., Skolnick, M. S., Whittaker, D. M., Emam-Ismail, M., Tartakovskii, A. I., *et al.* (2000). *Phys. Rev. Lett.* **85**, 3680.
19. Kavokin, A. and Gil, B. (1998). *Appl. Phys. Lett.* **72**, 2880.
20. For a review see Savona, V. *et al.*, (1999). *Phase Transitions* **68**, 169.
21. The term that has come from the atomic physics, and surprisingly has become very popular with microcavity physicists.
22. Müller, M., *et al.* (2000). *Phys. Rev. B* **62**, 16886.
23. Yamamoto, Y. (2000). *Nature* **405**, 629.
24. Baumberg, J. J., *et al.* (2000). *Phys. Rev. B* **62**, R16247.
25. Ciuti, C., *et al.* (2000). *Phys. Rev. B* **62**, R4825.
26. Tassone, F., *et al.* (1997). *Phys. Rev. B* **56**, 7554.
27. Rapaport, R., *et al.* (2000). *Phys. Rev. Lett.* **84**, 1607.
28. Benoit à la Guillaume, C., Debever, J-M., and Salvon, F. (1969). *Phys. Rev.* **177**, 567.
29. Haug, H. and Koch, S. (1977). *Phys. Stat. Sol.* (b), **82**, 531.
30. Koch, S. W., *et al.* (1978). *Phys. Stat. Sol.* (b) **89**, 43.
31. Ciuti, C. *et al.* (1998). *Phys. Rev. B* **58**, 7926.
32. Wu, Q., Zhang, X.-C. (1996). *Appl. Phys. Lett.* **68**, 1604.

13 Optoelectronic devices based on low-dimensional nitride heterostructures

Andreas Hangleiter

13.1 Introduction

Group-III-nitride-based optoelectronic devices have been the key driving force in the recent rapid development of nitride technology. Historically, the first light-emitting nitride device was a metal-insulator-semiconductor light-emitting diode (MIS-LED) developed by Jacques Pankove and coworkers [1] in the seventies, with dim pn-junction LEDs following in the late eighties [2]. Until 1992 the brightness of nitride LEDs reached only the millicandela range, still unable to surpass the SiC-based blue LEDs of those days. Only in 1993, were Nakamura and coworkers [3] able to report significant progress in GaN-based blue LED brightness, reaching the one candela level in 1994 [4]. While these first bright GaN-LEDs were still based on donor–acceptor transitions in thick GaN layers, super-bright violet, blue, and green LEDs utilizing the reduced dimensionality in GaInN/GaN quantum wells were reported only shortly thereafter [5, 6].

Until about the same time, most of the effort aiming at developing blue and green laser diodes was concentrated on II–VI semiconductors like CdTe and ZnSe. Due to the relatively weak bonds inherent to these materials, device lifetimes improved only very slowly, reaching less than 1000 hours in 1998 [7]. In early 1996, Nakamura *et al.* [8, 9] reported the first pulsed operation of a nitride laser diode, with continuous wave operation at room temperature achieved only one year later [10]. These laser diodes were also based on GaInN/GaN multiple quantum well structures and were taking advantage of their low-dimensional character.

In this chapter, we will review the present knowledge on low-dimensional nitride heterostructures as applied to optoelectronic devices, together with an account of the state of the art regarding device performance.

13.2 Peculiarities of nitride heterostructures

III-nitride semiconductors are different from conventional III–V semiconductors in many respects. Most basically, they exhibit a larger bandgap energy and therefore much larger effective masses. They prefer to crystallize in the wurtzite structure rather than in the zinc-blende structure.

Moreover, nitrides are highly polar materials, which leads to a strong piezo-electric and even spontaneous polarization, due to the low symmetry wurtzite structure.

13.2.1 Hetero-epitaxy and defects

Even though nitride research and development has reached an impressive level of maturity, the growth of bulk GaN crystals is still at an early stage. Only small pieces with dimensions of millimeters can be produced by direct bulk growth methods [11]. Even thick quasi-substrates grown by hydride vapour phase epitaxy on dissimilar substrates are only now becoming practical [12].

Today, all practical nitride-based devices are based on hetero-epitaxial GaN layers grown on substrates such as sapphire or silicon carbide. Other foreign substrates, such as silicon, gallium arsenide, or even zinc oxide have also been tried [13, 14]. The problems common to the hetero-epitaxial approach is the large mismatch of the lattice constant, which leads to a large number of misfit dislocations, and a large mismatch in thermal expansion coefficients. In fact, epitaxial growth on such dissimilar substrates works only if a two-step growth [15] procedure is applied, where a low-temperature nucleation step is followed by high-temperature coalescence and growth steps.

Nevertheless, the large lattice mismatch together with the nucleation procedure leads to a large number of threading dislocations [16] penetrating through the whole epitaxial layer stack. In typical GaN-based device structures, the defect densities are of the order 10^8–$10^{10}\,cm^{-2}$. This is in strong contrast to other III–V materials such as GaAs, where light-emitting devices cease to work at defect densities beyond $10^4\,cm^{-2}$ because of non-radiative recombination. Surprisingly, GaInN-based light-emitting diodes do work at such defect densities, with only a very weak dependence of the efficiency on defect density (c.f. Section 13.3.3).

Even without using GaN substrates, device properties can be dramatically improved by using lateral growth schemes. So-called 'epitaxial lateral over-growth (ELO)' [17, 18, 19] is starting from a GaN buffer layer e.g. on a sapphire substrate, where a silicon dioxide mask is deposited on parts of the still highly defective buffer layer. Further growth now proceeds, starting from the window regions and extends sidewards onto the mask regions without creating new defects there. After a few microns of growth, coalescence of the wings over the masks occurs, leading to a GaN layer with a low defect density over the masked regions and a high defect density over the window regions. Over the mask regions, a reduction of the defect density by about three orders of magnitude has been achieved. Variants of this method include growth on pre-patterned buffer layers, where lateral growth starts from the sidewalls of the patterned areas [20].

Further improvement of the layer properties can be achieved by growing a thick GaN layer on top of such a 'ELO' layer, possibly repeating the 'ELO' defect reduction step [21].

13.2.2 Optical properties

13.2.2.1 *Low carrier density (LEDs)*

The most unusual property of nitride heterostructures has already been mentioned: their quantum efficiency is large even though there exists a large number of potentially non-radiative defects. In a more detailed picture, there is a number of peculiar features in the optical properties of low-dimensional nitride heterostructures, which affect device operation and performance.

1. In typical nitride quantum wells, there is a large Stokes-like shift between emission and absorption, i.e. the emission appears at longer wavelength than the absorption edge [22]. This shift is larger for wide quantum wells and for large In mole fractions in GaInN/GaN structures and Al mole fractions in GaN/AlGaN structures, respectively [23]. For GaN/AlGaN structures, it is obvious that the quantum well emission sometimes appears even below the bulk GaN bandgap [24].
2. With increasing excitation level by either optical pumping or electrical injection, the emission is subject to a considerable blue-shift, particularly for wide quantum wells [25].
3. The emission decay times after pulsed excitation become rather large for wide wells and for large In or Al mole fractions. The decay is generally non-exponential and is accompanied by a red-shift of the emission peak [26, 27].
4. The emission linewidth of GaInN/GaN quantum wells is fairly large and increases with increasing In mole fraction [28] or increasing quantum well width [29].
5. Due to the low symmetry of the wurtzite structure, nitride heterostructures exhibit a considerable birefringence [30, 31].

Some of the early results along those lines together with high resolution electron microscopy led to speculations on carrier localization [22] or even quantum dot formation [32] due to phase separation in GaInN. This picture was also advertised as an explanation for the exceptional quantum efficiency of GaInN QWs, where localization at potential minima would save carriers from non-radiative recombination at dislocations [33].

However, it was later recognized that polarization effects such as piezoelectric fields and spontaneous polarization have a number of consequences for optical properties of nitride quantum wells [27, 28, 34]. In particular, strong piezoelectric fields of the order MV/cm [35] arise from the biaxial strain due to the lattice mismatch between well and barrier material. These internal fields lead to a strong quantum confined Stark effect (QCSE) [36] in low-dimensional heterostructures, as indicated in Fig. 13.1. Both in GaInN/GaN and in GaN/AlGaN quantum wells, internal fields give rise to a strong red-shift of the emission with respect to the absorption [29, 34], a strongly reduced oscillator strength due to carrier separation [27, 28], a blue-shift under

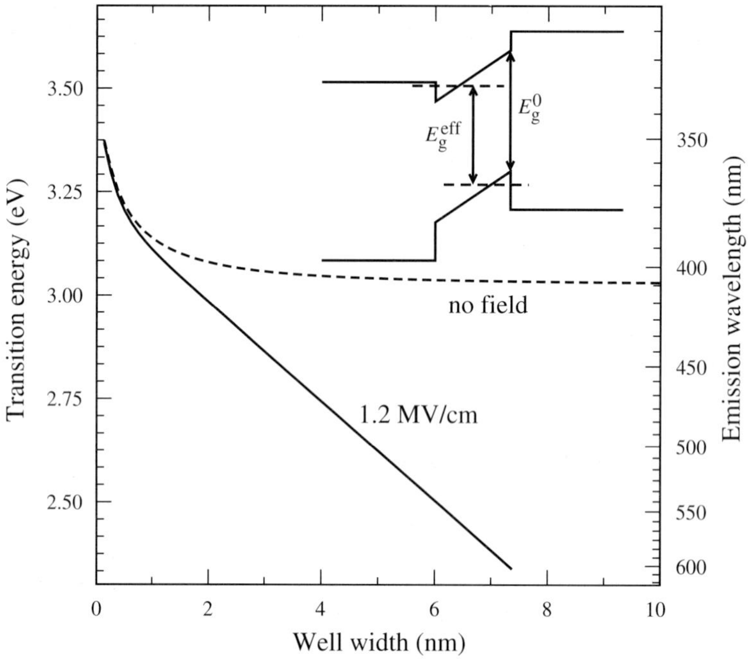

Fig. 13.1 Quantum confined Stark effect in piezoelectric quantum wells. Due to the internal field, the effective bandgap may become much lower than the real bandgap of the well material.

high injection levels or by background doping due to field screening [37, 38], and an enhancement of the emission linewidth [29].

Moreover, the internal field breaks the symmetry of an otherwise symmetric quantum well, which has been studied using deliberately asymmetric AlGaN/GaInN/GaN structure [39]. In multiple quantum well structures, Stark-ladder-like interwell transitions can be observed if the barriers are thin enough to allow overlap of the electron and hole wave functions of neighboring wells [40, 41]. Even if the well material is unstrained but the barriers are strained as in case of GaN/AlGaN QWs grown on a GaN buffer layer, Fermi-level-induced redistribution of fields [42, 43] leads to a strong field inside quantum wells and to a strong QCSE.

The effect of internal electric fields is equally important in deliberately made quantum dot structures. Since such structures can accommodate even larger strain than pseudomorphic quantum wells, the fields become giant and the red shift of the emission below the bulk bandgap becomes even more pronounced [44, 45].

Nevertheless, there are certainly compositional fluctuations in GaInN and AlGaN, as in any ternary material. Such compositional fluctuations, together with width fluctuations in quantum wells, are the origin of broadening and finite linewidth of optical transitions. From an analysis of the emission linewidth, it is possible to separate these two contributions quantitatively [29]. In this way it was found that typical GaInN quantum wells exhibit a relative compositional fluctuation of about 10%. Given these numbers, is not clear to date if such fluctuations are strong enough to protect carriers from recombining non-radiatively at room temperature.

13.2.2.2 Optical gain (lasers)

Highly sophisticated calculations of the optical gain in nitride single quantum well structures have already been reported [46, 47], which are based on a $k \cdot p$ model and which take into account the internal fields in such structures in a self-consistent manner. However, particularly in the presence of internal fields, multiple quantum wells are likely to behave considerably differently from single quantum wells, due to potentially inhomogeneous field distributions. Moreover, it has been shown that in the case of wurtzite nitrides, sub-band interactions in quantum wells are only of minor importance [48]. Therefore, a somewhat simplified model [49] concentrates on self-consistently calculating the effects of polarization fields on optical properties for multiple quantum wells by iteratively solving the Schrödinger and Poisson equations. The complications arising from the complex valence band structure are deliberately neglected. On the other hand, the complete multiple quantum well structure, including the commonly used AlGaN electron barrier is taken into account.

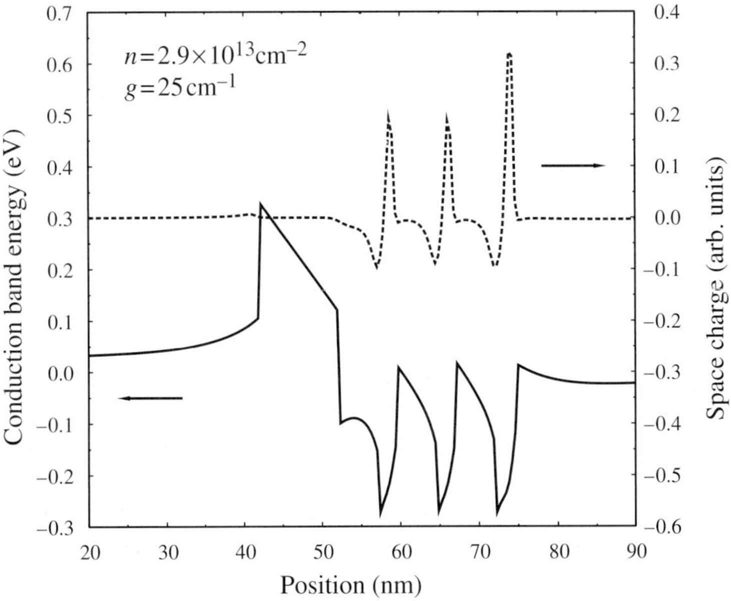

Fig. 13.2 Conduction band diagram and space charge distribution of a triple quantum well at laser threshold [49]. Clearly, the carrier distribution is strongly inhomogeneous.

As shown in Fig. 13.2, even under lasing conditions, the quantum well next to the substrate holds considerably more positive charge (i.e. holes) than the others. Here, a triple quantum well structure was studied, with a GaInN well bandgap of 3 eV, a well width of 2.5 nm, a barrier width of 5 nm, and a AlGaN electron barrier of 10 nm. As can be seen from the conduction band diagram, the overall polarization across the MQW is largely screened, but there remains a considerable field within the QWs. Accordingly, the oscillator strength is considerably lower than in the field-free case.

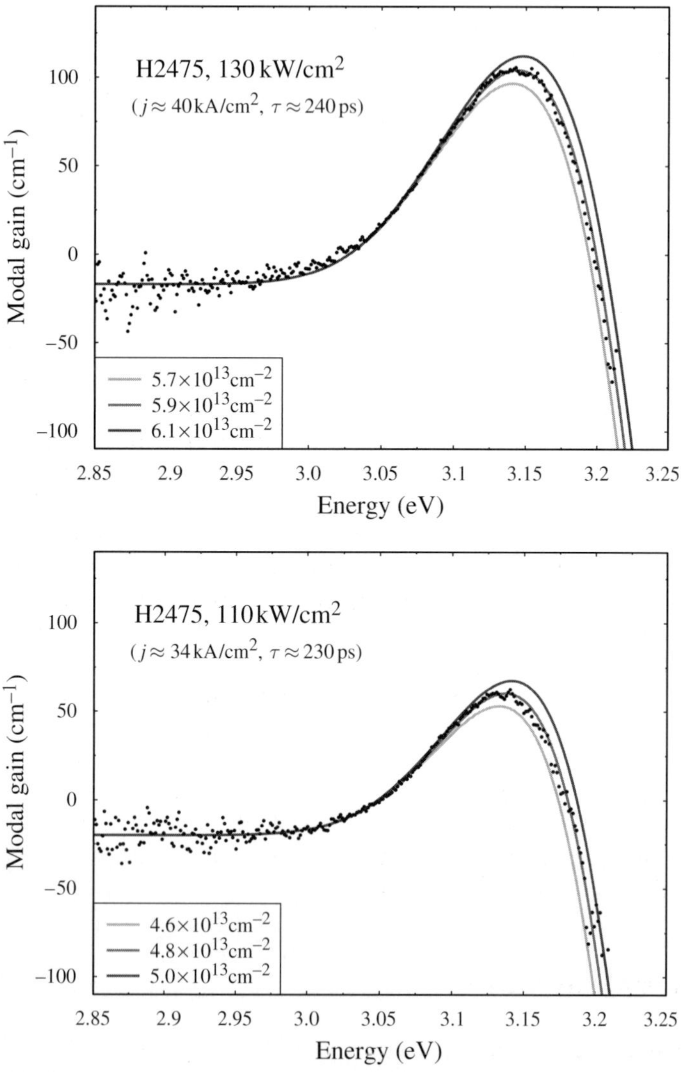

Fig. 13.3 Comparison of measured and calculated optical gain spectra, allowing the determination of the carrier density, the carrier lifetime, and the optical losses [49].

A critical test of any optical gain model for semiconductor structures is the direct comparison of calculated gain spectra with measured ones. As shown in Fig. 13.3, gain calculations performed using the very structural parameters of the real laser structures, nicely reproduce the experimental gain spectra.

For the particular example shown in Fig. 13.3, the carrier density in the structure is $4.8\times10^{13}\,cm^{-2}$ and $5.9\times10^{13}\,cm^{-2}$, respectively. These values are fairly insensitive to variation of the inhomogeneous broadening parameter. Together with the (absorbed) pump power densities of 110 and $130\,kW/cm^2$ used in the experiment, this translates into a carrier lifetime of 240 ps. This has to be compared to a radiative carrier lifetime of 5.8 ns and 4.8 ns determined from the calculations. Consequently, this laser structure is still dominated by non-radiative recombination losses associated with the high defect density.

Moreover, the optical losses in such structures typically are in the range $10–30\,cm^{-1}$ which is surprisingly low given the large defect density in heteroepitaxial nitride layers. However, the present result is similar to earlier data obtained for simple double heterostructures [50].

13.2.3 Doping and transport properties

Control of doping has been a notorious problem for most wide-gap semiconductors for quite some time. For GaN, where unintentional conductivity is n-type, n-type doping with silicon turned out to be simple and easy to control. For AlGaN problems similar to other Al-containing compounds like AlGaAs arise: At Al mole fractions around 50% n-type doping becomes ineffective and the material is insulating [51].

P-type doping, on the other hand, was a major obstacle until 1989, when Amano *et al.* [52] discovered that Mg acceptors can be activated by low energy electron irradiation. A little later, Nakamura *et al.* [53] introduced a thermal treatment to activate acceptors. It was later recognized that hydrogen, which is ubiquitous during growth, passivates Mg acceptors but can be driven out of the layers using one of the treatments mentioned above.

However, even though the chemical concentration of Mg reaches the upper $10^{19}\,cm^{-3}$ range, the actual hole density is at most $10^{18}\,cm^{-3}$, due to the large acceptor ionization energy for Mg of about 230 meV [54]. Moreover, the hole mobility in III-nitrides is extremely small, i.e. about $10–15\,cm^2/Vs$ [55], mostly due to the extremely large hole effective mass [56, 57]. Taken together, this leads to a minimum specific resistivity of about $0.2\,\Omega\,cm$. Also, so far all attempts to find acceptors with a lower ionization energy failed.

13.3 Light-emitting diodes

Highly efficient light-emitting diodes (LEDs) have marked the starting point of the nitride revolution. In 1994 blue nitride LEDs commercialized by Nichia Chemicals passed the one candela barrier and by far surpassed the brightness of previous SiC-based blue LEDs. Since then, blue, green, amber, and ultraviolet LEDs based on nitride quantum well heterostructures have been developed towards external efficiencies of up to 32% [58]. Moreover, even white light and tuneable colours are now being delivered by nitride-based LEDs [59, 60] thanks to the use of colour conversion schemes using inorganic or organic fluorescent materials.

13.3.1 Basic structure of nitride LEDs

Today's commercial nitride LEDs are fabricated either on sapphire or on SiC substrates. The conducting SiC substrate allows a vertical current flow and a backside n-contact, whereas the insulating sapphire substrate requires access to the n-contact from top.

Early blue nitride LEDs used donor–acceptor (DA) transitions for light emission by doping the active region with zinc, which has a deep acceptor level at about 330 meV [61]. Today's green, blue, and ultraviolet LEDs are based on GaInN quantum wells, which offer still higher quantum efficiency.

However, even though the details of the buffer structure differ for those substrates, the actual light-emitting quantum well region is essentially identical, and looks as in Fig. 13.4. On top of the n-type GaN buffer/contact layer follows a quantum well active region with a single or multiple GaInN/GaN quantum well. This is followed by a thin p-AlGaN cap layer and a p-GaN contact layer. Typically, the n-type buffer is a few microns thick, the quantum wells are in the range 2–4 nm, the AlGaN cap layer is a few ten nanometers, and the p-GaN contact layer is a few hundred nanometres thick.

13.3.2 Important concepts

The characteristics of nitride LEDs are determined by a number of design parameters and concepts, which will be discussed in the following.

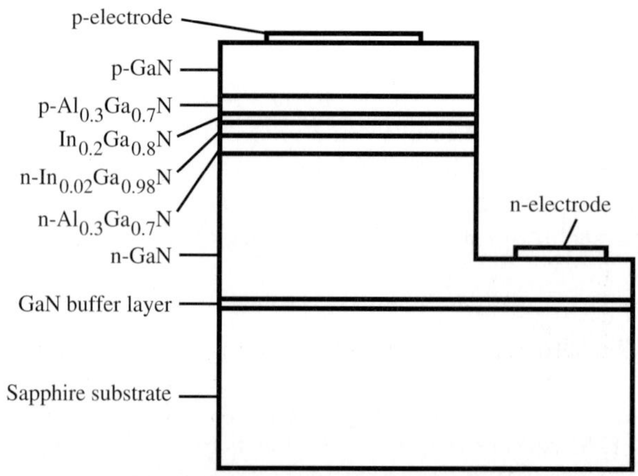

Fig. 13.4 Typical layer structure of a quantum-well-based nitride LED [5].

The emission wavelength (colour) of the LED is governed by the choice of the active material (i.e. its bandgap), its composition, the width of quantum wells, and the amount of strain built into the structure. The latter controls the magnitude of built-in piezoelectric fields, which can have a huge impact on emission wavelength via the quantum confined Stark effect. Most prominently, this is reflected in the properties of experimental red-emitting nitride LEDs [62]. As shown in Fig. 13.5, a strong blue shift under current injection of about 50 nm provides an unambiguous hint towards field screening by injected carriers. A more general picture is shown in Fig. 13.6, where the calculated emission wavelength is plotted as a map in well-width/In-content space. As one can see, for example, an emission wavelength of 500 nm can either be reached by a 1 nm GaInN QW with 27% In or by a 5 nm GaInN QW with only 12% In.

The internal efficiency is primarily a matter of the importance of non-radiative recombination, i.e. the influence of defects. As is well known, heteroepitaxial nitride structures contain a large density of threading dislocations, which penetrate through the active layer. As mentioned earlier, it is speculated that localization in potential fluctuations prevents carriers from recombining non-radiatively at the defects. Leakage currents across the active region by poor carrier confinement are an additional source of losses. This is particularly

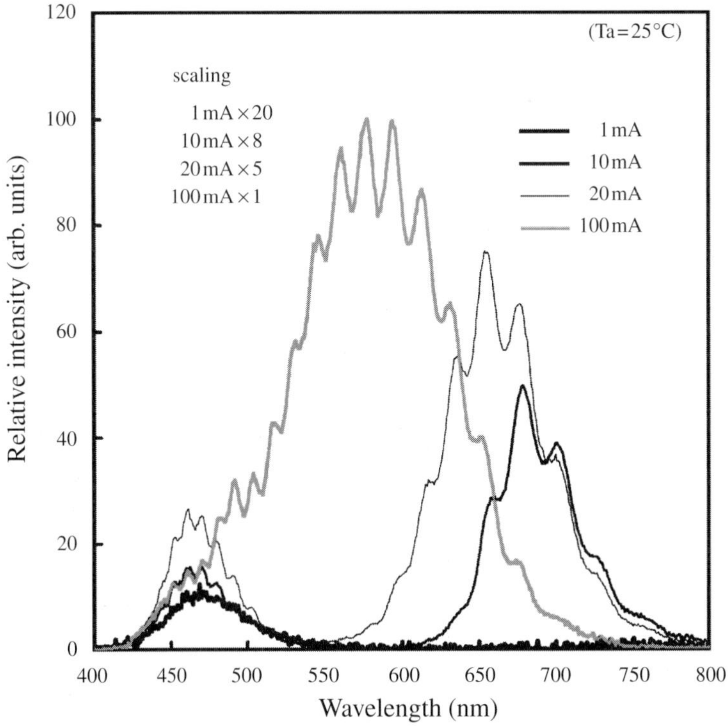

Fig. 13.5 Strong red-shift of the emission under current injection as observed in a red-emitting nitride LED [63].

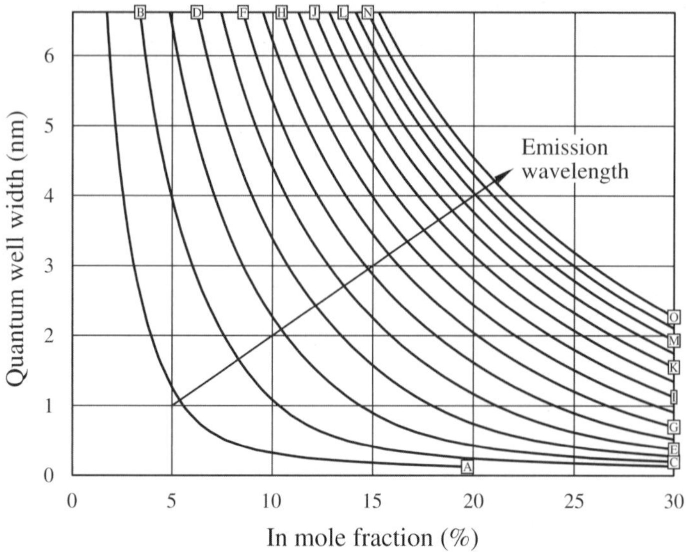

Fig. 13.6 Map of emission wavelength contours as a function of In content and well width in GaInN/GaN quantum wells. Contours start at 380 nm (A) and end at 660 nm (O), with a 20 nm increment.

important in case of shallow quantum wells, as, for example, used for near UV LEDs in the range 370–400 nm. In order to prevent electron leakage currents [64], most practical LED structures contain a special AlGaN electron barrier on the p-side next to the quantum wells. Moreover, enlarged AlGaN barriers instead of GaN on the n-side are also becoming necessary to inhibit hole overflow in case of such near UV LEDs [65].

The external efficiency is a notorious problem for all kinds of semiconductor-based LEDs. Due to the large refractive index of semiconductors, the angle of total reflection for rays leaving the device is rather small and is of the order 20° (Fig. 13.7). Those rays being reflected at the surface run a significant risk of being reabsorbed within the LED structure and thus get lost (Fig. 13.7). Besides the use of a plastic encapsulant with an intermediate refractive index, several ways of intelligent shaping of the LED itself have been developed. Among these are microstructured surfaces, photonic bandgap structures, and the overall shaping of the LED towards reducing the average angle of emitted rays onto the surface. For red- and orange-emitting LEDs an external quantum efficiency of more than 50% has been achieved by shaping the LEDs as an inverted pyramid [66], for infrared LEDs the use of photonic bandgap structures led to 70% external efficiency [67], and for blue GaInN/GaN LEDs grown on SiC a post-like shape with inclined lower surfaces as shown in Fig. 13.8 roughly doubled the extraction efficiency [68].

A major step towards a broad application of LEDs as general light sources has been the development of colour conversion LEDs [59]. The light emitted from a blue or UV LED chip is converted to white or any other colour using one or more inorganic or organic fluorescent luminescence conversion materials. A white LED, for example, has been realized using a blue LED chip combined with a yellowish YAG:Ce luminescence converter. This concept opens the door to a wealth of applications requiring individually tuned colours. The quantum efficiency of the colour conversion can be of the order 90%.

13.3.3 Typical properties and materials issues

Actual device results differ considerably, depending on emission wavelength and active material. While visible LEDs based on GaInN perform very

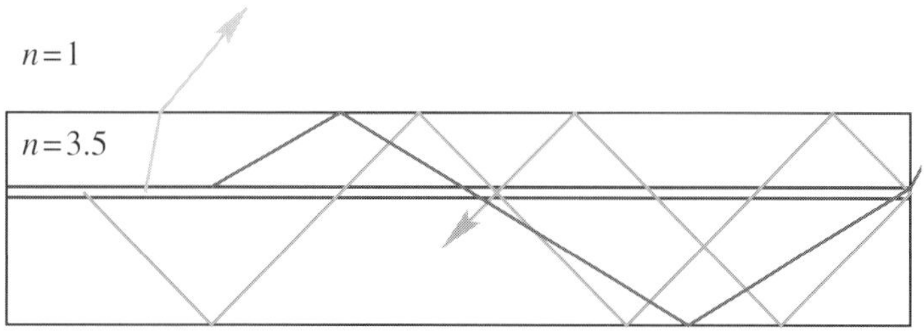

Fig. 13.7 Schematic view of emitted rays being transmitted or totally reflected at the outer surface of a LED.

Fig. 13.8 LED chip with enhanced light extraction efficiency due to the inclined lower surfaces [68].

favourably, UV LEDs based on GaN/AlGaN structures provide much less light output.

Light-emitting diodes in the violet, blue, and green spectral range based on GaInN quantum wells have a number of properties in common. External quantum efficiencies in excess of 10% are obtained, the emitted power at 20 mA reaching more than 10 mW [63]. The forward voltage of such LEDs is as low as 3.1 V [4]. Typically, there is a significant spectral blue-shift of the emission with increasing forward current [69]. This is most likely due to screening of the built-in piezoelectric field, but has also been attributed to filling of localized states [69]. There is also a characteristic dependence of the efficiency on forward current, as shown in Fig. 13.9: At low current an increase of efficiency is observed, whereas the efficiency starts to drop at current levels higher than about 1 mA.

Interestingly, there appears to be a sharp decrease in efficiency when going from InGaN/AlGaN structures with low In content to In-free GaN/AlGaN structures [70], as can be seen from Fig. 13.10. While the external quantum efficiency of InGaN/AlGaN LEDs at an emission wavelength of 370 nm can be of the order 10%, the efficiency drops to less than 2% for GaN/AlGaN LEDs [65]. For LEDs emitting at even shorter wavelengths, efficiencies of less than 10^{-4} are reported [71, 72].

Obviously, LEDs based on In-containing compounds are fairly insensitive with respect to defects, as evident from a comparison of LEDs on normal

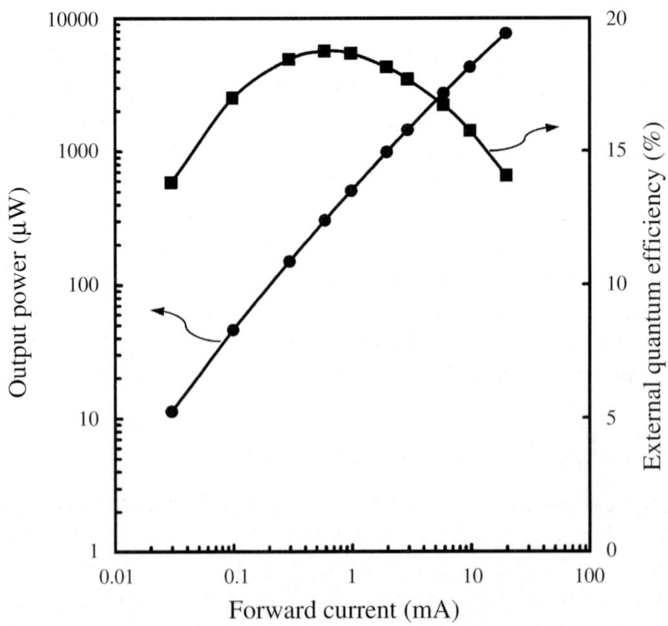

Fig. 13.9 Output power and quantum efficiency of a blue SQW LED on sapphire as a function of injection current [63].

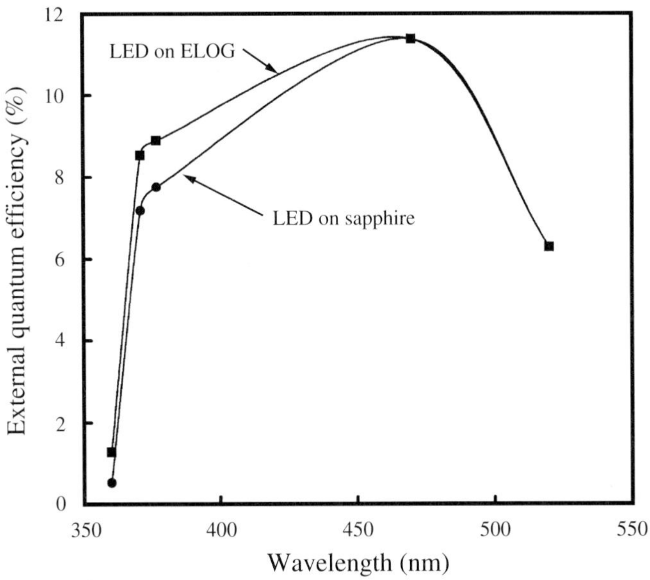

Fig. 13.10 Quantum efficiency of GaInN/AlGaN and GaN/AlGaN LEDs at 20 mA vs. emission wavelength [65].

sapphire substrates and LEDs on ELOG substrates [69], where no significant difference in performance was found. This can be seen very directly in Fig. 13.11, where a blue ELOG-based LED exhibits only a slight modulation in brightness between adjacent almost defect-free and highly defective stripes. In contrast, In-free GaN/AlGaN-based LEDs exhibit a strong improvement in efficiency, when fabricated on ELOG substrates [65], where the defect density is reduced by orders of magnitude.

It has been proposed that carrier localization due to phase separation [22] occurring only in the ternary InGaN active material, keeps carriers from recombining non-radiatively at dislocations. Even though bandgap fluctuations certainly exist in a ternary material, it has not been clear until now if the localization is strong enough to be effective at room temperature, particularly in case of emission wavelengths close to the GaN bandgap.

13.3.4 Current performance and limitations

At present, the external quantum efficiency of GaInN-based LEDs reaches up to 30% at violet–blue wavelengths. At the standard current level of 20 mA the emitted power is up to 21 mW in the blue (Cree Lighting, [58]) and up to 7 mW in the green [63]. The luminosity reaches 27 lm/W in the blue [74] and 60 lm/w in the green [63] spectral region. Their reliability is very high, easily reaching more than 50 000 h. In fact the most severe problem concerning reliability is degradation of the plastic encapsulant, which tends to degrade under UV illumination [75].

Fig. 13.11 Photograph of an ELOG-based blue LED under current injection. There is only a slight modulation of intensity between almost defect-free and highly defective stripes [73]. (See also color Plate IV.)

13.3.5 Applications and future trends

The rapid development of high-brightness LEDs both for the blue–green and for the red part of the spectrum during the past few years has opened the road towards a wealth of high-volume applications. Among these are large full-colour displays with millions of LEDs, signals and traffic lights, automotive applications such as taillights, and backlighting of LCD displays as used in mobile phones. Most of these applications make use of the high single-colour efficiency of high-brightness LEDs of up to 50% [66] for red LEDs and up to 20% for green and blue LEDs [63]. In comparison, filtering a conventional halogen light bulb to a spectral width equivalent to a green nitride LED gives an efficiency of about 2%.

A hot issue for the near future is the potential of white LEDs for general lighting applications. Such white LEDs are realized by using blue or UV-emitting LEDs and a luminescence conversion medium such as inorganic [59] or organic phosphors [60]. The performance of white LEDs has reached 15 lm/W at a total luminous flux of 15 lm for a single device [74].

13.4 Laser diodes

While light-emitting diodes are useful for signals, display, and even for lighting, it needs laser diodes to address applications such as high-density optical storage, high-density printing, or full-colour projection displays. For the red part of the visible spectrum, AlGaInP-based lasers have reached a high level of sophistication with wavelengths down to 630 nm and output powers up to 1.5 W at 670 nm [76]. On the other hand, lasers based on II–VI semiconductors such as ZnSe suitable for the blue–green spectral region are still being plagued by their short device lifetime due to defect generation attributed to the weak bonds of these materials [77].

A dramatic development has led to the commercial availability of violet and blue GaN-based laser diodes. The first pulsed violet laser diode being presented in 1995 [78], it took about two years to develop a CW laser with a commercially viable device lifetime [79]. Since that time a large number of groups have developed laser diodes on sapphire [80, 81, 82, 83], SiC [84, 85, 86], and GaN substrates [12]. In the following we will review the basic ideas and the important steps towards such laser diodes.

13.4.1 Basic structures (vs. substrate) and concepts

Most modern semiconductor laser diodes are based on separate confinement quantum well heterostructures, which take advantage of the reduced dimensionality of electrons and holes in quantum wells and of the waveguiding properties of heterostructures due to their refractive index profile. In case of the nitrides, this requires heterostructures comprising GaInN, GaN, and AlGaN layers. Due to the aforementioned lack of bulk GaN substrates, sapphire, silicon carbide, and HVPE-grown GaN are used as a substrate for laser diode structures.

13.4.1.1 *Basic layer structure*

The basic layer structure of a nitride laser diode, as shown in Fig. 13.12, is built around the active gain medium, i.e. the GaInN/GaN quantum well region. Typically, the quantum wells for violet–blue lasers with emission wavelengths between 400 and 450 nm consist of GaInN quantum wells with 10–15% In and with a well thickness of 2–4 nm and GaN or GaInN barriers 5–10 nm thick. On the p-side of the quantum wells follows a typically 20 nm-thick 20%-AlGaN electron blocking layer [28, 64], which is meant to provide a barrier to electrons trying to escape from the quantum wells into the p-waveguide.

This active region is surrounded by either doped or undoped GaN waveguiding layers ($\approx$100 nm on either side) and by p- and n-type AlGaN cladding layers. A p-GaN contact layer on top and a n-GaN contact layer below this stack complete the basic structure.

13.4.1.2 *Optical waveguiding*

In order to provide a well-defined optical mode and to maximize its overlap with the active gain medium, an optical waveguide is formed by an appropriate sequence of low-index cladding layers and high-index waveguiding layers. Typical layer structures, including their refractive index profile, being used

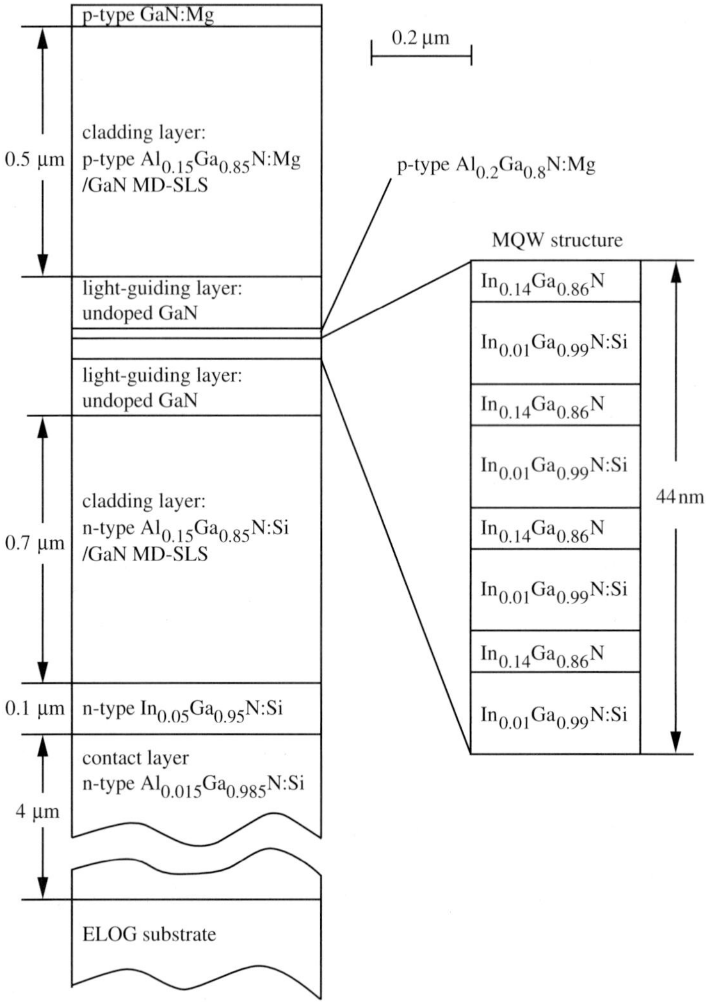

Fig. 13.12 Typical layer structure of a GaInN/GaN/AlGaN laser diode grown on sapphire [62].

for the two most common substrates, namely sapphire and SiC, are shown in Figs 13.13 and 13.14. The core waveguide structure is formed by the AlGaN cladding layers, the GaN waveguiding layers, and the multiple-quantum-well active gain region.

The most important difference (besides differences in growth and defect density for different substrates) is the refractive index of the substrate, which is lower than that of the waveguide layers in case of sapphire but similar in the case of SiC. In addition, in the case of sapphire one typically uses a micrometer-thick GaN buffer layer below the actual waveguide, whereas in case of SiC the lower AlGaN cladding layer usually extends down to the substrate.

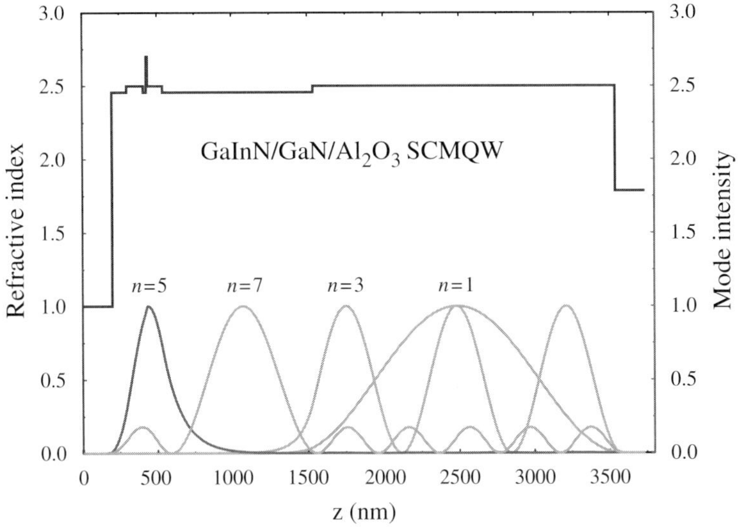

Fig. 13.13 Refractive index profile and optical modes in a GaInN/GaN/AlGaN SCMQW laser structure on sapphire. Due to the secondary waveguide formed by the GaN buffer, the lasing mode is not the fundamental mode of the structure [90].

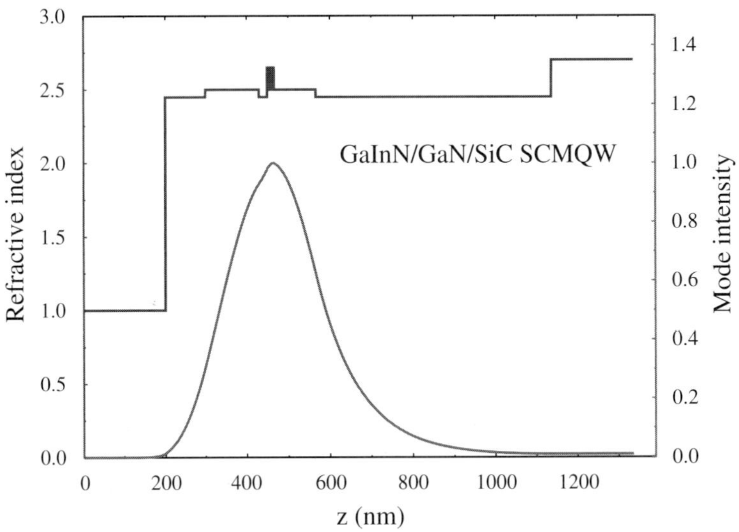

Fig. 13.14 Refractive index profile and optical modes in a GaInN/GaN/AlGaN SCMQW laser structure on SiC. Due to the large refractive index of SiC, the structure is single-mode [92].

As a consequence, usual laser structures on sapphire substrates actually represent two competing coupled waveguides, where the coupling is controlled by the thickness of the AlGaN cladding layer in between [87, 88]. Since the thickness of this AlGaN layer is limited to about 0.5 μm due to strain and crack formation [89] this leads to excitation of higher order modes and to multi-lobe vertical far-field patterns [90, 91], as shown in Fig. 13.15. Actually, the detailed mechanism of coupling into higher order modes relies on the birefringence of the wurtzite nitrides [92], which also leads to a spectral resonance behaviour. While for a conventional waveguide based on isotropic material, phase matching for TE and TM modes is never achieved and therefore there is no effective coupling between those modes, a birefringent waveguide allows coupling of modes. Mode coupling in typical nitride waveguide is mediated by the defects present in large numbers. Improvements to this situation have been found by using modulation-doped superlattices for the cladding layer [89] a thick AlGaN buffer layer instead of the GaN buffer [93] or by using thick HVPE-grown GaN substrates [12, 91].

In contrast to sapphire-based structures, SiC-based laser structures do not suffer from higher order modes, unless the waveguide is not properly designed [92]. Due to the fact that the refractive index of SiC is slightly higher than that of the nitrides and that SiC is absorbing wavelengths shorter than about 450 nm, the waveguide structure can be made truly single mode.

13.4.1.3 *Laser mirrors*

Traditionally, the mirrors forming the laser resonator in semiconductor lasers are obtained by cleaving along a crystal plane. For GaN-based laser diodes,

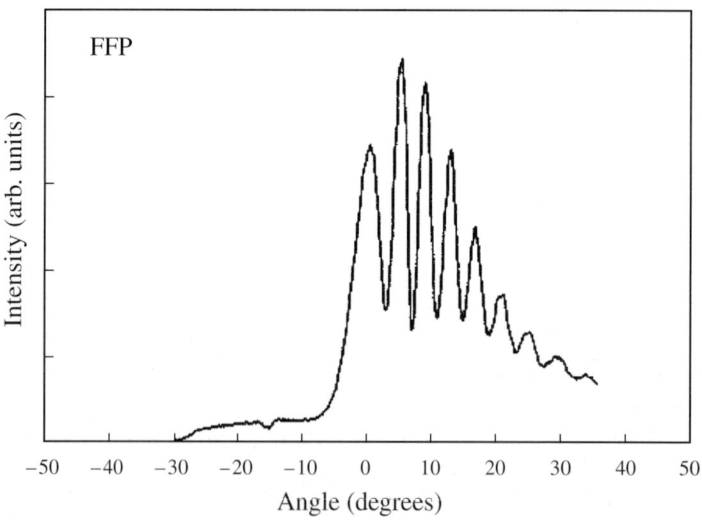

Fig. 13.15 Weird example of the higher-order far-field pattern caused by higher-order mode lasing in sapphire-based laser structures [91].

cleaving is difficult at least for structures grown on *c*-plane sapphire [78], where dry-etched mirrors are used by most groups. Such dry-etched mirrors exhibit a relatively large roughness of 5 nm [82] and therefore large optical losses. By using *a*-plane sapphire as substrate [9] or by using thick HVPE-grown GaN substrates [79] or 6H-SiC or 4H-SiC [94] substrates cleaving can be accomplished and the mirror quality can be greatly improved.

Besides etched or cleaved mirrors, distributed feedback (DFB) and distributed Bragg reflector (DBR) lasers have been realized. Beginning with optically pumped structures [95, 96, 97], electrically pumped DFB [98] and DBR [99] laser diodes have been realized by now. Such DFB laser diodes exhibit a stable single longitudinal mode emission [95], as compared to the multiple longitudinal modes emitted by a usual Fabry–Perot-type laser.

As an attractive alternative to edge-emitting lasers, vertical cavity surface emitting lasers (VCSELs) are being studied. Recent modelling results [100] underline the problems inherent to this approach as applied to nitrides:

(i) Due to the small refractive index contrast and large strain between GaN and AlN it is difficult to realize DBR mirrors with large reflectivity, (ii) the small p-conductivity in GaN makes electrical pumping of vertically emitting structures a real challenge. Nevertheless, there has been some progress recently toward vertically emitting lasers. Lasing emission in the vertical direction has been achieved under optical pumping [101, 102] by using dielectric mirror stacks on one or on both sides.

13.4.2 Material problems

The threshold current density even of the best nitride laser diodes is higher than 1.2 kA/cm^2 [103], which is much higher than that of longer wavelength red AlGaInP lasers [104] and that of blue–green II–VI lasers [75]. On the fundamental side, the large effective masses of electrons and holes associated with the wide bandgap of GaInN, GaN, and AlGaN are decisive. With an electron effective mass of about $0.2\,m_0$ [105, 106] and a hole effective density-of-states mass of about $2\,m_0$ [56, 107] the transparency carrier density is of the order $1 \times 10^{19}\,\mathrm{cm}^{-3}$ [108].

The lowest threshold current density has been obtained with two QWs for shorter wavelength lasers [103] and for a single QW for longer wavelength lasers [109] as shown in Fig. 13.16. This is reproduced by calculations [49] in the case of low optical losses (Fig. 13.17), while at somewhat larger optical losses the minimum threshold is expected for two quantum wells.

In addition to fundamental issues, the large defect density typical for heteroepitaxial nitride structures severely affects the threshold current density. Defect reduction schemes like epitaxial lateral overgrowth (ELOG, see below) clearly tend to reduce the threshold current density [103].

Defects have also been a major concern with respect to optical losses within the laser waveguide. Surprisingly, the optical losses of the epitaxial material are only about 10–20 cm^{-1} in good laser structures [49, 50]. For ridge stripe laser diodes, the losses may increase somewhat to about 50 cm^{-1} [8].

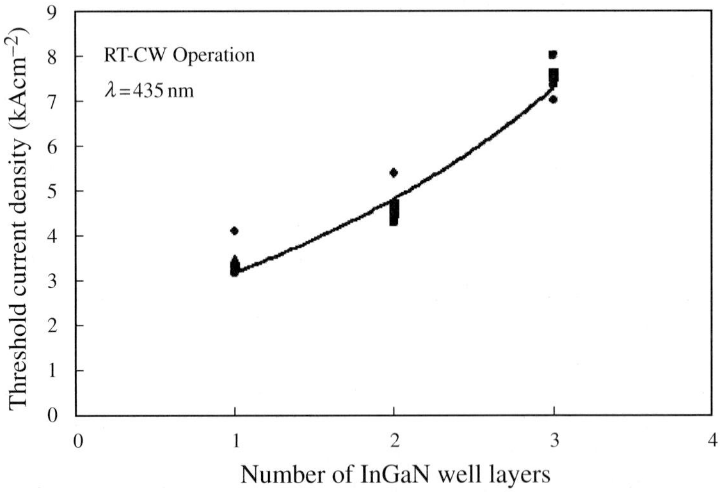

Fig. 13.16 Threshold current density of laser diodes emitting at 435 nm with 2, 3 and 4 quantum wells [108].

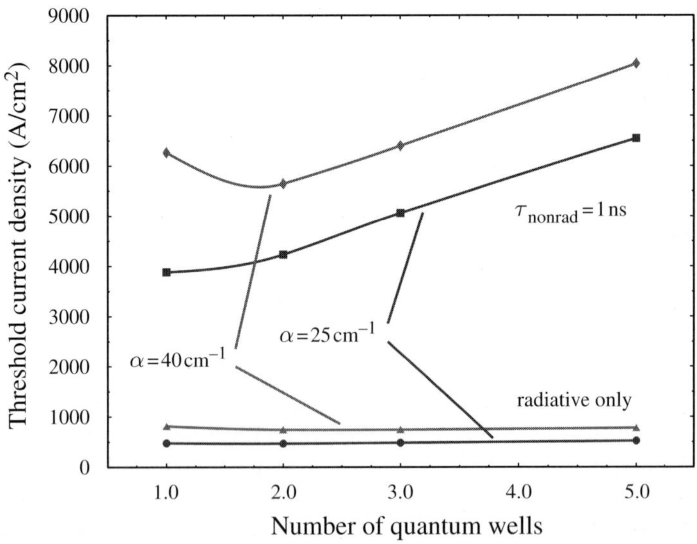

Fig. 13.17 Calculated dependence of the threshold current density for low and moderate optical losses [49].

Up to now, laser diodes have been realized for the wavelength range 376 [80] to 450 nm [109], with the lowest threshold current density around 410 nm [103]. As shown in Fig. 13.18 the calculated threshold current density [49] shows a steep increase both towards shorter and towards longer wavelength, consistent with the device results. On the long wavelength side, a deterioration of GaInN material quality at higher In content is often made responsible for the increase in threshold current [109]. On the other hand, at high In mole fractions the large piezoelectric polarization field, which has to be overcome by screening, is certainly a major factor [49].

On the short wavelength side, carrier confinement within the active quantum wells is becoming an issue below 400 nm. Since GaN is typically used in the waveguiding layer, the confinement energy is quickly becoming comparable with kT at room temperature. As will be discussed below, the use of higher-bandgap AlGaN in the waveguide is not without problems.

The route towards shorter wavelength lasers thus requires higher aluminium contents in the cladding layers and in the waveguide. Even for existing violet laser diodes, it has already been a challenge to make the cladding layers thick enough to provide sufficient optical confinement without generating cracks. The use of strained-layer superlattice structures has been a significant step forward in this context [110]. It might prove increasingly difficult to realize thick enough cladding layers at higher Al content.

In order to be commercially successful, laser diodes are required to operate for more than 10 000 hours under CW conditions. The first nitride laser diodes died after a few hours of CW operation. Closer inspection revealed that electromigration of contact metal along the core of threading defects provided an important mechanism of degradation [111]. By developing the technique of epitaxial lateral overgrowth (ELOG) [17, 19, 110], it was possible to reduce the density of threading dislocations by about two orders of magnitude [19].

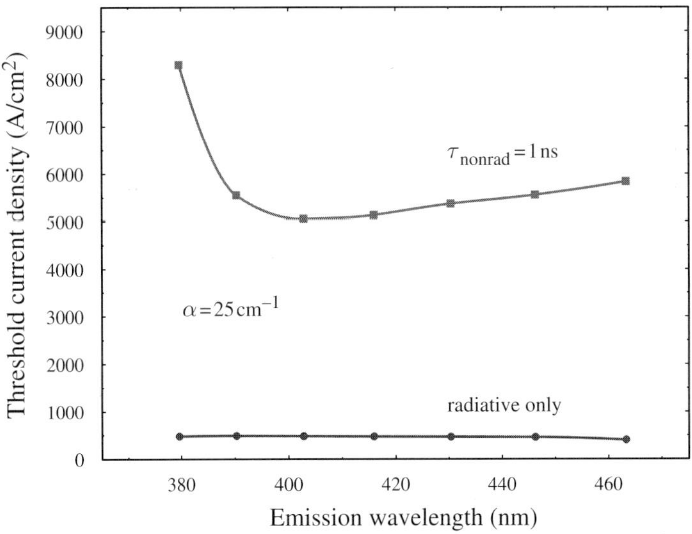

Fig. 13.18 Calculated threshold current density vs. emission wavelength. When non-radiative recombination is dominating, a clear minimum at 405 nm is observed [49].

In this way, it was possible to extend the operating lifetime of laser diodes up to more than 10 000 hours [21].

13.4.3 Present performance

Today's nitride laser diodes have been perfected to a level such that continuous operation for 15 000 h at an output power of 30 mW and at a case temperature of 60 °C has been achieved [21]. The threshold current density of nitride lasers can be as low as 1.2 kA/cm^2. Laser operation has been reported for the wavelength range 376–450 nm [80, 109]. A maximum output power as high as 420 mW has been observed for a single laser diode [103]. The far-field of nitride laser diodes has been improved through the use of GaN or SiC substrates to exhibit true Gaussian shape (Fig. 13.19) and angles of 11.2° in the lateral and 29.9° in the vertical direction [21].

13.4.4 Future trends and applications

As violet and blue nitride-based laser diodes are becoming commercially available at the present time, a large number of applications is awaiting them. Most prominently, ultra-high density optical storage [digital versatile disk (DVD)] is likely to be one of the driving forces for future applications. Whereas DVDs based on 650 nm red lasers provide a capacity of about 4.5 GB, UDVDs based on violet lasers have the promise of almost 20 GB capacity [112]. In order to fulfil the requirements for this application, laser diodes with a good far-field pattern, i.e. true single-mode waveguides both in the lateral and in the vertical direction, which allow a small spot size after focusing, are needed.

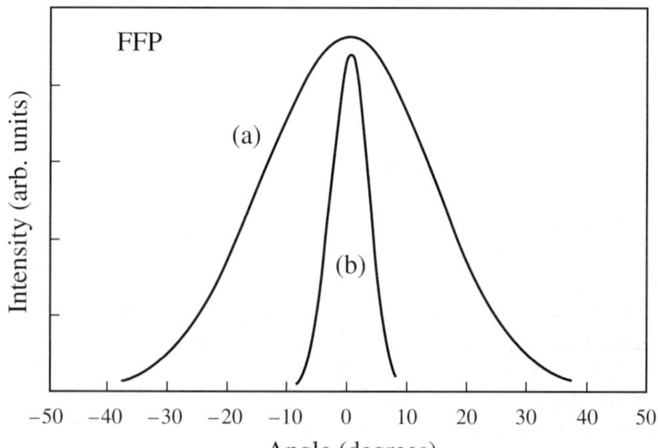

Fig. 13.19 Measured far-field distribution for a laser diode grown on a thick GaN substrate [91].

Similar requirements in terms of beam quality apply to applications for high-density high-speed printing. In addition to good beam quality, high output power is also needed for such applications.

Potential applications for full-colour projection displays require truly blue-emitting laser diodes, i.e. at wavelengths longer than 450 nm. Power levels of the order 1 W are needed for such applications, in addition to good beam quality. Up to now, the maximum output power reported for nitride lasers was 420 mW [103]. Broad stripe lasers or trapezoidal-shaped waveguides may be necessary to obtain even higher output power levels.

The potential of deliberately made quantum dots for lasers in the nitride material system is presently being explored [113]. Like in other material systems, such quantum dots may lead to improved laser performance.

13.5 Conclusions

Low-dimensional nitride heterostructures have been the key to the success of group-III-nitrides for optoelectronic devices. The unexpectedly large quantum efficiency of LEDs and the decent performance of nitride lasers are certainly strongly related to the properties of such structures. Quite a few of the unusual properties of nitride heterostructures are not completely understood and require further intensive study.

References

1. Pankove, J. I. and Norris, P. E. (1972). Luminescence from GaN MIS diodes. *RCA Review* **33**(2), 377–82.
2. Akasaki, I. K. H., Amano, H., and Kito, M. (1991). Photoluminescence of Mg-doped p-type GaN and electroluminescence of GaN p-n junction LED. *J. Lumin.* **48–49**, 666–70.
3. Nakamura, S., Senoh, M., and Mukai, T. (1993). High-power InGaN/GaN doubleheterostructure violet light emitting diodes. *Appl. Phys. Lett.* **62**(19), 2390–2.
4. Nakamura, S., Mukai, T., and Senoh, M. (1994). Candela-class high brightness In GaN/AlGaN double-heterostructure blue-light-emitting diodes. *Appl. Phys. Lett.* **64**(13), 1687.
5. Nakamura, S., Senoh, M., Iwasa, N., and Nagahama, S. I. (1995a). High-power InGaN single-quantum-well-structure blue and violet light-emitting diodes. *Appl. Phys. Lett.* **67**(13), 1868.
6. Nakamura, S., Senoh, M., Iwasa, N., Nagahama, S. I., Yamada, T., and Mukai, T. (1995b). Super-bright green InGaN single-quantum-well-structure light-emitting diodes. *Jpn. J. Appl. Phys.* **34**(10B), L1332.
7. Kato, E., Noguchi, H., Nagai, M., Okuyama, H., Kijima, S., and Ishibashi, A. (1998). Significant progress in II–VI blue-green laser diode lifetime. *Electron. Lett.* **34**, 282.
8. Nakamura, S., Senoh, M., Nagahama, S., Iwasa, N., Yamada, T., Matsushita, T., *et al.* (1996a). Optical gain and carrier lifetime of InGaN multi-quantum well structure laser diodes. *Appl. Phys. Lett.* **69**(11), 1568–70.

9. Nakamura, S., Senoh, M., Nagahama, S., Iwasa, N., Yamada, T., Matsushita, T., *et al.* (1996b). Ridge-geometry InGaN multi-quantum-well-structure laser diodes. *Appl. Phys. Lett.* **69**(10), 1477–9.
10. Nakamura, S., Senoh, M., Nagahama, S., Iwasa, N., Yamada, T., Matsushita, T., *et al.* (1996c). Room-temperature continuous-wave operation of InGaN multi-quantum-well structure laser diode. *Appl. Phys. Lett.* **69**(26), 4056–8.
11. Porowski, S. (1997). Growth and properties of single crystalline GaN substrates and homoepitaxial layers. *Mater. Sci. Eng. B* **44**(1–3), 407.
12. Kuramoto, M., Kimura, A., Yamaguchi, A. A., Sunakawa, H., Kuroda, N., Nido, M., *et al.* (1999). Room-temperature continuous-wave operation of InGaN multi-quantum-well laser diodes grown on an n-GaN substrate with a backside n-contact. *Jpn. J. Appl. Phys.* **38**(2B), L184–6.
13. Wang, J. W., Chen, Q., Sun, C. J., Lim, B., Anwar, M. Z., Khan, M. A., and Temkin, H. (1996a). InGaN-GaN based light-emitting diodes over (111) spinel substrates. *Appl. Phys. Lett.* **69**(3), 369.
14. Wang, J. W., Sun, C. J., Chen, Q., Anwar, M. Z., Khan, M. A., Nikishin, S. A., *et al.* (1996b). High quality GaN-InGaN heterostructures grown on (111) silicon substrates. *Appl. Phys. Lett.* **69**(23), 3566–8.
15. Amano, H., Sawaki, N., Akasaki, I., and Toyoda, Y. (1986). Metalorganic vapor phase epitaxial growth of a high quality GaN film using an AlN buffer layer. *Appl. Phys. Lett.* **48**, 353–5.
16. Lester, S. D., Ponce, F. A., Craford, M. G., and Steigerwald, D. A. (1995). High dislocation densities in high efficiency GaN-based light-emitting diodes. *Appl. Phys. Lett.* **66**(10), 1249.
17. Usui, A., Sunakawa, H., Sakail, A., and Yamaguchi, A. A. (1997). Thick GaN epitaxial growth with low dislocation density by hydride vapor phase epitaxy. *Jpn. J. Appl. Phys.* **36**, L899.
18. Sakai, A., Sunakawa, H., and Usui, A. (1997). Defect structure in selectively grown GaN films with low threading dislocation density. *Appl. Phys. Lett.* **71**(16), 2259–61.
19. Nam, O. H., Bremser, M. D., Zheleva, T. S., and Davis, R. F. (1997). Lateral epitaxy of low defect density GaN layers via organometallic vapor phase epitaxy. *Appl. Phys. Lett.* **71**(18), 2638–40.
20. Linthicum, K., Gehrke, T., Thomson, D., Carlson, E., Rajagopal, P., Smith, T., *et al.* (1999). Pendeoepitaxy of gallium nitride thin films. *Appl. Phys. Lett.* **75**(2), 196–8.
21. Nagahama, S. I., Iwasa, N., Senoh, M., Matsushita, T., Sugimoto, Y., Kiyoku, H., *et al.* (2000). High-power and long-lifetime InGaN multi-quantum-well laser diodes grown on low-dislocation-density GaN substrates. *Jpn. J. Appl. Phys.* **39**(7A), L647–50.
22. Chichibu, S., Azuhata, T., Sota, T., and Nakamura, S. (1996). Spontaneous emission of localized excitons in InGaN single and multiquantum well structures. *Appl. Phys. Lett.* **69**, 4188.
23. Im, J. S., Heppel, S., Kollmer, H., Sohmer, A., Off, J., Scholz, F., and Hangleiter, A. (1998a). Evidence for quantum-dot-like states in GaInN/GaN quantum wells? *J. Cryst. Growth* **189–190**(1–4), 228–9.
24. Im, J. S., Kollmer, H., Off, J., Sohmer, A., Scholz, F., and Hangleiter, A. (1998c). Effects of piezoelectric fields in GaInN/GaN and GaN/AlGaN heterostructures and quantum wells. In FA Ponce, S.P DenBaars, B.K Meyer, S. Nakamura, and S. Strite, editors, *Nitride Semiconductors*, volume 482 of *MRS. Symp. Proc.* pp. 513–18. Materials Research Society, Pittsburgh.
25. Riblet, P., Hirayama, H., Kinoshita, A., Hirata, A., Sugano, T., and Aoyagi, Y. (1999). Determination of the photoluminescence mechanism in InGaN quantum wells. *Appl. Phys. Lett.* **75**(15), 2241–3.

26. Narukawa, Y., Kawakami, Y., Fujita, S., Fujita, S., and Nakamura, S. (1997a). Recombination dynamics of localized excitons in $In_{0.20}Ga_{0.80}N/In_{0.05}Ga_{0.95}N$ multiple quantum wells. *Phys. Rev. B* **55**(4), R1938–41.
27. Im, J. S., Kollmer, H., Off, J., Sohmer, A., Scholz, F., and Hangleiter, A. (1998d). Reduction of oscillator strength due to piezoelectric fields in $GaN/Al_xGa_{1-x}N$ quantum wells. *Phys. Rev. B* **57**(16), R9435–8.
28. O'Donnell, K. P., Breitkopf, T., Kalt, H., van der Stricht, W., Moerman, I., Demeester, P., and Middleton, P. G. (1997). Optical linewidths of InGaN light emitting diodes and epilayers. *Appl. Phys. Lett.* **70**(14), 1843.
28. Hangleiter, A., Im, J. S., Kollmer, H., Heppel, S., Off, J., and Scholz, F. (1998). The role of piezoelectric fields in GaN-based quantum wells. *MRS Internet J. Nitride Semicond. Res.* **3**, 15.
29. Hangleiter, A., Im, J. S., Off, J., and Scholz, F. (1999). Optical properties of nitride quantum wells: How to separate fluctuations and polarization field effects. *Phys. Stat. Sol.* (b) **216**, 427.
30. Heppel, S., Wirth, R., Off, J., Scholz, F., and Hangleiter, A. (1999b). Mode coupling. In *Proceedings of the 3rd International Conference on Nitride Semiconductors.* Montpellier, To be published in *Phys. Stat. Sol.*
31. Ciplys, D., Gaska, R., Shur, M. S., Rimeika, R., Yang, J. W., and Khan, M. A., (2000). Propagation of guided optical waves in thick GaN layers grown on (0001) sapphire. *Appl. Phys. Lett.* **76**(16), 2232–4.
32. Narukawa, Y., Kawakami, Y., Funato, M., Fujita, S., Fujita, S., and Nakamura, S. (1997b). Role of self-formed InGaN quantum dots for exiton localization in the purple laser diode emitting at 420 nm. *Appl. Phys. Lett.* **70**(8), 981.
33. Nakamura, S. (1998). The roles of structural imperfections in InGaN-based blue light-emitting diodes and laser diodes. *Science* **281**(5379), 956–61.
34. Takeuchi, T., Sota, S., Katsuragawa, M., Komori, M., Takeuchi, H., Amano, H., and Akasaki, I. (1997). Quantum-confined Stark effect due to piezoelectric fields in GaInN strained quantum wells. *Jpn. J. Appl. Phys.* **36**(4A), L382–L385.
35. Bernardini, F., Fiorentini, V., and Vanderbilt, D. (1997). Spontaneous polarization and piezoelectric constant of III-V nitrides. *Phys. Rev. B* **56**(16), R10024–7.
36. Mendez, E., Bastard, G., Chang, L., Esaki, L., Morkoc, H., and Fischer, R. (1982). Effect of an electric field on the luminescence of GaAs quantum wells. *Phys. Rev. B* **26**, 7101.
37. Im, J. S., Kollmer, H., Off, J., Scholz, F., and Hangleiter, A. (1998b). Piezoelectric field effect on optical properties in GaN/GaInN/AlGaN quantum wells. *MRS Internet J. Nitride Semicond. Res.* **4S1**, G6.20.
38. della Sala, F., Bernardini, F., Fiorentini, V., Scholz, R., Jancu, J. M., di Carlo, A., and Lugli, P. (1999). Free-carrier screening of polarization fields in wurtzite GaN/InGaN laser structures. *Appl. Phys. Lett.* **74**(14), 2002–4.
39. Im, J. S., Kollmer, H., Off, J., Scholz, F., and Hangleiter, A. (1999). Carrier confinement in GaInN/AlGaN/GaN quantum wells with asymmetric barriers: direction of the piezoelectric field. *Mat. Sci. Eng. B* **59**, 315.
40. Kollmer, H., Im, J. S., Off, J., Scholz, F., and Hangleiter, A. (1999). Intra- and interwell transitions in GaInN/GaN multiple quantum wells with built-in piezoelectric fields. *Appl. Phys. Lett.* **74**(1), 82.
41. Gil, B., Lefebvre, P., Allègre, J., Mathieu, H., Grandjean, N., Leroux, M., Massies, J., *et al.* (1999). Observation of long-lived oblique excitons in GaN-AlGaN multiple quantum wells. *Phys. Rev. B* **59**(15), 10246–50.
42. Langer, R., Simon, J., Ortiz, V., Pelekanos, N. T., Barski, A., André, R., and Godlewski, M. (1999). Giant electric fields in unstrained GaN single quantum wells. *Appl. Phys. Lett.* **74**(25), 3827–9.

43. Leroux, M., Grandjean, N., Laügt, M., Massies, J., Gil, B., Lefebvre, P., and Bigenwald, P. (1998). Quantum confined Stark effect due to built-in internal polarization fields in (Al, Ga)N/GaN quantum wells. *Phys. Rev. B* **58**(20), R13371–4.
44. Widmann, F., Simon, J., Daudin, B., Feuillet, G., Rouvière, J. L., Pelekanos, N. T., and Fishman, G. (1998). Blue-light emission from GaN self-assembled quantum dots due to giant piezoelectric effect. *Phys. Rev. B* **58**(24), R15989–92.
45. Hirayama, H., Tanaka, S., Ramvall, P., and Aoyagi, Y. Intense photoluminescence from self-assembling InGaN quantum dots artificially fabricated on AlGaN surfaces. (1998). *Appl. Phys. Lett.* **72**(14), 1736–8.
46. Chow, W. W., Amano, H., Takeuchi, T., and Han, J. (1999). Quantum-well width dependence of the threshold current density in InGaN lasers. *Appl. Phys. Lett.* **75**(2), 244–6.
47. Park, S. H., Chuang, S. L., and Ahn, D. (1999). Piezoelectric effects on many-body optical gain of zinc-blende and wurtzite GaN/AlGaN quantum-well lasers. *Appl. Phys. Lett.* **75**(10), 1354–6.
48. Uenoyama, T. and Suzuki, M. (1995). Valence subband structures of wurtzite GaN/AlGaN quantum wells. *Appl. Phys. Lett.* **67**(17), 2527.
49. Hangleiter, A., Heppel, S., Off, J., Kuhn, B., Scholz, F., Bader, S., *et al.* (2001). Analysis of the threshold current in nitride-based lasers. *J. Cryst. Growth* **230**(3–4), 522–6.
50. Frankowsky, G., Härle, V., Scholz, F., and Hangleiter, A. (1996). Optical gain in GaInN/GaN heterostructures. *Appl. Phys. Lett.* **68**(26), 3746.
51. Stampfl, C. and de Walle, C. G. V. (1998). Doping of $Al_xGa_{1-x}N$. *Appl. Phys. Lett.* **72**(4), 459–61.
52. Amano, H., Kito, M., Hiramatsu, K., and Akasaki, I. (1989). p–type conduction in Mg-doped GaN treated with low-energy electron beam irradiation (LEEBI). *Jpn. J. Appl. Phys.* **28**, L2112.
53. Nakamura, S., Iwasa, N., Senoh, M., and Mukai, T. (1992). Hole compensation mechanism of p-type GaN films. *Jpn. J. Appl. Phys.* **31**, 1258.
54. Fischer, S., Wetzel, C., Haller, E. E., and Meyer, B. K., (1995). On p-type doping in GaN-acceptor binding energies. *Appl. Phys. Lett.* **67**(9), 1298.
55. Nakamura, S., Senoh, M., and Mukai, T. (1991). Highly p-typed Mg-doped GaN films grown with GaN buffer layers. *Jpn. J. Appl. Phys.* **30**, L1708.
56. Kim, K., Lambrecht, W. R. L., Segall, B., and van Schilfgaarde, M. (1997). Effective masses and valence-band splittings in GaN and AlN. *Phys. Rev. B* **56**(12), 7363–75.
57. Im, J. S., Härle, V., Scholz, F., and Hangleiter, A. (1996). Radiative lifetime of excitons in GaInN/GaN quantum wells. *MRS Internet J. Nitride Semicond. Res.* **1**, 37.
58. Cree Lighting (2001). http:/www.cree.com/cree lighting.
59. Schlotter, P., Schmidt, R., and Schneider, J. (1997). Luminescence conversion of blue light emitting diodes. *Appl. Phys. A* **64**(4), 417–18.
60. Hide, F., Kozodoy, P., DenBaars, S. P., and Heeger, A. J. (1997). White light from InGaN/conjugated polymer hybrid light-emitting diodes. *Appl. Phys. Lett.* **70**(20), 2664.
61. An, H. Y., Cha, O. H., Kim, J. H., Yang, G. M., Lim, K. Y., Suh, E. K., and Lee, H. J. (1999). Thermal treatment effect of the GaN buffer layer on the photoluminescence characteristics of the GaN epilayer. *J. Appl. Phys.* **85**(5), 2888–93.
62. Nakamura, S. (1999). InGaN-based violet laser diodes. *Semicond. Sci. Technol.* **14**, R27–40.
63. Mukai, T., Yamada, M., and Nakamura, S. (1999). Characteristics of InGaN-based UV/blue/green/amber/red light-emitting diodes. *Jpn. J. Appl. Phys.* **38**(7A), 3976–81.
64. Domen, K., Tanahashi, T., Soejima, R., and Kuramata, A. (1998). Electron overflow to the AlGaN p-cladding layer in InGaN/GaN/AlGaN MQW laser diodes. *MRS Internet J. Nitride Semicond. Res.* **3**, 2.

65. Mukai, T. and Nakamura, S. (1999). Ultraviolet InGaN and GaN single-quantum-well-structure light-emitting diodes grown on epitaxially laterally overgrown GaN substrates. *Jpn. J. Appl. Phys.* **38**(10), 5735–9.
66. Krames, M. R., Ochial-Holcomb, M., Höfler, G. E., Carter-Coman, C., Chen, E. I., Tan, I. H., *et al.* (1999). High-power truncated-inverted-pyramid (AlGa)InP/GaP light-emitting diodes exhibiting ø50% external quantum efficiency. *Appl. Phys. Lett.* **75**(16), 2365–7.
67. Boroditsky, M., Krauss, T. F., Coccioli, R., Vrijen, R., Bhat, R., and Yablonovitch, E. (1999). Light extraction from optically pumped light-emitting diode ba thin-slab photonic crystals. *Appl. Phys. Lett.* **75**(8), 1036–8.
68. Härle, V. (2001). Osram enhances brightness of blue InGaN LEDs. *Compound Semiconductor* **7**(1), 7.
69. Mukai, T., Takekawa, K., and Nakamura, S. (1998a). InGaN-based blue light-emitting diodes grown on epitaxially laterally overgrown GaN substrates. *Jpn. J. Appl. Phys.* **37**(7B), L839–41.
70. Mukai, T., Yamada, M., and Nakamura, S. (1998b). Current and temperature dependences of electroluminescence of InGaN-based UV/blue/green light-emitting diodes. *Jpn. J. Appl. Phys.* **37**(11B), L1358–61.
71. Otsuka, N., Tsujimura, A., Hasegawa, Y., Sugahara, G., Kume, M., and Ban, Y. (2000). Room temperature 339 nm emission from $Al_{0.13}Ga_{0.87}N/Al_{0.10}Ga_{0.90}N$ double heterostructure light-emitting diode on sapphire substrate. *Jpn. J. Appl. Phys.* **39**(5B), L445–8.
72. Nishida, T., Saito, H., and Kobayashi, N. (2001). Milliwatt operation of AlGaN-based single-quantum-well light emitting diode in the ultraviolet region. *Appl. Phys. Lett.* **78**(25), 3927–8.
73. Nakamura, S., Iwasa, N., Matushita, T., Mukai, T., Senoh, M., and Nagahama, S. (1999). InGaN/GaN/AlGaN-based LEDS and laser diodes. *MRS Internet J. Nitride Semicond. Res.* 4S1, ??,
74. Wierer, J. J., Steigerwald, D. A., Krames, M. R., Oshea, J. J., Ludowise, M. J., Christenson, G., *et al.* (2001). High-power AlGaInN flip-chip light-emitting diodes. *Appl. Phys. Lett.* **78**(22), 3379–81.
75. Osinski, M., Zeller, J., Chiu, P. C., Phillips, B. S., and Barton, D. L. (1996). AlGaN/InGaN/GaN blue light emitting diode degradation and pulsed current stress. *Appl. Phys. Lett.* **69**(7), 898.
76. Lichtenstein, N., Schweizer, H., Weiss, S., Hutter, M., Reichl, H., Winterhoff, R., and Scholz, F. (2000). The impact of LOC structures on 670-nm (Al)GaInP high-power lasers. *IEEE J. Sel. Topics Quantum Electron.* **6**(4), 564–70.
77. Nurmikko, A. and Gunshor, R. L. (1997). Blue and green semiconductor lasers: a status rcport. *Semicond. Sci. Technol.* **12**, 1337–47.
78. Nakamura, S., Senoh, M., Nagahama, S. I., Iwasa, N., Yamada, T., Matsushita, T., *et al.* (1996d). InGaN-based multi-quantum-well-structure laser diodes. *Jpn. J. Appl. Phys.* **35**(1B), L74.
79. Nakamura, S., Senoh, M., Nagahama, S., Iwasa, N., Yamada, T., Matsushita, T. *et al.*, (1998b). High-power, long-lifetime InGaN/GaN/AlGaN-based laser diodes grown on pure GaN substrates. *Jpn. J. Appl. Phys.* **37**(3B), 309–12.
80. Akasaki, I., Sota, S., Sakai, H., Tanaka, T., Koike, M., and Amano, H. (1996). Shortest wavelength semiconductor laser diode. *Electron. Lett.* **32**(12), 1105–6.
81. Mack, M. P., Abare, A., Aizcorbe, M., Kozodoy, P., Keller, S., Mishra, U. K., *et al.* (1997). Characteristics of indium-gallium-nitride multiple-quantum-well blue laser diodes grown by MOCVD. *MRS Internet J. Nitride Semicond. Res.* **2**, 41.
82. Kneissl, M., Bour, D. P., Johnson, N. M., Romano, L. T., Krusor, B. S., Donaldson, R., *et al.* (1998). Characterization of AlGaInN diode lasers with mirrors from chemically assisted ion beam etching. *Appl. Phys. Lett.* **72**(13), 1539–41.

83. Kimura, Y., Miyachi, M., Takahashi, H., Tanaka, T., Nishitsuka, M., Watanabe, A., Ota, H., and Chikuma, K. (1998). Room-temperature pulsed operation of Gan-based laser diodes on a-face sapphire substrate grown by low-pressure metalorganic chemical vapor deposition. *Jpn. J. Appl. Phys.* **37**(10B), L1231–3.
84. Bulman, G. E., Doverspike, K., Sheppard, S. T., Weeks, T. W., Kong, H. S., Dieringer, H. M., *et al.* (1997). Pulsed operation lasing in a cleavedfacet InGaN/GaN MQW SCH laser grown on 6H-SiC. *Electron. Lett.* **33**, 1556–7.
85. Kuramata, A., Domen, K., Soejima, R., Horino, K., Kubota, S. I., and Tanahashi, T. (1997). InGaN laser diode grown on 6H-SiC substrate using low-pressure metal organic vapor phase epitaxy. *Jpn. J. Appl. Phys.* **36**(8AB), L1130.
86. Bader, S., Hahn, B., Lugauer, H. J., Lell, A., Weimar, A., Brüderl, G., *et al.* (2000). First European GaN-based violet laser diode. *Phys. Stat. Sol.* (a) **180**(1), 177–82.
87. Bougrov, V. E. and Zubrilov, A. S. (1997). Optical confinement and threshold currents in III-V nitride heterostructures: simulation. *J. Appl. Phys.* **81**(7), 2952–6.
88. Bergmann, M.J. and H.C., Casey, J. (1998). Optical-field calculations for lossy multiplelayer Al/sub x/Ga/sub 1- x/N/In/sub x/Ga/sub 1–x/N laser diodes. *J. Appl. Phys.* **84**(3), 1196–203.
89. Nakamura, S., Senoh, M., Nagahama, S., Iwasa, N., Yamada, T., Matsushita, T., *et al.* (1997). InGaN/GaN/AlGaN-based laser diodes with modulation-doped strained-layer superlattices. *Jpn. J. Appl. Phys.* **36**(12A), L1568.
90. Hofstetter, D., Bour, D. P., Thornton, R. L., and Johnson, N. M. (1997). Excitation of a higher order transverse mode in an optically pumped $In_{0.15}Ga_{0.85}N/In_{0.05}\ Ga_{0.95}N$ multiquantum well laser structure. *Appl. Phys. Lett.* **70**(13), 1650–2.
91. Nakamura, S., Senoh, M., Nagahama, S., Iwasa, N., Yamada, T., Matsushita, T., *et al.* (1998a). Continuous-Wave operation of InGaN/GaN/AlGaN-based laser diodes grown on GaN substrates. *Appl. Phys. Lett.* **72**(16), 2014–6.
92. Heppel, S., Wirth, R., Off, J., Scholz, F., and Hangleiter, A. (1999a). Mode conversion in GaN-based laser structures on sapphire due to the birefringence of the nitrides. *Phys. Stat. Sol.* (a) **176**, 73.
93. Takeuchi, T., Detchprohm, T., Iwaya, M., Hayashi, N., Isomura, K., Kimura, K., *et al.* (1999). Improvement of far-field pattern in nitride laser diodes. *Appl. Phys. Lett.* **75**(19), 2960–2.
94. Kuramata, A., Kubota, S., Soejima, R., Domen, K., Horino, K., Hacke, P., and Tanahashi, T. (1999). Continuous wave operation at room temperature of InGaN laser diodes fabricated on 4H-SiC substrates. *Jpn. J. Appl. Phys.* **38**(5A), L481–3.
95. Hofmann, R., Gauggel, H. P., Griesinger, U. A., Gräbeldinger, H., Adler, F., Ernst, P., Bolay, H., Härle, V., Scholz, F., Schweizer, H., and Pilkuhn, M. H. (1996). Realization of optically pumped second-order GaInN distributed-feedback lasers. *Appl. Phys. Lett.* **69**(14), 2068–70.
96. Hofstetter, D., Thornton, R. L., Kneissl, M., Bour, D. P., and Dunnrowicz, C. (1998a). Demonstration of an InGaN/GaN-based optically pumped multiquantum well distributed feedback laser using holographically defined third-order gratings. *Appl. Phys. Lett.* **73**(14), 1928–30.
97. Hofstetter, D., Bour, D. P., Kneissl, M., Romano, L. T., and Paoli, T. L. (2000). Realization of a complex-coupled InGaN/GaN-based optically pumped multiple-quantum-well distributed-feedback laser. *Appl. Phys. Lett.* **76**(17), 2337–9.
98. Hofstetter, D., Thornton, R. L., Romano, L. T., Bour, D. P., Kneissl, M., and Donaldson, R. M. (1998b). Room-temperature pulsed operation of an electrically injected InGaN/GaN multi-quantum well distributed feedback laser. *Appl. Phys. Lett.* **73**(15), 2158–60.
99. Cho, J., Chae, S., Sone, C., Nam, O. H., Lee, J. W., Park, Y., *et al.* (2000). InGaN/GaN multi-quantum well distributed Bragg reflector laser diode. *Appl. Phys. Lett.* **76**(12), 1489–91.

100. Mackowiak, P., and Nakwaski, W. (2000). Designing guidelines for possible continuous-wave-operating nitride vertical-cavity surface-emitting lasers. *J. Phys. D* **33**(6), 642–53.
101. Someya, T., Werner, R., Forchel, A., Catalano, M., Cingolani, R., and Arakawa, Y. (1999). Room temperature lasing at blue wavelengths in gallium nitride microcavities. *Science* **285**, 1905.
102. Song, Y. K., Zhou, H., Diagne, M., Nurmikko, A. V., Schneider, R. P. J., Kuo, C. P., *et al.* (2000). A quasicontinuous wave, optically pumped violet vertical cavity surface emitting laser. *Appl. Phys. Lett.* **76**(13), 1662–4.
103. Nakamura, S., Senoh, M., Nagahama, S., Iwasa, N., Yamada, T., Matsushita, T., *et al.* (1998c). InGaN/GaN/AlGaN-based laser diodes grown on GaN substrates with a fundamental transverse mode. *Jpn. J. Appl. Phys.* **37**(9A/B), L1020–2.
104. Raisch, P., Winterhoff, R., Wagner, W., Kessler, M., Schweizer, H., Riedl, T., *et al.* (1999). Investigations on the performance of multiquantum barriers in short wavelength (630 nm) AlGaInP laser diodes. *Appl. Phys. Lett.* **74**(15), 2158–60.
105. Meyer, B. K., Volm, D., Graber, A., Alt, H. C., Detchprohm, T., Amano, K., and Akasaki, I. (1995). Shallow donors in GaN – the binding energy and the electron effective mass. *Solid State Commun.* **95**(9), 597.
106. Perlin, P., Litwin-Staszewska, E., Suchanek, B., Knap, W., Camassel, J., Suski, T., *et al.* (1996). Determination of the effective mass of GaN from infrared reflectivity and Hall effect. *Appl. Phys. Lett.* **68**(8), 1114.
107. Im, J. S., Moritz, A., Steuber, F., Härle, V., Scholz, F., and Hangleiter, A. (1997). Radiative carrier lifetime, momentum matrix element, and hole effective mass in GaN. *Appl. Phys. Lett.* **70**, 631.
108. Park, S. H. and Ahn, D. (1997). Many-body effects on optical gain in strained hexagonal and cubic GaN/AlGaN quantum well lasers. *Appl. Phys. Lett.* **71**(3), 398.
109. Nakamura, S., Senoh, M., Nagahama, S., Iwasa N., Yamada, T., Matsuhita, T., *et al.* (1998e). Violet InGaN/GaN/AlGaN-based laser diodes with an output power of 420 mW. *Jpn. J. Appl. Phys.* **37**(6A), L627–9.
109. Nakamura, S., Senoh, M., Nagahama, S., Iwasa, N., Matsushita, T., and Mukai, T. (2000). Blue InGaN-based laser diodes with an emission wavelength of 450 nm. *Appl. Phys. Lett.* **76**(1), 22–4.
110. Nakamura, S., Senoh, M., Nagahama, S., Iwasa, N., Yamada, T., Matsushita, T., *et al.* (1998d). InGaN/GaN/AlGaN-based laser diodes with modulation-doped strained-layer superlattices grown on an epitaxially laterally overgrown GaN substrate. *Appl. Phys. Lett.* **72**(2), 211–3.
111. Barton, D. L. and Osinski, M. (1998). Degradation mechanisms in GaN/AlaN/InGaN LEDs and LDs. In *Semiconducting and Insulating Materials 1998. Proceedings of the 10th Conference on Semiconducting and Insulating Materials*, pp. 259–62. IEEE, Piscataway, NJ, USA.
112. Iwata, K., Nakano, E., Hosoda, A., Oishi, K., Ohshima, K., Ito, F., *et al.* (2001). High-density rewritable optical disk using groove recording. *Jpn. J. Appl. Phys.* **40**(3B), 1637.
113. Tachibana, K., Someya, T., Arakawa, Y., Werner, R., and Forchel, A. (1999). Room-temperature lasing oscillation in an InGaN self-assembled quantum dot laser. *Appl. Phys. Lett.* **75**(17), 2605–7.

14 GaN-based modulation doped FETs

Hadis Morkoç, Aldo Di Carlo, and Roberto Cingolani

Because of their large bandgap, large high-field electron velocity, large breakdown field, large thermal conductivity, and robustness, GaN and its heterostructures with InGaN and AlGaN have recently attracted a good deal of attention for high power/high temperature electronics applications, and for use in optoelectronic devices operative in the UV and visible wavelengths. Specifically, the heterostructures based on GaN have paved the way to AlGaN/GaN MODFETs with CW power levels of about 6 W at 10 GHz in devices with 1 mm gate periphery which are comparable to power densities extrapolated from smaller devices. This means that the 1-mm devices are not thermally limited yet. On the noise figure front, a minimum noise figure of 0.85 dB with an associated gain of 11 dB at 10 GHz has been obtained. The expected electron mobility in GaN has recently been upgraded from a maximum of about 900 $cm^2V^{-1}s^{-1}$ to about 2000 $cm^{-2}V^{-1}s^{-1}$, which bodes very well for electronic device applications. While the experimental device effort is progressing well, basic issues such as polarization-induced charge has gained prominence as it is inextricable from charge induced/controlled by the gate. Last but not least, the GaN-based material system, though far from being perfected, has already demonstrated remarkable performance in optical emitters optical detectors, which are not discussed in this review.

14.1 Introduction

Semiconductor nitrides such as aluminum nitride (AlN), gallium nitride (GaN), and indium nitride (InN) are very promising materials for their potential use in optoelectronic devices (both emitters and detectors), and high power/temperature electronic devices as have been treated in length and reviewed recently [1–7]. These materials and their ternary and quaternary alloys cover an energy bandgap range of 1.9 to 6.2 eV, suitable for band-to-band light generation with colors ranging from red (potentially) to ultraviolet (UV) wavelengths. Specifically, nitrides are suitable for such applications as surface acoustic wave devices [8], UV detectors [9, 10], Bragg reflectors [11], waveguides, UV and visible light-emitting diodes (LEDs) [12–14], and laser diodes (LDs) [15] for digital data read–write applications. During the last several decades, lasers and LEDs have expanded remarkably both in terms of the range of emission wavelengths available and brightness. The nitride semiconductor-based LEDs have proven to be reliable in such applications as displays, lighting, indicator lights, advertisements, and traffic signs/signals. Additional possible applications include use in agriculture as light sources for accelerated photosynthesis, and in health care for diagnosis and treatment. Lasers, as coherent sources, are crucial for high-density optical read and write technologies because the diffraction-limited optical storage density increases approximately quadratically in the ideal case as the probe laser wavelength is reduced. Nitride-based coherent UV sources are attracting a good deal of attention for optical storage devices. Optical storage would enable the storage and retrieval of an inordinate number of images and vast quantities of text with untold efficiency. Other equally attractive applications envisioned include printing and surgery.

When used as UV sensors in jet engines, automobiles, and furnaces (boilers), the devices would allow optimal fuel efficiency and control of effluents for a cleaner environment. Moreover, UV sensors that operate in the solar blind region (260 to 290 nm) would have high detectivity because the ozone layer absorbs solar radiation at those wavelengths, thus virtually eliminating the radiation noise. Consequently, these detectors are expected to play a pivotal role in threat recognition aimed against aircraft and other vehicles [9, 10, 16]. GaN photodiodes [17] exhibited zero-bias responsivities of about 0.21 A/W at 356 nm that decreased by more than three orders of magnitude for wavelengths longer than 390 nm. The noise equivalent power (NEP) at a reverse bias of 10 V is ($f > 100$ Hz) 6.6×10^{-15} W/Hz$^{1/2}$ which is extremely small [18]. Detector speed while affected in terms of uniformity by the sheet resistance of the p-layer, which suffers from the notoriously low doping levels, is in the picosecond range [19]. Finally, the GaN-based detectors with AlN mole fractions approaching the solar blind region of the spectrum have been fabricated into arrays for imaging. Detector arrays with pixel sizes of 32×32 have been fabricated and tested already [20].

GaN's large bandgap, large dielectric breakdown field, fortuitously good electron transport properties [21–23] (an electron mobility possibly in excess of 2000 cm^{2}V^{-1}s^{-1} and a peak velocity approaching 3×10^7 cm^{-1} at room temperature), and good thermal conductivity are trademarks of high power/temperature electronic devices [24]. Sheppard *et al.* [25] have reported that 0.45 μm gate, high-power modulation doped FETs (MODFETs) on SiC substrates exhibited a power density of 6.8 W/mm in a 125-μm-wide device and a total power of 4 W (with a power density of 2 W/mm) at 10 GHz. Other groups have also reported on the superior performance of GaN-based MODFETs on SiC and sapphire substrates with respect to competing materials, particularly at X band and higher frequencies [26–29]. What is astounding is that researchers at HRL Laboratories have recently demonstrated GaN/AlGaN MODFETs prepared by MBE on SiC substrates, which exhibited a total power level of 6.3 W at 10 GHz from a 1-mm-wide device. What is more astounding is that the power level is not really thermally limited as the power density extrapolated from a 0.1-mm device is 6.5 W. Equally impressive is the noise figure of 0.85 dB at 10 GHz with an associated gain of 11 dB. The drain breakdown voltages in these quarter micron gate devices are about 60 V, which are in part responsible for such a record performance. [30].

Applications of high power GaN based MODFETs include amplifiers operative at high-power levels, high temperatures and in unfriendly environments such as radar, missiles, satellites as well as low-cost compact amplifiers for wireless base stations. Much of these applications are currently met by pseudomorphic modulation doped FETs [31].

Nitride semiconductors have been deposited by vapor phase epitaxy (i.e. both hydride VPE [32] [HVPE] which has been developed for thick GaN layers and organometallic VPE [33] [OMVPE] which has been developed for heterostructures), and in a vacuum by a slew of variants of molecular beam epitaxy (MBE) [16]. All the high performance light-emitters, which require high quality InGaN, have been produced by OMVPE. On the other hand, MBE has been very successful in producing structures, which do not require InGaN. Some examples are FETs and detectors. With its innate refined control of growth parameters, *in situ* monitoring capability, and uniformity, MBE is well suited for depositing heterostructures and gaining insight to deposition/incorporation mechanisms. MBE's control over growth parameters is such that any structure can be grown in any sequence. The structures based on conventional compound semiconductors such as IR lasers for CD players, surface emitting vertical cavity lasers, and high-performance pseudomorphic MODFETs have all been produced very successfully, most of them commercially, by MBE. Nitride growth, however, requires much higher temperatures than those used in producing conventional Group III–V semiconductors for which the MBE systems were designed. In addition, it has proved difficult to provide active N species at sufficiently high rates for nitride growth. Despite these mechanical/engineering limitations and its relatively late entry, with appropriate modifications, MBE has already played a key role on a number of fronts, such as high performance GaN-based MODFETs and fast solar blind detectors, which are the topics of discussion in this article.

14.2 Electron transport properties in GaN and GaN/AlGaN heterostructures

Electron mobility is a key parameter in the operation of n-channel FETs in that it affects the access resistances as well as the rate with which the carrier velocity increases with field. Consequently, we will treat the low field mobility in GaN and its dependence on various scattering events first. This will be followed by treating the two-dimensional electron gas mobility as occurs in modulation doped field effect transistors which are the types that are of technological importance hitherto. Ultimately, electron mobility is limited by the interaction of electrons with phonons, and in particular with optical phonons. This holds for bulk mobility as well as that in AlGaN/GaN modulation doped field effect transistors.

The room-temperature electron mobility values in bulk GaN grown with HVPE to a thickness of 60-μm was reported to be 950 cm^2/Vs [34]. Freestanding

GaN templates grown by HVPE exhibited room temperature electron mobilities approaching $1400\,cm^2/Vs$ [35]. Recent HVPE layers exhibit much higher mobilities on the surface of the layers, which approach that of free-standing templates. Those reported for Metal Organic Chemical Vapor Deposition (MOCVD) grown layers were also in excess of $900\,cm^2/Vs$ [36], though the temperature dependence of mobility in this particular sample was rather unique. Early MBE layers exhibited mobilities as high as $580\,cm^2/Vs$ on SiC substrates, which at that time were not as commonly used as in recent times [37]. Typically, however, the MBE-grown films produce much lower mobility values of $100–300\,cm^2/Vs$ [38]. The lower mobilities have been attributed to both high dislocation densities [38–40] and elevated levels of point defects [41, 42].

Dislocations are considered by some to be an important scattering mechanism in films having dislocation densities above $1\times10^8\,cm^{-2}$ [38, 39]. Depending on the particulars of the growth and substrate preparation, GaN films grown by MBE typically have dislocation densities in the range of $5\times10^9-5\times10^{10}\,cm^{-2}$ [38]. With refined procedures, however, dislocation densities in the $8\times10^8-2\times10^9\,cm^{-2}$ can be obtained when grown directly on sapphire substrates with AlN or GaN buffer layers. Dislocation reduction, and other scattering centers that are inherently related to dislocations, is really the key to achieving high mobility GaN which goes to the heart of buffer layer and or early stages of growth. Based on the premise that the [002] X-ray diffraction is affected by screw dislocations and the [104] peak by edge dislocations and the fact that RF-nitrogen grown MBE layers produce excellent [002] peaks (in the 40–120 arcsec range) while the [104] peaks are wider and weaker (in the 180 to 300 arcsec) one can conclude that the majority of the dislocations in MBE layers are that of propagating edge type.

The strength of MBE, namely producing 2D growth does not bode well in dislocation reduction as the edge dislocations which propagate along the *c*-axis go right through the sample. The detailed picture, though, is somewhat dependent on the particulars of the growth such as pitted or pit free growth which depends on the Group V/III ratio. Some sort of 3D growth at the early stages of the growth, as in the case of growth from vapor, followed by a smoothing layer would help reduce dislocations. The other option is to use HVPE or MOCVD buffer layers for MBE growth. This approach led to record or near record bulk ($1{,}150\,cm^2/Vs$ RT, even higher of late) [43] and 2DEG (53500 at 4.2 K) mobilities [44]. It is clear that the buffer layers grown by the vapor phase epitaxy method helps eliminate the main problem associated with MBE; that is, the poor quality of the buffer layer. The other long-standing obstacle for MBE, difficulties associated with sapphire and SiC substrate preparation, has been eliminated. In the case of sapphire, a high temperature anneal in O_2 environment produces atomically smooth and damage free surface [45]. In the case of SiC, some form of H_2 etching at elevated temperatures removes the surface damage caused by polishing [46] as in the case of sapphire. Controlling the Ga/N ratio and substrate temperature causes the dislocation density across the homoepitaxial interface to remain constant [47, 48].

While the above two works are related to RF MBE, ammonia MBE has also produced very high electron mobilities (as high as 70 000 cm^2/Vs at 4 K) when grown on bulk GaN wafer grown under high pressure and temperature conditions [49].

Electron mobility is one of the most important parameters associated with the material with great impact on devices. The temperature dependence of mobility and the carrier concentration can be used to extract fundamental information regarding scattering mechanisms [50, 51]. As compared to the other III–V semiconductors such as GaAs, GaN possesses many unique material and physical properties [1]. However, the lack of high-quality material, until very recently, prevented detailed investigations of carrier transport. The earlier transport investigations had to cope with poor crystal quality and low carrier mobility, well below predictions [52–55].

We should point out that unintentionally doped GaN exhibits n-type conduction with a typical electron concentration of $\sim 10^{17} cm^{-3}$, with heavy compensation. Typical compensation ratios observed for MOCVD and MBE grown films are about 0.3, though a lower ratio of ~ 0.24 was reported for HVPE grown crystals [56–58]. Compensation reduces the electron mobility in GaN for a given electron concentration. Another point that should be kept in mind is that GaN layers are often grown on foreign substrates with very different properties. The degenerate layer at the interface (caused by extended defects and impurities), spontaneous polarization at heterointerfaces, and piezoelectric effects all should be considered. Experiments show that, even for thick GaN grown by HVPE, the degenerate interfacial layer has an important contribution to the Hall conductivity, especially at low temperatures where freeze-out occurs for the donors in bulk, leading to domination by the interfacial layer [34, 59–61]. In these cases, the measured data must be corrected to extract meaningful data [56]. The typical extended defect density of GaN grown by various techniques is $\sim 10^9 cm^{-2}$ [62]. In many cases, the dislocation and defect scattering may also limit the carrier mobility, especially at low temperatures [63, 64]. Finally, many material and physical parameters of GaN were not available for some of the previous simulations where those parameters were treated as adjustable. Needless to say, reliable parameters are required in the calculation of the electron mobility and in the interpretation of experimental results accurately.

14.3 Bulk mobility

To treat bulk mobility, we made use of free-standing high-quality GaN templates grown by HVPE at Samsung where the questionable interface layer had been chemically removed. A quantitative comparison with theoretical calculations demonstrates that the one-layer and one-donor conductance model is sufficient to account for the measured data in the entire temperature range without considering any dislocation scattering and any adjustable parameter other than the acceptor concentration [65]. The sample shows a low impurity concentration, a low compensation ratio, negligible dislocation scattering, and high electron mobilities derived from the Hall measurements show the high-quality transport properties associated with the sample.

A thick ~300 μm GaN film was first deposited by HVPE on the *c*-plane of sapphire. It was then thermally decomposed at the film/substrate interface and lifted off by scanning a laser beam with a photon energy larger than that corresponding to the GaN band gap. Both sides of the freestanding GaN crystal were mechanically polished, and the Ga-face was dry etched to yield a smooth surface. Wet chemical etching was then applied to the N-face to remove some 30 μm of material, the region containing the high conductance degenerate layer. The final thickness of the GaN template is about 200 μm. The sample has also been characterized by X-ray diffraction, atomic force microscopy, and photoluminescence [66].

The Raman spectra were measured for this sample to derive the Debye temperature and the optical dielectric constant as literature values of these parameters vary depending on sample quality. The phonon energies measured in this sample are: A1(LO)=737.0 cm^{-1}; A1(TO)=532. 5 cm^{-1}; E1(LO)= 745.0 cm^{-1}; E1(TO)=558.5 cm^{-1} with an accuracy of +/-0.5 cm^{-1}. These phonon energies are very close to those measured by Azuhata *et al.* [67] for a thick GaN film grown on sapphire. The optical phonon modes were measured in a variety of scattering configurations as indicated. As usual, X=(100), Y=(010), and Z=(001). Here Z was taken to be parallel to the growth direction. Shown in Fig. 14.1(a) and (b) are the A1(LO) and A1(TO) phonon spectra where the half-width of the phonon modes is only 8 cm^{-1}. E1(TO) and E2 phonon modes also appear in Fig. 14.1(b).

Variable temperature Hall effect measurements were performed in the dark in a temperature range of 26.5 to 273 K using the van der Pauw geometry. Ohmic contacts were formed on the Gaface with Ti/Al/Ti/Au metallization followed by rapid thermal annealing at 900 °C for 30 s [68]. Good ohmic contacts were verified over the measured temperature range inside the cryostat. Extreme care was taken for the accuracy of the experimental conditions such as the magnetic field, electrical current, and the sample temperature.

The measured Hall mobility, and carrier concentration are shown in Figs 14.2 and 14.3, respectively, as a function of temperature. A mobility of 1425 cm^2/Vs was observed near room temperature (273 K) with the peak mobility being 7386 cm^2/Vs at 48 K. The temperature dependence of both the Hall mobility and electron concentration (lack of saturation) point to the absence of a degenerate interfacial layer after etching. Therefore, only the single-layer model with a single donor was successfully applied to the analysis of the data.

The temperature dependence of the electron mobility was calculated by solving the Boltzmann transport equation (BTE) iteratively [50]. The following scattering processes were included: acoustic deformation potential scattering, piezoelectric scattering, polar optical phonon scattering, and ionized impurity scattering. Only one donor and one acceptor were assumed. The dislocation

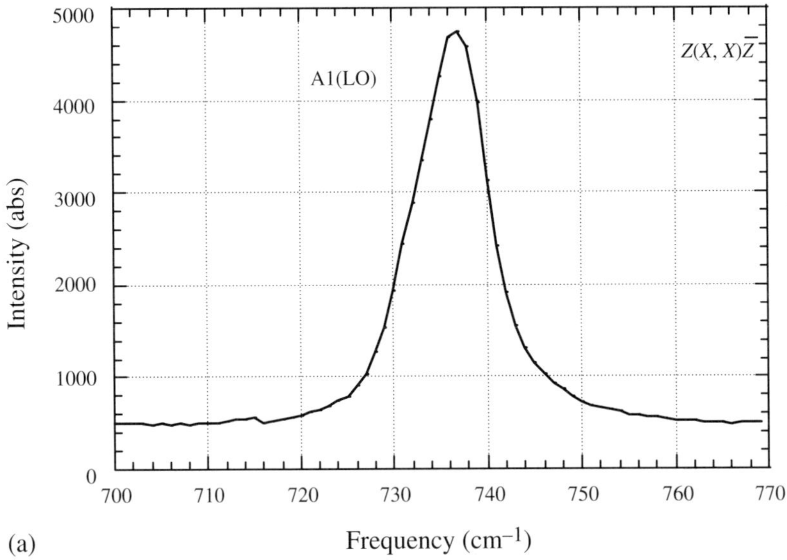

(a)

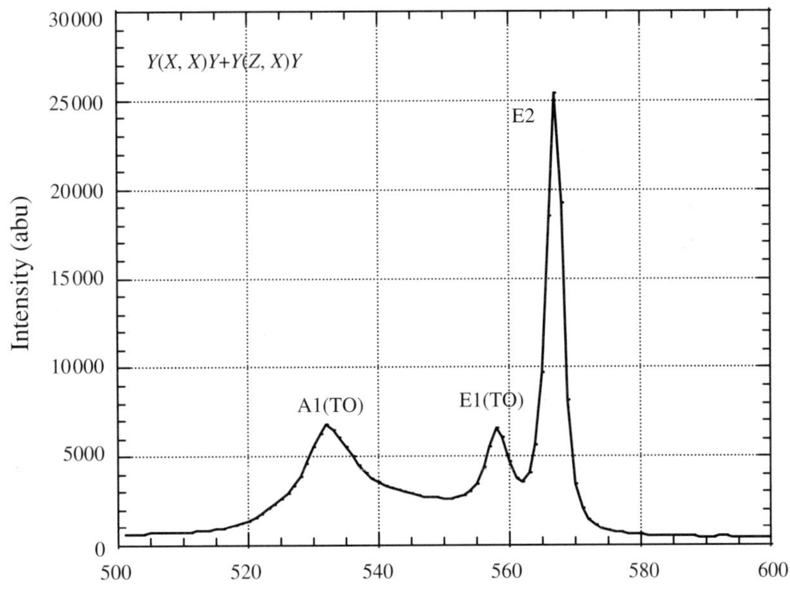

(b)

Fig. 14.1 (a) A1(LO) phonon spectrum of the sample investigated. The scattering configuration is shown in the upper right corner. (b) A1(TO) phonon spectrum of the sample investigated. The *E* phonon spectra are also shown. The scattering configuration is shown in the upper left corner.

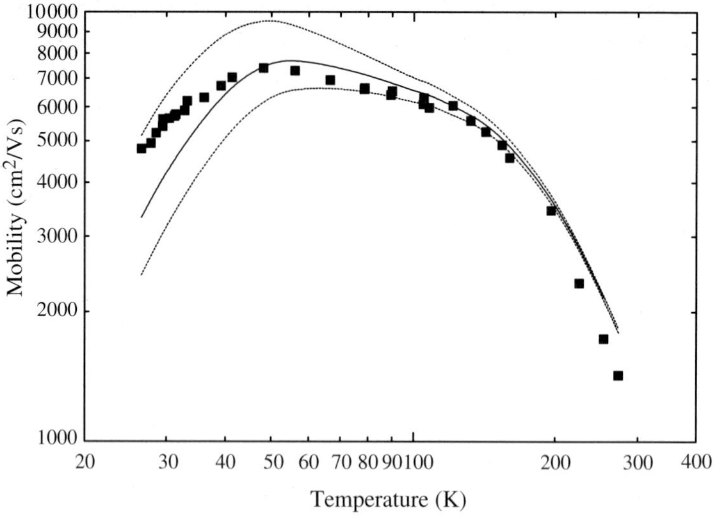

Fig. 14.2 The measured Hall mobility data (solid squares) from the GaN template grown by HVPE as a function of temperature. The solid line is the calculated result using $N_a = 1.4 \times 10^{15}\,\text{cm}^{-3}$, representing the best fit to the measured results. The upper and lower dot lines are the calculated results using $N_a = 1.4 \times 10^{15}$ and $3.4 \times 10^{15}\,\text{cm}^{-3}$.

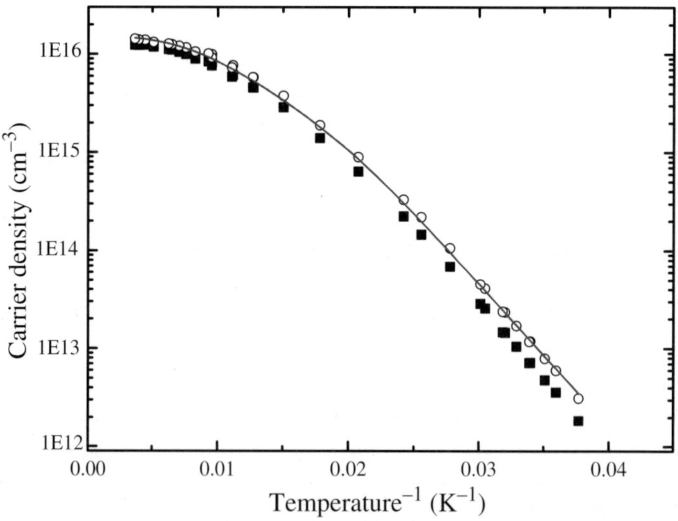

Fig. 14.3 The measured Hall densities n_H (solid squares) as a function of reciprocal temperature from the GaN template grown by HVPE. The open circles represent the carrier density corrected by the Hall factor, $n = n_H r_H$. The solid line is the fit to the theoretical expression of charge balance with hole and neutral acceptor densities neglected.

scattering was not considered because the dislocation density in our sample ($<10^6\,cm^{-2}$) [66] is much less than that where dislocation scattering affects the transport properties ($10^8\,cm^{-2}$) [63, 64]. Likewise, neutral-impurity scattering was not included since it is insignificant. The material parameters used in the calculations are listed in Table 14.1.

The high-frequency dielectric constant for this sample was calculated using the Lyddane–Sachs–Teller relation using the phonon frequencies of A modes, which should represent a benchmark value due to the high quality of the sample. Importantly, instead of allowing the acoustic deformation potential to be an adjustable fitting parameter, the recent unscreened acoustic deformation potential ($E_{DS}=8.54\,eV$) was used. This value is deduced from recent high mobility 2DEG system mobilities at low temperatures where the acoustic phonon scattering is important [69]. If screening is included, a value of 12 eV describes experimental 2DEG mobilities better [70]. However, this is not necessarily applicable here in part because the sample under investigation has a very low electron concentration, and in part due to freeze-out at low temperatures. Therefore, the deformation potential without screening was used in our simulations as a fixed parameter. The other parameters were kept the same as those reported previously [50, 1] and shown in Table 14.1.

The theoretical calculation was fitted to the measured data in concert with the temperature dependence of the electron concentration and charge neutrality. The donor and acceptor concentrations are used as the fitting parameters. The solid line in Fig. 14.2 represents the best fit to the measured data. The dotted lines give the estimated upper and lower bounds for the acceptor concentration. As shown, quantitative agreement with the measured mobility in the entire temperature range is obtained to within about 30%. The acceptor concentrations obtained from the fitting is $2.4\times10^{15}\,cm^{-3}$.

The carrier concentration as a function of temperature is also calculated using the charge neutrality condition with hole and neutral acceptor density being neglected. The actual value and the temperature dependence of the Hall factor are implicit in calculations and are shown in Fig. 14.4, and a degeneracy factor of 2 was assumed. The solid line shown in Fig. 14.3 represents the fit to the measured electron concentration corrected by the Hall factor. The impurity

Table 14.1 Material parameters of GaN used in the calculations.

Parameter	Symbol (units)	Value
High frequency dielectric constant	ε_∞	5.43
Low frequency dielectric constant	ε_0	10.4
Polar phonon Debye temperature	θ_{LO} (K)	1060
Mass density	ρ (kg/m^3)	6.10×10^3
Sound velocity	v_s (m/s)	6.59×10^3
Piezoelectric coefficient	P	0.118
Acoustic deformation potential	E_{DS} (eV)	8.54
Effective mass	m^* (kg)	$0.22\,m_0$

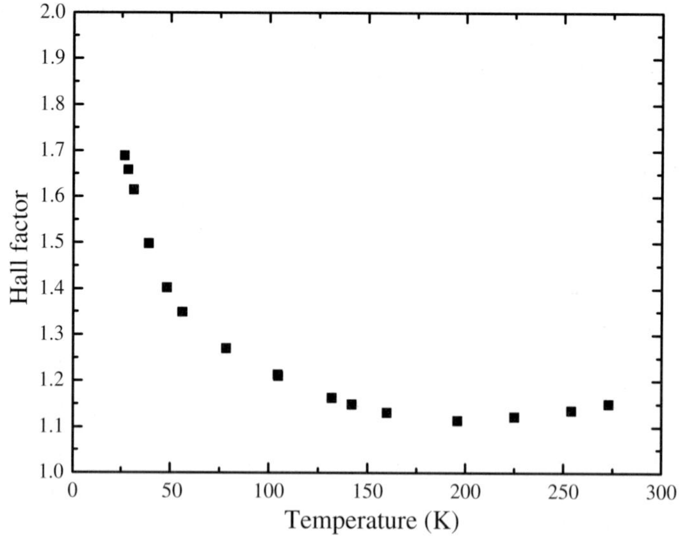

Fig. 14.4 Temperature dependence of the calculated Hall factor, r_H.

concentrations and the activation energy used in the fitting are N_D= $1.76\times100\,\text{cm}^{-3}$, N_A=$2.40\times10^{15}\,\text{cm}^{-3}$, and E_D=25.2 meV. The agreement between the N_D and N_A values derived from the mobility and the carrier concentration data demonstrates the self-consistency of the results with the assumed model.

Considering the screening effect, the donor binding energy E_{D0} can be calculated from $E_{D0}=E_D+\alpha N_D^{1/3}=30.7$ meV, where the screening factor $\alpha=2.1\times10^{-5}$ meVcm. This result is close to the measured value from infrared absorption (29.0 meV) [71]. We note that the donor binding energy obtained here is higher than the 27.5 meV derived from the HVPE sample with higher donor concentration of $1.25\times10^{17}\,\text{cm}^{-3}$ reported in reference [34]. The discrepancy may be introduced from the two-layer conductance in the latter case. As shown earlier [34, 59–61], the interface conductance mainly affects the low-temperature data. Since the donor activation energy can only be accurately determined from the low temperature region in which the ionized-impurity scattering is dominant, any high-conductivity interface contribution would make it difficult to deduce an accurate donor binding energy.

The measured impurity concentrations N_D and N_A as well as the compensation ratio (N_A/N_D) of 0.14 are among the lowest values reported for bulk GaN grown by any technique. The peak mobility of 7386 cm^2/Vs and the value near room temperature are also among the highest for bulk GaN. The quantitative agreement between the calculations and the measured data demonstrates that the single-layer and single-donor model amply describes the conduction process. The results also show that dislocation scattering is not important for the sample under investigation in the entire measured temperature range, demonstrating the high crystal quality.

14.4 Polarization effects, mobility, and electron concentration in 2DEG systems

As mentioned in the introduction section, AlGaN/GaN heterostructures have been the subject of many recent investigations because of their potential for use in high-temperature, high-power devices [3, 24, 72, 73]. Due to the large band discontinuities and polarization-induced screening charge, interface bound two-dimensional electron gas (2DEG) concentrations exceed $10^{13} cm^{-2}$. While polarization effects cause a redistribution of weakly bound and free charges, they cannot directly produce free electrons to form a 2DEG [74–76]. In GaN-based systems, issues dealing with heterointerfaces must include the discussion of polarization. Polarization induces field, which in turn affects the interface charge through screening in that mobile carriers move to where the fixed polarization charge with opposite polarity is. Since, nitrides are large bandgap materials, they tend to be n-type, and the hole concentration is extremely low, mobile carriers are electrons donated by intentional donors or donor-like defects.

The AlGaN barrier [77] or the AlGaN surface [78] have been suggested as the source of electrons. Positive surface charge has also been forwarded [79, 80] to account for the experimental observations in the form of dependence of the 2DEG density on the thickness and/or alloy composition of the AlGaN barrier. Typically, the AlGaN barrier is grown on a relatively thick GaN layer to form the basis for the 2DEG system in a Ga-polarity sample. The inherent lattice mismatch causes a biaxial tensile strain, and the thermal mismatch causes a biaxial compressive strain in the growth plane. The resultant strain induces a macroscopic electric field in the polar material. In addition, due to the particular crystal structure of the wurtzite lattice, a spontaneous polarization field is also found in both materials even in the absence of strain. In most current heterostructures, where the growth takes place along the (0001) direction, both the spontaneous and induced polarizations are directed opposite to the growth direction. The effect of polarization field on the position of the band-edges has been calculated by several groups [74, 81–84]. Polarization-induced fields in Ga-polarity samples, just as any negative gate voltage-induced field in FETs, increase the conduction band edge in the AlGaN barrier with distance from the interface. In the presence of free, weakly bound and surface charge, the internal polarization field is screened by a redistribution of these charges. The surface states may be in the form of donor-like states, which donate their electrons to the lowest unoccupied energy states at the AlGaN/GaN interface.

The holy grail of GaN/AlGaN heterostructures is the debate on the origin of the carriers, which end up at the interface. The observed dependence of the

2DEG density on the thickness and composition of the AlGaN barrier has been linked to surface donor states, the binding energy of which is roughly equal to the Schottky barrier height in n-type GaN [78]. Though this may point to the same surface states being responsible for a possible and weak Fermi level pinning, which is not that notable on the surface of GaN. In addition, pinning of the surface Fermi level in n-type GaN requires electrons to be transferred from bulk donors to surface acceptor states, whereas an excess of surface donors is required to form 2DEG in an AlGaN/GaN heterostructure. This inconsistency can be resolved, however, by assuming that the surface defects are amphoteric (i.e. they can act as either acceptors or donors depending on the circumstances) [85]. The polarization fields present in nitride heterostructures are strong enough to shift the Fermi energy at the surface of the AlGaN barrier below the charge transition level of the surface defects for Ga-face growth. This causes the surface defects on the barrier to transform from being acceptor-like to donor-like surface defects, which can provide the electrons for the 2DEG at the AlGaN/GaN interface [85]. Hsu and Walukiewicz [69] elaborated on the surface donor-like defect that is likely to form at growth temperature and its manifestation as a source of carriers confined at the underlying interface between the AlGaN top layer and GaN below it. The model calculations appeared to be somewhat insensitive to parameters such as donor formation energy and surface Fermi level. One sensitive parameter is the strength of the polarization field, which will be discussed below.

The wurtzite Group III nitrides are tetrahedrally coordinated and lack the inversion symmetry along the *c*-direction. Because of this low symmetry structure and large ionicity associated with the covalent metal nitrogen bond, a large spontaneous polarization oriented along the hexagonal *c*-axis occurs. This spontaneous polarization is noteworthy particularly when heterointerfaces between two nitride semiconductors with varying ionicity are involved. Moreover, Group III nitride semiconductors have large piezoelectric coefficients. The strain-induced piezoelectric and spontaneous polarization charges have profound effects on device structures. In heterojunction devices such as MODFETs where strain and heterointerfaces are present, the polarization charge is present and is inextricably connected to free carriers, which are indeed present. As such, polarization affects device operation in all nitride-based devices, particularly MODFETs, and thus must be taken into consideration in device design. As mentioned above, polarization charge arises from two sources: piezoelectric (PE) effects and the difference in spontaneous polarization between AlGaN and GaN, even in the absence of strain. These charges exist in all compound semiconductors to varying degrees unless self-cancelled by the symmetry of the particular orientation under consideration.

Spontaneous polarization has only recently been fully understood. As pointed out earlier [86], nitrides lack inversion symmetry and exhibit piezoelectric effects when strained along the [0001] direction. Piezoelectric coefficients in nitrides are almost an order of magnitude larger than in many of the traditional Group III–V semiconductors [87]. In addition, wurtzite GaN has a

Table 14.2 Piezoelectric constants and spontaneous polarization charge in nitride semiconductors

	AlN	GaN	InN
e_{33}* [C/m^2**]	1.46	0.73	0.97
e_{31}* [C/m^2]	−0.60	−0.49	−0.57
P_0 [C/m^2]	−0.081	−0.029	−0.032
[e_{31}-[C_{31}/C_{33}***]e_{33}]	−0.86	−0.68	−0.90

*e_{31} and e_{33} are piezoelectric constants
**C/m^2 is coulombs per square meter
***C_{31} and C_{33} are elastic constants

unique axis, thus allowing spontaneous polarization (P_0, whose values are given in Table 14.2) to be present even in the absence of any strain. This can manifest itself as polarization charge at hetero-interfaces. The magnitude of the polarization charge, converted to number of electrons can be in mid $10^{13}\,\mathrm{cm}^{-2}$ level for AlN/GaN heterointerfaces, which is huge by any standards. For comparison, the interface charge in the GaAs/AlGaAs system used for MODFETs is less than 10% of this figure. An excellent review of the polarization effects can be found in ref. 88.

Let us compare the relative importance of spontaneous polarization to piezoelectric polarization. For a biaxially strained layer, the effective piezoelectric polarization is given by

$$P_z^{\mathrm{piezo}} = [e_{31} - (C_{31}/C_{33})e_{33}]\varepsilon_\perp, \tag{14.1}$$

where $\varepsilon_\perp = \varepsilon_{xx} + \varepsilon_{yy}$ is the in-plane strain and C_{31} and C_{33} are elastic constants.

For $Al_xGa_{1-x}N$ coherently strained on a relaxed GaN substrate, the strain $\varepsilon_\perp$ is expected to be proportional to x and given by $\varepsilon_\perp = 2(a_{\mathrm{GaN}} - a_{\mathrm{AlGaN}})/a_{\mathrm{AlGaN}}$, which is $0.051x$ and is tensile. The piezoelectric polarization is then $P^{\mathrm{piezo}} = -0.044x$, i.e. pointing in the $[000\bar{1}]$ direction. The corresponding difference in spontaneous polarization between $Al_xGa_{1-x}N$ and GaN is also expected to be proportional to x, the AlN mole fraction, and is given by $\Delta P^{\mathrm{spon}} = -0.052x$. Consequently, the two are in the same direction for this particular orientation, and are comparable in magnitude. The total polarization for AlN/GaN interface, which is defined in this case as the sum of the piezoelectric polarization and the differential polarization-charge, is $-0.096x$. Note that these are all in C/m^2 and that $1\,\mathrm{C/m^2} = 0.624 \times 10^{15}$ electrons/cm^2. Thus, for x of the order of 0.1, we are dealing with total polarization charge on the order of mid-$10^{12}\,\mathrm{cm}^{-2}$.

For a coherently strained $In_xGa_{1-x}N$ layer on relaxed GaN, the difference in spontaneous polarization is much smaller, $\Delta P^{\mathrm{spon}} = -0.003x$. Furthermore, the $In_xGa_{1-x}N$ layer on GaN would be under compressive strain $\varepsilon_\perp = 0.203x$, and

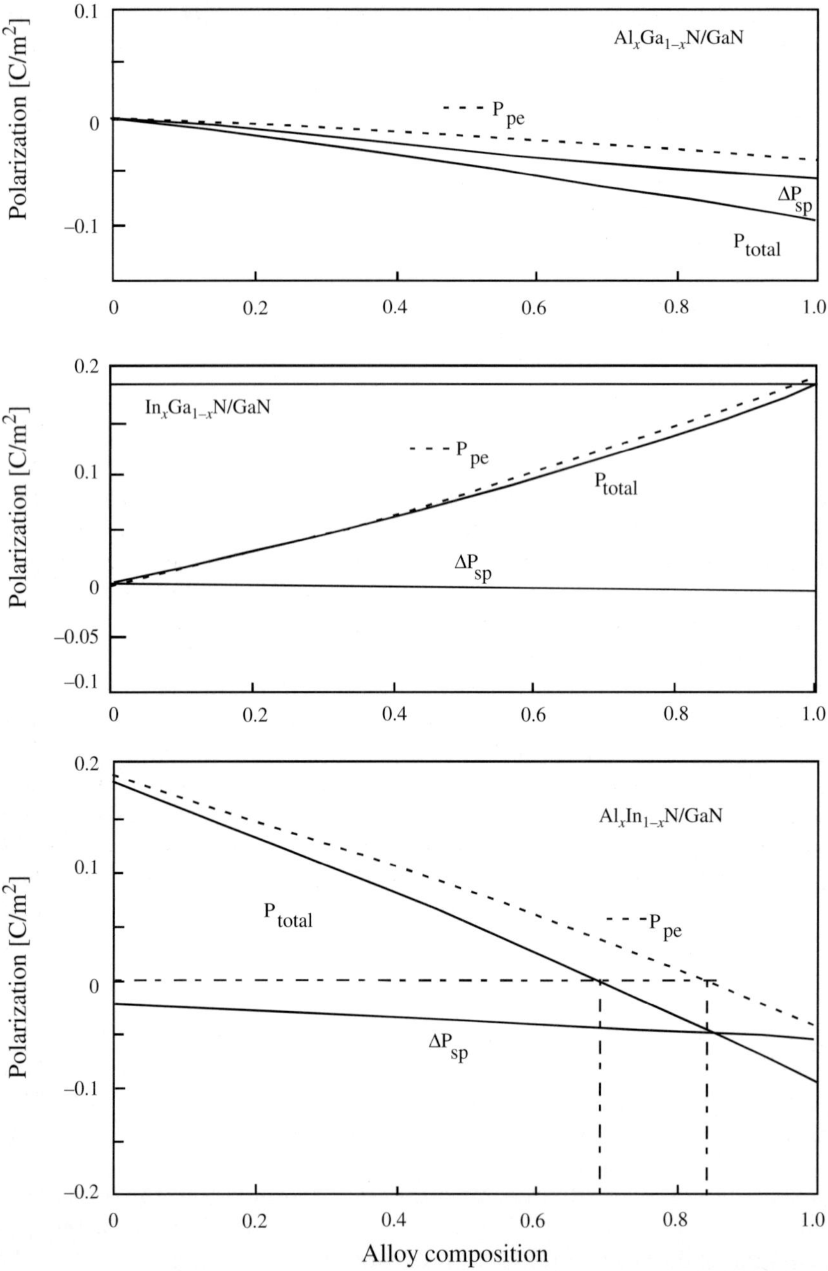

Fig. 14.5 Piezoelectric, spontaneous, and total polarization charge of coherently strained AlGaN, InGaN and AlInN alloys grown on completely unstrained Ga-polarity GaN buffer layers versus the alloy composition. The polarization values were determined by linear extrapolation of the physical properties from the binary compounds. The figure is similar to that reported by Ambacher *et al.* [90], but the parameter used are consistent with our previous publications.

$P^{\text{piezo}} = +0.183x$. Here the piezoelectric polarization dominates and is opposite in direction to the spontaneous polarization charge, but even larger in absolute magnitude. In the case of coherently strained $Al_xIn_{1-x}N$ layer on a relaxed GaN layer, the situation is unique in that for $x=0$ we revert to the InN on GaN case, and for $x=1$ we revert to the AlN on GaN case.

Numerical figures can be generated for the total polarization charge following the expressions outlined above by using a linear extrapolation of the strain and differential spontaneous polarization. The total polarization at the interface is the sum of the piezoelectric and differential spontaneous polarization, $P_{\text{total}} = \Delta P_{\text{sp}} + P_{\text{PE}}$. Taking the normal modulation doped structures where the GaN buffer layer is assumed completely relaxed and the AlGaN barrier layer is assumed coherently strained, one arrives at the plot shown in Fig. 14.5 for the total polarization charge at the interface. Additionally, data for InGaN and AlInN on GaN are shown. For tensile-strained $Al_xGa_{1-x}N$ or $Al_xIn_{1-x}N$ (for large x values) on GaN, the piezoelectric and the spontaneous polarization are negative and point in the same direction, thus they add up. The spontaneous and piezoelectric polarizations oppose one another for compressively strained $In_xGa_{1-x}N$ or $Al_xIn_{1-x}N$ (for small x values) layers. To calculate the differential spontaneous and piezoelectric polarization associated with alloys, one can employ a linear interpolation for the spontaneous polarization, piezoelectric and elastic constants from the binary compounds [89]. As discussed above, the piezoelectric polarization in coherently strained $Al_xGa_{1-x}N$ layers grown on GaN increases to $-0.044\,\text{C/m}^2$ for $x=1$. For InGaN layers, the piezoelectric polarization increases up to $+0.183\,\text{C/m}^2$ for $x=1$. The ternary $Al_{0.82}In_{0.18}N$ can be grown lattice matched to GaN and the piezoelectric polarization vanishes. For lower Al-concentrations, i.e. $x < 0.82$, the piezoelectric polarization increases due to increasing biaxial compressive strain. For higher Al-concentrations, i.e. $x > 0.82$, the layer is under tensile strain and the piezoelectric polarization

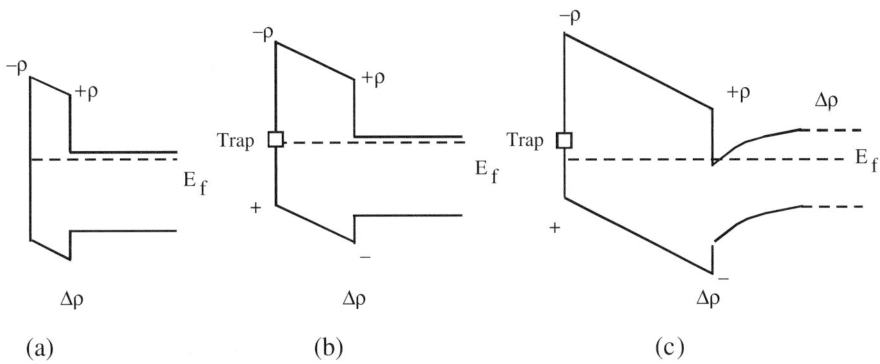

Fig. 14.6 Schematic representation of very simple-minded band structure of an AlGaN/GaN structure with varying AlGaN thickness, as one goes from (a) to (c), which demonstrates how the surface charge could participate in the screening of the polarization induced charge or the field. Patterned after ref. [78].

becomes negative (Fig. 14.5). Opposing strain and spontaneous polarization charge in $Al_xIn_{1-x}N$ cancel one another for a mole fraction of about $x=0.7$ with the underlying assumption that the $Al_xIn_{1-x}N$ is coherently strained and GaN layer on which it is grown is completely relaxed [90]. The discussion of more accurate effective mass calculations can be found in the MODFET modeling section.

Some further words of caution about the above estimates are needed. If the AlGaN layers are not pseudomorphic but partially relaxed (by misfit dislocations, for example), then the piezoelectric effect would be reduced but the spontaneous polarization would still be present. If the interfaces are not atomically sharp but exhibit a certain degree of inter-diffusion, the differences in spontaneous polarization would be reduced as well. Finally, if domains with inverted polarity exist, the overall polarization effects may be washed out. Also note that in an inverted structure with nitrogen (N) polarity towards the surface, it may be possible to create a two-dimensional hole–gas (2DHG) at the AlGaN/substrate GaN interface, providing that free holes are available. However, if an n-type GaN layer is placed on top, a 2DEG may form on top of the AlGaN layer.

Let us now consider a hypothetical case in which no free carriers exits. In this case, the polarization charge causes a linear band bending in $Al_xGa_{1-x}N$ with no change in the band of the underlying GaN as shown in Fig. 14.6. Let us also assume that there are surface states below within the bandgap as shown. As the AlGaN thickness is increased, the band bending would be such that the surface states would become ionized. The released electrons in the process would wind up at the interface, which tend to screen the polarization charge. If the surface state concentration were not sufficiently high, the surface Fermi level comes very close to the valence band. This implies that holes are present within the AlGaN layer near the surface. This positive charge needs to be balanced by a negative charge at the interface, the source of which could be surface states. In reality, the semiconductor system contains defects and free carriers that would make the picture somewhat more complicated. For one thing, the Fermi level may not be able to come as close to the valence band as that shown in the figure. The surface defect supposition has recently been forwarded with experimental backing [91].

With Ga polarity, the conduction band-edge of the $Al_xGa_{1-x}N$ will slope up towards the surface where the $Al_xGa_{1-x}N$ layer is. The band diagram shown would hold in this hypothetical case only if the $Al_xGa_{1-x}N$ layer thickness is such that the Fermi level at the surface is still a few kTs away from the valence band edge. If the $Al_xGa_{1-x}N$ thickness were made larger, the Fermi-level near the surface would be very close to the valence band, causing a positive hole-charge (positive) at the surface as shown in Fig. 14.6. A negative image charge would then form at the heterointerface tending to screen the polarization charge. This charge is different from the polarization charge in that it is mobile and can partake in current flow in addition to altering the band diagram. In short, the polarization-charge is a fixed charge. However, it would induce mobile charges to screen it if they are present. The recent device literature is confusing, as some authors are implying that polarization-charge alone is capable of doping a semiconductor. Misleading nomenclature with no physical basis such as 'piezo doping' has already been coined. To make the point in another way, polarization charge causes band bending and any free carriers, if present in the system, would tend to screen the polarization charge. The screening charge represents the mobile carriers whose source could be native defects, impurity dopants (intentional or unintentional) and surface states.

Returning to modulation doped structures with $Al_xGa_{1-x}N$ barriers, the sign of the polarization is such as to produce a potential energy for electrons sloping down from the Ga face towards the N face. Thus, for a structure in which the Ga face is turned towards the surface, the potential will slope down from the $Al_xGa_{1-x}N$ surface towards the AlGaN/GaN interface and helps to drive carriers towards the two-dimensional electron gas (2DEG) forming at this interface. For example, if there is an ohmic metal contact on the $Al_xGa_{1-x}N$ surface, electrons will flow towards the 2DEG below that layer.

The most favorable situation for enhancing sheet carrier concentration would occur for an InGaN quantum well on top of relaxed n-GaN and below a AlGaN barrier with the whole structure having cation polarity towards the surface. In that case, the field will slope down towards the InGaN/AlGaN interface in the quantum well and will help localize the carriers in the 2DEG. Note that the piezoelectric polarizations estimated here are based on the theoretical values for a perfectly insulating material. The free carriers that are present in each layer screen the field. For example, if free carriers are provided from metal contacts and they flow towards the 2DEG, this process sets up a screening field, which counters the polarization-induced field. Under equilibrium conditions, if they are reached, the net field is determined by the condition that the chemical potential for electrons (the Fermi level) must be constant throughout the structure. This depends on the doping and band bending in the substrate and, possibly, in each of the layers. At the very least, one may expect these fields to be reduced by a factor corresponding to the macroscopic dielectric constant, i.e. a factor of order 10 but possibly larger if the conductivity of the layers increase by the free carriers in the system. Consequently, a more realistic expectation for the effect on sheet carrier concentration is on the order of 10^{11}–10^{12} electrons/cm^2.

The difference between these and traditional device structures without polarization effects is that for uniform dopant concentrations, one obtains parabolically varying potentials with distance, whereas here the linear terms come from polarization on top of the parabolic terms. These linear terms lead to variations of the potential over a shorter distance scale determined by the thickness of the layers, whereas the parabolic terms correspond to the space-charge regions. Thus, the linear terms may help to localize carriers if the polarity of the structure is chosen properly.

The immediate impact of this polarization is that the field generated by this process must be considered together with that induced by the applied voltage and charge redistribution. Moreover, as alluded to earlier, free carriers can

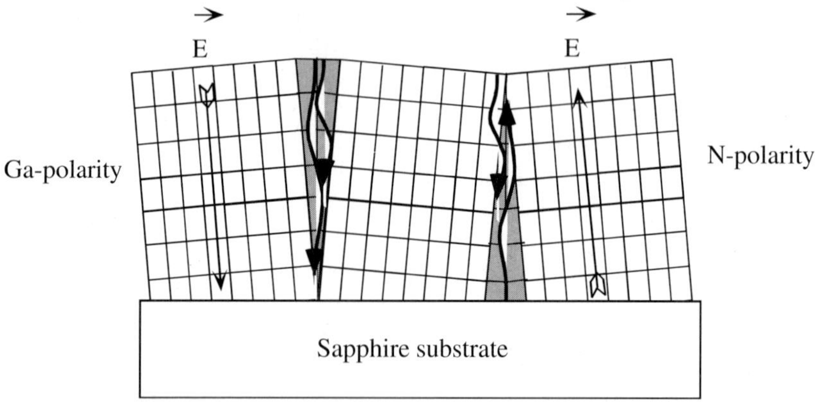

Fig. 14.7 Domains in GaN, with N-polarity (nitrogen surface layer) on the right and Ga polarity (with Ga on surface) on the left side under compressive residual strain. The arrows show the direction of the piezoelectric field in each of the domains.

also be drawn from any shallow and weakly bound impurities and metal in contact with the semiconductor. In any case, the free carriers would tend to screen the piezoelectric-induced polarization field. An additional complicating factor in nitrides in relation to polarization is that the semiconductor tends to twist and tilt in a columnar mode, in an effort to minimize strain as shown in Fig. 14.7. Multiple polarity has been confirmed in epitaxial GaN layers on sapphire substrates by convergent beam electron diffraction [92–94]. These columns do not necessarily have the same cation/anion ordering polarity as shown in Fig. 14.7. In addition, stacking faults would also lead to mixed polarity samples. In the presence of strain, Ga-polarity domains and N-polarity domains would have opposite polarization, causing increased scattering.

Figure 14.8 is a schematic representation of an ideal inversion domain boundary formed in growth along the [0001] direction. On the left of the boundary, the growth initiates with N, and on the right it begins with Ga. On the left side, the bond along the [0001] direction is from Ga to N; this is called Ga polarity. On the right side, the [0001] bonds are from N to Ga; this is called N-polarity. In N-polarity and under tensile strain, the PE field generated points toward the surface, whereas that for the Ga polarity region points in from the surface. When the strain is compressive, the direction of the field changes. Yet an additional complicating factor is the asymmetry in the apparent (measured) barrier discontinuities between GaN and its binary and ternaries caused by polarization, [95–97] which we have not really discussed here.

To reiterate, the spontaneous polarization arises simply because of the ionicity of the bonds and the low symmetry in wurtzite. In fact, Bernardini *et al.* [98] showed that the field that occurs in quantum wells is almost completely determined by the difference in spontaneous polarization between the two bulks, and their PE contribution. The field (i.e. the slope of the potential) is

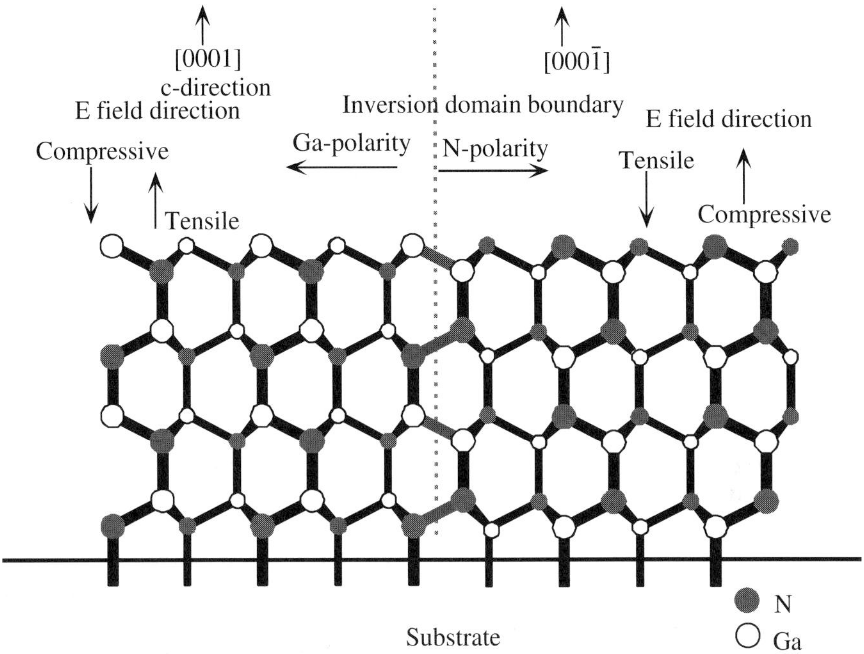

Fig. 14.8 Schematic representation of an ideal inversion domain boundary formed in along the *c*-axis.

quite independent of the offset (i.e. the dipole discontinuity that occurs at the interface between the two materials). Recent experimental results obtained in the authors' laboratory with colleagues from the University of Lecce support this premise that the spontaneous polarization component is dominant.

In considering a Normal MODFET (N-MODFET) structure where the larger bandgap AlGaN donor layer is deposited on top of a GaN channel layer, (see Fig. 14.9) both the spontaneous polarization and piezoelectric polarization must be accounted for. For an N-MODFET structure with Ga polarity, the potential will slope down from the surface of the AlGaN layer towards the AlGaN/GaN interface and will help to drive free electrons towards the interface forming a 2DEG as shown in Fig. 14.9. For example, if there is an ohmic metal contact on the AlGaN surface, electrons will flow towards the 2DEG below the AlGaN layer from contacts. Since nitride semiconductors in question have large bandgaps, thermal generation rates are minuscule and the role played by thermally generated carriers can be ignored.

A case of importance in GaN based semiconductors is the polarity of the epitaxial layer, i.e. whether the Ga or the N plane forms the surface. It is therefore essential that the N-face case also be considered. The case of an AlGaN (tensile strained)/GaN (relaxed) heterostructure with nitrogen polarity for an

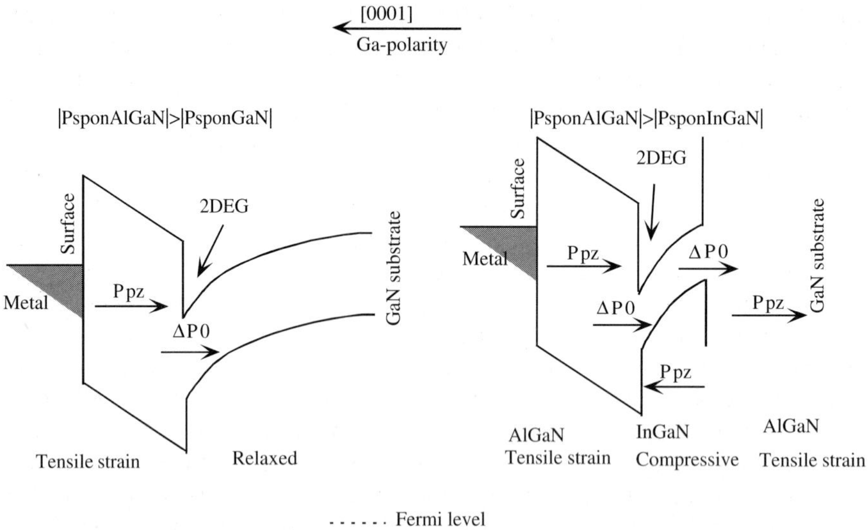

Fig. 14.9 GaN-based normal modulation doped structures with Ga polarity and GaN and InGaN active layers. If the sign of strain were to change, to compressive from tensile, then the direction of the piezoelectric polarization would change. In that case the spontaneous and piezoelectric polarization charges would oppose one another, and the larger one would determine whether hole or electron accumulation is favored at the interface.

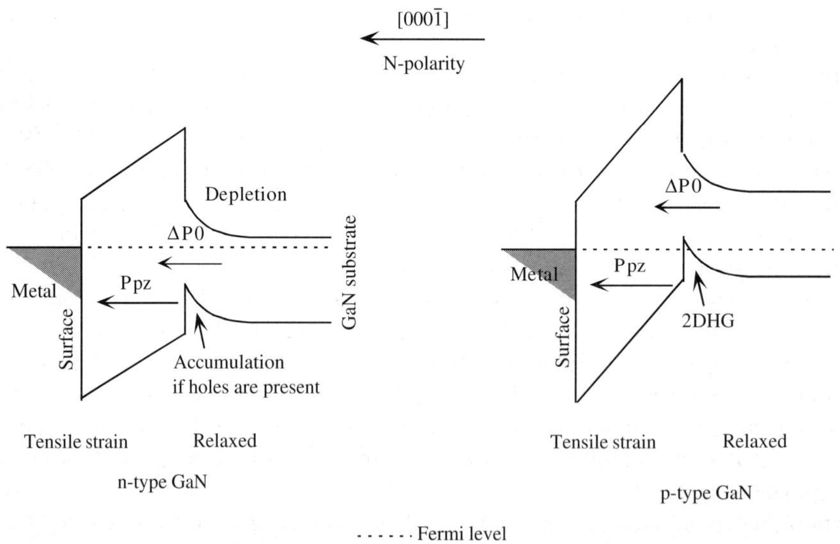

Fig. 14.10 AlGaN/GaN based normal modulation doped structures with N-polarity for two cases, one for n-type and the other for p-type buffer layer. If the sign of strain were to change, to compressive from tensile, then the direction of the piezoelectric polarization would change. In that case the spontaneous and piezoelectric polarizations would oppose one another, and the larger one would determine whether hole or electron accumulation is favored at the interface.

n-type GaN and for a p-type GaN buffer layers is shown in Fig. 14.10 as in the cases depicted in Fig. 14.9 where the piezoelectric polarization and spontaneous polarization charges support one another. Unlike the Ga-polarity case, the polarization charge is such that screening charge will be made of holes if they are present in the film. If holes constitute the minority charge in the film, then the thermal process is the means by which they would be created. However, this process in a wide bandgap semiconductor such as GaN is very slow and equilibrium condition may not be attained. If the strain in AlGaN were compressive, the direction of the piezoelectric polarization vector would change, causing the piezoelectric polarization to counter the spontaneous polarization. This would actually represent the case when the epitaxial films are relaxed at the growth temperature and upon cooling to room temperature, the film would be under compressive strain if on sapphire substrates. This is due to the expansion coefficient of sapphire being larger than that of GaN. In such a case, the larger of the two would dominate and determines whether hole or electron accumulation would be favored. If on the other hand the film is grown on a SiC substrate, the strain due to thermal expansion would be tensile. This would lead to the case where the piezoelectric polarization and spontaneous polarization would support one another.

Inverted modulation doped structures can also be used to interrogate the picture in effect and perhaps take advantage of the unique features presented. In such a case, the AlGaN layer precedes the GaN top layer where the charge accumulation would occur. The interface between the AlGaN layer and the bottom GaN layer which is referred to as the buffer layer would be graded to avoid a normal interface.

Figure 14.11 shows, schematically, AlGaN/GaN based inverted modulation doped structures with Ga, and N-polarities. As can be seen, in the case of Ga-polarity and tensile strain in AlGaN, both the piezoelectric and spontaneous polarization vectors support each other leading to a hole accumulation at the interface if holes are present in the system. Since the thermal generation rate is very small, the semiconductor structure cannot be expected to reach equilibrium by this means at room temperature in a reasonable period of time. On the other hand, with N-polarity and tensile strain in AlGaN, the structure favors electron accumulation at the interface. If the sign of strain were to change, to compressive from tensile, then the direction of the piezoelectric polarization would change. In that case the spontaneous and piezoelectric polarization charges would oppose one another, and the larger one would determine whether hole or electron accumulation is favored at the interface.

Ambacher et al. [90] employing Hall effect and capacitance-voltage profiling measurements, measured the sheet carrier concentration and its profile in modulation doped structures at room temperature, the results of which are shown in Fig. 14.12. As seen in the figure, the sheet carrier concentrations are consistently larger than those expected from piezoelectric polarization alone. Since, the GaN buffer contribution was verified to be negligible, some additional sources, namely spontaneous polarization was invoked to explain the

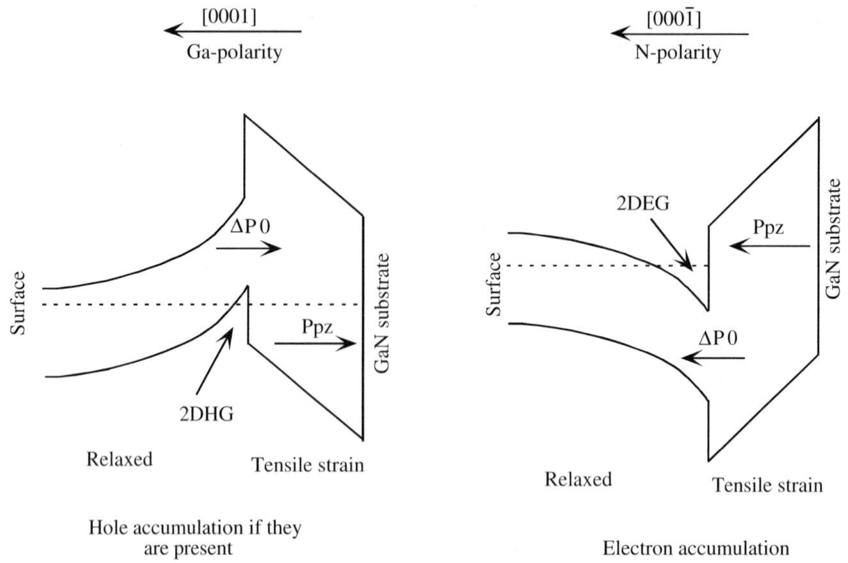

Fig. 14.11 AlGaN/GaN-based inverted modulation doped structures with Ga, and N-polarity. As can be seen, in the case of Ga polarity and tensile strain in AlGaN, both the piezoelectric and spontaneous polarization vectors support each other leading to hole accumulation at the interface if holes are present in the system. The other source, thermal generation rate is very small and the semiconductor structure cannot be expected to reach equilibrium by this means at room temperature in a reasonable period of time. On the other hand, with N-polarity and tensile strain in AlGaN, the structure favors electron accumulation at the interface. If the sign of strain were to change, to compressive from tensile, then the direction of the piezoelectric polarization would change. In that case the spontaneous polarization and piezoelectric polarization would oppose one another, and the larger one would determine whether hole or electron accumulation is favored at the interface.

observations. The contribution from the unintentionally doped barrier layer also was found to be negligible as the measure of 2DEG concentration with increasing barrier thickness up to 650 Å ($x=0.25$) did not appear to affect the results. The highest measured sheet carrier concentrations for $0.2<x<0.45$ are in good agreement with the calculated values of total polarization charge taking spontaneous polarization and the measured strain relaxation of AlGaN into account. The maximum sheet carrier concentration for undoped AlGaN/GaN (and also for InGaN/GaN) heterostructures is limited to about $2\times10^{13}\,\mathrm{cm}^{-2}$ due to strain relaxation of the top alloy layer. Not in good agreement are the calculated and measured sheet carrier concentrations of undoped MODFET structures with alloy compositions $x<0.2$. The measured sheet carrier concentrations are up to $5\times10^{12}\,\mathrm{cm}^{-2}$ smaller than the predicted ones. In addition, the large scattering of the measured sheet-carrier concentrations for heterostructures with similar barrier thicknesses and alloy compositions is unexpected and much larger than error in experiments. For a more accurate calculation of the

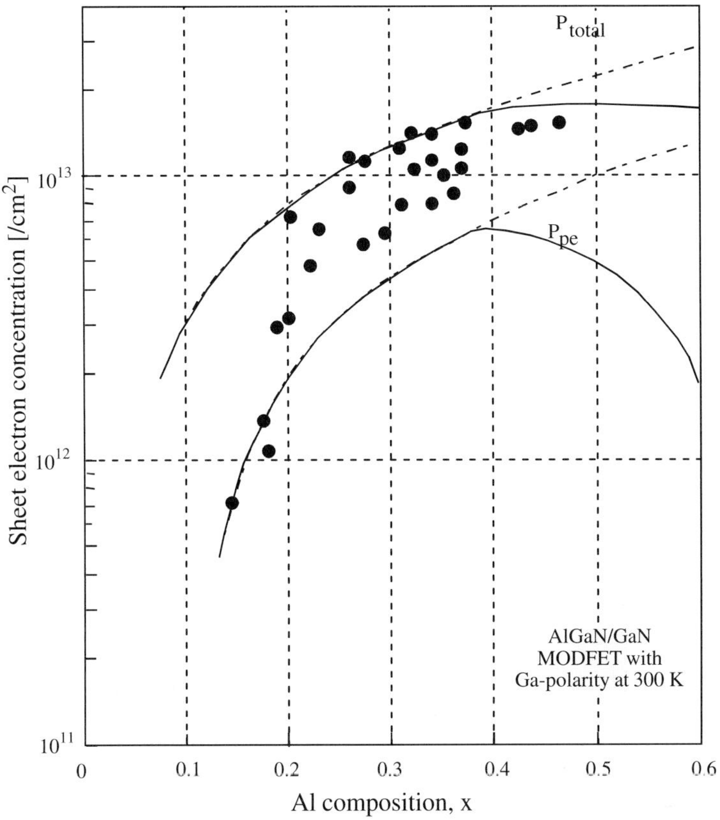

Fig. 14.12 Comparison of calculated and measured electron sheet carrier concentrations in Ga-face AlGaN/GaN heterostructures. The dashed lines are calculated results for pseudomorphic structures. The solid lines are indicative of the calculated results, which take the measured partial strain relaxation into account (for more details, see ref. [89, 90]).

polarization charge in Ga-polarity modulation doped structure, see the section on MODFET modeling.

The physics literature appears to be fairly clear in that the polarization charge is a bound charge and that any free carriers act only to screen it. However, the device reports on GaN FETs have gone so far as to suggest that piezoelectric effect, in and of itself, is sufficient to provide the free carriers needed for devices. Polarization effects, particularly spontaneous polarization, have immense impact on measured band discontinuities. For example, the dependence of measured band discontinuities on the order in which the larger and the smaller band-gap semiconductors are grown, is one that can be attributed to polarization effects [96, 97].

To reiterate, as a result of polarization, the static potential at the GaN/AlN interface is different from that at the AlN/GaN interface that gives rise to interface charge larger than the charge densities used in devices. A substantial level of effort has been expended toward determining band discontinuities, but the field is in desperate need of more in-depth investigations in improved structures. The observed asymmetry in AlN/GaN and GaN/AlN interfaces caused by spontaneous polarization is within the experimental errors of Martin *et al.* [95–97]. Inversion domains, see Fig. 14.8 combined with any strain in semiconductor nitrides lead to flipping PE fields, (see Fig. 14.14) with adverse effects on our ability to characterize the films, let alone exploit this phenomenon for devices. Such flipping fields would also cause increased scattering of carrier as they traverse in the *c*-plane. Simply put, identical device structures with different polarity layers would have widely differing performance underscoring the importance that these issues will have to be investigated and reconciled. The polarity mixing causes the PE-induced electric field to flip from one domain to the next, causing a variation in the sheet carrier concentration along the channel of an FET-like device. The same polarity mixing would have deleterious effects in the base of an HBT as well and depending on the polarity the induced field would either aid or impede minority carrier transit.

14.5 Partial strain relaxation

The heretofore discussion dealt only with coherently strained AlGaN top layer and relaxed GaN buffer layer in the case of normal modulation doped structures and relaxed AlGaN and coherently-strained GaN top layer in the case of inverted modulation doped structures. This is, of course, an assumption whose validity has not been confirmed. In reality, a mostly likely scenario is that a partial relaxation would take place coupled with a residual thermal strain. Ambacher *et al.* treated this issue of partial relaxation [90]. As pointed out by them, calculations of the critical thickness of InGaN and AlGaN grown on GaN show that for a typical AlGaN barrier thickness of about 20 nm, strain relaxation should occur for alloy compositions above $x=0.2$. At the outset, we should mention that the critical thickness, the thickness above which the film begins to relax, is difficult to determine because of the large density of dislocations in this material system. Nevertheless, proceeding as if it can be done does indeed give an estimate, which is very useful. In order to determine the alloy composition and relaxation of the barrier, the lattice constants $a(x)$ and $c(x)$ were measured by high resolution X-ray diffraction (HRXRD) [89]. For barrier thicknesses of about 25 nm, macroscopic strain relaxation was observed for alloy compositions of $x_c=0.20$ and 0.38 for InGaN and AlGaN, respectively. In both cases the degree of relaxation increased nearly linearly with increasing alloy composition for $x>x_c$. Naturally, for a barrier with a fixed alloy composition, the piezoelectric polarization and the sheet charge induced by that polarization decrease linearly with an increasing level of an relaxation, r. The polarization induced charge at the interface for a Ga-face AlGaN/GaN heterostructure is shown versus Al-concentration for different degrees of strain relaxation in Fig. 14.13. The maximum sheet charge caused by

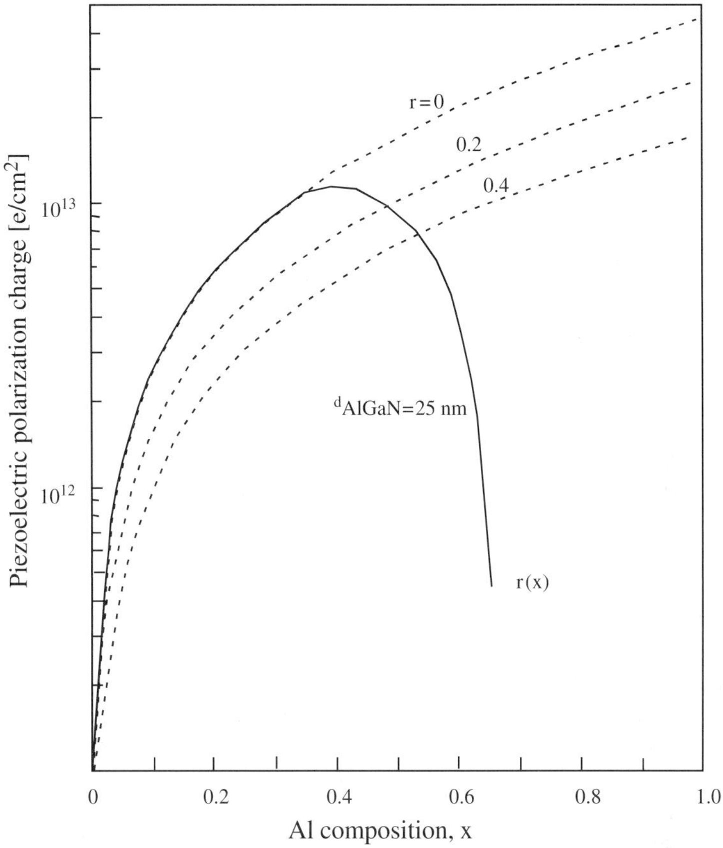

Fig. 14.13 Piezoelectric polarization induced sheet charge at the interface of a Ga-face AlGaN/GaN heterostructure for different degrees of relaxation of the barrier layer (dashed lines). The solid line was calculated taking into account the measured degree of relaxation for AlGaN top layers with thicknesses of 25±5 nm. The figures have been modified to be consistent with the piezoelectric and polarization charge figures used in our previous publications.

piezoelectric polarization of a strained 30 nm thick barrier was calculated using the measured degree of relaxation. As can be seen, for a barrier with $x=0.4$ the total sheet charge induced by spontaneous and piezoelectric polarization decreases if the AlGaN becomes partially or completely relaxed.

Hsu and Walukiewicz [69] incorporated issues related to polarization into a normal modulation doped structure calculating the dependence of parameters such as the sheet-carrier concentration on structural parameters. While this issue will be treated in detail in the FET section, we will discuss the treatment by Hsu and Walukiewicz since it relates the sheet concentration to structural parameters and discusses the electron mobility.

Figure 14.14(a) shows the dependence of the 2DEG density as a function of the Al content of an $Al_xGa_{1-x}N$ barrier with a thickness of 31 nm. Parametric

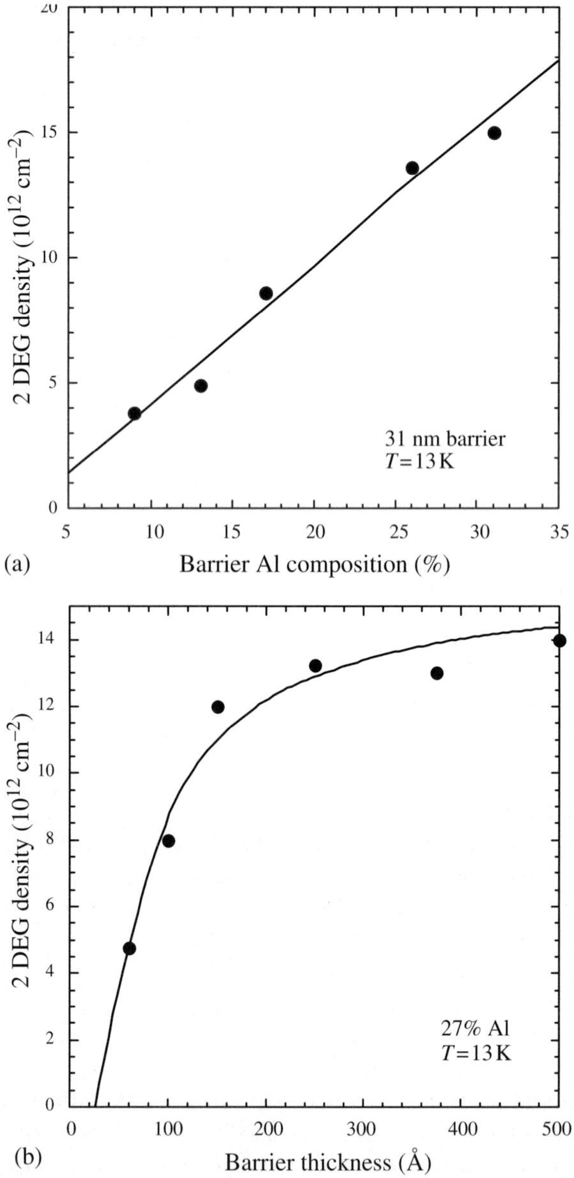

Fig. 14.14 (a) 2DEG densities as functions of AlGaN barrier Al composition. (After ref. [69]). (b) 2DEG densities as a function of thickness of a $Al_{0.27}Ga_{0.73}N$ barrier. (After ref. [69].)

values of $A = 7.9 \times 10^4$ V/cm and $B = 1.15 \times 10^7$ V/cm in $P(x) = A + Bx$ describing the dependence of polarization on AlN mole fraction led to good quantitative agreement with experiments. In principle, the parameter A is related to the uncompensated, spontaneous polarization field in GaN. Its relatively small

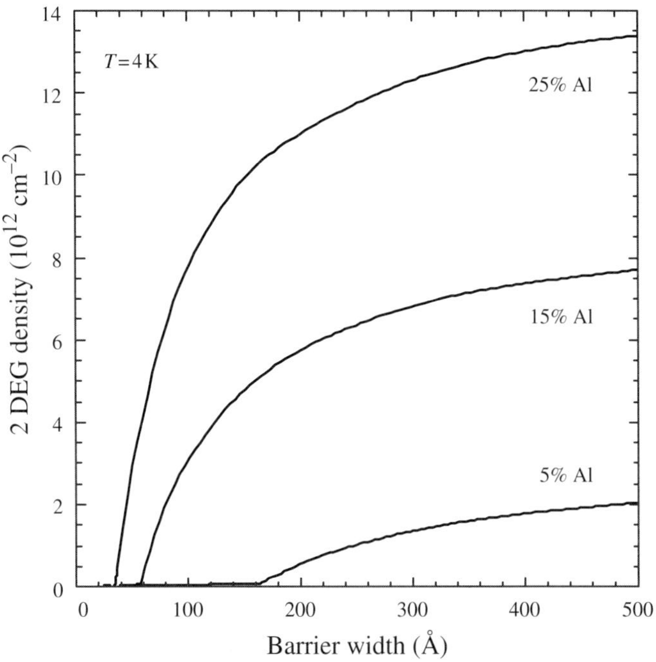

Fig. 14.15 Calculated 2DEG densities as a function of barrier width for three different AlGaN/GaN heterostructures with barrier of different Al compositions. (After ref. [69].)

value indicates either that this spontaneous polarization is very small, or that it is well compensated by charges from unintentional dopants within the AlGaN layer. Figure 14.14(b) shows the dependence of the 2DEG density as a function of the thickness of an $Al_{0.27}Ga_{0.73}N$ barrier.

Figure 14.15 shows the calculated dependence of the 2DEG density on the barrier thickness for several alloy compositions at 4 K, formed by surface donors in an unintentionally doped structure. For very thin barriers (<30 Å for a barrier with x=25% and <160 Å for x=5%), the defect level is below the Fermi energy and so the surface defects are neutral. In this case, only the electrons from any dopant in the barrier are transferred to the GaN well and if one assumes a donor concentration of 5×10^{16} cm^{-3} the density remains below 10^{11} cm^{-2}. Once the barrier thickness increases to the point of bringing the defect level above the Fermi energy, electrons are transferred from these surface donors and the 2DEG density increases rapidly.

At very low alloy compositions (5%), the 2DEG density rises continuously at an approximately constant rate throughout the entire range of barrier widths. The cause for this is that, as the barrier width increases, the donor charge transition state energy also increases, due to the polarization field. This leads to an increase in the energy gain from transferring electrons from the donors to GaN. Consequently, the donor formation energy decreases and the surface defect concentration increases.

For larger barrier alloy compositions, the 2DEG density increases very rapidly at first, followed by a leveling for barrier thicknesses greater than 250 Å. The initial increase is due to the effect of the polarization field on the donor level in that larger Al fractions in the barrier result in larger polarization fields and a more rapid increase in defect concentration. As the 2DEG density approaches 10^{13} cm^{-2}, the opposing field due to the transfer of electrons becomes comparable in magnitude to the polarization field. This field acts to reduce the rate at which the defect formation energy is lowered. For a barrier of sufficient thickness, this charge transfer field could completely cancel the polarization field and thereafter, the 2DEG density would be independent of barrier thickness assuming that the AlGaN barrier remains coherently strained and is not relaxed.

14.6 Low field transport in 2DEG systems

Calculations of the electron mobility at the AlGaN/GaN interface have been performed [69] using methods which have been described previously [77]. The scattering mechanisms considered were acoustic phonons, Coulomb scattering from both the donor-like defects on the AlGaN barrier surface and unintentional dopants in the GaN, and alloy disorder scattering. In addition, interface roughness scattering was also included.

Figure 14.16 shows calculated 2DEG mobilities at low temperature as a function of barrier width for several heterostructures with different Al compositions. Calculations show three distinct regions [69]. For very small barrier widths, the mobility is quite low and increases slowly with increasing barrier width. Comparing these curves with those in Fig. 14.15, one can see that the 2DEG density is very small in this region, and is due to unintentional dopants in the bulk of the AlGaN barrier. Consequently, Coulomb scattering by charged impurities in the GaN well and the AlGaN barrier keeps the mobility low. In the region with a sudden increase of the mobility, the vast majority of the 2DEG electrons originate from donors at the surface of the AlGaN barrier. In this region, the high electron density reduces the efficacy of Coulomb scattering, which results in much higher mobilities. Moreover, most of the Coulomb scattering centers are at the surface of the AlGaN barrier, away from the carriers at the AlGaN/GaN interface. Finally, when the 2DEG density becomes high enough, the alloy disorder scattering, which varies with the square of the electron concentration, is the dominant scattering mechanism and the mobility decreases with increasing 2DEG density. Maximum mobilities are generally achieved for barrier thicknesses between 50 Å ($x = 0.25$) and 200 Å ($x = 0.05$). Fig. 14.15 indicates that maximum mobilities are obtained for 2DEG densities in the range of about 3.5×10^{12} cm^{-2} ($x = 0.25$) to 5×10^{11} cm^{-2} ($x = 0.05$).

Figure 14.17 shows the overall and individual mobilities as a function of barrier thickness for a 2DEG at the interface of a $Al_{0.07}Ga_{0.93}N$/GaN heterostructure. This is the Al alloy fraction which would produce the highest

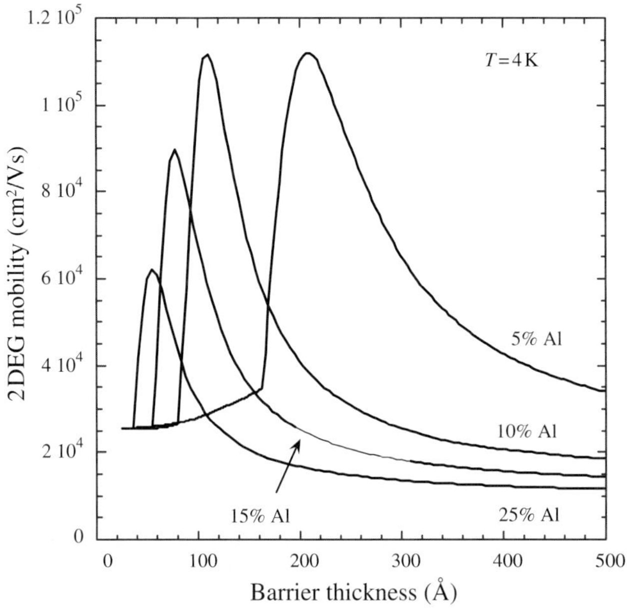

Fig. 14.16 Low temperature 2DEG mobilities as a function of barrier thickness for four different AlGaN/GaN heterostructures with different Al barrier compositions. (After ref. [69].)

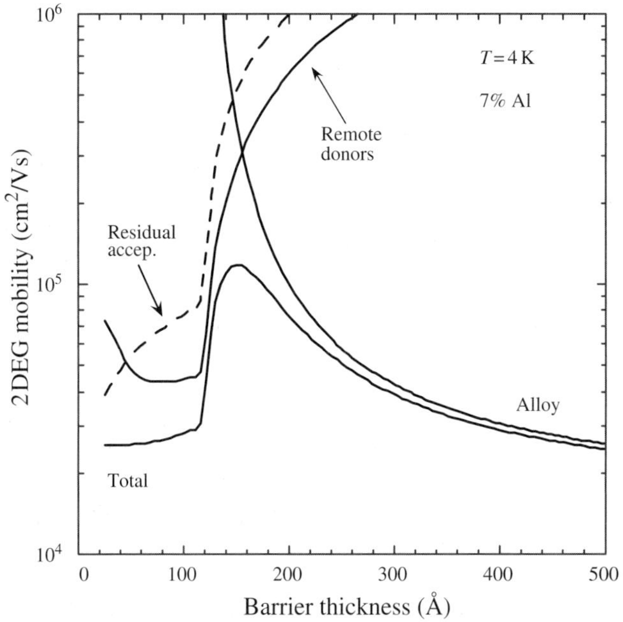

Fig. 14.17 DEG mobility as a function of AlGaN barrier thickness for a $Al_{0.07}Ga_{0.93}N$/GaN heterostructure. (After ref. [69].)

possible low temperature mobilities for the indicated unintentional doping levels [69]. The component mobility curves illustrate the dominant scattering mechanisms in each of the three different regions mentioned above. At a barrier thickness of 110 Å, the surface defect ionization energy is pushed above the Fermi energy and there is a rapid increase in both the mobility and the 2DEG concentration. For barriers that are thicker than 150 Å, alloy disorder becomes the dominant scattering mechanism. A maximum electron mobility of somewhat above 105 cm^2/Vs is predicted for a barrier width of about 150 Å and a corresponding 2DEG density of $1.8\times10^{12} cm^{-2}$. These calculations indicate that for a standard, undoped AlGaN/GaN heterostructure the maximum low-temperature mobility is only slightly larger than $10^5 cm^2/Vs$. The principal reasons for the lower values here as compared to those obtained in GaAs/AlGaAs heterostructures are stronger alloy disorder scattering (resulting from the much higher 2DEG density and larger conduction band offsets found in AlGaN/GaN heterostructures) and to a smaller extent relatively high levels of residual impurities in the nitride layers.

The calculated mobilities agree with the experimental values well for barrier Al compositions less than about 15%, as shown in Fig. 14.18 [69]. However, for higher Al fractions, the measured mobilities are a nearly constant factor of 2 to 2.5 lower. There are two possible reasons. First, as can be seen in Fig. 14.14(a), the electron density exceeds $6\times10^{12} cm^{-2}$ for $x>0.15$. This corresponds roughly to the electron density at which the higher sub-bands in the GaN quantum well begin to be occupied, which causes increased scattering. Second, the lattice mismatch at the interface increases with increasing x, leading possibly to increased roughness and thereby reduced electron mobility at the interface. In the end, technological issues may come to be the culprit as was the case in the early development of the GaAs/AlGaAs heterointerface.

The effect of interface roughness scattering on the total mobility has been estimated using 15 Å as the parameter for the characteristic lateral extent of the islands of roughness and adjusting the height of the islands in order to obtain the best fit with experimental data [69]. A constant island height of 0.9 Å produced the fit shown in Figs 14.18(a) and 14.18(b) [69]. One possible interpretation of the data is a sudden degradation of a near perfect interface for heterostructures with a barrier of greater than 15% Al.

The above arguments indicate that the highest electron mobilities are expected from heterostructures with 2DEG densities in the range of a few times $10^{12} cm^{-2}$. Since the polarization-induced charge transfer increases with increasing width and Al composition of the barrier, the highest mobilities are limited to heterostructures with relatively thin barriers with low Al content for undoped structures. A maximum electron mobility of about $10^5 cm^2/Vs$ can be expected from heterostructures with a 15-nm-thick $Al_{0.07}Ga_{0.93}N$ barrier [69].

Figure 14.19 compares the temperature dependent mobility, both calculated and measured for an $Al_{0.09}Ga_{0.91}N$/GaN heterostructure [69]. The calculations were performed for a constant 2DEG density of $2.3\times10^{12} cm^{-2}$. The low

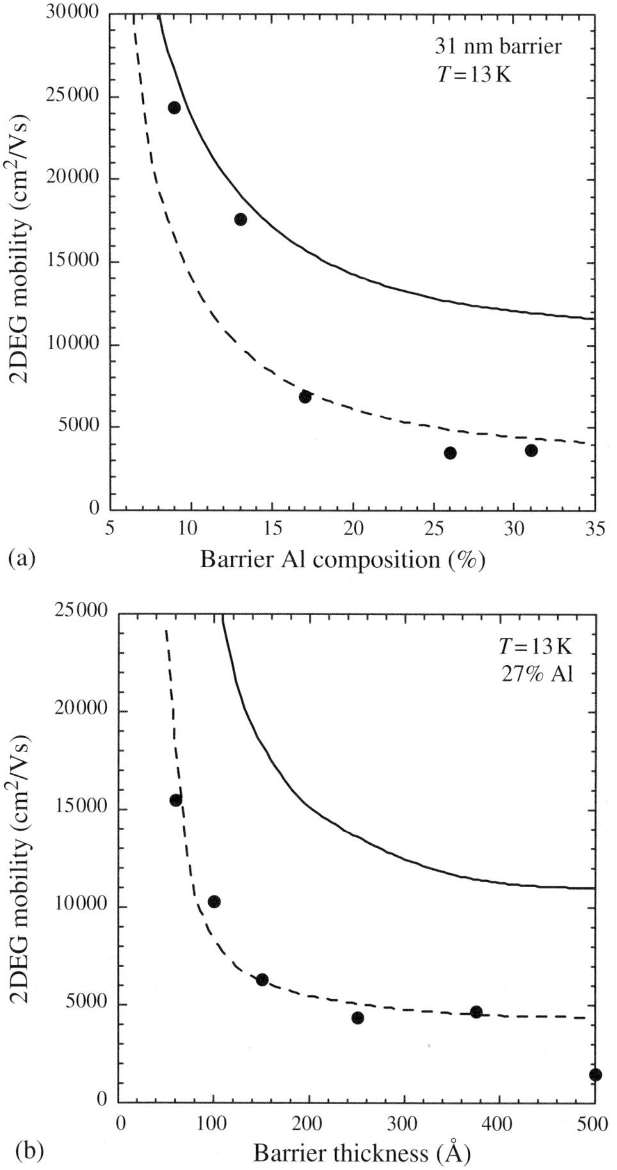

Fig. 14.18 (a) 2DEG mobilities as a function of Al composition of the barrier. The points are experimental measurements taken from ref. [7] and correspond to the samples in Fig. 14.14(a). The solid line is the calculated mobility neglecting interface roughness scattering. The dashed line is the calculated mobility including interface roughness scattering. (After ref. [69].). (b) 2DEG mobilities as a function of thickness of a $Al_{0.27}Ga_{0.73}N$ barrier along with the experimental data points and correspond to the samples in Fig. 14.14(b). The solid line is the calculated mobility neglecting interface roughness scattering. The dashed line is the calculated mobility including interface roughness scattering. (After ref. [69].)

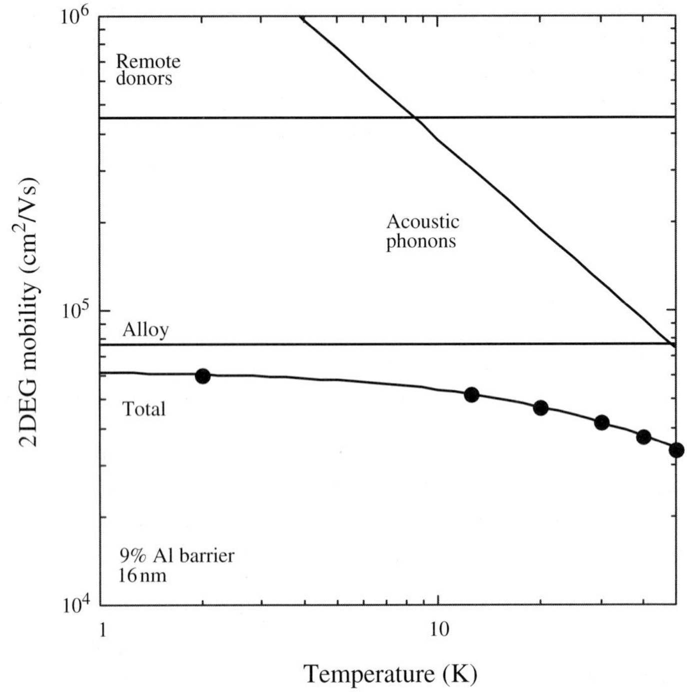

Fig. 14.19 2DEG mobilities as a function of temperature in an $Al_{0.09}Ga_{0.91}N$/GaN heterostructure with a 16-nm-thick barrier. (After ref. [69].)

temperature mobility is dominated by alloy disorder scattering with only small contributions from ionized centers and acoustic phonon scattering. Above 10 K, however, scattering by acoustic phonons becomes important and it is responsible for a weak but still clearly discernible temperature dependence of the mobility. The acoustic phonon component of the mobility varies roughly as 4.02×10^6 (1/T) cm^2/Vs and is about an order of magnitude larger in AlGaN/GaN heterostructures than in AlGaAs/GaAs systems.

The calculations reviewed above indicate that scattering by ionized impurities is important only for heterostructures in which the 2DEG density is 10^{12} cm^{-2} or smaller. Consequently, if one wishes to increase the electron mobility one should reduce the effect of alloy disorder and interface roughness scattering along with improved AlGaN technology. The penetration depth of the 2DEG into an $Al_{0.09}Ga_{0.91}N$ barrier of height $\Delta E_c = 0.2$ eV is $z_p = 0.9$ nm leading to the assertion that the alloy scattering is the dominant mechanism [69]. One could envision a composite barrier design consisting of a very thin layer of AlN, grown at the interface, followed by a thicker $Al_{0.07}Ga_{0.93}N$ layer. Because the penetration depth of electrons into AlN is only 3 Å and the electron wave function decays exponentially, a layer of AlN of only a few angstroms thick would sufficiently confine the electrons in the binary compounds and nearly eliminate alloy scattering. The thin AlN film at the interface should not significantly affect the charge transfer, which is determined mostly by the composition and thickness of the much thicker remaining part of the barrier. Thus, for a barrier composed of 5 Å of AlN followed by 120 Å of $Al_{0.07}Ga_{0.93}N$, one could expect to obtain a 2DEG with a density between 2 and $3 \times 10^{12}\,cm^{-2}$, but with negligible alloy disorder scattering.

14.7 High field transport

FETs, by their nature rely on transport under high electric fields. As such, high field effects are an integral part of any short channel FET, realizing that low field effects too are important in that they determine many of the parasitic resistances with detrimental effects on device performance. Consequently, high field properties of any semiconductor contemplated for use in FETs must be scrutinized in terms of its high field transport, and semiconductor nitrides are no exception. High field transport, in nitride semiconductors has recently been reviewed [99]. However, a short treatement of results are deemed necessary for inclusion in this review for completeness. While the bulk of the experimental, and to some extent the calculations, have focused on GaN, the transport properties from the point of electron mobility and velocity of InN are much more conducive for FETs. What is holding InN back, among others causes, is its poor qualtiy and high unintentional electron concentration. Borrowing from the GaAs system, it is very likely that some composition or a range of compositions of InGaN will be used for FET channels while taking full advantage of the GaN and AlGaN in the rest of the structure. Consequently, transport properties, inclusive of low and high field, of nitride semiconductors InN, GaN, AlN and their alloys will be briefly reported following ref. [99].

The steady-state electron drift velocity versus electric field has been calculated for the nitride binaries and ternaries at different temperatures and various doping concentrations [99]. The results for 300 K and an electron concentration of $10^{17}\,cm^{-3}$ will be shown here. Fig. 14.20 shows the calculated electron steady-state drift velocity versus applied electric field, for GaN, $Al_{0.2}Ga_{0.8}N$, $Al_{0.5}Ga_{0.5}N$, $Al_{0.8}Ga_{0.2}N$, and AlN. This set of calculations is made assuming the maximum alloy scattering rate. The alloy scattering was observed to be the dominant scattering mechanism when the random alloy potential was set to the conduction band offset between GaN and AlN. A significantly lower drift velocity is observed when alloy scattering is present which is simply a result of the higher total scattering rate. Moreover, in the presence of alloy scattering, the peak velocity occurs at a higher applied field compared to the case when alloy scattering is neglected. This is also due to the fact that in the presence of alloy scattering, the total scattering rate is higher; thus a higher field is required to heat the carriers prior to the onset of intervalley transfer.

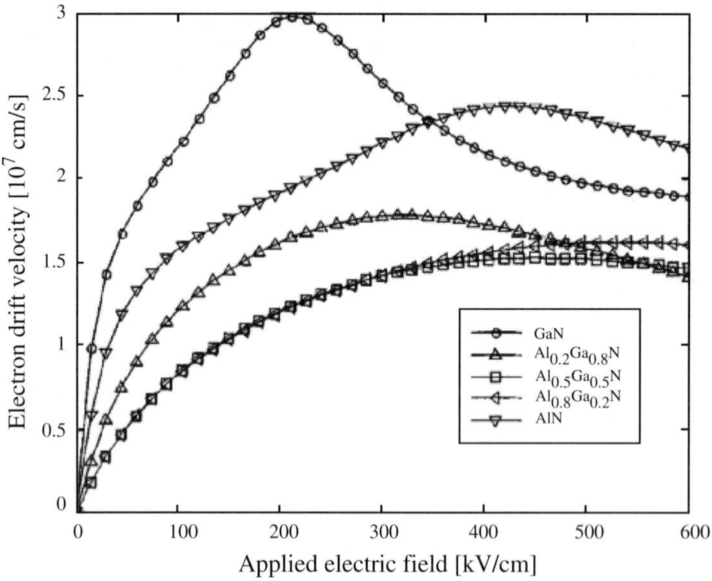

Fig. 14.20 Calculated electron drift velocity versus applied electric field for GaN, $Al_{0.2}Ga_{0.8}N$, $Al_{0.5}Ga_{0.5}N$, $Al_{0.8}Ga_{0.2}N$, and AlN. For this calculation, the random alloy potential was set equal to conduction band offsets. Lattice temperature is at 300 K, and electron concentration is equal to 10^{17} cm^{-3}. (After ref. [99].)

Figure 14.21 shows the calculated electron drift velocity versus applied electric field for the GaN, $In_{0.2}Ga_{0.8}N$, $In_{0.5}Ga_{0.5}N$, $In_{0.8}Ga_{0.2}N$, and InN. The random alloy potential is set equal to the conduction band offsets. It should be mentioned that for $In_xGa_{1-x}N$ compounds the random alloy scattering potential calculated from J. C. Phillips' theory of electronegativity differences results in values on the same order as the conduction band offsets.

14.8 Modulation doped field effect transistors (MODFETs)

With its reduced impurity scattering and unique gate capacitance-voltage characteristics, the MODFET has become the dominant high frequency device. Among the MODFET's most attractive attributes are close proximity of the mobile charge to the gate electrode and high drain efficiency. As in the case of emitters, the GaN-based MODFETs have quickly demonstrated record power levels at high frequencies with very respectable noise performance and large drain breakdown voltages.

In MODFETs, the carriers that form the channel in the smaller bandgap material are donated by the larger bandgap material, ohmic contacts or both. Since the mobile carriers and their parent donors are spatially separated, short-range ion scattering is nearly eliminated, which leads to mobilities that are characteristic of nearly pure semiconductors. A Schottky barrier is then

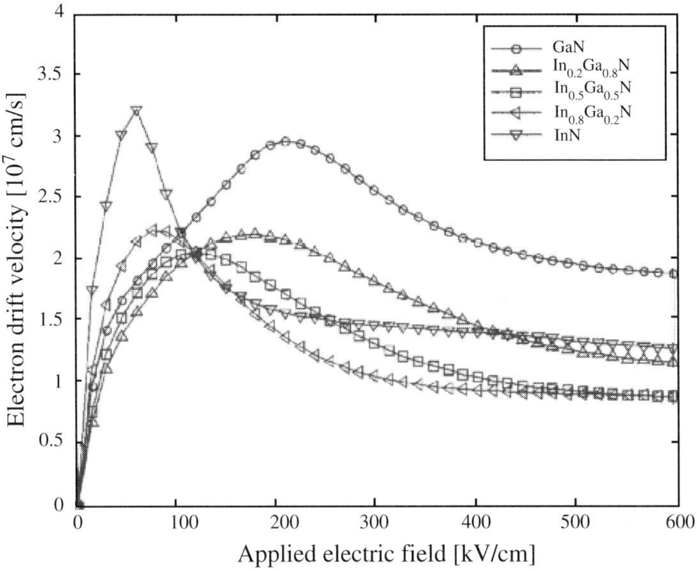

Fig. 14.21 Calculated electron drift velocity versus applied electric field for GaN, $In_{0.2}Ga_{0.8}N$, $In_{0.5}Ga_{0.5}N$, $In_{0.8}Ga_{0.2}N$, and InN. For this calculation, the random alloy potential was set equal to conduction band offsets. Lattice temperature is at 300 K, and electron concentration is equal to 10^{17} cm^{-3}. (After ref. [99].)

used to modulate the mobile charge that in turn causes a change in the drain current. Because of this heterolayer construction, the gate can be placed very close to the conducting channel, resulting in large transconductances [100]. Fig. 14.22 presents a schematic representation of a GaN/AlGaN MODFET heterostructure in which the carriers are provided by the donors in the wider bandgap AlGaN. In a MODFET device under bias, the carriers can also be provided by the source contact.

14.9 MODFET model

The Schrödinger's equation and Poisson equation can be used self-consistently in order to study the channel formation and of current flow mechanisms in GaN based MODFETs [100, 101]. Several approaches have been used to define the system Hamiltonian used in the Schrödinger equation, namely effective masses [102], **k.p** expansion [103] and tight-binding expansion [104, 105]. The use of sophisticated models such as *k.p* or tight-binding is justified, perhaps even made necessary, by the complex wurtzite band structure, particularly for determining the valence band states. Thus, calculations of optical processes involving band to band transitions must consider the details of the band structure beyond the simple effective mass approximation (EMA) [106, 107].

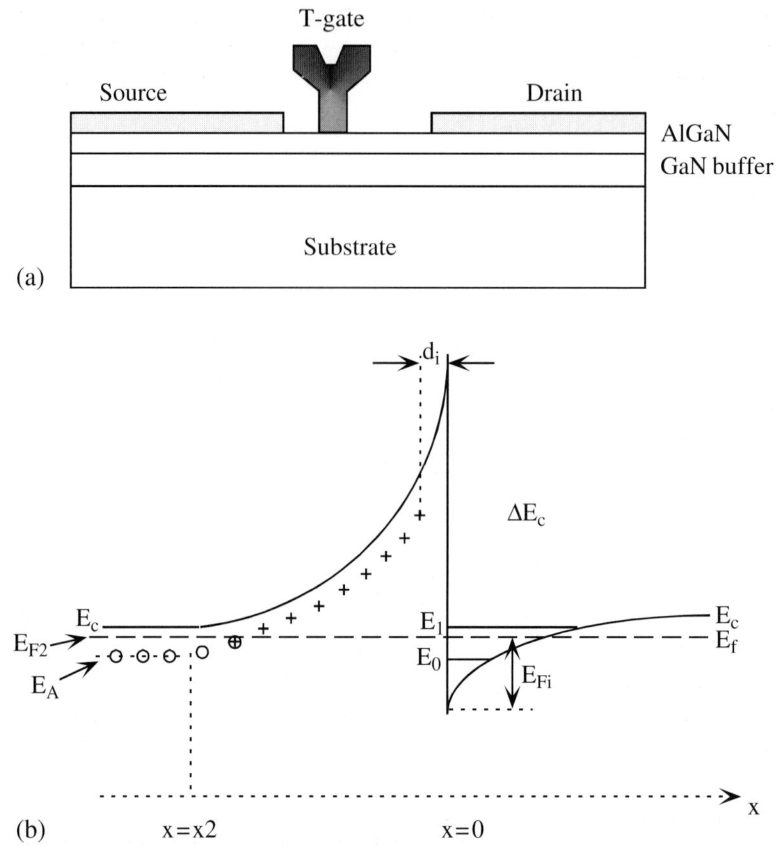

Fig. 14.22 (a) Schematic representation of an AlGaN/GaN modulation doped field effect transistor (MODFET). (b) Schematic band structure of an AlGaN/GaN modulation doped heterostructure in which the free carriers are provided to the GaN layer by the dopant impurities placed in the larger bandgap AlGaN barrier layer.

However, even when only the conduction band processes are of interest, EMA is still a very accurate means of determining the properties of interest. In fact, nitride based semiconductors in the wurtzite structure posses a conduction band with a Γ minimum, which can be described reasonably well within such an approximation. Within the effective mass theory Schrödinger's equation takes the form [100, 102, 108]

$$-\frac{h^2}{2}\frac{d}{dz}\left(\frac{1}{m(z)}\frac{d}{dz}\right)\varphi = \mathrm{e}V(z)\varphi = E\varphi, \qquad (14.2)$$

where $m(z)$ is the (position dependent) effective mass, V the electrostatic potential, φ the electron wave function and E the energy level. Non-parabolicity may induce deviations from the simple parabolic band model; however, this will not substantially change the results that we will present. In the nitride semiconductors with wurtzite phase, spontaneous and piezoelectric polarization effects are present [109], which necessitates that the Poisson equation be solved for the displacement field, $D(z)$

$$\frac{d}{dz}D(z) = \frac{d}{dz}\left(-\varepsilon(z)\frac{d}{dz}V(z) + P(z)\right) = e(p(z) - n(z) + N_D^+ - N_A^-) \quad (14.3)$$

where ε is the dielectric constant, P the total polarization, $n(p)$ the electron(holes) charge concentration and N_D^+ (N_A^-) the ionized donor (acceptor) density. In the self-consistent procedure, potential V is obtained using eqn (14.3) from an initial guess of the mobile charge concentration, and then inserted into the Schrödinger's equation, eqn (14.2), which is solved to get the energy levels and wavefunctions of the systems. The new electron charge density is obtained by applying Fermi statistics as follows,

$$n_{2D}(z) = \frac{m(z)k_B T}{\pi h^2}\sum_i |\varphi_i(z)| \ln[1 + e^{(E_F - E_i/k_B T)}] \quad (14.4)$$

where E_F is the Fermi level, E_i the energy of the ith quantized level, T the temperature and k_B the Boltzman constant. The calculated density is than plugged into Poisson equation (eqn 14.3) and the iteration repeated until convergence is achieved. Convergence of the self-consistent algorithm can be improved adopting special relaxation techniques. Here we have used a first order expansion of the model reported in ref [110].

In the following, we consider two MODFET structures, namely a single heterojunction AlGaN/GaN Normal Modulation Doped FET (NMODFET) where the AlGaN donor layer is grown on top of the GaN channel layer, and an 'inverted' GaN/AlGaN/GaN MODFET (IMODFET) where the channel layer is grown on top of the AlGaN donor layer. The NMODFET structure consists (from the gate to the substrate) of a 150 Å n-doped ($n=10^{18}\,cm^{-3}$) AlGaN, 50 Å unintentionally doped AlGaN layer and a thick GaN buffer. The IMODFET consists (from the gate to the substrate) of a 300 Å unintentionally doped GaN layer, 50 Å unintentionally doped AlGaN, 150 Å n-doped

Table 14.3 Bandgap and conduction band discontinuities with respect to GaN of the $Al_xGa_{1-x}N$ layer.

x (Al)	E_G [eV]	ΔE_C [eV]
0.1	3.62	0.17
0.2	3.85	0.33
0.3	4.09	0.51
0.4	4.35	0.69

($n=10^{18}\,cm^{-3}$) AlGaN, 300 Å unintentionally doped AlGaN layer, and a thick GaN layer.

We consider a residual doping of 1017 cm^{-3} in both GaN and AlGaN layers. We use a Schottky barrier (ϕ_B) of 1.1 eV for the metal-GaN interface and a $\phi_B=1.2$ eV for the metal–AlGaN interface. Calculations have been performed for $Al_xGa_{1-x}N$ with Al concentrations of $x=0.1$, 0.2, 0.3, 0.4. Both [0001] and [000$\bar{1}$] growth directions are considered. In the simulations, we have used an effective mass of 0.19 for electrons and 1.8 for holes in both GaN and AlGaN layers. The band gaps and band discontinuities of the AlGaN layers used are tabulated in Table 14.3.

As discussed previously the presence of polarization is quite important in the nitride based N-MODFET. The conduction band-edge profile for the NMODFET grown in [000$\bar{1}$] direction is depicted in Fig. 14.23 for the cases (i) with both spontaneous and piezoelectric polarization fields, (ii) without considering the polarization fields, and (iii) with only the piezoelectric polarization fields. The difference in piezoelectric and spontaneous polarization between AlGaN and GaN layer manifests itself as a fixed 2D-charge density at the interface between the two materials. For the [0001] growth direction considered in figure the difference in polarization between the two materials induces a positive charge ($\sigma=+1.12\times10^{13}\,cm^{-2}$) at the $Al_{0.2}Ga_{0.8}N$/GaN interface. Electrons are attracted by this positive charge, tending to accumulate at the interface and thus forming a conductive channel. Moreover, the high electric field due to the interface-charge, favors the build-up of a large channel density and of a strong channel confinement. Within the lGaN layer,

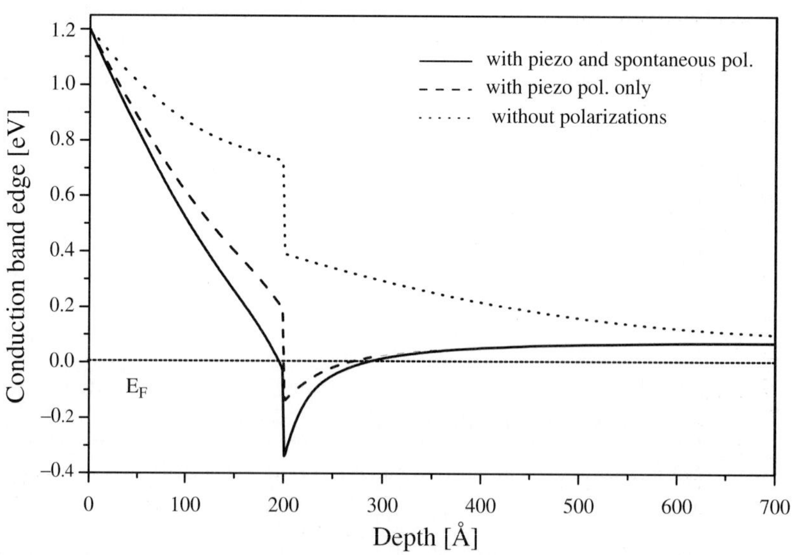

Fig. 14.23 Conduction band edge for the N-MODFET structure grown in the [0001] direction for $V_G=0$ with and without polarization fields.

the strong electric field compensates the space-charge contribution coming from the ionized donors. Consequently, it prevents the appearance of the parasitic channel that would otherwise form in the doped AlGaN layer [103, 111].

The comparison reported in Fig. 14.23 between the three cases with different contribution of the polarization fields shows the importance of considering both spontaneous and piezoelectric polarizations in GaN-based device modeling. In fact, neglecting the spontaneous polarization, as was done recently [112–115], the channel electron density is underestimated [103]. Clearly, the sign of the polarization charge is crucial. For the same N-MODFET structure grown in the $[000\bar{1}]$ direction, the resulting polarization charge would be negative (with the same magnitude) and electrons would be repelled from the channel as shown in Fig. 14.24.

The distribution of the free electron charge in the channel is shown in Fig. 14.25 for several values of the Al concentration of the AlGaN layer. Increasing the Al content induces a larger polarization charge at the GaN/AlGaN interface and consequently a higher channel electron concentration.

The calculations we have shown so far are obtained by considering only the polarization charge at the AlGaN/GaN interface. In reality, however, polarization charges that form at the metal–AlGaN and at the end of the GaN buffer region should be accounted for. The metal–AlGaN charge is completely screened by the charges induced on the metal surface and can therefore be neglected. On the other hand, the charges at the end of the buffer region may induce large deviation with respect to the situation depicted above. Oberhuber *et al.* [103] have considered $a-\sigma/2$ charge at the interface between

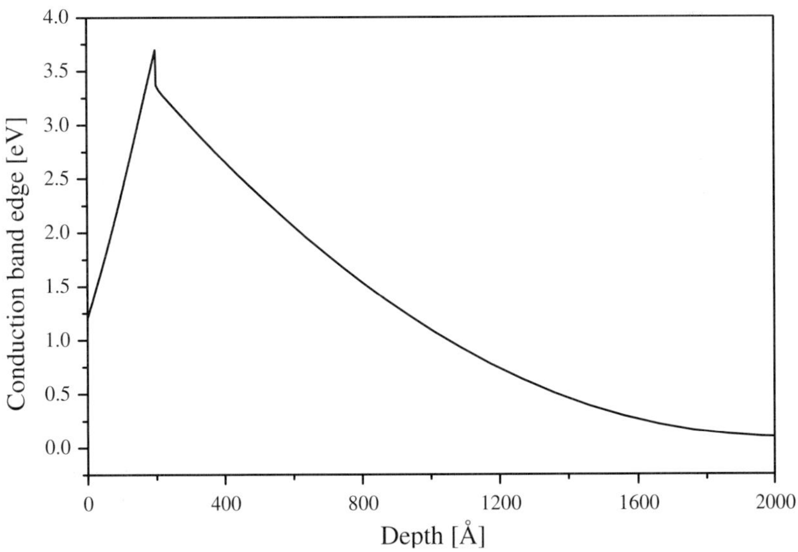

Fig. 14.24 Conduction band edge for the N-MODFET structure grown in the $[000\bar{1}]$ direction for $V_G=0$.

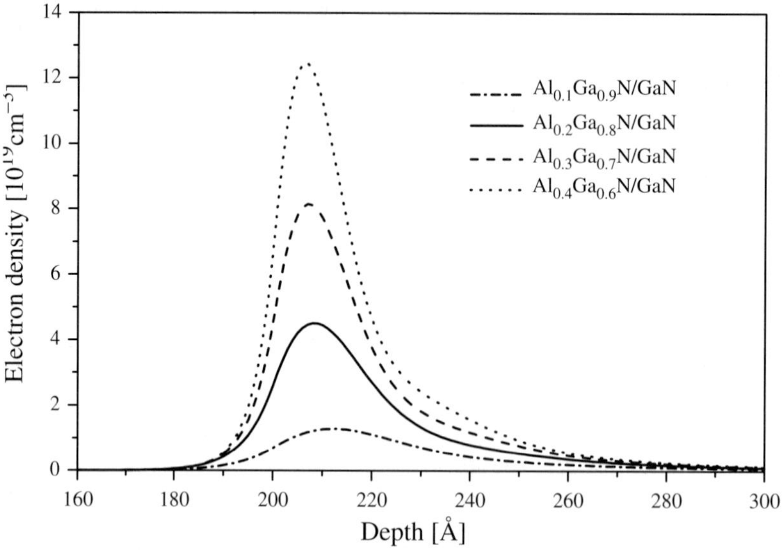

Fig. 14.25 Electron density distribution in the channel of the [000$\bar{1}$] grown N-MOD-FET for V_G=0 for several Al contents of the $Al_xGa_{1-x}N$ layer.

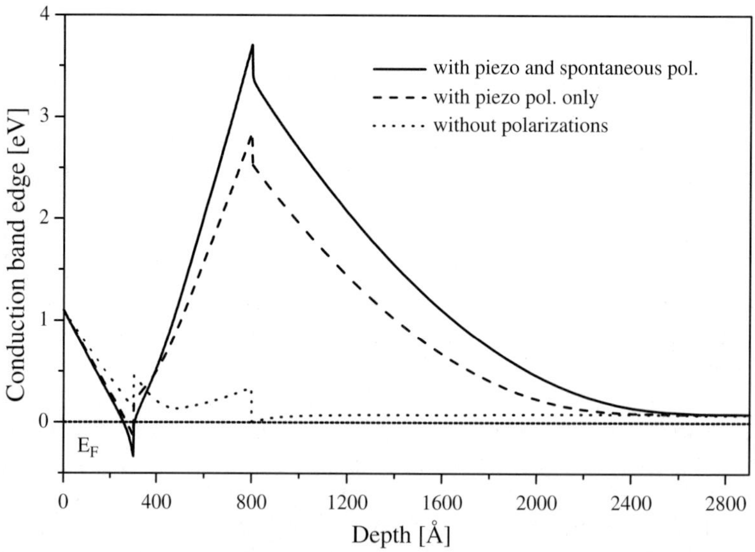

Fig. 14.26 Conduction band-edge for the I-MODFET structure grown in the [000$\bar{1}$] direction for V_G=0 with and without polarization fields.

the GaN and a nucleation region. The exact amount of such charge depends, however, on the morphology of the heterojunction and may differ from the theoretical value $\sigma=\Delta P/e$. The situation is less critical if the bottom interface is far away from the main AlGaN/GaN heterojunction. In this case, the

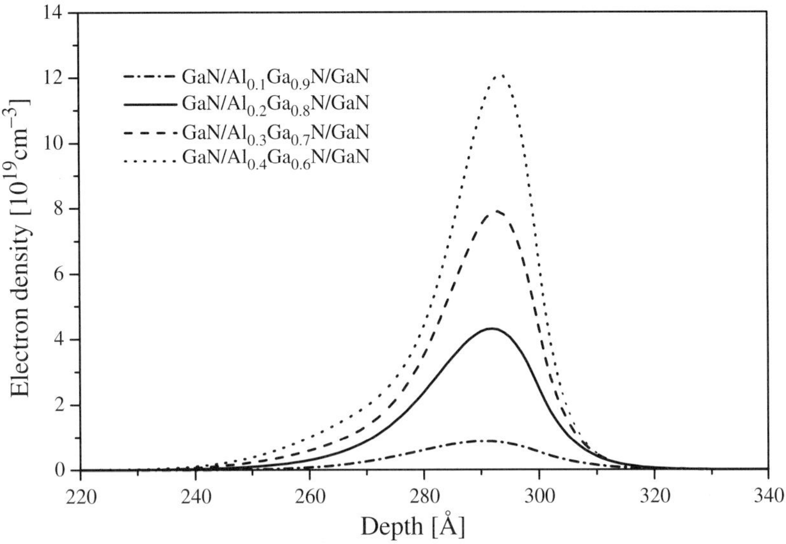

Fig. 14.27 Electron density distribution in the channel of the $[000\bar{1}]$ grown I-MODFET for $V_G=0$ for several Al contents of the $Al_xGa_{1-x}N$ layer.

polarization charge that arises can be completely screened by the residual doping of the GaN substrate. On the contrary, if such interface is close to the AlGaN/GaN heterojunction, the polarization charge can completely deplete the channel. In our simulations we have considered a thick GaN substrate. Thus, the effect of the polarization charge at the end of the GaN substrate is completely screened.

The band-edge profile and electron densities for the I-MODFET grown in the $[000\bar{1}]$ direction are shown in Figs. 14.26 and 14.27 respectively. A comparison of the conduction band-edges with and without polarization charges is also plotted. As for the N-MODFET, the presence of the fixed and positive polarization charge at the GaN/AlGaN interface induces the formation of a channel not present in absence of the polarization charge. For the I-MODFET $a-\sigma$ polarization charge is also present at the end of AlGaN region (i.e. at the AlGaN/GaN interface). Similar to the [0001] grown N-MODFET, a larger Al content of the AlGaN layer induces a larger polarization charge at the GaN/AlGaN interface and consequently an increasing of electron concentration in the channel. Naturally, for [0001] orientation the interface-charge forms below the AlGaN layer, which is not what is desired for an I-MODFET. What is desired is the formation of the electron sheet layer on top of the AlGaN layer, which is possible when the $[000\bar{1}]$ orientation is employed. The structure in its present shape, i.e. the [0001] polarity would show FET performance providing that the AlGaN layer is completely depleted but with small transconductance. If the AlGaN is not depleted, then the device would

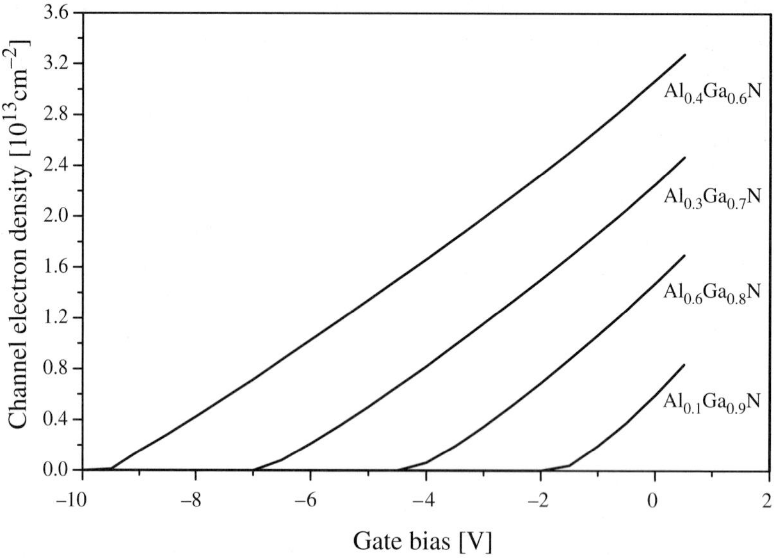

Fig. 14.28 Channel electron density as a function of Al and V_G for the [000$\bar{1}$] grown N-MODFET.

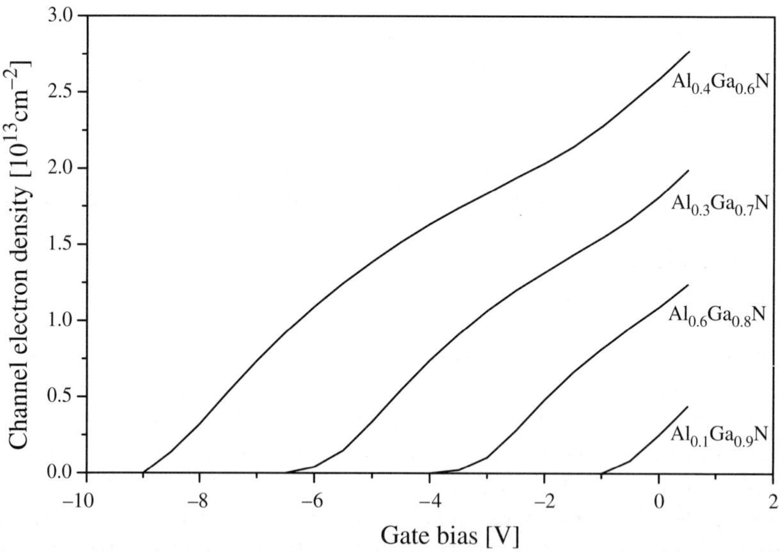

Fig. 14.29 Channel electron density as a function of Al and V_G for the grown I-MODFET.

function as a MESFET dominated by transport in the AlGaN layer unless the gate potential is large enough to deplete the AlGaN layer. To eliminate the formation of an interface electron charge at the bottom of the AlGaN layer, the bottom heterointerface should be graded substantially. In that case, the [0001] polarity would cause the band diagram to accumulate holes at the top interface if they are present. That top interface would accumulate electrons in the $[000\bar{1}]$ polarity.

The channel charge density is therefore controlled by two factors: (i) the Gate bias as in traditional N-MODFET device, and (ii) the Al content of the AlGaN layer, which tailors the polarization field. Charge control in nitride-based devices can be achieved by adjusting two independent parameters and thus offers a wide degree of flexibility with respect to traditional devices. This can be seen from the sheet charge concentration in the channel as obtained by integrating the electron density distribution along the z-direction. Considering the explicit dependence of the sheet charge density on the gate voltage V_G, we have:

$$n_s(V_G) = \int n(V_G,z)dz. \tag{14.5}$$

Figures 14.28 and 14.29 show the sheet electron density in the channel as a function of gate bias for several Al contents in the AlGaN layer for both $[000\bar{1}]$ grown N-MODFET and $[000\bar{1}]$ grown I-MODFET, respectively. Naturally, the channel electron density increases for larger Al contents of the AlGaN layer. We note also that the density is higher for the N-MODFET with respect to the I-MODFET because the particulars relating to the band bending on the top interface of the I-MODFET. As mentioned earlier, the I-MODFET structure is intended to be used for investigative purposes only as the body of work in the AlGaAs/GaAs system showed the N-MODFET to be desired device structure.

14.10 Drain current model in MODFETs

We have implemented a quasi-2D [116, 117, 118] model for the calculation of the current-voltage characteristics of the nitride MODFETs. This model makes use of the exact value of the sheet charge density in a MODFET device channel, obtained from the self-consistent Schrödinger–Poisson solution presented above.

We assume a FET model shown in Fig. 14.30 where the x-axis is along the channel and the z-axis is along the growth direction. The model also considers the presence of a drain (R_D) and source (R_S) resistance. When a drain bias (V_D) is applied, the potential along the channel may be considered as varying gradually from the source bias (V_S) to V_D. In this situation it is still possible to calculate the sheet-charge density n_s at every section grid, provided that one considers the $V(x)$ potential (on the top surface) for each point of the channel. Since for n-channel devices, V_D is positive and V_S is zero, $V(x)$ contributes to the channel depletion and the sheet charge density n_s for the generic x section of the FET will therefore be:

$$n_s(x) = n_s[V_G - V(x)]. \tag{14.6}$$

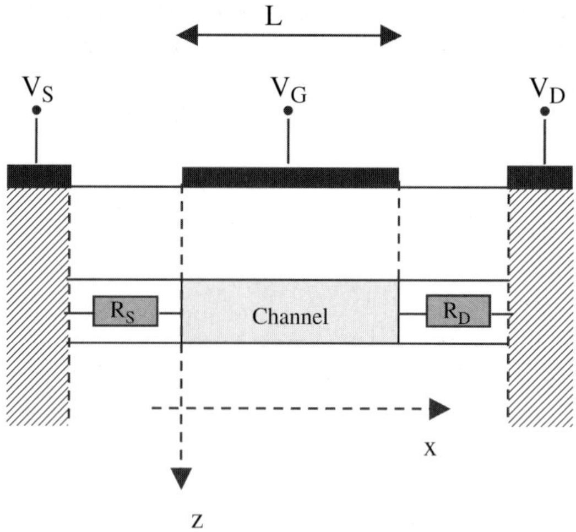

Fig. 14.30 Schematic representation of the quasi-2D FET model used.

By neglecting diffusion contributions, the source-to-drain current I_{DS} is given by:

$$I_{DS} = -qWv(x)n_s(x), \tag{14.7}$$

where W is the gate width and $v(x)$ the electron mean velocity, supposed independent from transverse coordinate.

The dependence of the drift velocity on the longitudinal electric field is empirically given by:

$$v(x) = \frac{\mu_0 F(x)}{1 + [F(x)/F_C]}, \tag{14.8}$$

where $F(x)$ is the electric field $(=-dV(x)/dx)$, μ_0 is low field mobility and $F_c = v_{sat}/\mu_0$ is the electric field at saturation. Parasitic components are included explicitly through the drain and source resistances (R_D, R_S)

$$\begin{aligned} V_S^e &= V_S + I_{DS}R_S \\ V_D^e &= V_D - I_{DS}R_D \end{aligned} \tag{14.9}$$

where V_D^e and V_S^e represent the effective bias boundaries of the gate region on the drain and source sides, respectively. For a certain value of I_{DS}, we can calculate the V_D by solving eqn (14.7).

The explicit equation for the current is:

$$I_{DS} = qW\frac{\mu_0 F(x)}{1 + [F(x)/F_C]}n[V_G - V(x)]. \tag{14.10}$$

The numerical solution is based on the discretization of this expression into N sections, each with amplitude h, so that $Nh=L$, where L is the gate length. Given the $(i-1)$th section potential, the ith potential $V_i=V_{i-1}+F_ih$ where F_i is the ith section electric field. We have then the N relations:

$$I_{DS} = qW\frac{\mu_0 F_i}{1 + (F_i/F_C)}n(V_G - V_{i-1} - F_ih). \qquad (14.11)$$

Since the $(i-1)$th section potential is known from the previous step, this is a nonlinear equation in the unknown F_i. Solving iteratively for all the N sections, one obtains the value of the drain voltage V_D consistent with the assumed current.

Repeating this procedure for a suitable range of values of I_{DS}, one obtains the set of corresponding values of V_{DS} and thus the MODFET I–V characteristics, which are elaborated on below.

14.11 I–V characteristics

In this section we discuss the simulated I–V characteristic of the normal and inverted MODFET, obtained for a gate length of $L=0.3\,\mu m$. We have chosen a drain and source contact resistivity of about $1\,\Omega\,mm$ which is consistent with experimentally measured values on these types of devices [119]. We use a saturation velocity of $2.5\times10^7\,cm/s$ [22], while for the low field mobility we choose a value of $\mu_0=1100\,cm^2/Vs$, slighter higher than the GaN bulk value, according to the experimental and theoretical results for similar devices [22, 103, 120, 121].

In Fig. 14.31, we show the I_{DS} vs. V_{DS} for the [0001] polarity MODFET for several gate (V_{GS}) voltages. The results are presented for both $x=0.2$ [Fig. 14.31(a)] and $x=0.4$ [Fig. 14.31 (b)] Al concentration of the top layer. For $x=0.2$, the MODFET reaches pinch-off for a bias voltage of $V_{GS}=-4.4\,V$ while for $x=0.4$ the pinch-off is reached at $V_{GS}=-9.5\,V$. On the other hand, the saturation drain current for $x=0.2$ is $I_{DS}=2.4\,A/mm$ at $V_{GS}=0$ and it increases up to 5.76 A/mm for an Al content of $x=0.4$. Thus, the current flowing in the devices depends strongly on the Al contents of the top layer. This is essentially due to the increasing of the channel electron density induced by the increasing of the polarization charge going from an Al content of 0.2 up to 0.4. This peculiarity of the MODFET should be considered in the design of these devices since fluctuation of the alloy composition of the top layer may induce large variation with respect to nominal electrical values of the device. It should also be pointed out that the gate leakage would determine the extent of gate voltage that can be applied to the gate. For a gate bias of 9.5 V and AlGaN layer thickness of 20 nm, the vertical field under the gate near the source can reach 4.75 MV/cm. This means that MODFETs utilizing large mole fractions of Al may require thin AlGaN layers or recessed gates to keep the gate voltage smaller.

A similar situation is obtained for the I-MODFET with the GaN/AlGaN/GaN structure grown in the $[000\bar{1}]$ direction, meaning with N-polarity.

The calculated I_{DS} vs. V_{DS} characteristics are reported in Fig. 14.32(a) and (b) for $x=0.2$ and $x=0.4$ Al composition of the AlGaN layer, respectively. Also in this case the pinch-off bias depends critically on the Al composition and varies from −3.9 V for $x=0.2$ up to −9.0 for $x=0.4$. Saturation currents are lower for the I-MODFET at $x=0.2$ with respect to the equivalent MODFET structure. Such difference, however, is negligible for the case with $x=0.4$.

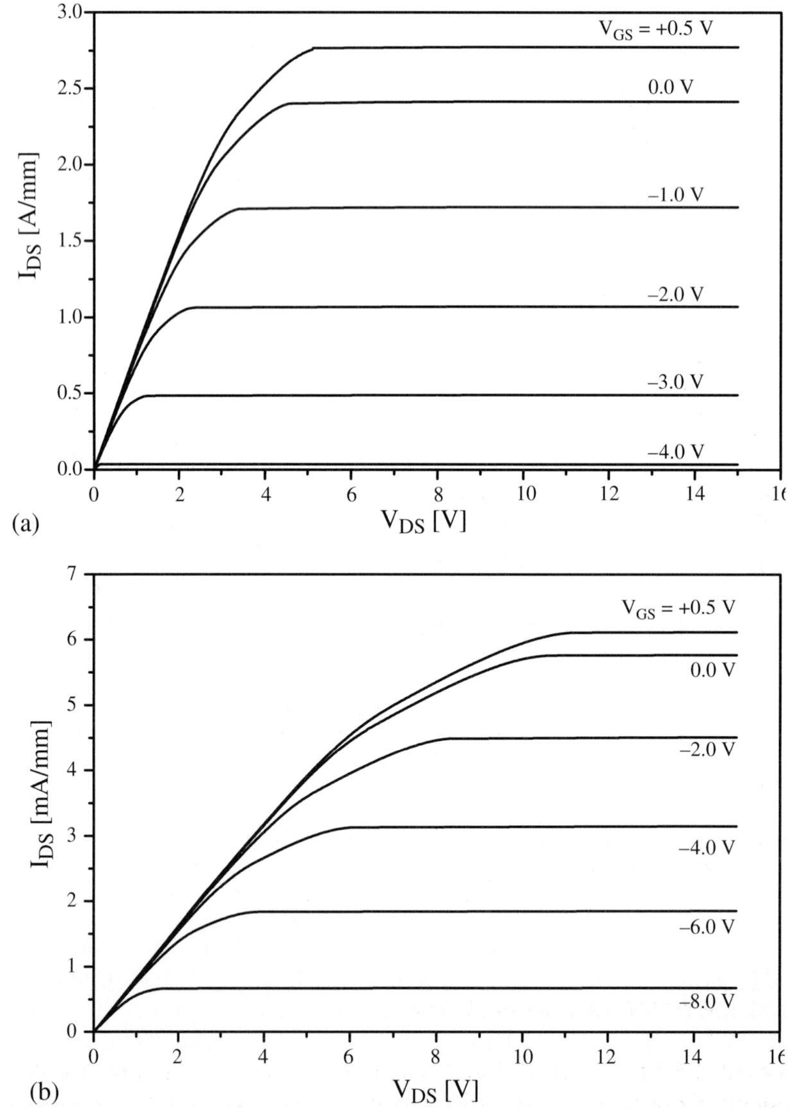

Fig. 14.31 (a) I_{DS} vs. V_{DS} for the [0001] polarity Normal MODFET for several gate (V_{GS}) voltages and for $x=0.2$. (b) I_{DS} vs. V_{DS} for the [0001] polarity Normal MODFET for several gate (V_{GS}) voltages and for $x=0.4$.

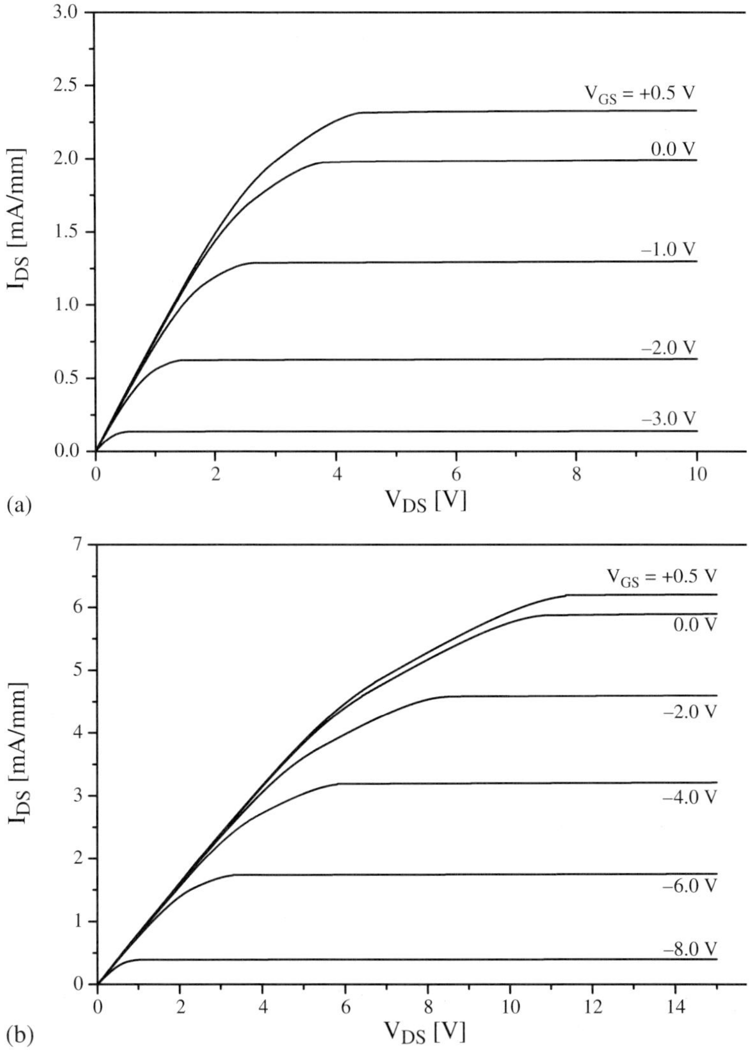

Fig. 14.32 (a) I_{DS} vs. V_{DS} for the $[000\bar{1}]$ polarity Inverted MODFET for several gate (V_{GS}) voltages and for $x=0.2$. (b) I_{DS} vs. V_{DS} for the $[000\bar{1}]$ polarity Inverted MODFET for several gate (V_{GS}) voltages and for $x=0.4$.

14.12 Experimental considerations

GaN-based FETs are intended primarily for power application at high frequencies. Consequently, traditional small signal considerations have to be augmented by large signal specific issues. The main parameter facing a power device is the maximum power level that can be obtained and the associated gain. In many applications the noise figure of the device must also be considered. In simple terms if the device has large drain breakdown voltage, high gain at high frequencies, and high drain efficiency, the stage is set for a desirable device. Even in a well-designed device in a semiconductor with all the accolades, the thermal wall is a very formidable one. Thus, it is imperative that the effect of temperature on device performance is accounted for accurately. As in small signal modeling, the first step in power modeling is to establish the basic device geometrical factors that are needed to calculate the current voltage characteristics. Once these are known, the output characteristics superimposed with the load line can be used to estimate the power level that can be obtained from the device provided that it is not limited by the input drive as shown in Fig. 14.33. In Class-A operation, the maximum power that can be expected from the drain circuit of a device is given by

$$P_{\mathrm{max}} = \frac{I_{\mathrm{dson}}(V_b - V_{\mathrm{knee}})}{8} \tag{14.12}$$

where I_{dson} is the maximum drain current (this is the drain current with a small positive voltage on the gate electrode), V_b is the drain breakdown voltage, and V_{knee} is the knee voltage as shown in Fig. 14.33. The allowable positive gate voltage ($\approx 1\,V$) will depend on the channel doping and the work function of the gate metal. The positive gate voltage is limited by the onset of forward Schottky-diode current. The DC load line shown in Fig. 14.33 would be used in a Class-A RF amplifier with the maximum drain voltage $V_D = V_b/2$. The slope of the load line is $1/R_L$ where R_L is the value of the load resistance

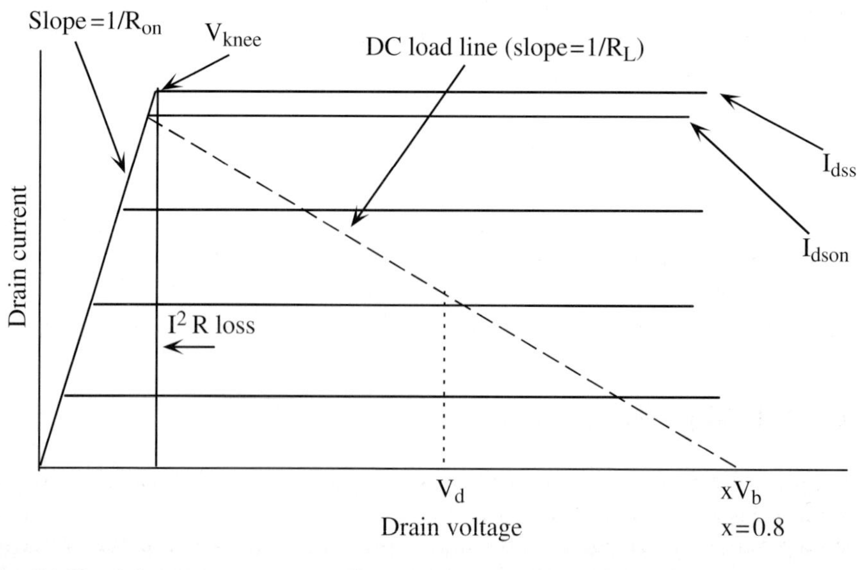

Fig. 14.33 Schematic representation of I–V characteristics with a load line for a Class-A operation.

at the output of the FET. What can be gleaned from eqn (14.12) is that V_b and I_{dson} must be made as large as possible. The utility of wide bandgap semiconductors such as GaN in this juncture is that the drain breakdown voltage is larger than that in conventional Group III–V semiconductors. In general, the drain can be swung to voltages up to within 80% of the drain breakdown for a 20% margin of safety. It should be pointed out that the maximum drain current in nitride semiconductor-based MODFETs is in the same ballpark as that of more conventional semiconductors. This implies that increased power handling capability is a direct result of large breakdown voltages and thermal conductivity and the fact that higher junction temperatures can be tolerated. Ability to increase drain bias increases the load resistance and makes it easier to impedance match, particularly in devices with large gate widths.

In power devices, power dissipation within the device increases the junction temperature and alters the output characteristics. On the one hand, higher junction temperatures with respect to the case temperature would enhance the heat dissipation to the power of four of the temperature differential, but along with it comes reduced current and increased series resistances, which in turn increase the heat dissipation. Moreover, the thermal conductivity of the semiconductor decreases with increased temperature, exacerbating the situation. Consequently, the effect of junction temperature on the output characteristics must be taken into consideration. Temperature-dependent material parameters, if known, can be used to calculate the output characteristics with respect to temperature. However, a more pragmatic approach can be taken, particularly when the aforementioned parameters and or models required are not available, in which one measures the output characteristics of the device under consideration as a function of temperature. The junction temperature is critically dependent on the substrate thermal conductivity that is available for various substrates including GaN [1]. The functional dependence of thermal conductivity on temperature is

$$\chi(T) = (T_0)\,(T/T_0)^{-r} \tag{14.13}$$

where the coefficient r is 0.559, 0.443, 0.524, and 0.544 for Si, GaAs, SiC, and sapphire, respectively [122]. Thermal conductivity of sapphire, SiC, GaAs, and Si as a function of temperature are shown in Fig. 14.34. In Fig. 14.34, $\chi(T_0)$ has also been appropriately reduced to account for the doping of the substrate material.

14.13 Schottky barriers for gates

Any semiconductor device requires metal contacts, and MODFETs are no exception. These devices require ohmic source and drain contacts as well as a rectifying Schottky barrier for controlling the charge in the channel. Schottky barrier-related processes for GaN-based devices are nascent, but rapid progress is being made. Until recently it has been difficult to fabricate good

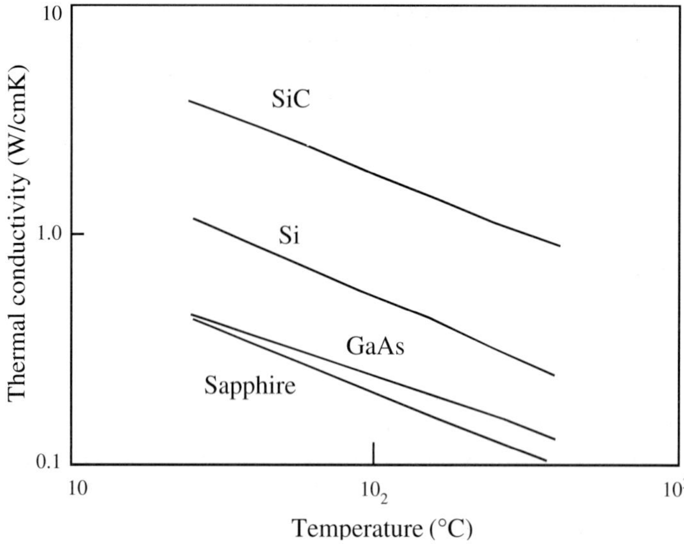

Fig. 14.34 Thermal conductivity vs. temperature for SiC and sapphire. After ref. [122].

quality single-crystal films on which a Schottky metal could be deposited, and upon which the properties of Schottky barriers could be studied. However, considerable progress has been made with Pt-GaN Schottky barriers [123, 124] that have been successfully implemented in GaN-based MODFETs.24'-[125, 126].

Recent successes in growing good-quality single-crystal Group III–V GaN layers prompted the fundamental electrical property studies of metal–semiconductor barriers on GaN. In order to determine the properties of only the metal–semiconductor junction, one must be able to model the semiconductor. Semiconductors with large defect concentrations are notorious for exhibiting parasitic processes in current-voltage and capacitance-voltage characteristics that cloud the picture. Consequently, good epitaxial layers as well as good metal–semiconductor interfaces are imperative. During the evolutionary period, while the sample quality is acceptable, temperature and frequency dependence of the capacitance-voltage characteristics and temperature- dependent current-voltage characteristics are measured and analyzed for determining effective metal–semiconductor barrier height. To get a large Schottky barrier height for rectifying metal contacts on GaN, which is imperative for low leakage, metals with large work functions such as Au and Pt [123] have been explored. Hacke *et al.* [127] have studied Schottky barriers made of Au on unintentionally doped *n*-GaN grown by HVPE. The forward current ideality factor was n_{idl} ~1.03 and the reverse bias leakage current was $<10^{-10}$ A at a reverse bias of −10 V. While the current-voltage measurement indicated the barrier height to be 0.844 eV, the capacitance measurements led to a value of 0.94 eV.

Suzue *et al.* [123] and Mohammad *et al.* [124] have studied the Pt Schottky barriers on unintentionally doped *n*-GaN. Temperature-dependent current-voltage and capacitance-voltage characteristics in the range of –195 °C to 42 °C were studied to gain insight about the current conduction mechanism. Any excess current observed is traditionally attributed to defects (generation recombination centers) and surface leakage current. The ensuing current is called the Shockley–Read–Hall (SRH) recombination current resulting from the mid-gap states. If one neglects this excess current, a barrier height of about 0.8 eV is deduced as opposed to about 1 eV deduced from the CV measurements. Because of the effect of excess current on the slope of the I–V curve, CV measurements in this particular case may represent the metal barrier height. An examination of the CV plots, however, indicated that under reverse bias condition, the capacitance depended insignificantly on the sample temperature and the signal frequency as shown in Fig. 14.35(a) and (b). This leads one to conclude that the density of traps has been lowered. The curves corresponding to all temperatures were largely linear, which yielded barrier heights ranging between 0.95 and 1.05 eV. Reduced capacitance with decreasing temperature is consistent with relatively deep donors. Binari *et al.* [128] determined Ti Schottky barrier heights to be 0.58 and 0.59 eV from the current-voltage and capacitance measurements, respectively. The ideality factor n_{idl} is approximately 1.28. The diode series resistance (R_S) is 100 Ω.

The ternary $Al_xGa_{1-x}N$ is an essential component of nitride-based MODFETs, which makes the investigation of metal $Al_xGa_{1-x}N$ contacts imperative. M. A. Khan *et al.* [129] reported the fabrication of a Cr/Au Schottky barrier on *n*-AlGaN. Moreover, M. R. H. Khan *et al.* [130] studied the Schottky barrier characteristics of the Au-$Al_xGa_{1-x}N$ system. A typical current-voltage characteristic of an $Al_{0.14}Ga_{0.86}N$ Schottky diode had an ideality factor of 1.56 under reverse bias and a threshold voltage of about 0.9 V at 0.1 A. The reverse bias leakage current was recorded to be marginally low (10^{-10} A) for a reverse bias of –10 V. By using the current-voltage method, the barrier height and the electron affinity were determined to be 0.94 eV and 4.16 eV, respectively. From the C^{-2} vs. V plot, the same barrier height and the electron affinity were deduced to be 1.3 ± 0.05 eV and 3.8 eV, respectively. As the AlGaN quality increases, periodic as well as more in-depth investigations will become necessary to get an accurate picture of intrinsic parameters. In short, the current conduction mechanism in metal–semiconductor structures is strongly affected by surface and bulk states. Deviations from an ideal ideality factor, such as is the case here, indicates such states. The situation gets more complicated with AlGaN and gets worse as the AlN mole fraction is increased. Likewise, capacitance-voltage measurements also are affected by states that are charged, either by interface state or by bulk state. As is the case in many facets of research and development, insights into the metal–nitride contacts will be gained in an evolutionary manner hinging upon the developments in nitride layers.

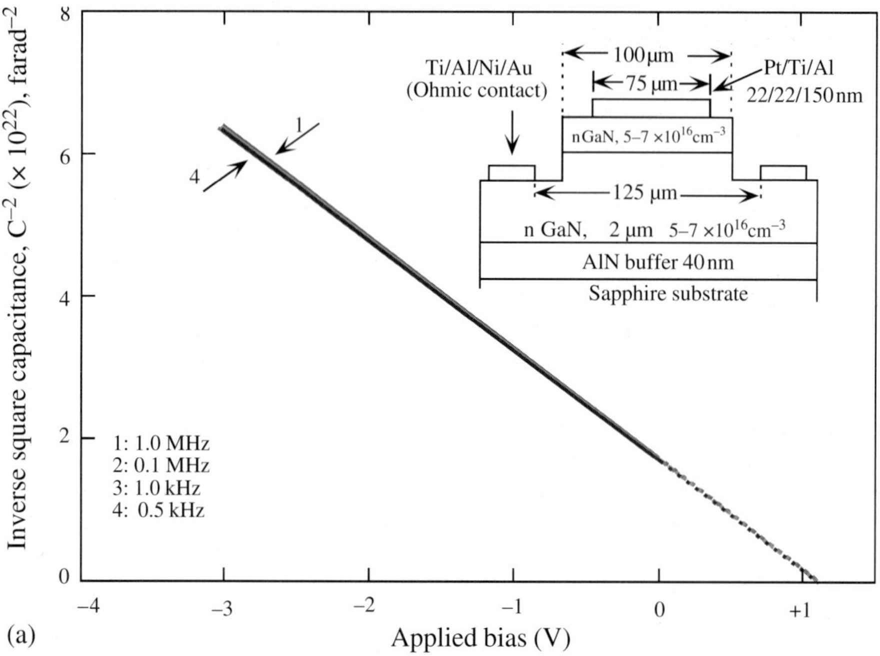

(a)

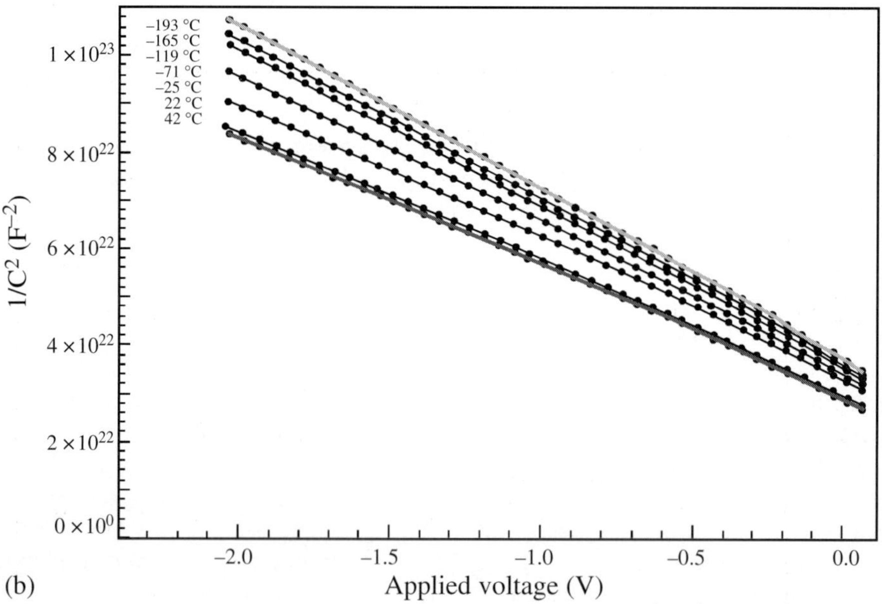

(b)

Fig. 14.35 (a) Variation of inverse square capacitance, C^{-2} vs. applied bias V for various signal frequencies ω in Pt-GaN Schottky barriers. Curves 2 and 3 lie in between curves 1 and 4, respectively; (b) Temperature dependence of the C^{-2} vs. applied bias V in Pt-GaN Schottky barriers.

14.14 Contacts to GaN

Ohmic contacts in power devices are extremely important because they affect their efficiency as well as heat dissipation. Initial inferior results helped fuel concerns that GaN-based electronic devices may not perform well. Early specific contact resistivities on *n*-type GaN using Al and Au metallizations [131] were in the range of 10^{-4} and $10^{-3}\,\Omega\,\mathrm{cm}^2$. Major improvements were realized by using Ti/Au [132] and TiAl [133], in that specific contact resistivities in the high $10^{-6}\,\Omega\,\mathrm{cm}^2$ were obtained with the latter. Carrying the TiAl contact work one step further, Wu *et al.* [134] confirmed that, except at very high annealing temperatures, the ohmic contact suggested by Lin *et al.* [133] functions very effectively. At very high temperatures, Al of the metal contact melts and tends to ball up, resulting in rough surfaces and increased ohmic contact resistances as pointed out already by Lin *et al.* [133]. In an attempt to circumvent this difficulty, Wu *et al.* [134] designed a separate layer-metallization method where a realignment and deposition of a second thin Ti layer, and a 2000 Å Au overlayer were carried out. Specific contact resistivities were in the range of $3.0\times10^{-6}\,\Omega\,\mathrm{cm}^2$ and $5.5\times10^{-6}\,\Omega\,\mathrm{cm}^2$, depending on the doping concentration in the semiconductor.

In an attempt to obtain improved ohmic contacts, Fan *et al.* [135] have designed a multilayer ohmic contact method. By using a composite metal layer of Ti/Al/Ni/Au (150 Å / 2200 Å / 400 Å / 500 Å), they obtained very low contact

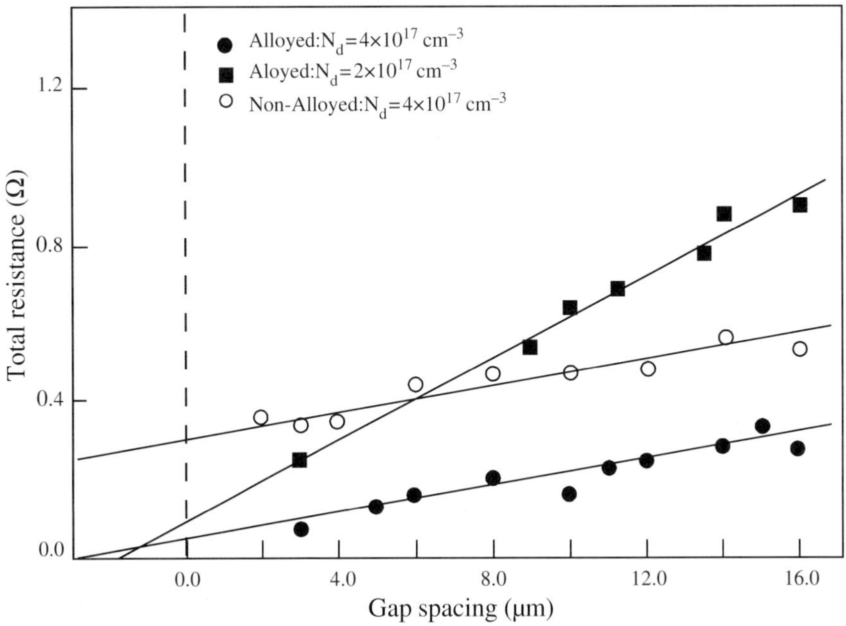

Fig. 14.36 Least-squares linear regression of the total resistance between the two adjacent ohmic contact pads in multiple-layer Ti/Al/Ni/Au ohmic contacts on GaN.

resistivities. Specifically, for *n*-GaN with doping levels between $2\times10^{17}\,cm^{-3}$ and $4\times10^{17}\,cm^{-3}$, they obtained specific contact resistivities in the range of $\rho_S=1.19\times10^{-7}\,\Omega\,cm^2$ and $8.9\times10^{-8}\,\Omega\,cm^2$. The resistance R_T between the two contacts was measured at 300 K using a four-point probe arrangement. The contact resistivity ρ_S was derived from a plot of R_T vs. gap length. The method of least squares was used to fit a straight line to the experimental data. These straight lines, and the actual experimental results for both alloyed and non-alloyed contacts are shown in Fig. 14.36. Calculation of the contact resistivity was based on the assumption that the semiconductor sheet resistance underneath the contacts remains unchanged, which is not true for non-alloyed contacts. As for the current conduction mechanism in these ohmic contacts, the large metal–semiconductor barriers diminish the possibility of thermionic-emission-governed ohmic contacts to GaN. The alternative mechanism is some form of tunneling that may take place if GaN is so heavily doped as to cause a very thin depletion region. Tunneling is possible if, due to annealing (for example, at 900 °C for 30 s), Al and Ti along with Ni undergo substantial interaction with each other and GaN. Investigations showed that Ti receives N from GaN, forming a metallic layer, while the lack of N on GaN provides the desired benefit of increased electron concentration through N vacancy formation [136]. Aluminum passivates the surface and also possibly reacts with Ti to form TiAl.

14.15 AlGaN/GaN MODFETs

To reiterate, MODFET's performance is due to the conduction channel that allows large sheet carrier concentrations to be maintained and its unique capacitance-voltage relationship [137]. Moreover, spatial separation of scattering centers (such as ionized donors) from the electrons leads to low field transport void of ionized impurity scattering. As discussed in detail above, what is somewhat unique to GaN and its alloys is the spontaneous and the strain-induced piezoelectric polarization that causes redistribution of mobile and weakly bound charge and charge collected from metal contacts.

14.16 Experimental performance of GaN MODFETs

Initial GaN MODFETs utilized the background donors in the AlGaN layer, the density of which is not controllable, to say the least, and any other free and weakly bound electrons drawn to the interface. Congruent with the early stages of development and the defect-laden nature of the early GaN and $Al_xGa_{1-x}N$ layers, the MODFETs exhibited very low transconductances (on the order of 20 mS/mm), large on resistances. In addition, they also exhibited a low-resistance state, which was relatively high to begin with, and a high-resistance state before and after the application of a high drain voltage (20 V). As in the case of GaAs/AlGaAs MODFETs, hot electron trapping in the larger bandgap material at the drain side of the gate is primarily responsible for the current collapse. The negative electron charge accumulated because of this trapping causes a significant depletion of the channel layer, more probably a pinch-off, leading to a drastic reduction of the channel conductance and the decrease of the drain current. This continues to be effective until the drain–source bias is substantially increased, leading to a space-charge injection and giving rise to an increased drain–source current.

With improvements in the materials quality available, the transconductance, current capacity, and drain breakdown voltage are all increased to the point where GaN-based MODFETs are now strong contenders in the arena of high power devices/amplifiers, particularly at X band and higher frequencies. As is the case for FET device structure, improved and high resistivity buffer layers have once again played a pivotal role. For chronological purposes, a brief review of the latest class of MODFETs with high transconductances and current levels is given later in this article.

The first breakthrough for N-MODFETs based on GaN came in the 1994–1995 time-scale in the author's (HM) laboratory [125]. These devices with a gate length of 3 µm and gate width of 40 µm, exhibited transconductances or about 120 mS/mm with low resistances as they sported doped AlGaN donor layers and low resistance ohmic contacts. The I–V characteristics of a typical device of the kind are shown in Fig. 14.37. Shortly thereafter, devices with a gate length of 2 µm, gate width of 40 µm, and the drain–source separation of 4 µm exhibited drain currents of approximately 500 mA/mm and extrinsic transconductances of approximately $g_{em}=185$ mS/mm. The drain breakdown voltage for 1 µm gate to drain spacing was approximately 100 V, the exact value depending on the layer design and quality of the layered structure. Soon thereafter, other laboratories achieved similar results in similar structures. What is unique to AlGaN/GaN MODFETs as compared to their GaAs varieties is the polarization charge discussed earlier. The polarization issues particularly the piezoelectric are used rather liberally. Terms such as piezoelectric doping have been coined. Very high sheet carrier concentrations observed have been ascribed to piezoelectric polarization. We have to recognize that ultimately, regardless of the source of the carriers, the strength of the electric field that can be accommodated by the semiconductor under the gate without excessive leakage will set an upper limit on the number of carriers at the interface. Use of multi-2DEG structures is one obvious method to increase the current capability of MODFETs, and they have been employed. In those cases, the GaN layer is straddled by two doped AlGaN molecules that donate electrons to the channel, thus increasing the number of electrons available for current conduction. By Hall effect measurement, the mobility and sheet carrier densities in the 2DEG were about 304 cm^2/Vs and 3.7×10^{13} cm^{-2}, respectively, at room temperature. The sheet carrier concentration may have been affected by a piezoelectric effect. A number of double hetero-channel MODFETs (DHCMODFETs) with gate lengths of 1.5 to 1.75 µm and a gate width of 40 µm have been reported.

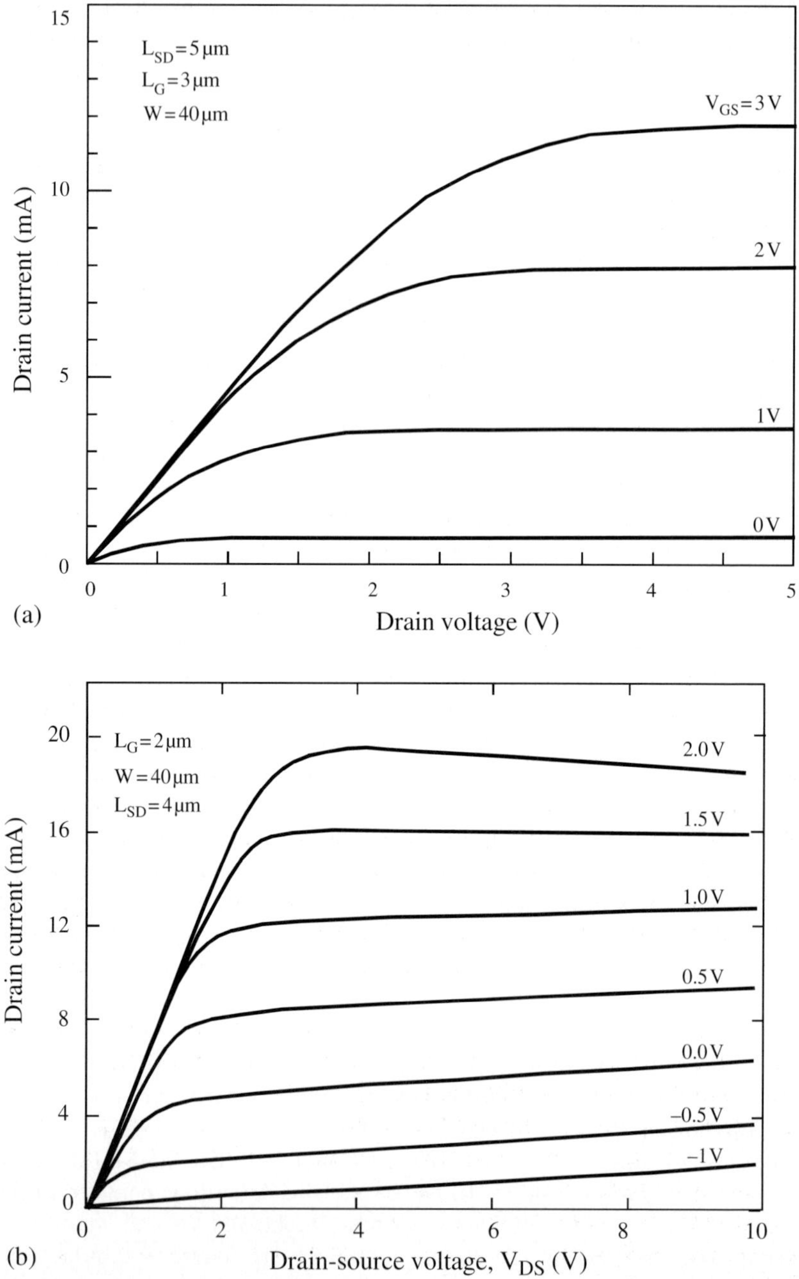

Fig. 14.37 (a) Output I–V characteristics of the first GaN MODFET, which exhibited respectable performance. The 3 μm gate device had a gate length of 3 microns and sported low resistance ohmic contacts and low leakage Schottky barriers. (b) Output I–V characteristics of a GaN MODFET on sapphire with a 2 μm gate, which exhibits negative output conductance due to thermal effects associated with the relatively low thermal conductivity of sapphire.

The maximum drain saturation current I_{DS} corresponding to a drain-source voltage V_{DS}=7 V, and gate source voltage V_{GS}=3.5 V in a DHCMODFET is about 1100 mA/mm, which is important because, in high power devices, the input is momentarily forward biased. The DHCMODFET has a room temperature extrinsic transconductance of g_m=270 mS/mm. The value of the total resistance R_T extracted from the linear region of the I–V curves is 4 Ω/mm. Near pinch-off, the drain breakdown voltage is about 80 V, indicating excellent power potential of the device. These measurements were made in a nitrogen-pressurized container to avoid possible oxidation of the contacts and probes. The maximum drain–source current and extrinsic transconductance of the DHCMODFET are 500 mA/mm and 120 mS/mm, respectively. These devices maintain reasonable output characteristics at temperatures as high as 500 °C with maximum drain current and extrinsic transconductance values of 380 mA/mm and 70 mS/mm, respectively. Cooling to room temperature restored the characteristics, which demonstrates the robustness of this material system and of the metallization employed. It should be noted, however, that high power operation requires large drain breakdown voltages with the added benefit of having large output resistances, which ameliorates impedance matching.

The heat dissipation is a major problem, however, in GaN MODFETs on sapphire substrates as the thermal conductivity of this substrate is about 0.3 W/cmK, and may be even somewhat lower. To make matters worse, the thermal conductivity decreases rapidly as the temperature increases. Consequently, devices show a decreasing drain current (negative differential output conductance) as the drain bias is increased, and needless to say, the power performance is degraded. To overcome this, one must either remove the sapphire substrate followed by mounting the structure on a substrate with better thermal conductivity, employ flip-chip mounting, or grow the structure on a substrate with better thermal conductivity. Among the substrates with better thermal conductivity are Si and in particular SiC. Layers on Si, however, are not of as high quality as one would like, which leaves SiC substrates, which are expensive and suffer from inferior surface—(characteristics or smoothness) due to the hardness of SiC.

The early attempts in the authors' laboratory to grow GaN layers on SiC met with difficulty due to the surface damage roughness, though occasionally very high mobility could be obtained [138].

Two approaches can be employed to remove the surface damage. One is the mechanical chemical polish, which is very slow in coming, and the other is etching in H and Cl environment at very high temperatures such as 1500 °C. The H cleaning process has been adopted as a standard procedure for MBE growth of GaN on SiC [139] with cleaning temperature of about 1700 °C. The authors, in collaboration with the group of E. Janzen at Linköping University, were able to H etch Leyl SiC followed by MODFET growth. These devices did not exhibit the negative differential resistance characteristic of the sapphire substrates. However, SiC substrates prepared by the sublimation method did not appear to survive this high temperature H etching process. Researcher have exploited the *in situ* H etching process [140] and HCl etching process [141].

Reports detailing these processes and their effects have appeared in the open literature already. The I–V characteristics of the particular device prepared in 1996 on SiC substrates are shown in Fig. 14.38. These results were reported in meetings dealing with the development of high power devices and the case was made for SiC substrates as intrinsically being better for GaN power MODFET applications. Several groups participating in those meetings expended a good deal of effort on SiC substrates with initially comparable high power performance to that on sapphire. Other groups later propelled GaN MODFETs on SiC substrate to their pinnacle with outstanding performance as will be discussed below. It should be pointed out that the pitch of gates for a power FET on a substrate with very good thermal conductivity can be made smaller than on a substrate with inferior thermal conductivity. Consequently, the chip size can be made much smaller in addition to other advantages.

MODFETs have progressed to a point where microwave measurements have been performed on a variety of devices with gate length as wide as 2 μm and as narrow as about 0.2 μm. A typical MODFET structure with 2 μm gate lengths has been tested for small-signal S-parameters performed at bias conditions used for the power measurements (i.e. 15 V, –2.5 V, and 20 mA for the drain voltage, gate voltage, and drain current, respectively). The unity current gain cutoff frequency (f_T) and maximum frequency of oscillation (f_{max}) were 6 GHz and

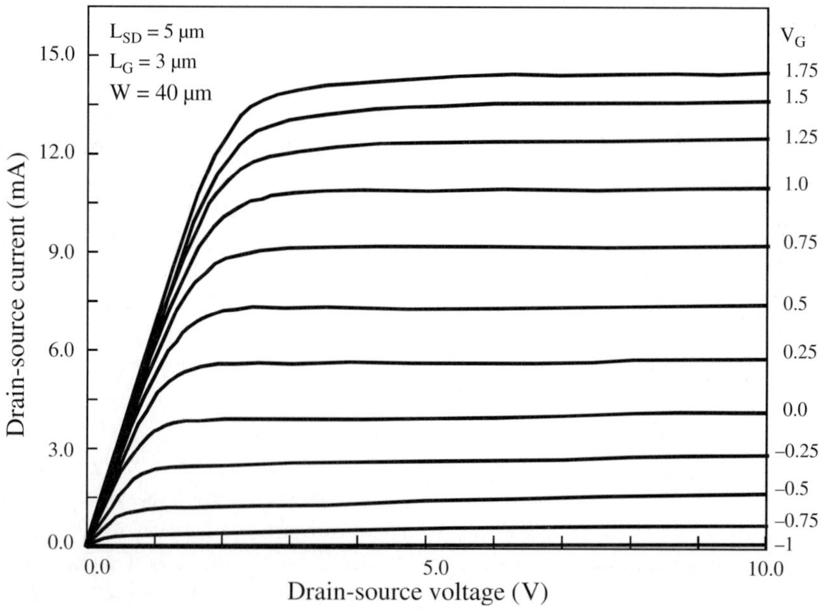

Fig. 14.38 Output characteristics of a 3 μm gate AlGaN/GaN MODFET grown on Leyl SiC substrate which is void of the output negative conductance. However, the Leyl substrates are highly conductive and not well suited for FETs due to RF shorting/loading. Nevertheless, experiments of this kind serve to prove the point that the negative output conductance observed in devices on sapphire is most likely thermal oriented.

11 GHz, respectively, at both 15 and 30 V bias. Values of f_T and f_{max} in excess of 50 GHz and 100 GHz have been reported for short channel (about 0.2 μm) devices, respectively. As touched upon earlier, devices on sapphire substrates suffer from the low thermal conductivity of sapphire substrates and exhibit negative differential resistance in the output characteristics. Remedies include better heat sinking by flip-chip mounting and the use of high resistivity 4H-SiC substrates, which provide good thermal conductivity but are hard to obtain.

GaN MODFET devices that have been prepared in the authors' laboratory on conducting 6H-SiC substrates exhibited output characteristics that lacked the negative resistance (i.e. they exhibited good heat sinking). There have subsequently been a few reports of MODFET power devices on high resistivity SiC [25, 28, 29] and *p*-type SiC [27] substrates with phenomenal improvement in power handling capability notwithstanding the rapid progress on sapphire substrates. On sapphire, recent 0.7-μm gate-length $Al_{0.5}Ga_{0.5}N$/GaNMODFETs exhibited a current density of 1 A/mm, three-terminal breakdown voltages up to 200 V, and CW power densities of 2.84 and 2.57 W/mm at 8 and 10 GHz, respectively, representing a marked performance improvement for GaN-based FETs.

Outstanding power levels are being achieved with near-half-micron gate lengths. With 0.7-μm gate length devices on SiC substrates, where the gate-source spacing and gate-drain spacing were 0.5 and 0.8 μm, respectively, a total output power of 2.3 W in a device with a 1.28 mm gate periphery has been obtained [28]. The power gain at the 2.3 W output power point was 3.6 dB with

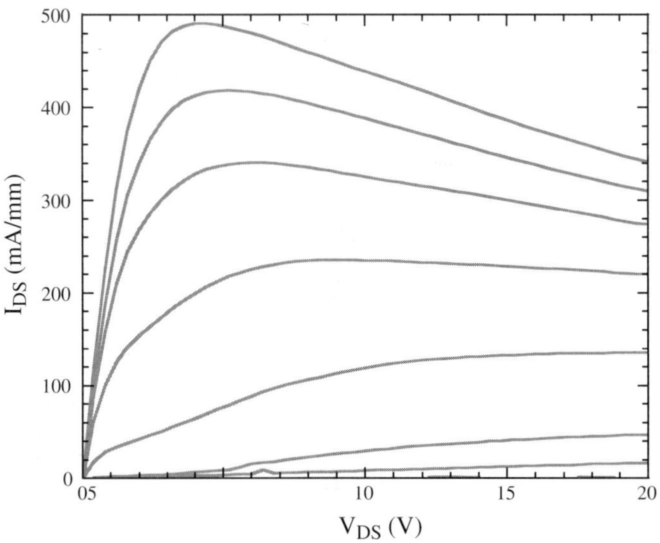

Fig. 14.39 The output I–V characteristics of a 0.25 μm gate AlGaN MODFET on sapphire fabricated at HRL laboratories. Note the negative differential output conductance due to the poor thermal conductivity of sapphire substrate. (Courtesy of Drs N. X. Nguyen and C. Nguyen.)

a PAE of 13.3%. The power measurement was conducted at a drain voltage of 33 V. The current and power gain cutoff frequencies were 15 and 42 GHz. The contact resistance, though not the best, was between 2.6 and 3.5 Ωmm. The maximum normalized transconductance was 270 mS/mm and the drain current was 293 mA/mm.

Though, excellent results have been reported by many laboratories on SiC and sapphire substrates, the results that represent the pinnacle are those obtained recently by Nguyen *et al.* at HRL in devices grown by MBE. While, the I–V characteristics, small signal RF performance, and noise performance of devices on sapphire have been made available to the authors, only the maximum power performance of 1-mm-wide devices on SiC are available to the authors at this time. The output I–V characteristics of a 0.25 μm AlGaN MODFET is shown in Fig. 14.39 and the small signal current and power gains are shown as a function of frequency in Fig. 14.40. DC characteristics include 600 mA/mm current performance and >60 V drain breakdown voltage. The current gain cutoff and maximum power gain cutoff frequencies measured were about 40 and 90 GHz, respectively. The minimum noise figure and the associated gain of the same device are shown in Fig. 14.41 showing a minimum noise figure of 0.85 dB at 10 GHz with an associated gain of 11 dB, which is simply remarkable. We should mention, however, that even better results are on the horizon.

The power performance of devices on SiC substrates of the HRL devices is simply outstanding. At HRL laboratories 6.3 W of output power was obtained at 10 GHz from a 1-mm-wide transistor device. This is the highest output power density reported for a-mm or larger power device. More importantly,

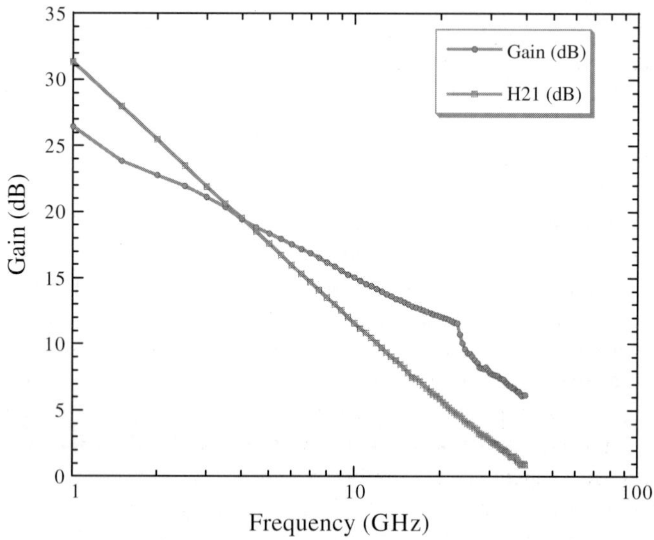

Fig. 14.40. Small signal current and power gains as a function of frequency of a 0.25 μm gate AlGaN MODFET on sapphire fabricated at HRL laboratories. (Courtesy of Drs N. X. Nguyen and C. Nguyen.)

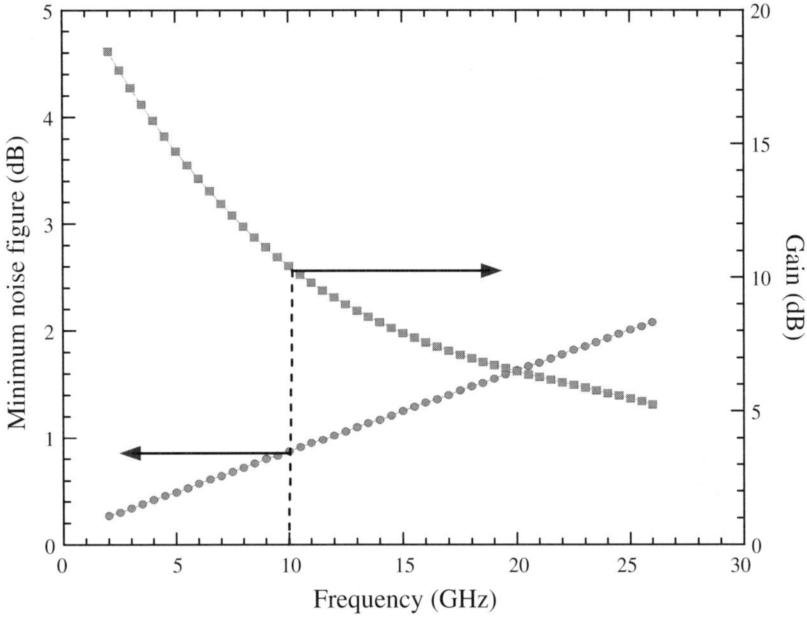

Fig. 14.41 The minimum noise figure and the associated gain of a 0.25 μm gate AlGaN MODFET on sapphire fabricated at HRL laboratories. (Courtesy of Drs N. X. Nguyen and C. Nguyen.)

the power density remains nearly constant as the device size is scaled upward from 0.1 mm width, where the device exhibits 6.5 W/mm, to 1.0 mm. These record-setting transistors were epitaxially grown AlGaN/GaN heterostructures on semi-insulating SiC (silicon carbide) substrates. HRL has developed a growth process using molecular beam epitaxy (MBE) which has virtually eliminated material defects common to other reported GaN devices, thereby enabling the scaling. MBE growth also produces device characteristics with less than 5% standard deviation over the 2-inch diameter SiC substrate, a six-fold improvement over previously reported results.

In power devices, the thermal limitation can never be eliminated completely as is the case in nitride devices as well, particularly when fabricated on sapphire substrates with a thermal conductivity of only approximately 0.3 W/cmK. Inclusion of thermal limitations leads to results shown in Fig. 14.42 for devices that compete in the high-power device arena [24, 122]. Since new device developments do in general compete with existing and alternative technologies, a brief account of competing technologies for the power arena is given below. The Si metal–semiconductor FET (MESFET) analytical curve, modeled for its simplicity, is slightly above the SiC analytical curve and indicates a maximum power density of 0.35 W/mm at $V_{\mathrm{dS}}=7\,\mathrm{V}$ which is slightly lower than 0.39 W/mm. Since Si RF MESFETs are unavailable, commercial Si RF metal–oxide semiconductor FET (MOSFET) results were used for comparison

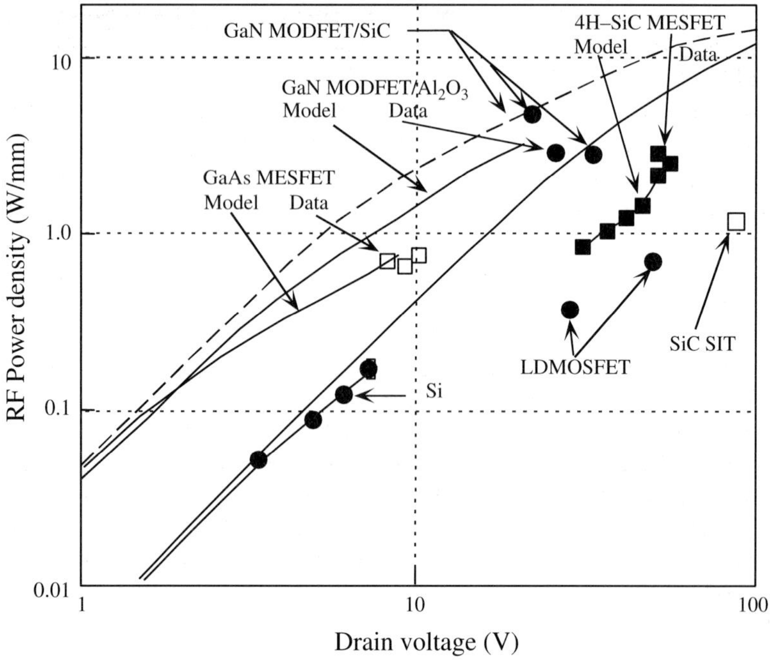

Fig. 41.42 Simulated and experimental RF power density data for Si, GaAs, SiC, and GaN FETs. (After ref. [122].)

instead. At low voltages, the Si MOSFET data parallel the analytical curve suggesting the validity of the functional dependence of power density on drain voltage. Also shown are two higher power density data points 0.4 W/mm, V_{DS}=28 V and 0.87 W/mm, V_{DS}=48 V. These higher power densities were obtained with specially designed RF power MOSFETs that incorporate a lightly doped drain and field plates that significantly increase the breakdown voltage.

The GaAs analytical curve shows the highest power density of all of the devices at the lowest voltages primarily because of the higher electron mobility of GaAs. However, the low breakdown field limits the GaAs MESFET's drain voltage to about 8 V and power density to 0.63 W/mm including thermal effects. Typical commercially available GaAs MESFET power densities are below 1 W/mm. However, high-performance GaAs FETs with more complex device cross-sections have achieved power densities as high as 1.4 W/mm at 18 V. At 100 V, the SiC MESFET has calculated maximum power densities of 7.96 W/mm with thermal effects and 9.7 W/mm without thermal effects. The highest demonstrated CW power density 3.3 W/mm (V_{DS}=50 V) for a SiC MESFET [142] is also shown for comparison. Additional SiC data again illustrate the functional dependence of power density on drain voltage. The GaN analytical results are highly dependent on the thermal conductivity of the substrate. With a sapphire substrate, the device is severely thermally limited to 2.24 W/mm at 30 V with a resulting channel temperature of over 400 °C. With a SiC substrate, however, the analysis predicts that a GaN MODFET could achieve 15.5 W/mm at 100 V with a channel temperature of about 300 °C. We should caution that while the power density figure can be used during the evolution process, eventually the total power figure must prevail. Normalized power density measurements, though frequently reported are often misleading because smaller gate widths naturally lead to larger power densities. This experimental datum point is actually slightly higher than the simulated result, possibly because of the very small size of the experimental device (100 μm width).

The GaN results of analytical models are highly dependent on the thermal conductivity of the substrate. With a sapphire substrate, the device is severely thermally limited to 2.24 W/mm at 30 V with a resulting channel temperature over 400 °C. However with a SiC substrate [122], the analysis predicts that a GaN MODFET could achieve 15.5 W/mm at 100 V while keeping the channel temperature at about 300 °C. The key to further improvements lies with our ability to control the polarity of the films, to prepare domain free material, and to reduce defects. If the past few years are any indication, substantial progress is in the wings.

14.17 Anomalies in GaN/AlGaN MODFETs

Field effect transistors in general and modulation doped field effect transistors in particular exhibit anomalies in their output I–V characteristics. Among the causes of these anomalies are channel carriers being trapped in the wide bandgap material and bulk, meaning the buffer layers. In addition, surface states, if present and not passivated, could cause reduction of the sheet conducting charge, particularly between the gate and the drain region of the FETs [143]. This is depicted schematically in Fig. 14.43. The AlGaAs/GaAs MODFET,

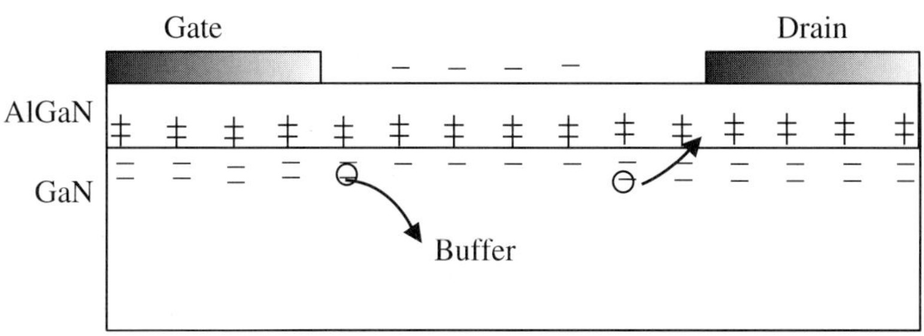

Fig.14.43 Schematic representation of an AlGaN/GaN MODFET indicating how the surface charge and charge injection into traps in the buffer layer and defects in AlGaN could serve in depleting conducting channel carriers. For simplicity, all the 2DEG is assumed to be due to donors placed in AlGaN. In addition, the regions from which carrier injection takes place are limited to the spacing between the gate and drain electrodes, the exact location is arbitrarily chosen.

which is the predecessor to the current AlGaN/GaN MODFET, exhibited what was then referred to as 'current collapse' [144]. This is caused by carrier injection from the channel to the AlGaAs at reasonably high fields where they are trapped at low temperatures. With light excitation, increasing temperature and exchange of the source and drain terminals, the effect could be eliminated. The GaAs buffer layer for the AlGaAs/GaAs based device is of such high quality that its trapping effect was not dominant. The surface states in the AlGaAs/GaAs device were not deemed to have as profound an effect on the current voltage characteristics. However, it is always a prudent approach to passivate the surface states, as was done in the AlGaAs/GaAs device, as they affect the device operation with time greatly. In the AlGaAs/GaAs variety, the trapping effect in the AlGaAs barrier was attributed to DX levels, which are caused by lattice distorting defects leading to massive change to the bandgap of the semiconductor at the local level, and their behavior could be described by a lattice coordination diagram. Since this effect reduced with a decrease in the AlAs mole fraction, AlGaAs/InGaAs pseudomorphich modulation doped FETs were developed [145], which are the dominant compound semiconductor FETs in industry at the moment, to mitigate the effect of DX centers.

In the AlGaN/GaN system, the surface states and or defects play a much more important role due to polarization fields as the layers are on polar surfaces. Anomalous characteristics go back to the early development of the devices in conjunction with somewhat detailed investigation of the devices [146]. Now that these devices are strong contenders in the market place for

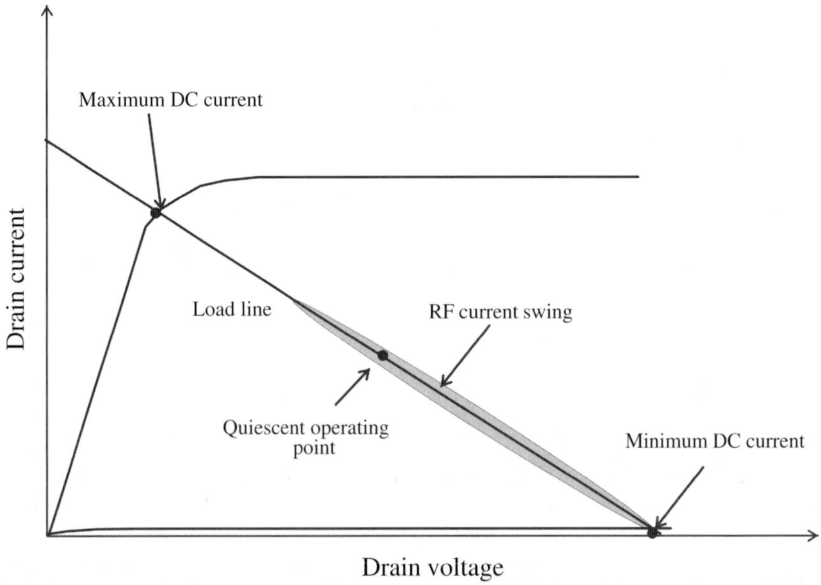

Fig. 14.44 Schematic representation of RF current lag superimposed on top of DC Drain I–V characteristics with a load line.

systems applications, they are getting a good deal of attention. The drain current lag which is basically a failure on the part of drain current to keep up with the gate bias voltage in response to a high-frequency large signal gate modulation [147] is attributed to surface states. Filled surface states pinch the channel in the region between the gate and drain. This is shown schematically in Fig. 14.44 where the load line and quiescent operating conditions for class A operation is shown. Also shown are the extremes of DC current as governed by the load line and RF current, (shaded) which fails to follow the gate bias, with expected higher drain currents. Recently, lattice distortions linked to the minimization of the polarization field have been proposed as a likely cause of current lag [148]. This current lag can be measured as a function of frequency at RF frequencies with an appropriate load line. Since the drain current does not follow the input stimulus due to surface traps, the term 'lag' has been coined to describe the phenomenon. The surface must be appropriately passivated to avoid this degradation. The effective methods so far have been the use of low temperature AlN [149] or Si_3N_4 [150] post-growth and fabrication passivation layers. If passivation alone is sufficient to eliminate the current lag, the issue of lattice distortion becomes an interesting one in that it raises the question whether the surface states are involved and if so whether passivation layers alter the strain picture also.

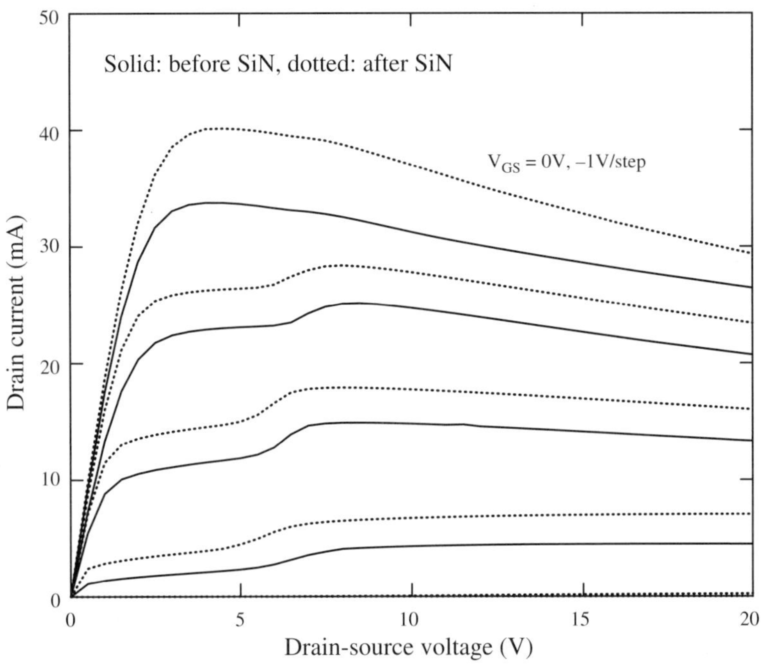

Fig. 14.45 Drain characteristics for a AlGaN/GaN MODFET with a 250Å AlGaN layer and a width of 50 μm. The dotted lines are for $V_{DS} < 10$ V and the solid lines are for V_{DS} up to 20 V. (After ref. [143].)

The drain characteristics for a GaN MODFET with a 250 Å Si-doped AlGaN layer that exhibit the aforementioned anomalies are shown in Fig. 14.45 where two sets of characteristics are included for the same device. The characteristics indicated by the dashed lines are the result when the maximum V_{DS} is limited to 10 V. On the other hand, the solid lines are for those measured when the maximum V_{DS} is 20 V. By comparing these characteristics, a reduction in drain current for $V_{DS} < 8$ V is noted. This reduction in current after the application of a high drain voltage is referred to as current collapse. This effect is similar to that reported for GaN MESFETs and is attributed to hot electron injection and trapping in the GaN buffer layer [151, 152]. As mentioned above, at high drain voltages, electrons are injected into the GaN buffer layer, where they are trapped. This trapped charge depletes the 2DEG from beneath the active channel and results in a reduction in drain current for subsequent V_{DS} traces. The trapped charge can be released through illumination or thermal emission. The gradual reduction in current for $V_{DS} > 10$ V, seen in Fig. 14.45, is attributed to self-heating due in part to the sapphire substrate with low thermal conductivity.

The effect of SiN passivation on the drain characteristics is shown in Fig. 14.46 for the same device before and after passivation. The drain current went up as a result of the increase in n_{sh}. It can be seen that the reduction in current associated with the current collapse phenomenon is unaffected. This is consistent

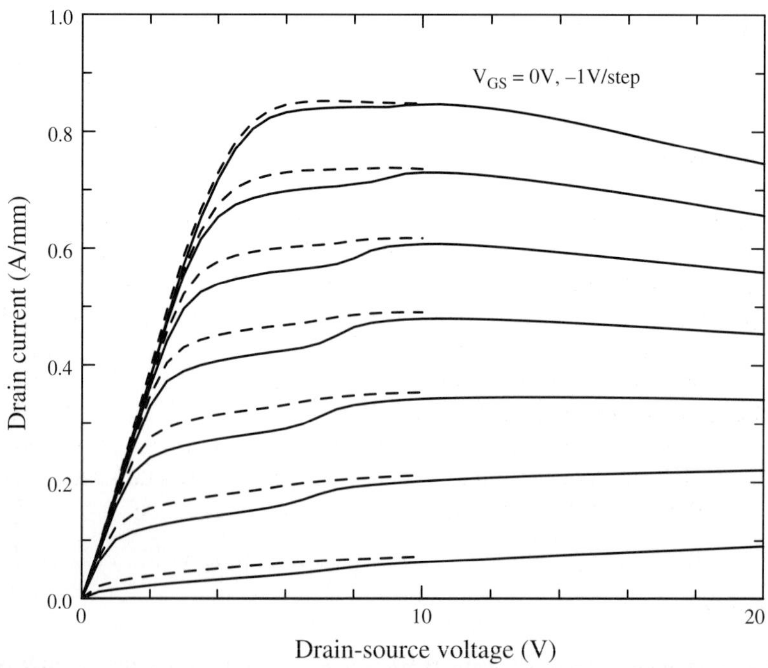

Fig. 14.46 Measured drain characteristics of the device of Fig. 14.45 before and after SiN passivation. (After ref. [143].)

with the proposed mechanism for current collapse, i.e. hot electron injection and trapping in the buffer layer without surface involvement.

14.18 Conclusions

Modulation doped field effect transistors (MODFETs) based on the GaN material system have been discussed. Unlike the GaAs based MODFETs on (100) surfaces, polarization induced charge in GaN varieties on the polar (0001) surfaces is quite large. Consequently, even undoped structures contain sheet electron concentrations in the 10^{13} cm^{-2}. An in-depth discussion of polarization effects for normal and inverted MODFETs has been presented. The charge is most likely provided by donor-like surface defects. Experimental data and calculation results have been provided on the particulars of the interface charge in relation to parameters such as the AlGaN mole fraction and thickness. In addition, calculation results for electron distribution and current-voltage characteristics or MODFETs have been presented. An elaboration of details surrounding the AlGaN/GaN MODFETs, with CW power levels of about 20 W at 10 GHz in devices with 2 mm gate periphery, has also been presented. A minimum noise figure is 0.85 dB with an associated gain of 11 dB at 10 GHz has been obtained. Finally, a discussion of the current collapse of lag occurring in GaN-based MODFETs has been discussed. In closing, GaN-based MODFETs have made great strides and are continuining to do so despite the less than ideal materials properties. It may be only a matter of time for inclusion of these devices in systems.

Acknowledgments

This work was supported by the Air Force Office of Scientific Research Office, the Naval Research, and the National Science foundation and monitored by G. L. Witt and D. Johnstone (AFOSR) C. E. Wood, Y.S. Park, and M. Yoder (ONR), and Verne Hess (NSF). The authors appreciate fruitful discussions with and the assistance of Profs. Paolo Lugli, W. Lambrecht and F. Bernardini, and Drs. Fabio Sacconi, C. Nguyen, X. Nguyen, O. Ambacher, John Albrecht, and W. Walukiewicz provided many of their reprints and in some cases their preprints. Finally, the authors would like to thank their colleagues at VCU and University of Rome and elsewhere for their contributions throughout the evolution of much of the work reported here.

References

1. Morkoç, H. (1999). *Nitride semiconductors and devices*. Springer-Verlag, Heidelburg.
2. Mohammad, S. N. and Morkoç, H. (1996). Progress and prospects of Group III–V nitride semiconductors. *Progress in Quantum Electronics* **20**(5–6), 361–525.

3. Mohammad, S. N., Salvador, A., and Morkoç, H. (1995). Emerging GaN based devices. *Proc. IEEE* **83**, 1306–55.
4. Morkoç, H., Strite, S., Gao, G. B., Lin, M. E., Sverdlov, B., and Burns, M. (1994). A review of large bandgap SiC, III–V nitrides, and ZnSe based II–VI semiconductor structures and devices. *J. Appl. Phys. Rev.* **76**(3), 1363–98.
5. Strite, S. T. and Morkoç, H. (1992). GaN, AlN, and InN: a review. *J. Vac. Sci. Technol. B* **10**, 1237–1266.
6. Ambacher, O. (1998). Growth and applications of Group III-nitrides. *J. Phys. D: Appl. Phys*. **31**, 2653–710.
7. Pearton, S. J., Zolper, J. C., Shul, R. J., and Ren, F. (1999). GaN: processing, defects, and devices. *J. Appl. Phys*. **86**, 1–78.
8. Duffy, M. T., Wang, C. C., O'Clock, G'D., McFarlane III, S. H., and Zanzucchi, P. J. (1973). Epitaxial growth and piezoelectric properties of AlN, GaN, and GaAs on sapphire or spinel. *J. Elect. Mater*. **2**, 359.
9. Razeghi, M. and Rogalski, A. (1996). Semiconductor ultraviolet detectors. *J. Appl. Phys.* **79**, 7433–73.
10. Xu, G. Y., Salvador, A., Kim, W., Fan, Z., Lu, C., Tang, H., Morkoç, H. *et al.* (1997). Ultraviolet photodetectors based on GaN p-i-n and AlGaN(p)-GaN(i)-GaN(n) structures. *Appl. Phys. Lett*. **71**(15), 2154–6.
11. Fritz, I. J. and Drummond, T. J. (1995). AlN-GaN quarter-wave reflector stack grown by gas-source MBE on (100) GaAs. *Electron. Lett.* **31**, 68–9.
12. Nakamura, S., Mukai, T., and Senoh, M. (1994). Candela-class high-brightness InGaN/AlGaN double-heterostructure blue light-emitting diodes. *Appl. Phys. Lett.* **64**, 1687–1689.
13. For a review see: Morkoç, H. and Mohammad, S. N. (1995). High luminosity gallium nitride blue and blue-green light emitting diodes. *Science Magazine* **267**, 51–5.
14. For a review, see: Morkoç, H. and Mohammad, S. N. (1999). Light emitting diodes. In *Wiley Encyclopedia of electrical engineering and electronics engineering* (Ed. J. Webster), Wiley, New York.
15. Nakamura, S., Senoh, M., Nagahama, N., Iwara, N., Yamada, T., Matsushita, T. *et al.* (1997). InGaN/GaN/AlGaN-based laser diodes with modulation-doped strained-layer superlattices. *Jpn. J. Appl. Phys*. **38**, L1578.
16. Morkoç, H. (1998). Wurtzite GaN based heterostructures by molecular beam epitaxy. *IEEE J. Selected topics in quantum electronics* (Eds R. Miles and I. Akasaki), **4**(3), 537–49.
17. Schetzina, J. North Carolina State University, Private communication.
18. Temkin, H. Texas Tech University, Private communication.
19. Campbell, J. C. University of Texas at Austin, Private communication.
20. Schetzina, J. *et al.* In Proceedings of the International Conference on Silicon Carbide and Related Materials, October 10–15, 1999, Research Triangle Park, NC USA.
21. Ridley, B. K. Oct. (1998). Exact electron momentum-relaxation times in GaN associated with scattering by polar-optical phonons. *J. Appl. Phys*. **84(**7), 1, 4020–21.
22. Bhapkar, U. V. and Shur, M. S. (1997). Monte Carlo calculation of velocity-field characteristics of wurtize GaN. *J. Appl. Phys*. **82**, 1649.
23. Kolnik, J., Oguzman, I. H., Brennan, K. F., Wang, R., Ruden, P. P., and Wang, Y. (1995). *J. Appl. Phys.* **78**, 1033–8.
24. Morkoç, H. (1998). Beyond SiC! III–V nitride based heterostructures and devices. In *Sic materials and devices* (Ed. Y. S. Park). Academic Press, Willardson and Beer Series (Eds Willardson and Weber), **52**(8), 307–94.
25. Sheppard, S. T., Doverspike, K., Pribble, W. L., Allen, S. T., Palmour, J. W., Kehias, L. T., and Jenkins, T. J. (1999). High power microwave GaN/AlGaN HEMTs on semi-insulating silicon carbide substrates. *IEEE Electron. Dev. Lett.* **20**(4), 161.

26. Wu, Y.-F., Keller, B. P., Fini, P., Keller, S., Jenkins, T. J., Kehias, L. T. *et al.* (1998). High Al-content AlGaN/GaN MODFET's for ultrahigh performance. *IEEE Electron. Dev. Lett.* **19**(2), 50–3.
27. Ping, A. T., Chen, Q., Yang, J. W., Khan, M. A., and Adesida, I. (1998). DC and microwave performance of high-current AlGaN/GaN heterostructure field effect transistors grown on p-type SiC substrates. *IEEE Electron. Dev. Lett.* **19**(2), 54–56.
28. Sullivan, G. J., Chen, M. Y., Higgins, J. A., Yang, J. W., Chen, Q., Pierson, R. L., and McDermott, B. T. (1998). High-power 10-GHz operation of AlGaN HFET's on insulating SiC. *IEEE Electron. Dev. Lett.* **19**, 198–199.
29. Binari, S., Redwing, J. M., Kelner, G., and Kruppa, W. (1997). AlGaN/GaN HEMTs grown on SiC substrates. *Electron. Lett.* **33**(3), 242–3.
30. Nguyen, N. and Nguyen, C. and of HRL Laboratories, private communication.
31. Henderson, T., Aksun, M. I., Peng, C. K., Morkoç, H., Chao, P. C., Smith, P. M. *et al.* (1986). Microwave performance of a quarter micron gate low noise pseudomorphic InGaAs/AlGaAs modulation doped field effect transistor. *IEEE Electron. Dev. Lett.* **EDL-7**, 649–51.
32. Molnar, R. J., Goetz, W., Romano, L. T., and Johnson, N. M. (June 1997). Growth of gallium nitride by hydride vapor-phase epitaxy. *J. of Crys. Growth* **178**(1–2), 147–56.
33. Yamaguchi, S., Kariya, M., Nitta, S., Takeuchi, T., Wetzel, C., Amano, H., and Akasaki, I. (June 1999). Structural properties of InN on GaN grown by metal-organic vapor-phase epitaxy. *J. Appl. Phys.* **85**(11), 1, 7682–8.
34. Look, D. C., Reynolds, D. C., Hemsky, J. W., Sizelove, J. R., Jones, R. L., and Molnar, R. J. (1997). *Phys. Rev. Lett.* **79**, 2273.
35. Huang, D., Yun, F., Visconti, P., Reshchikov, M. A., Wang, D., Morkoç, H., *et al.* Hall mobility and carrier concentration in GaN free-standing templates grown by hydride vapor phase epitaxy with high quality. *Solid State Electronics*, in press.
36. Nakamura, S., Mukai, T., and Senoh, M. (1992). *J. Appl. Phys.* **71**, 5543.
37. Lin, M. E., Sverdlov, B., Zhou, G. L., and Morkoç, H. (1993). *Appl. Phys. Lett.* **62**, 3479.
38. Ng, H. M., Doppalapudi, D., Moustakas, T. D., Weimann, N. G., and Eastman, L. F. (1998). *Appl. Phys. Lett.* **73**, 821.
39. Look, D. C. and Sizelove, J. R. (1999). *Phys. Rev. Lett.* **82**, 1237.
40. Weimann, N. G., Eastman, L. F., Doppalapudi, D., Ng, H. M., and Moustakas, T. D. (1998). *J. Appl. Phys.* **83**, 3656.
41. Fang, Z. Q., Look, D. C., Kim, W., Fan, Z., Botchkarev, A., and Morkoç, H. (1998). *Appl. Phys. Lett.* **72**, 2277.
42. Wook, K., Botohkarev, A. E., Morkoç, H., Fang, Z. Q., Look, D. C., and Smith, D. J. (1998). *J. Appl. Phys.* **84**, 6680.
43. Heying, B., Smorchkova, I., Poblenz, C., Elsass, C., Fini, P., and Den Baars, S. *et al.* (October 30, 2000). Optimization of the surface morphologies and electron mobilities in GaN grown by plasma-assisted molecular beam epitaxy. *Appl. Phys. Lett.* **77**(18), 2885–7.
44. Manfra, M. J., Pfeiffer, L. N., West, K. W., Stormer, H. L., Baldwin, K. W., Hsu, J. W. P. *et al.* (October 30, 2000). High-mobility AlGaN/GaN heterostructures grown by molecular-beam epitaxy on GaN templates prepared by hydride vapor phase epitaxy. *Appl. Phys. Lett.* **77**(18), 2888–90.
45. Cui, J., Sun, A., Reshichkov, M., Yun, F., Baski, A., and Morkoç, H. Preparation of sapphire for high quality III-nitride growth. MRS Internet J. The URL for the front page is http://nsr.mij.mrs.org/5/7/.
46. Powell, J. A., Larkin, D. J., and Trunek, A. J. (1998). Use of gaseous etching for the characterization of structural defects in silicon carbide single crystals. *Silicon Carbide, III-nitrides, and related materials* (Eds G. Pensl, H. Morkoc, B. Monemar, and E. Janzen), Sweden, Trans Tech Publications. **264–8**; 421–4.

47. Tarsa, E. J., Heying, B., Wu, X. H., Fini, P., DenBaars, S. P., and Speck, J. S. (1997). *J. Appl. Phys*. **82**, 5472.
48. Heying, B., Averbeck, R., Chen, L. F., Haus, E., Riechert, H., and Speck, J. S. (2000). *J. Appl. Phys.* **88**, 1855.
49. Grandjean, N. CHREA-CNRS, Valbonne, France. Private communication.
50. Rode, D. L. (1975). *Semiconductors and semimetals* (Eds R. K. Willardson and A. C. Beer). Academic press, New York. Vol. **10**, pp. 1–90.
51. Seeger, K. (1982). *Semiconductor physics*, 2nd edn. Springer-Verlag, Berlin.
52. Chin, V. W. L., Tansley, T. L., and Osotchan, T. (1994). *J. Appl. Phys*. **75**, 7365.
53. Rode, D. L. and Gaskill, D. K. (1995). *Appl. Phys. Lett.* **66**, 1972.
54. Jain, S. C., Willander, M., Narayan, J., and Van Overstraeten, R. (2000). *J. Appl. Phys.* **87**, 963.
55. Ridley, B. K., Foutz, B. E., and Eastman, L. F. (2000). *Phys. Rev. B*. **61**(24), 1682–9.
56. Dhar, S. and Ghosh, S. (1999). *J. Appl. Phys*. **86**, 2668.
57. Chin, V. W. L., Tansley, T. L., and Osotchan, T. (1994). *J. Appl. Phys*. **75**, 7365.
58. Look, D. C., Reynolds, D. C., Hemsky, J. W., Sizelove, J. R., Jones, R. L., and Molnar, R. J. (1997). *Phys. Rev. Lett.* **79**, 2273.
59. Cheong, M. G., Kim, K. S., Oh, C. S., Namgung, N. W., Yang, G. M., Hong, C. H. *et al.* (2000). *Appl. Phys. Lett.* **77**, 2557.
60. Götz, W., Romano, L. T., Walker, J., Johnson, N. M., and Molnar, R. J. (1998). *Appl. Phys. Lett.* **72**, 1214.
61. Joshi, R. P. (1994). *Appl. Phys. Lett.* **64**, 223.
62. Visconti, P., Jones, K. M., Reshchikov, M. A., Cingolani, R., Morkoç, H., and Molnar, R. J. (2000). *Appl. Phys. Lett.* **77**, 3532.
63. Ng, H. M., Doppalapudi, D., Moustakas, T. D., Weimann, N. G., and Eastman, L. F. (1998). *Appl. Phys. Lett.* **73**, 821.
64. Zhu, Q. S. and Sawaki, N. (2000). *Appl. Phys. Lett.* **76**, 1594.
65. Huang, D., Yun, F., Visconti, P., Reshchikov, M. A., Wang, D., Morkoç, H. *et al.* (June 2001). *Solid State Electronics* **45**(5), 711–15.
66. Visconti, P., Jones, K. M., Reshchikov, M. A., Yun, F., Cingolani, R., Morkoç, H. *et al.* (2000). *Appl. Phys. Lett.* **77**, 3743.
67. Azuhata, T., Sota, T., Suzuki, K., and Nakamura, S. (1995). *J. Phys.: Condensed Matter* **7**, L129.
68. Lin, M. E., Ma, Z., Huang, F. Y., Fan, Z. F., Allen, L. H., and Morkoç, H. (1994). *Appl. Phys. Lett.* **64**, 1003.
69. Hsu, L. and Walukiewicz, W. (2001). Effect of polarization fields on transport properties in AlGaN/GaN heterostructures. *J. Appl. Phys.* **89**, 1783.
70. Walukiewicz, W. Private communication.
71. Wang, Y. J., Kaplan, R., Ng, H. K., Doverspike, K., Gaskill, D. K., Ikedo, T. *et al.* (1996). *J. Appl. Phys.* **79**, 8007.
72. Özgür Aktas, Kim, W., Fan, Z., Mohammad, S. N., Botchkarev, A., Salvador, A. *et al.* (1995). *Electron. Lett.* **31**(16), 1389–90.
73. Binari, S. C., Redwing, J. M., Kelner, G., and Kruppa, W. (1997). *Electron. Lett.* **33**, 242.
74. Hadis Morkoç, Aldo Di Carlo, and Cingolani, R. (2001). *Condensed Matter News* (Ed. Patrick Bernier), Vol. 8, issue 2, pp. 4–46.
75. Hadis Morkoç, Roberto Cingolani, and Bernard Gil, October (October 1999). *Solid State Electronics*, **43(**10), 1909–27.
76. Hsu, L. and Walukiewicz, W. (1998). *Appl. Phys. Lett.* **73**, 339.
77. Hsu, L. and Walukiewicz, W. (1997). *Phys. Rev. B* **56**, 1520.
78. Smorchkova, I. P., Elsass, C. R., Ibbetson, J. P., Vétury, R., Heying, B., Fini, P. *et al.* (1999). *J. Appl. Phys*. **86**, 4520.

79. Shur, M. S., Bykhovsky, A. D., and Gaska, R. (2000). *Solid-State Electron.* **44**, 205.
80. Asbeck, P. M., Yu, E. T., Lau, S. S., Sun, W., Dang, X., and Shi, C. (2000). *Solid-State Electron.* **44**, 211.
81. Hsu, L. and Walukiewicz, W. (1998). *Appl. Phys. Lett.* **73**, 339.
82. Shur, M. S., Bykhovsky, A. D., and Gaska, R. (2000). *Solid-State Electron.* **44**, 205.
83. Oberhuber, R., Zandler, G., and Vogl, P. (1998). *Appl. Phys. Lett.* **73**, 818.
84. Maeda, N., Nishida, T., Kobayashi, N., and Tomizawa, M. (1998). *Appl. Phys. Lett.* **73**, 1856.
85. Walukiewicz, W. (1989). *Appl. Phys. Lett.* **54**, 2094.
86. Bykhovski, A. D., Kaminski, V. V., Shur, M. S., Chen, Q. C., and Khan, M. A. (1996). Piezoresistive effect in wurtzite n-type GaN. *Appl. Phys. Lett.* **68**, 818–19.
87. Bernardini, F., Fiorentini, V., and Vanderbilt, D. (1997). Spontaneous polarization and piezoelectric constants in III-V nitrides. *Phys. Rev. B* **56**, R10024.
88. For a review see, Resta, R. (1994). Macroscopic polarization in crystalline dielectrics: the geometric phase approach. *Rev. Mod. Phys.* **66**, 899, and references therein.
89. Ambacher, O., Smart, J., Shealy, J. R., Weimann, N. G., Chu, K., Murphy, M. *et al.* (1999). *J. Appl. Phys.* **85**, 3222.
90. Ambacher, O., Dimitrov, R., Stutzmann, M., Foutz, B., Murphy, M., Smart, J. *et al.* *J. Appl. Phys. Submitted.*
91. Smotchkova, I. P., Elsass, C. R., Ibbetson, J. P., Vétury, R., Heying, B., Fini, P. *et al.* (1999). *J. Appl. Phys.* **86**, 4520–6.
92. Daudin, B., Rouviere, J. L., and Arley, M. (1996). Polarity determination of GaN films by ion channeling and convergent beam electron diffraction. *Appl. Phys. Lett.* **69**, 2480.
93. Vermaut, P., Ruterana, P., Nouet, G., Salvador, A., and Morkoç, H. (1997). Structural defects due to interface steps and polytypism in III-V semiconducting materials: a case study using high resolution electron microscopy of the 2H-AlN/6H-SiC interface. *Philos. Mag. A* **75**, 239–59.
94. Potin, V., Ruterana, P., Nouet, G., Salvador, A., and Morkoç, H. (1997). The atomic structure of the {10-10} inversion domains in GaN/sapphire layers. *Proc. of Gallium Nitride and Related Materials II Symposium* (Eds C. R. Abernathy, H. Amano, and J. C. Zolper), Vol. 468, pp. 323–28, Materials Research Society, Pittsburgh, PA, USA.
95. Martin, G. A., Strite, S., Botchkarev, A., Agarwal, A., Rockett, A., Morkoç, H. *et al.* (1994). *Appl. Phys. Lett.* **65**, 610.
96. Martin, G. A. (1996). Ph.D. thesis, Semiconductor electronic band alignment at heterojunctions of wurtzite AlN, GaN and InN, Department of Physics, University of Illinois.
97. Martin, G. A., Botchkarev, A., Agarwal, A., Rockett, A., and Morkoç, H. (1996). Valence-band discontinuities of wurtzite GaN, AlN, and InN heterojunctions measured by X-ray photoemission spectroscopy. *Appl. Phys. Lett.* **68**, 2541.
98. Bernardini, F. and Fiorentini, V. (1998). Macroscopic polarization and band offsets at the nitride heterojunctions. *Phys. Rev. B.* **57**(16), 1–4.
99. Maziar Farahmand, Carlo Garetto, Enrico Bellotti, Kevin, F., Brennan, Michele Goano, Enrico Ghillino *et al.* March 2001. Monte Carlo simulation of electron transport in the III-nitride wurtzite phase materials system: binaries and ternaries, *IEEE Transactions on Electron Devices.* **48**, 535–42. and references therein.
100. Morkoç, H., Ünlü, H., and Ji, G. (1991). *Fundamentals and technology of MODFETs*, Vols. I and II. Wiley, Chichester, West Sussex, UK.
101. Fabio Sacconi, Aldo Di Carlo, Lugli, P., and Hadis Morkoç (2001). Spontaneous and piezoelectric polarization effects on the output characteristics of AlGaN/GaN

heterojunction modulation doped FETs. *IEEE Trans on Electron Devices*, special issue (Eds U. K. Mishra and J. Zolper), *IEEE Trans. on Electron Devices*, TED-**48**(3), 450–7.

102. Bastard, G. (1987). In Wave mechanics applied to semiconductor heterostructures. Edition de Physique, Paris, France.
103. Oberhuber, R., Zandler, G., and Vogl, P. (1998). *Appl. Phys. Lett.* **73**, 818.
104. Di Carlo, A., Pescetelli, S., Paciotti, M., Lugli, P., and Graf, M. (1996). *Solid State Commun.* **98**, 803.
105. Di Carlo, A. (2000). *Phys. Stat. Solidi.* **217**, 703.; Della Sala, F., Di Carlo, A., Lugli, P., Bernardini, F., Fiorentini, V., Scholz, R., and Jancu, J. M. (1999). *Appl. Phys. Lett.* **74**, 2002.; Di Carlo, A., Della Sala, F., Lugli, P., Fiorentini, V., and Bernardini, F. (2000). *Appl. Phys. Lett.* **76**, 3950.
106. Cingolani, R., Botchkarev, A., Tang, H., Morkoç, H., Coli', G., Lomascolo, M. e*t al.* (2000). *Phys. Rev. B* **61**, 2711.
107. Bonfiglio, A., Lomascolo, M., Traetta, G., Cingolani, R., Di Carlo, A., Della Sala, F. *et al.* (2000). *J. Appl. Phys.* **87**, 2289.
108. Lugli, P., Paciotti, M., Calleja, E., Munoz, E., Sanchez-Rojas, J. J., Dessenne, F. *et al.* (1996). HEMT models and simulations. In *Pseudomorphic HEMTs: technology and applications*. (Eds R. Lee Ross, S. Swensson, and P. Lugli), pp.141–63, Kluwer Press, Dordrecht.
109. Fiorentini, V., Bernardini, F., Della Sala, F., Di Carlo, A., and Lugli, P. (1999). *Phys. Rev. B* **60**, 8849.
110. Trellakis, A., Halick, A. T., Pacelli, A., and Ravaioli, U. (1997). *J. Appl. Phys.* **81**, 7880.
111. Lee, K., Shur, M. S., Drummond T. J., and Morkoç, H. (1984). Parasitic MESFET in (Al,Ga)As/GaAs modulation doped FETs and MODFET characterization. *IEEE Trans. Electron. Dev.* ED-**31**, 29–35.
112. Yu, E. T., Sullivan, G. J., Asbeck, P. M., Wang, C. D., Qiao, D., and Lau, S. S. (1997). Measurement of the piezoelectrically induced charge in GaN/AlGaN heterostructure field-effect transistors. *Appl. Phys. Lett.* **71**, 2794.
113. Ramvall, P., Aoyagi, Y., Kuramata, A., Hacke, P., and Horino, K. (1999). Influence of a piezoelectric field on the electron distribution in a double GaN/$Al_{0.14}Ga_{0.86}N$ heterojuction. *Appl. Phys. Lett.* **74**, 3866.
114. Gaska, R., Yang, J. W., Osinsky, A., Bykhovski, A. D., and Shur, M. S. (1997). Piezoeffect and gate current in AlGaN/GaN high electron mobility transistors. *Appl. Phys. Lett.* **71**, 3673.
115. Gaska, R., Yang, J. W., Osinsky, A., Bykhovski, A. D., Shur, M. S., Kaminski, V. V., and Soloviov, S. M. (1998). The influence of the deformation on the two-dimensional electron gas density in GaN-AlGaN heterostructure. *Appl. Phys. Lett.* **72**, 64.
116. Carnez, B., Cappy, A. *et al.* (1980). Modeling of a submicrometer gate field effect transistor including effects of nonstationary electron dynamics. *J. Appl. Phys.* **51**, 784–90.
117. Sandborn, P. A. *et al.* (1987). Quasi-two-dimensional modelling of GaAs MESFETs, *IEEE Trans. Electron. Dev.* ED-**34**, 985–91.
118. Snowden, C. M. *et al.* (1989). Quasi-two-dimensional modelling MESFET simulation for CAD. *IEEE Trans. Electron. Dev.* ED-**36**, 1564–73.
119. Morkoç: H. (1999). GaN-based modulation doped FETs and UV detectors. *Naval Research Reviews* **51**, 28–45.
120. Murphy, M. J., Foutz, B. E., Chu, K. *et al.* (1999). *MRS Internet J. Nitride Semicond. Res.* **4S1**, G8.4.
121. Yu, Y. F., Keller, B. P., Keller, S., Kapolnek, D., Kozodoy, P., Denbaars, S. P., and Mishra, U. K. (1996). *Appl. Phys. Lett.* **69**, 1438.

122. Weitzel, C., Pond, L., Moore, K., and Bhatnagar, M. (1998). Effect of device temperature on RF FET power density. Proc. of Silicon Carbide, III-Nitrides and Related Materials, ICSI, August 1997, Stockholm, Sweden, Trans Publications, Ltd., Materials Science Forum, Vols. 264–268, pp. 907–12.
123. Suzue, K., Mohammad, S. N., Fan, Z. F., Kim, W., Aktas, Ö., Botchkarev, A. E., and Morkoç, H. (1996). Electrical conduction in platinum-gallium nitride schottky diodes. *J. Appl. Phys.* **80**(6), 4467–78.
124. Mohammad, S. N., Zhifang Fan, Botchkarev, A. E., Kim, W., Aktas, Ö., Salvador, A., and Morkoç, H. (1996). Near ideal platinum-GaN Schottky diodes. *Electron. Lett.* **32**, 598.
125. Aktas, Ö., Kim, W., Fan, Z., Mohammad, S. N., Botchkarev, A., Salvador, A. *et al.* (1995). High transconductance-normally-off GaN MODFETs. *Electron. Lett.* **31**(16), 1389–90.
126. Aktas, Ö., Fan, Z., Botchkarev, A., Roth, M., Jenkins, T., Kehias, L. T., and Morkoç, H. (1997). Microwave performance of AlGaN/GaN Inverted MODFETs. *IEEE Electron. Dev. Lett.* **18**, 293–295.
127. Hacke, P., Detchprohm, T., Hiramatsu, K., and Sawaki, N. (1993). Schottky barrier on n-type GaN grown by hydride vapor phase epitaxy. *Appl. Phys. Lett.* **63**, 2676–8.
128. Binari, S. C., Dietrich, H. B., Kelner, G., Roland, L. B., Doverspike, K., and Gaskill, D. K. (1994). Electrical characterisation of Ti Schottky barriers on n-type GaN. *Electron. Lett.* **30**, 909–910.
129. Khan, M. A., Van Hove, J. M., Kuznia, J. N., and Olson, D. T. (1991). High electron mobility GaN/$Al_xGa_{1-x}N$ heterostructures grown by low-pressure metalorganic chemical vapor deposition. *Appl. Phys. Lett.* **58**, 2408–10.
130. Khan, M. R. H., Nakayama, H., Detchprohm, T., Hiramatsu, K., and Sawaki, N. (1997). A study on barrier height of Au-$Al_xGa_{1-x}N$ Schottky diodes in the range of $0<x<0.20$, in topical workshop on III-V nitrides Proc., Nagoya, Japan, 1995, *Solid State Electronics* **41**(2), 259–66.
131. Foresi, J. S. and Moustakas, T. D. (1993). Metal contacts to gallium nitride. *Appl. Phys. Lett.* **62**, 2859–861.
132. Khan, M. A., Kuznia, J. N., Bhattarai, A. R., and Olson, D. T. (1993). Metal semiconductor field-effect transistor based on single-crystal GaN. *Appl. Phys. Lett.* **62**, 1786–87.
133. Lin, M. E., Ma, Huang, F. Y., Fan, Z., Allen, L., and Morkoç, H. (1994). Low resistance ohmic contacts on wide band-gap GaN. *Appl. Phys. Lett.* **64**, 1003–5.
134. Wu, Y., Jiang, W., Keller, B., Keller, S., Kapolnek, D., Denbaars, S., and Mishra, U. (1997). Low resistance ohmic contacts to n-GaN with a separate layer method. *Progress. in Solid State Electronics* **41**(2), 75–78.
135. Fan, Z., Mohammad, S. N., Kim, W., Aktas, Ö., Botchkarev, A. E., and Morkoç, H. (1996). Very Low resistance multi-layer ohmic contact to n-GaN. *Appl. Phys. Lett.* **68**, 1672–74.
136. Ruvimov, S., Liliental-Weber, Z., Washburn, J., Duxstad, K. J., Haller, E. E., Mohammad, S. N. *et al.* (1996). Microstructure of Ti/Al and Ti/Al/Ni/Au ohmic contacts for n-GaN. *Appl. Phys. Lett.* **69**(11), 1556–8.
137. Moloney, M., Ponse, F., and Morkoç, H. (1985). Gate capacitance voltage characteristics of MODFETs: its effect on transconductance. *IEEE Trans. Electron. Dev.* ED-**32**(9), 1675–84.
138. Lin, M. E., Strite, S., Agarwal, A., Salvador, A., Zhou, G. L., Teraguchi, N. *et al.* (1993). GaN grown on hydrogen plasma cleaned 6H-SiC substrates. *Appl. Phys. Lett.* **62**(7), 702–4.
139. Feenstra, R. A. Private communication.

140. Powell, J. A., Larkin, D. J., and Trunek, A. J. (1998). Use of gaseous etching for the characterization of structural defects in silicon carbide single crystals. *Silicon Carbide, III-Nitrides, and Related Materials* (Eds G. Pensl, H. Morkoç, B. Monemar and E. Janzen), Sweden, Trans Tech Publications. 264–68: 421–4.
141. Powell, J. A., Larkin, D. J., Neudeck, P. G., Yang, J. W., and Pirouz, P. (1994). Investigation of Defects in Epitaxial 3C-SiC, 4H-SiC and 6H-SiC films grown on SiC substrates. *Silicon carbide and related materials* (Eds M. G. Spencer, R. P. Devaty, J. A. Edmond *et al.*), Bristol, IOP Publishing: 161–4.
142. Moore, K. E., Weitzel, C. E., Nordquist, K. J., Pond, III, L. L., Palmour, J. W., Allen, S., and Carter, Jr., C. H. (1997). *IEEE Electron. Dev. Lett.* **18**(2), 69–70.
143. Binari, S. C., Ikossi-Anastasiou, K., Roussos, J. A., Kruppa, W., Park, D., Dietrich, H. B. *et al.* (2001). Special Issue of *IEEE Electron. Dev.* **48**, 465–71.
144. Fischer, R., Drummond, T. J., Klem, J., Kopp, W., Henderson, T., Perrachione, D., and Morkoç, H. (1984). On the collapse of drain I-V characteristics in modulation doped FETs at cryogenic temperatures. *IEEE Trans. Electron. Dev.* **ED-31,** 1028–32.
145. Masselink, W. T., Ketterson, A., Klem, J., Kopp W., and Morkoç, H. (1985). Cryogenic operation of pseudomorphic AlGaAs/InGaAs single quantum well MODFETs. *Electron. Lett.* **21**, 937–9.
146. Kruppa, W., Binari, S. C., and Doverspike, K. (1995). Low-frequency dispersion characteristics of GaN HFETs. *Electron. Lett.* **31**, 1951–52.
147. Nguyen, C., Nguyen, N. X., and Grider, D. E. (1999). Drain current compression in GaN MODFETs under large-signal modulation at microwave frequencies. *Electron. Lett.* **35**, 1380–2.
148. Shur, M. S. *et al.* Private communication.
149. William, J. Schaff, Jeonghyun Hwang, Bruce Green, The effect of a MBE grown AlN surface layer on the DC and RF characteristics of GaN heterojunction field effect transistors, in preparation.
150. Vetury, R., Zhang, N. Q., Keller, S., and Mishra, U. K. (2001). Special Issue of *IEEE Electron. Dev.* **48**, 560–6.
151. Binari, S. C., Kruppa, W., Dietrich, H. B., Kelner, G., Wickenden, A. E., and Freitas, Jr., J. A. (1997). Fabrication and characterization of GaN FETs. *Solid-State Electronics* **41**, 1549–54.
152. Klein, P. B., Freitas, Jr., J. A., Binari, S. C., and Wickenden, A. E. (1999). Observation of deep traps responsible for current collapse in GaN metal-semiconductor field-effect transistors. *Appl. Phys. Lett.* **75**, 4016–018.

15 Mixed III–V–N semiconductors: a challenge for tomorrow?

Eric Tournié and Bernard Gil

15.1 Introduction

The first investigations on mixed III–V–N semiconductors date back almost 40 years [1–4]. III–V–N samples, however, have long been prepared by near-equilibrium crystal-growth techniques such as bulk growth or liquid- and vapour-phase epitaxy. The nitrogen content is then governed by thermodynamic considerations which predict an exceedingly low solubility of N in III–V compounds (e.g. $<10^{14}\,cm^{-3}$ in GaAs at 650 °C) under equilibrium conditions [5–7]. The nitrogen concentration in early III–V–N samples was thus limited to doping amounts. The problem had thus been entered from its fundamental aspect, by considering N as an isoelectronic impurity in III–V compounds.

The 80s, however, have seen a tremendous development of far-from-equilibrium epitaxy techniques, such as molecular-beam epitaxy (MBE) or metal-organic vapour phase epitaxy (MOVPE) which have progressively emerged as unique tools for the epitaxial growth of custom-designed semiconductor heterostructures. In addition, the efforts toward N-doping of wide bandgap II–VI compounds such as ZnSe on the one hand, and toward the growth of GaN-based wide bandgap materials on the other, have fostered the development of efficient N-sources for both MOVPE and MBE growth techniques.

In a permanent quest for new semiconductor systems as well as functionalities, the promises shown by GaN led naturally to consider the possibility to alloying GaN to conventional III–V compounds. The first results obtained with GaAs [8] and GaP [9] showed that in spite of GaN wide bandgap, incorporation of nitrogen in the alloy range results in a drastic reduction of the bandgap which indicates a giant bowing of the bandgap in these alloys. This most surprising result was soon supported by theoretical works using either the dielectric theory [10] or first-principles total-energy calculations [11]. It then became clear that III–V–N alloys could be the cornerstone of a totally new family of narrow bandgap semiconductors with many potential applications, as confirmed by the rapid development of GaInNAs (GINA)-based vertical-cavity surface-emitting lasers (VCSELs) operating near 1.3 μm (see Section 15.4). These findings have thus renewed the interest of the community for III–V–N semiconductors which appear now attractive from both fundamental as well as applied point of views.

The past few years have thus seen the resurgence of a massive research-activity on III–V–N semiconductors and their heterostructures. The aim of this chapter is to review the current knowledge. Section 15.2 deals with the band structure of III–V–N materials while their epitaxial growth is described in Section 15.3. Actual as well as potential applications are detailed in Section 15.4 before conclusions and perspectives are drawn in Section 15.5. We emphasize, however, that one should keep in mind the rapidly-evolving nature of the subject.

15.2 Band structure phenomena in III–V–N alloys

Both the size and the electronegativity of nitrogen differ markedly from the values of other group-V elements. This has strong implications on the band structure of these alloys and makes them rather singular among III–V compounds. In turn, this peculiarity is very attractive from both fundamental as well as applied perspectives, and nitrogen as an isoelectronic impurity which substitutes the group-V element of the III–V semiconductor compounds has been the subject of intense investigations beginning some four decades ago [1–4, 12]. We review successively the knowledge on the band-structure of phosphorus- and arsenic-based compounds and heterostructures. In the following, III–V:N refers to III–V compounds where N is incorporated in doping amounts while IIIVN refers to N-containing alloys.

15.2.1 Gallium phosphide and nitrogen: from doping to alloying

Zinc-blende GaP and its alloys GaAsP and GaInP have been originally the most investigated host materials. Although the fundamental bandgap structure of GaP is indirect in reciprocal space, changing the chemical composition of the GaAsP and GaInP solid solutions allows to tune the bandgap from indirect to direct [13, 14]. In GaInP alloys, nitrogen (which substitutes phosphorus) can be bound to various Ga_mIn_{m-4} $(m=0-4)$ 'species', while in GaAsP alloys it experiences the influence of structural disorder on the anion sublattice [15]. These alloys are thus complementary from the point of view of basic physics.

15.2.1.1 *N-doped Ga(In)P alloys*

In GaP:N, signature of the nitrogen bound exciton is easily obtained in absorption. It is observed at 2.317 eV at 2 K, an energy some 13 meV below the value of the indirect band gap of GaP and 6 meV higher than the recombination feature due to the sulphur-donor bound-exciton at 2.3108 eV (so-called C line) [4, 16]. Group theory arguments predict a fine structure splitting of the photoluminescence (PL) line of the nitrogen bound exciton caused by the short range electron–hole interaction. Indeed, the experimental spectra consist of a pair of PL peaks (so-called A and B lines) separated by 0.91 meV [16]. When increasing the nitrogen concentration, a series of additional peaks appears deeper in the band gap as shown in Fig. 15.1, that were attributed to excitons bound to

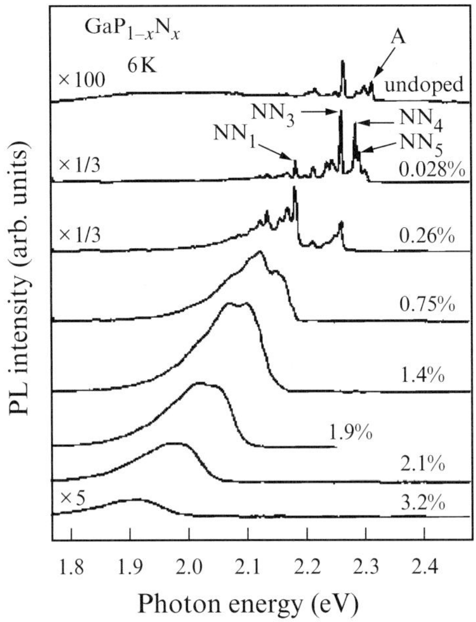

Fig. 15.1 Evolution of the 6 K photoluminescence spectrum of GaP:N and GaPN crystals. Note the distortion of the PL line towards the low energy for nitrogen compositions higher than 1%. (Figure reproduced from ref. [37].)

nitrogen pairs and are conventionally labelled NN_i. Careful examination of the excitonic fine structure of the NN_i excitons reveals in addition to the fine structure electron–hole interaction splitting an additional splitting due to the reduction of the zinc-blende crystal field in the vicinity of these defects [17]. Quantum mechanics textbook models have been invoked to explain these lines, where two nitrogen atoms are represented by a double well potential that binds the electron, the hole itself being bound by the long-range interaction with the electron [3]. The essence of these models is to discretely change the two-well inter-spacing, according to the substitution of phosphorus by nitrogen atoms that are located in first, second, . . . *i*th neighbour (for NN_i) configuration [18–21]. The result deduced from these initial calculations is that the closer the two impurities, the stronger the localization energy of the exciton.

However, more sophisticated theoretical models involving the multivalley structure of the GaP conduction band, and later lattice relaxation around nitrogen atoms, challenged their predictions and showed the importance of including L conduction states to the nitrogen wave function in GaP [22–25]. Further, uniaxial stress investigations of the symmetry of the defects revealed that, in fact, NN_i excitons are not trapped by nitrogen pairs in *i*th neighbour positions [26]. In particular, the NN_2 bound-exciton line detected between the NN_1 and NN_3 features in the PL spectra is due to trapping of the exciton to a nitrogen triplet defect [27, 28], even though the probability for the occurrence of such defect is extremely small if one assumes a random distribution of nitrogen in the crystal at doping concentrations allowing the observation of NN_2 ($[N] \geq 10^{18}\,cm^{-3}$). In addition, the symmetries of the other nitrogen pairs indicate that the N–N axes always lie in a (110) plane [26]. This shows that nitrogen does not incorporate randomly in GaP. Finally, elementary Slater–Koster calculation of the electron binding energies to NN_2 nitrogen triplets reveals that the three nitrogen atoms are in the (0,0,0), (1,1,0), and ($0,\bar{1},1$) positions in the crystal [28]. This results in a Cs symmetry of the defect, in agreement with the experimental data [26]. There are four possibilities to arrange three nitrogen atoms to form relevant triplets [28]. The highest localization energy is computed to arise when the three nitrogen atoms form a regular tetrahedron, a defect which is called NN_1N_1. Its corresponding PL signature is predicted to occur near 2.1 eV at 2 K [28].

By investigating crystals with nitrogen incorporation up to $3.5 \times 10^{19}\,cm^{-3}$ [29], Gil *et al.* observed that when the nitrogen concentration increases the distribution of the exciton population shifts from shallow to deeper levels [30]. This phenomenon, which is activated by temperature [31] and hydrostatic pressure [32], was called 'transfer mechanism' and has been detailed by Mariette [15]. When increasing hydrostatic pressure, the localization energy of the nitrogen-bound excitons decreases [33] and there is progressive ionization of the shallow and deep bound excitons. Ultimately, for pressures around 6 GPa the PL spectra of GaP : N samples can be totally freed from NN_i excitons [34–36]. However, none of these high-pressure experiments could reveal the signature of NN_1N_1 triplets for [N] up to $6 \times 10^{18}\,cm^{-3}$. The reasons for this remain unclear today. Experimental conditions such as the signal–noise ratio at a pressure of 6 GPa in a diamond-anvil cell could be considered as an argument. At the time these experiments were done, the comprehension of GaP:N saturated, probably due to the lack of sufficiently accurate theoretical examinations of the problem and of innovative ideas.

15.2.1.2 *Ga(In)PN alloys*

Using modern epitaxy techniques GaPN and GaAsPN alloys with nitrogen contents in the alloy-range have been obtained [37–50]. Compositions up to 4% have even been reported for GaPN-GaP multiple quantum wells (MQWs) [49].

The exciton transfer mechanism described above for GaP : N samples is exalted in GaPN alloys and a novel PL band can be unambiguously observed at the energy of the NN_1N_1 triplet, i.e. near 2.071 eV at 2 K. The most recent determination of the nitrogen composition required to observe this line is about 0.7% [48]. Below this value, the NN_i PL lines are no longer observed and we come back to the GaP:N behaviour. When increasing the nitrogen composition beyond 0.7%, the 2.071 eV line broadens towards the low-energy side (Fig. 15.1). As observed also for GaP:N samples [27–29], this low-energy tail arises from the influence of additional nitrogen atoms which, while being far from the NN_i defect, are still close enough to perturbate the excitons. They stimulate the migration of the NN_i bound-exciton towards an energetically more favourable NN_j bound-exciton configuration. Therefore the tail of the PL line corresponds to NN_iN_α triplets with a third atom far from the NN_i doublet [15, 31]. The higher the nitrogen concentration the higher the number of active NN_iN_α triplets, and the most extended the low energy PL tail [27–29], as can be clearly seen in Fig. 15.1 for GaPN alloys. Finally, the formation of GaN_4 tetrahedra cannot be ruled out, of course. In that case one would observe a marked PL band probably peaking at 1.92 eV at 2 K, as shown in Fig. 15.1 for a composition of 3%.

Absorption measurements in GaP:N and GaPN alloys reveal a red shift of the onset of absorption when the nitrogen concentration increases as shown in Fig. 15.2. Note that absorption peaks related to NN_i pairs are still detected for concentrations around 1% which indicates that a novel band-structure does not appear until [N]=1% as confirmed by PLE experiments [41]. Finally the evolution of the conduction band of GaPN alloys is shown in Fig. 15.3.

It is now clear that there is a continuity in the optical properties of GaP:N and GaPN alloys as evidenced in Fig. 15.4 by the dependence on the hydrostatic pressure of PL lines from GaP:N bound excitons and GaPN alloys. This is not a surprise, in fact: when increasing the number of nitrogen atoms in the crystal, their wave function hybridizes more and more efficiently with the wave functions of the gallium and phosphide atoms, up to forming some novel energy states. This, together with the absorption data shown above, is consistent with the prediction of Bellaiche *et al.* [51, 52] who computed that

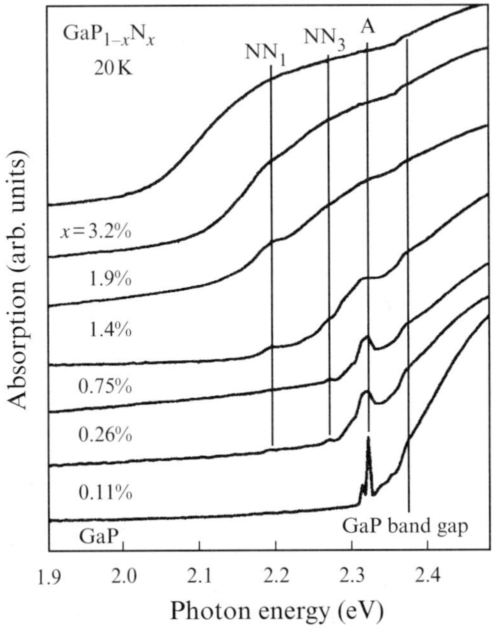

Fig. 15.2 20 K absorption spectra of GaPN alloys. (Figure reproduced from ref. [33].)

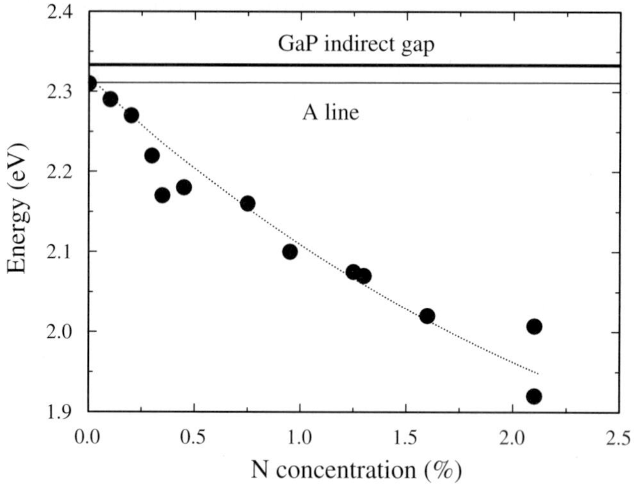

Fig. 15.3 Evolution of the bandgap of GaPN alloys with the nitrogen composition. (Figure reproduced from ref. [41].)

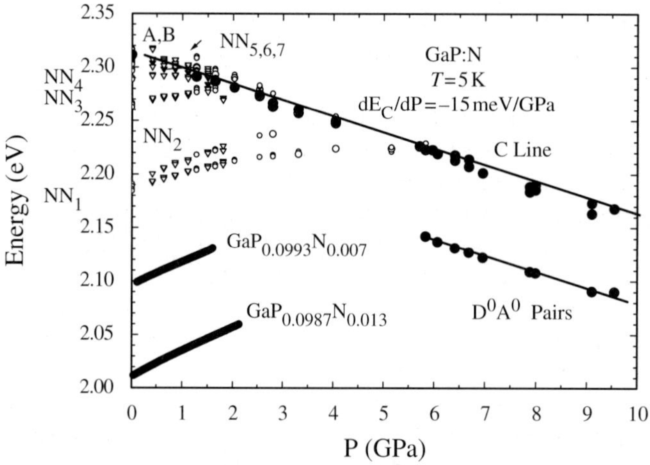

Fig. 15.4 Hydrostatic pressure-dependence of nitrogen bound excitons [33–35] and of GaPN alloys [47]. The GaPN data which were taken at 300 K in ref. [40] was blue-shifted such as that all the figure represents a 2 K situation.

the bandgap of GaPN becomes direct for a nitrogen composition of 3%. We believe that the optical probe used in these experiments, namely excitons, indicates that a true GaPN alloy exist for nitrogen compositions giving rise to a PL signature below ~1.92 eV at 2 K. For lower concentrations, the material behaves more probably like an amalgam of strongly inhomogeneous GaPN and GaP:N domains, and the notion of band structure does not strictly exist, at least from an exciton point of view.

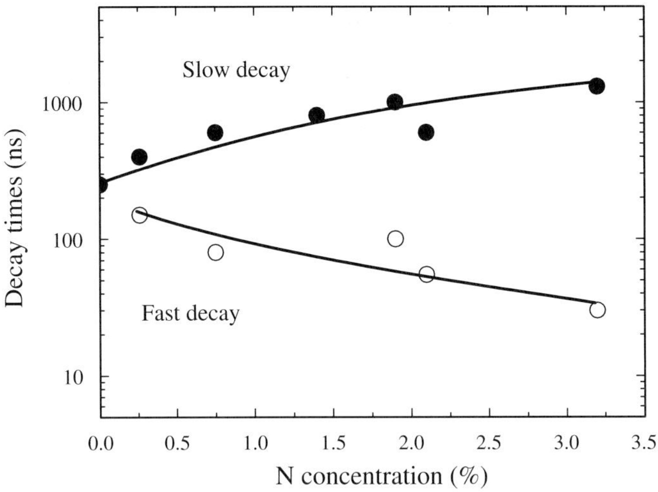

Fig. 15.5 Evolution of the slow and fast components of the photoluminescence decay times measured at 4.2 K for various GaPN samples. (Figure reproduced from ref. [42].)

Time-resolved PL spectra of GaPN alloys exhibit a fast- and a slow-decay-component as shown on Fig. 15.5. When the nitrogen concentration increases, the slow decay time (in the microsecond range) increases while on the other hand the fast decay time (in the 100 ns to 20 ns range) seems to decrease. However, this decrease is not monotonous and a maximum is found when PL peaks at 2.071 eV, i.e. when the NN_1N_1 triplet dominates the PL spectrum of GaPN. We interpret this non-monotonous behaviour as arising from the reduction of the non-radiative recombination channels for excitons efficiently localized on NN_1N_1 defects which means that the measured decay time is influenced by the amount of excitons localized on NN_1N_1 defects in the PL band under investigation. This model indicates that the fast decay time is radiative in nature and that there would be also an increase of the decay time when GaN_4 tetrahedra are formed with a reasonable probability, or, in other words, when PL peaks at 1.92 eV for a nitrogen composition of ~3%. Finally, we believe that the slow decay time is also radiative in nature but concerns these extended states that were computed by Kent and Zunger [53] (see Fig. 15.6).

15.2.2 Gallium arsenide and nitrogen: from doping to alloying

15.2.2.1 *N-doped GaAs*

The evolution of the nitrogen-related PL in GaAsP alloys [54] indicates that the nitrogen impurity in GaAs should give rise to a state which is resonant in the conduction band, serendipitously close to the energy of the L minimum. Wolford *et al.* [55] and Leroux *et al.* [56] could shift this state into the bandgap by applying hydrostatic pressure to GaAs. The hydrostatic pressure required to

Fig. 15.6 Isosurfaces of the conduction wave functions for (a) $x << x_c$ GaAsN showing nitrogen-localized cluster states within the gap (red), (b) $x \sim x_c$ showing the 'amalgamated state' of localized (red) and the delocalized (Bloch-like) states (green), and (c) $x >> x_c$ extended states (green). (Figure reproduced from ref. [53].) (See also color Plate V.)

detect a related PL line is around 2 GPa. On the other hand, Schwabe *et al.* [57] have also reported some nitrogen-related PL peaks in the 1.509 to 1.440 eV range at ambient pressure that they have ascribed to nitrogen on the basis that their GaAs crystal was intentionally N-doped. These apparently conflicting findings have been reconciled, however, by Liu *et al.* [58] who proposed that at ambient pressure nitrogen gives a resonant state, but the NN_1 pair gives a PL line in the bandgap of GaAs, at the energy position reported by Schwabe *et al.* [57]. Some identifications of these lines in the bandgap of GaAs in terms of specific orientation of the nitrogen pairs were later proposed by Makimoto *et al.* [59], and calculated by Zhang and Mascarenhas [60] and Kent and Zunger [25]. In the absence of investigation using symmetry breaking perturbation the identification is still pending. The formation of nitrogen clusters has also been postulated [61]. In fact, GaP:N and GaAs:N have, *mutatis mutandis*, a similar behaviour when doped with nitrogen as shown in Fig. 15.7 [62]. The appearance of an excitonic peak (at 1.47 eV) for a nitrogen concentration of 0.23% is an evidence for the onset of the formation of a band structure. This critical composition is also compatible with the recent hydrostatic pressure investigations of Klar *et al.* [63]. Note that in the case of pseudomorphic growth onto GaAs, the GaAsN layer is expected to experience a biaxial tension splitting the valence band into light-hole and heavy hole states. The fundamental bandgap should thus involve the light hole that is strain-induced coupled with the spin-orbit split-off hole. However, no splitting can be detected in Fig. 15.7 [62].

15.2.2.2 *Ga(In)NAs (GINA) alloys*

As stated above, the nitrogen state in GaAs is serendipitously very close to the L-conduction band minimum. Keeping in mind that the L Bloch states are important to explain the properties of GaP:N and GaPN alloys (cf. Section 15.2.1.1), one also expects a strong interaction of the nitrogen wave function with the Bloch states of GaAs, and strong consequences on the optical properties of

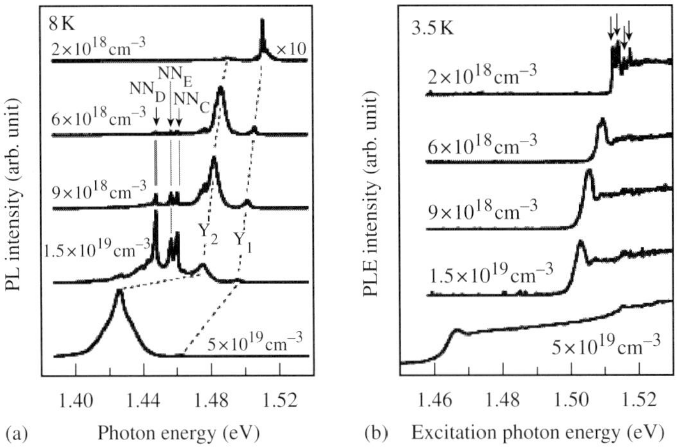

Fig. 15.7 (a) Photoluminescence spectra of GaAs:N as a function of the N concentration, (b) photoluminescence excitation spectroscopy spectra of GaAs:N as a function of the N concentration.

GINA alloys. In fact, the situation is much more complex and the reader is referred to the review article of Zunger for a complete description of the anomalous behaviour of the nitride alloys [64].

Many theoretical investigations [65–76] have predicted a red shift of the bandgap when increasing the nitrogen content in GaAsN and in other III–V III–N mixed compounds. A red shift as high as 100–180 meV/%N has in fact been reported [9, 77–95]. Fig. 15.8 shows electro-reflectance spectra taken at room temperature from a series of GaAsN samples and from a couple of GINA epilayers [81]. The striking point is that not only does the fundamental transition red shift, as expected, but an additional feature is also detected which, in contrast, blue-shifts when the nitrogen concentration increases. These transitions have been labelled E_- and E_+ respectively. Shan *et al.* [80] have followed the evolution of the E_- and E_+ transitions when such samples are submitted to an hydrostatic pressure, as shown in Fig. 15.9. In contrast to the behaviour of nitrogen-free GaAs and GaInAs [55, 56], the fundamental bandgap of the mixed nitride alloys shifts strongly nonlinearly with the pressure, as do the bound excitons in GaP : N and as does the bandgap of GaPN alloys. Further, the larger the nitrogen concentration, the stronger the nonlinear behaviour, and the higher the range of pressure which allows detection of a photoreflectance signal [80].

A two-band anticrossing model was then developed to explain this nonlinear behaviour as well as the evolution of the E_-–E_+ splitting with nitrogen concentration. This model involves the fundamental Ga(In)As conduction state and the nitrogen level. From normalization conditions of the wave functions, the coupling term varies like the square root of the nitrogen composition [72]. This model was, *mutatis mutandis,* recently extended to II–VI semiconductor ternary alloys [96]. However, elaborate theoretical calculations have

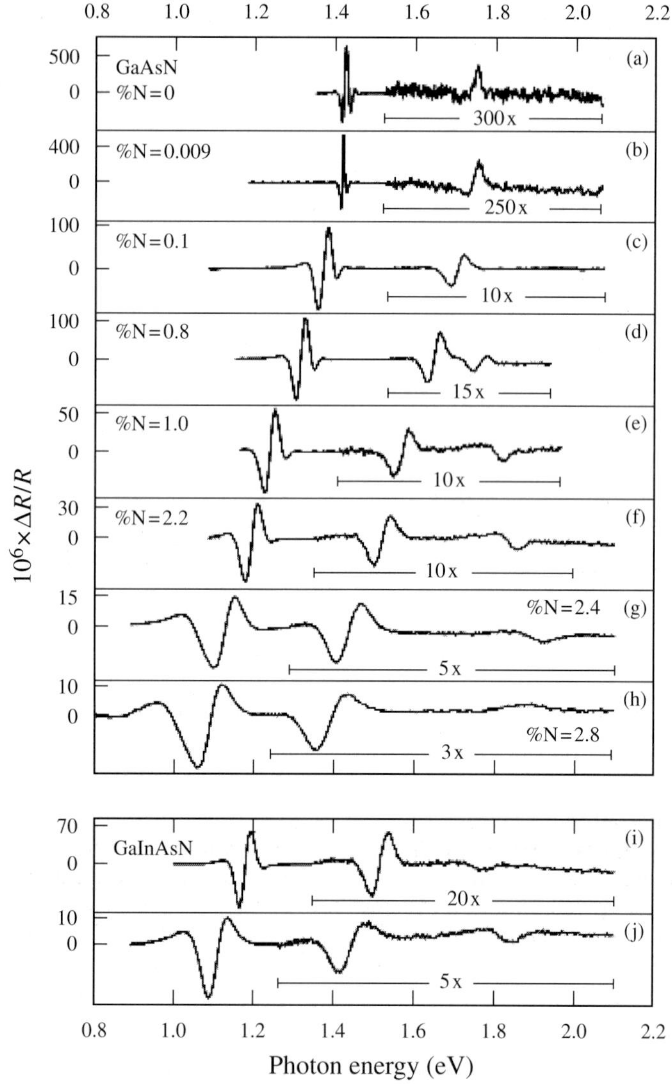

Fig. 15.8 Electroreflectance spectra for $GaAs_{1-x}N_x$ [(a)–(h)] and GINA [(i) and (k)]. Note the different scale used for each panel as well as the expanded scale within each panel used to more clearly display the above bandgap transitions. For the $GaAs_{1-x}N_x$ samples, the nitrogen content ranges from $x=0$ (a) to $x=0.028$ (h). For the GaInAsN samples, the compositions are $Ga_{0.95}In_{0.05}As_{0.987}N_{0.013}$ (i) and $Ga_{0.92}In_{0.08}As_{0.978}N_{0.022}$ (j). (Figure reproduced from ref. [81].)

challenged this representation which does not include interaction with the rest of the conduction band, namely X and L states, [69, 70] despite the importance of the interaction with these conduction band states. Further, the expansion of the conduction state along the Γ, X, L and IN> states drastically varies with the composition for GaAsN [70] as well as GINA [69] alloys as shown in

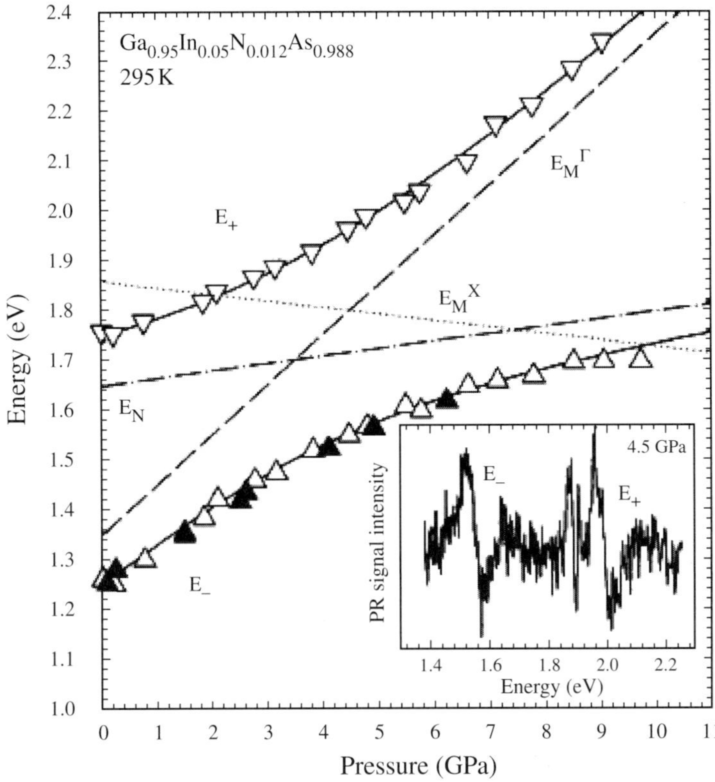

Fig. 15.9 Change of the E_- and E_+ transition energies in $Ga_{0.95}In_{0.05}N_{0.012}As_{0.988}$ as a function of applied pressure. The open triangles are photoreflectance data and the filled triangles are photomodulated transmission data. The dashed, dotted, and dot-dashed lines are the pressure dependence of the Γ and *X* conduction band edges of the $Ga_{0.95}In_{0.05}$ As matrix and the N level relative to the top of the valence band, respectively. The inset shows a PR spectrum taken at 4.5 GPa. The narrow PR spectral feature at energy below E_+ originates from the GaAs substrate. (Figure reproduced from ref. [80].)

Fig. 15.10. In addition the computed pressure-dependence of the PL line is in excellent agreement with the experiment (Fig. 15.11). The interesting point is that these calculations predict a reduction of the oscillator strength (Γ component) near the E_-–X_+ pressure-induced anticrossing as shown in Fig. 15.2 of ref. [70]. This renders quite impossible the detection of any photoreflectance signal at high pressure, as it is in fact experimentally observed [80]. Such a behaviour cannot be accounted for by the two-band anticrossing model. These theoretical calculations have been reproduced by Gil [73], by using a fully heuristic empirical approach which is quite tractable and in which the nitrogen content can be varied continuously. This phenomenological approach is based on well-known symmetry arguments: nitrogen which substitutes to the group V atom give an a_1 impurity level which couples with the

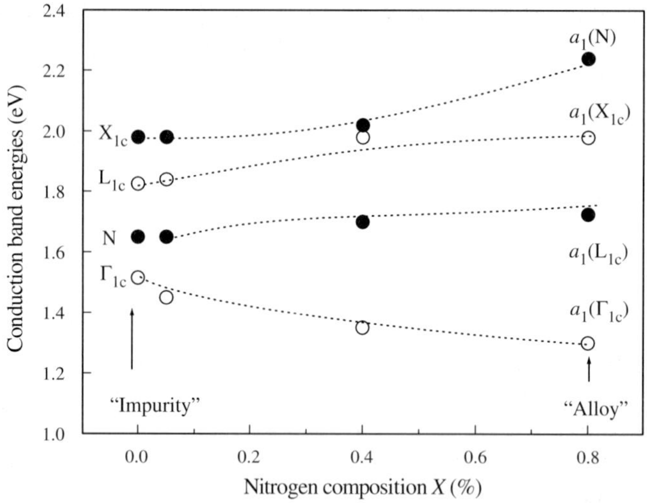

Fig. 15.10 The composition dependence of the energies of the lowest conduction band states in nitrogen-poor $GaAs_{1-x}N_x$ alloy. (Figure reproduced from ref. [71].)

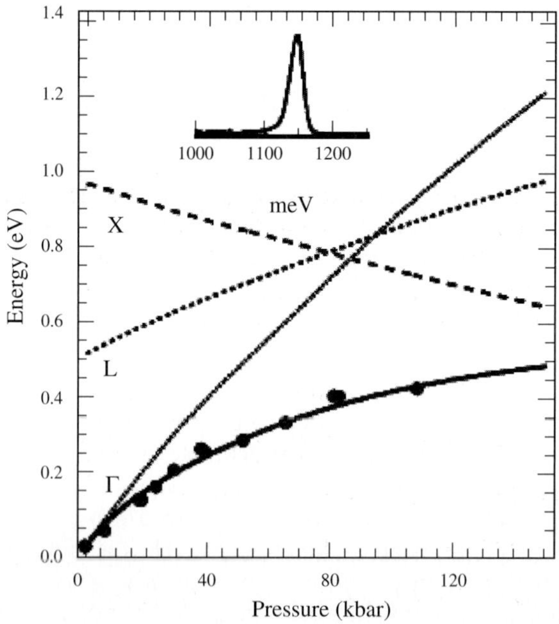

Fig. 15.11 Experimental (dots) and theoretical (solid line) dependence of the bandgap energy shift versus pressure at 4 K for 2% nitrogen in $In_xGa_{1-x}As_{1-y}N_y$. The dotted lines are the calculated pressure dependencies for the GaAs Γ, L, and X points. The inset shows the 4 K ambient pressure photoluminescence spectrum. (Figure reproduced from ref. [69].)

a_1 zone centre folded states that are fully symmetric linear combination of the contributions of the X or L conduction extrema. The conduction states of GaAsN alloys are obtained by using the following 4×4 matrix description:

$\|N>$	$\|a_1^\Gamma>$	$\|a_1^L>$	$\|a_1^X>$
E_N	$V_N\Gamma$	V_{NL}	V_{NX}
$V_N\Gamma$	Ea^Γ	0	0
V_{NL}	0	Ea^L	0
V_{NX}	0	0	Ea_1^X

where $|N>$ is a Wannier state, the energy of which is the limit value at the asymptotic nitrogen doping conditions. This matrix describes the zone center states, but the dispersion relations can be obtained very easily by assuming that the Bloch theorem still holds in the low nitrogen concentration range. To do that we simply have to rewrite the above equation with the following modifications (we restrict ourselves to the [001] axis):

$$Ea_1^\Gamma = Ea_1^\Gamma(0) + \frac{\hbar^2}{2m_\Gamma^*}k_Z^2;$$

$$Ea_1^X(k_z) = Ea_1^X(0) + \frac{\hbar^2}{6}\left(\frac{1}{m_l^{*X}} + \frac{2}{m_t^{*X}}\right)k_Z^2;$$

$$Ea_1^L(k_z) = Ea_1^L(0) + \frac{\hbar^2}{6}\left(\frac{1}{m_l^{*L}} + \frac{2}{m_t^{*L}}\right)k_Z^2.$$

For the four L minima the projections of the effective mass in the z direction are the same, but the Z [001] minimum deals with the longitudinal mass, while the X [100] and Y [010] extrema deal with the transverse mass. The fit to the data (evolution of the optically active states with nitrogen composition x, and pressure P) could be achieved by using the following set of matrix elements:

$$E_N(0)(\text{meV}) = 1675\text{–}2520\,x + 24(\text{meV/GPa})P(\text{GPa});$$

$$Ea_1^\Gamma(0)(\text{meV}) = 1430\text{–}1550\,x + 110(\text{meV/GPa})P(\text{GPa});$$

$$Ea_1^L(0)(\text{meV}) = 1745 + 55(\text{meV/GPa})P(\text{GPa});$$

$$Ea_1^X(0)(\text{meV}) = 1870\text{–}15(\text{meV/GPa})P(\text{GPa});$$

$$V_{N^\Gamma}(\text{meV}) = -2400\sqrt{x};\ V_{NL}(\text{meV}) = -400\sqrt{x};\ V_{NX}(\text{meV}) = -200\sqrt{x}.$$

The evolution of the different states with N are shown in Fig. 15.12. For nitrogen concentrations higher than 5%, this model collapses since the coupling terms become very high. Thus, mixing with upper conduction states as well as mixing of the conduction states with the valence band states may occur and any perturbation approach is then hopeless.

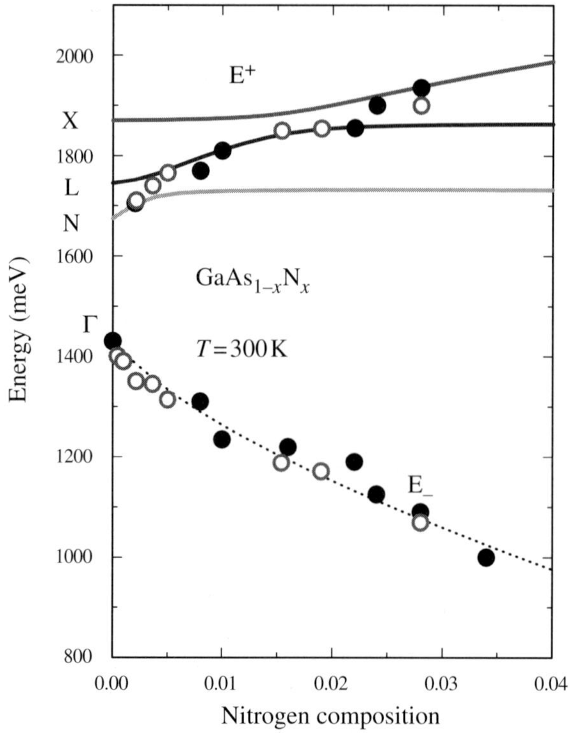

Fig. 15.12 Evolution of the conduction levels in GaAsN alloys. (Figure reproduced from ref. [73].)

The E'_0 bandgap does not vary much with the nitrogen content in GaAsN alloys because of the large energy separation between its value in GaAs and the energy of the nitrogen state [90]. A similar behaviour of the E_2 gap (at the X zone boundary of the Brillouin zone) is also obtained [90] since the coupling between N and X is small [69, 70, 73]. In contrast, the strong coupling of N with L is theoretically expected to influence the E_1 bandgap [69, 70, 73], as is experimentally observed [87, 90, 91]. Finally, the valence band shear deformation potential *b* has also been found to present a nonlinear behaviour [97] the origin of which remains to be explained.

The effective masses can be obtained from the $E(k)$ relations from the second derivative of the dispersion relation or by $m^*(k_z) = [(1/\hbar^2 k_z)(dE/dk_z)]^{-1}$.

To compute the dispersion relations, we have used $m^*_\Gamma = 0.067$, $m^{*X}_\ell = 1.3$, $m^{*X}_t = 0.23$, $m^{*L}_\ell = 1.8$, and $m^{*L}_t = 0.26$. The effective masses obtained by this crude approach are shown in Fig. 15.13. Experiments confirm that in spite of narrower bandgaps the effective masses in GINA alloys are much heavier than in their N-free Ga(In)As parent, a result which is predicted by both the two-band anticrossing model as well as the Γ, X, L and IN> mixing. Finally, the effective masses appear to be sensitive to the doping level [25, 84, 89, 98], as shown in Fig. 15.14. This issue merits further in-depth investigations.

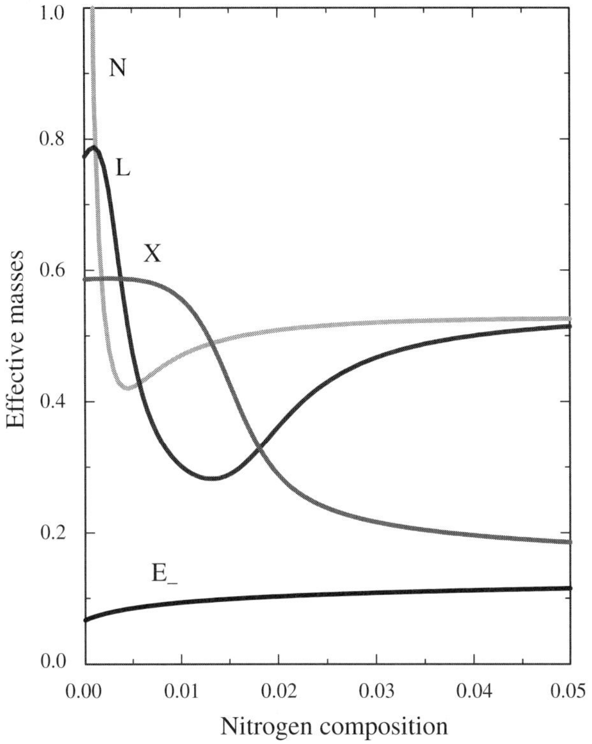

Fig. 15.13 Computed electron effective mass m^* for GaN_xAs_{1-x}.

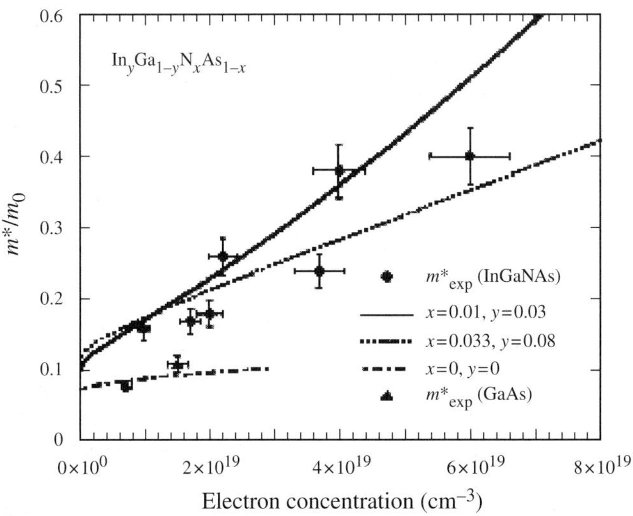

Fig. 15.14 Experimental electron effective mass m^* versus the electron density, n_e, for $In_yGa_{1-y}N_xAs_{1-x}$ samples with different alloy compositions. The lines are the theoretically calculated masses for the indicated compositions. (Figure reproduced from ref. [84].)

It appears that most, but not all, experimental findings can be interpreted within the two-band anticrossing model. However, a complete and more realistic picture of the band structure of Ga(In)NAs alloys has to take into account the mixing of the Γ, X, L and IN> bands [99].

15.2.3 Gallium nitride doped with arsenic

During the last few years there has been little experimental and theoretical interest in As-doped GaN [66–68, 79, 100–117]. As expected from a theoretical point of view a red shift of the GaN bandgap is experimentally observed [79].

In addition, a strong blue luminescence is observed near 2.6 eV in GaN : As. Its origin has been addressed by theoretical calculation [102] as well as time-resolved spectroscopy [118]. In principle, isoelectronic doping of GaN by As could be achieved by incorporation of As atoms on N sites during growth. This is predicted to be favoured under n-type conditions, in a Ga-rich environment [102]. Unfortunately, due to the much larger size of the As atom as compared to N, the formation energy of As_N is very high (4.2 eV), and almost independent of the position of the Fermi level (under Ga-rich conditions). The creation of large amounts of As_N is thus highly improbable. In n-type conditions, less Ga-rich conditions will raise the formation energy of As_N and will reduce the formation energy of As_{Ga}^{0}. In this context, Van de Walle and Neugebauer [102] calculated the formation energies of As_{Ga}^{0}, As_{Ga}^{+}, and As_{Ga}^{2+} defects. They found that the formation of an As_{Ga}^{2+} complex is energetically favourable. Consequently they associate the 2.6 eV luminescence to this double donor charge state. If this transition $\varepsilon(2+/+)$ is found at 2.7 eV, the $\varepsilon(+/0)$ transition is then computed at 2.2 eV, a behaviour characteristic of 'negative U centres'. Such negative U centres involve large lattice relaxation and should result in long decay times. Indeed, *long* decay times in the 50 ns to the 150 ns range have been measured for the 2.6 eV PL line [118] which support the interpretation that this PL line arises from the incorporation of arsenic on gallium sites.

Isoelectronic exciton binding energies of 310 meV [102], and 410 meV [103, 104] have been computed for the As_N configuration which is in reasonable agreement with the 287 meV value of the activation energy deduced in ref. [119]. This should give a PL emission near 3 eV. Large binding energies of 1.27 eV [104] and 1.9 eV [102] have been computed for As_N–As_N and As_N–As_{Ga} pairs, respectively. However, the corresponding deep levels could not be detected up to now in PL spectra. In fact, although high these binding energies remain unsufficient to compensate for the formation energy of As_N. Formation of isolated As_{Ga} is thus always favoured over formation of As_N–As_N or As_N–As_{Ga} pair defects [102].

Finally, an important conclusion to be drawn from these theoretical investigations is that since As_{Ga} behaves as a deep double donor, it acts as a compensating centre in p-type GaN and renders the p-type doping of N-rich GaNAs alloys basically impossible [102].

15.2.4 Other III–V–N compounds

Even though most work up to now has focused on the afore-described compounds a few papers report on other III–V—III–N mixed compounds. PL, absorption and photo-reflectance experiments performed on InPN alloys grown onto InP reveal a strong reduction of the bandgap with increasing the nitrogen content [119, 120]. Finally, the GaAsSbN quaternary alloys grown onto GaAs has been proposed as an alternative to GaIn-NAs. Its bandgap shrinks with the nitrogen content in a way similar to its counterpart [121].

15.2.5 The 'anomalous' temperature behaviour of the bandgap of III–V–Ns

GaAsN/GaAs and GINA/GaAs quantum wells have been recently investigated [122–133] which exhibit a type-I configuration of the band line-up [132]. Their optical properties are dominated by localized excitons [128, 133, 134]. The measured decay-times decrease with increasing temperature which reveals the presence of a high density of non-radiative recombination channels [133].

There has been several reports [45, 128, 129, 133–136] of an anomalous temperature-dependence of the bandgap of mixed III–V–N alloys and of their heterostructures. In fact, when the temperature increases the red shift of the bandgap is much smaller for III–V–N alloys and heterostructures than for their nitrogen-free III–V parent alloys. This has been interpreted in a quantum mechanics context by Kent and Zunger [53] in terms of thermally-induced detrapping between the localized cluster states (CS) and the delocalized perturbated host states(PHS). The genesis of the effect originates in the conduction band. For low-dimensional systems, the situation will be even much more complex due to confined valence states and to possible additional thermal population effects between light-hole and heavy-hole confined states which depend on parameters like the heavy-hole to light-hole splitting, i.e. on the thickness of the quantum well.

A brutal blue-shift of the PL line upon a temperature rise is reported in ref. [133]. Further, a comparison of the temperature-dependent PL of GaInAs, GaAsN and GINA QWs (Fig. 15.15) reveals that this singular behaviour is observed with GINA QWs only which shows that it is the concomitant presence of both N and In which results in a strong carrier localization in GINA quantum-wells [134]. The situation is somewhat to be compared to the case of GaInN wide bandgap quantum wells where a strong localization has also been detected and ascribed to the formation of InN-rich clusters [137, 138]. Indeed, recent theoretical calculations [139] support the assertion that the GINA alloy is not random, but shows a tendency for the formation of In–N bonds which is accompanied by a broadband tail of localized states. However, careful investigations of thermal annealing effects [135, 136] indicate that localization in GINA quantum-wells can be controlled if not eradicated. This suggests that the

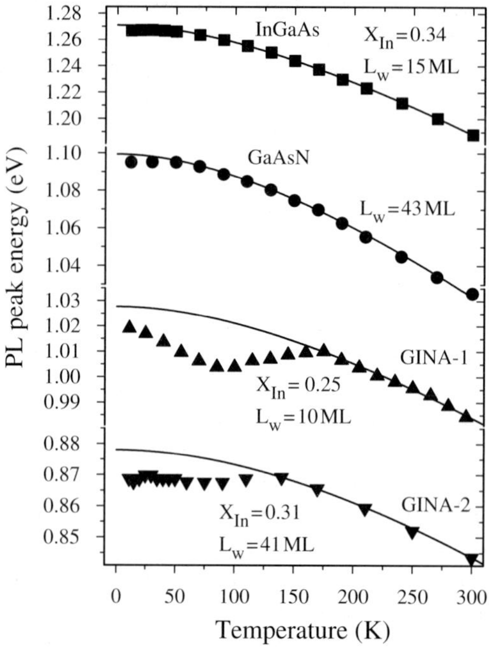

Fig. 15.15 Evolution with the temperature of the PL-peak energy for a set of GaInAs, GaAsN et GINA QWs. Strong localization is observed for GINA QWs only.

full technological processing chain may significantly affect the sample properties on a nanoscopic scale. In any case, the history of a sample should be taken into account before describing the temperature-dependence of its properties.

15.3 Epitaxial growth of III–V–N alloys and heterostructures

As mentioned in the Introduction, thermodynamics predicts an exceedingly low solubility of N in III–V compounds under equilibrium conditions [5–7]. Consequently, the N content in early III V N samples, which were prepared by near-equilibrium epitaxial techniques, was limited to the doping range. However, with the advent of far-from-equilibrium techniques such as molecular-beam epitaxy (MBE) or metal-organic vapour-phase epitaxy (MOVPE), thermodynamics can be (to some extent) circumvented. In these conditions kinetics as well as surface phenomena come into play and a greatly-enhanced nitrogen-solubility (in the % range) is theoretically predicted [140] as experimentally achieved.

We now summarize the main results obtained by MOVPE- and MBE-related epitaxial techniques. A few factors, common to both kinds of techniques, have emerged as the key for the successful growth of III–V–N alloys. Among others, one can cite the necessity to develop efficient N-sources, the use of much lower temperatures than for conventional III–V compounds, or the effectiveness of a post-growth annealing to improve the material quality. In the following we will restrict ourself to the growth of III–As–N alloys which are, however, representative of the problems faced during the growth of the whole family of III–V–N compounds.

15.3.1 Epitaxial growth of III–V–N alloys and heterostructures by MOVPE and related techniques

The crucial point in the MOVPE growth of III–V–N compounds is to find the most appropriate N-source. Ammonia (NH_3) is most commonly used for the epitaxy of wide bandgap GaN-based materials. Early experiments on the growth of GaAsN alloys have thus been carried out using NH_3 [8]. Even though [N] contents around 1% could be achieved, the PL spectra of these samples were plagued by strong deep-level emission. In addition, the most straightforward way to reduce the tendency to phase separation in thermodynamically unstable materials is to grow them at low temperatures. However, efficient cracking of NH_3 is obtained at high temperatures only [141]. These boundary conditions are thus marginally compatible. Consequently, a strong effort has been directed toward investigating other N-containing molecules. With a low cracking temperature and a high vapour-pressure [142], dimethylhydrazine (DMHy) rapidly emerged as a candidate of choice [143–145]. A comparison of several precursors has shown that no N incorporation could be measured when using tertiobutylamin, while significant incorporation was obtained when using hydrazine, even at lower nitrogen/III ratios than would be required using DMHy [146]. Still, hydrazine is a very explosive compound. By now DMHy has thus been adopted as the N-source by most groups involved in the field, whatever the conditions (atmospheric-pressure or low-pressure) used during MOVPE growth.

Furthermore, safety considerations during MOVPE-growth of As-based compounds had previously led to the search for alternative group-V sources as substitutes for the highly-toxic traditional III–V epitaxy precursors arsine (AsH_3) [147]. It turns out that the selected compound, tertiarybutylarsine (TBAs), has improved decomposition properties, allowing the use of lower growth temperatures and lower V/III ratios. These characteristics are being used by several groups to grow III–V–N compounds at substrate temperatures in the 500–550 °C range [143, 144, 148–153]. Fig. 15.16 shows the evolution with the growth temperature of the N-content in GaAsN epilayers for two fractional molar flows x_v (x_v = (DMHy/(DMHy + TBAs))) of DMHy [143]. In the low temperature region (500–530 °C), the N incorporation appears to be controlled only by the fractional flow of DMHy. In contrast, at higher temperatures the N-content decreases exponentially. Fig. 15.17 displays the effect of the DMHy/TBAs molar-flow ratio on the N-content for GaAsN layers grown at 530 °C [152]. It shows that the effect of the V/III ratio on the N composition is dependent on the fractional molar ratio x_v. When x_v is low, the nitrogen incorporation is fairly insensitive to the V/III ratio [150]. These figures show

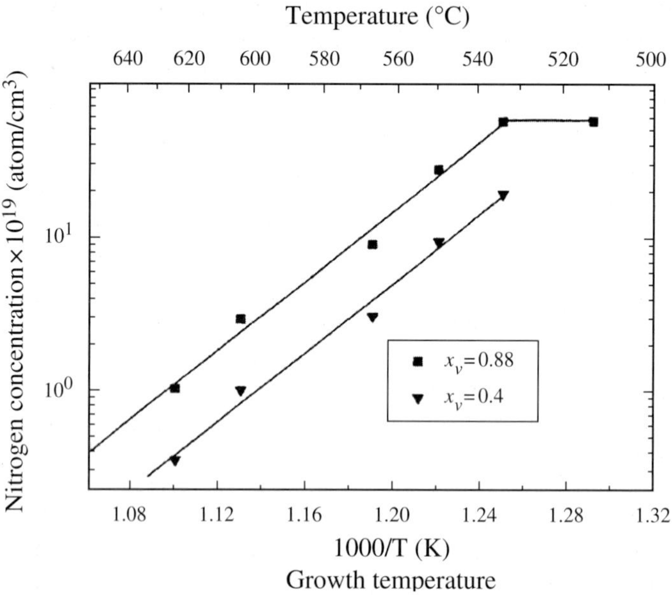

Fig. 15.16 Evolution of the nitrogen concentration (measured by SIMS) in GaAsN layers grown by MOVPE as a function of growth temperature for two fractional molar flows x_v of DMHy. (Figure reproduced from ref. [143].)

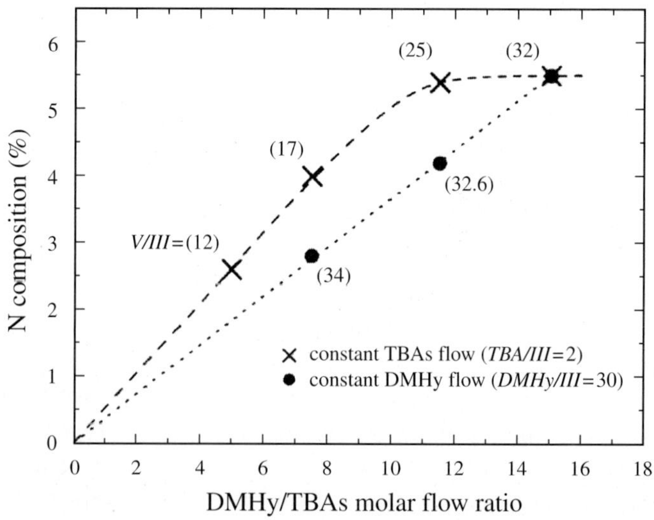

Fig. 15.17 Effect of DMHy/TBAs molar flow ratio on the N composition of GaAsN layers in MQW structures. The V/III ratio of each sample is shown in parantheses. The lines are guide to the eyes. (Figure reproduced from ref. [152].)

that the use of TBAs and DMHy sources allows growth conditions to be found for which the N content in GaAsN can easily be controlled up to around [N] = 5% since it varies linearly with the DMHy flow.

The situation appears to be much more complex during MOVPE growth of GINA alloys. In fact, several groups have reported that under otherwise identical conditions the N-incorporation in GINA is much lower than in GaAsN [145, 148, 153, 154]. Friedman *et al.* [154] have studied this problem in details. Figure 15.18 shows the N-content in GINA epilayers with different In contents grown onto GaAs substrates. For In contents above 1% there is a clear trend toward decreasing N content with increasing In content. The origin of this behaviour is not yet established. It appears to be connected neither to the strain of the layer, nor to a pre-reaction between the reagents. The influence of In segregation at typical GINA growth-temperatures (>500 °C) has been invoked as a possible explanation [154]. Indeed, segregation leads to an In-rich growing front [155] and the InNAs ternary alloy is known to be extremely difficult to synthetize, whatever the growth technique [156–158]. Another argument is that the equilibrium vapour pressure of N_2 over In-based compounds is much higher than over Ga-based compounds [159]. Consequently for the same amount of active N species, less N is incorporated in In-based than in Ga-based compounds [160].

GaAsN- and GINA-based heterostructures have also been grown by chemical beam epitaxy (CBE) and metal-organic molecular-beam epitaxy (MOMBE) using metal-organic group-III sources and various combinations of

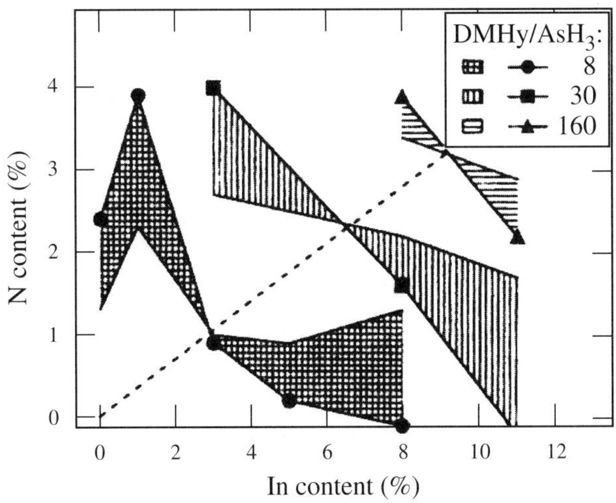

Fig. 15.18 N contents for a matrix of GaInNAs/GaAs epilayers grown by MOVPE at various DMHy and TMIn flows, at 5 µm/h and AsH_3/III = 4.4. The N contents are estimated from the lattice constants; the lines bounding the shaded area indicate the unstrained- and strained-epilayer limits of the estimate. The diagonal dashed line represents the compositions $Ga_{1-3x}In_{3x}N_xAs_{1-x}$ lattice-matched to GaAs. (Figure reproduced from ref. [154].)

As- (solid arsenic, AsH_3, tridimethylaminoarsenic) and N-sources (plasma-cracked N_2, DMHy, mono-MHy) [153, 161–164]. The only works having addressed the In/N incorporation-interplay used a plasma-cracked N_2 source [163, 164]. They concluded that the presence of In marginally impacts the N incorporation which could reveal that the reduced N-uptake observed during MOVPE-growth of GINA alloys is specific to the MOVPE environment. Note, however, that CBE and MOMBE growths are performed at much lower temperatures (450–500 °C) than MOVPE which does reduce In-segregation.

15.3.2 Epitaxial growth of III–V–N alloys and heterostructures by MBE and related techniques

Gas-source MBE (GS-MBE) [77, 160, 165–169] and solid-source MBE (SS-MBE) [121, 127, 170–175] are currently taking an increasing importance in the development of Ga(In)AsN compounds and heterostructures. For both techniques elemental Ga and In are used as group-III sources while plasma-cracked N_2 is the N-source. Arsenic is obtained from thermally cracked AsH_3 in GS-MBE and from evaporation of solid-arsenic in SS-MBE.

As for the N-source, most groups use radio-frequency (RF) plasma cells, although a few reports on the use of electron-cyclotron-resonance (ECR) [169], or DC [173] plasma-cells. The main differences between the type of sources lie in the nature and density of excited species produced in the plasma. In particular, ECR sources are known to produce more ions than RF sources [176], while DC cells give a low flux of excited species. Although it has not been clearly established which kind of species is preferrable, the common belief is that ions damage the growing crystal, which explains why RF cells are generally preferred.

Early experiments performed on GaP substrates showed that very high N contents, in the 10–20% range, could be obtained for GaAsN alloys grown by GS-MBE [160, 166] or SS-MBE [170]. Further, it has been reported that at high temperature (>650 °C) these alloys were phase separated [160, 170], as expected from thermodynamic considerations, while they can be preserved as single phases at much lower temperatures (<500 °C) [160]. The quality of such high N-content alloys is however probably very low, and most works have subsequently focused on growing high quality alloys with N contents in the % range.

A crucial parameter in the MBE growth of Ga(In)AsN alloys is, again, the growth temperature. We show in Fig. 15.19 the evolution, as a function of the growth temperature, of the PL intensity and linewidth from GINA/GaAs QW heterostructures grown by SS-MBE [177]. There is a clear optimum for the temperature which, in this MBE system, lies in the 420–440 °C range, i.e. around 50–100 °C below the typical temperatures for the growth GaInAs/GaAs QW heterostructures. Similar conclusions can be drawn for the growth of GaAsN/GaAs QW heterostructures [178]. There is a clear relationship between the surface reconstruction which is observed by *in situ* reflection high-energy electron diffraction (RHEED) during growth and the material

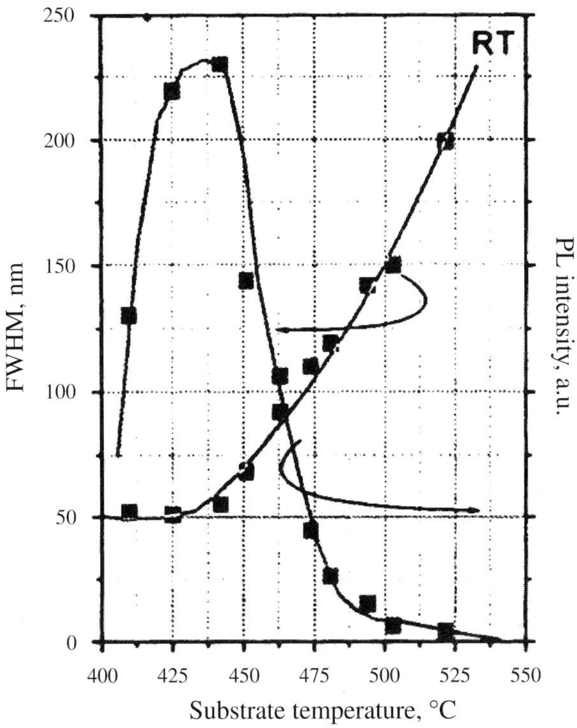

Fig. 15.19 Influence of growth temperature on the PL characteristics of GINA/GaAs QW heterostructures grown by SS-MBE. (Figure reproduced from ref. [177].)

quality [178]. When initiating Ga(In)AsN growth on a typical (2×4) GaAs-reconstruction in the 400–480 °C range, the surface exhibits an unclear, faint (2×1) or (2×3) reconstruction [179]. Epitaxy then proceeds in a two-dimensional growth-mode. High quality samples are obtained under these conditions. When increasing the growth temperature, the growth mode evolves progressively from a two-dimensional- toward a three-dimensional-mode. Finally, at high temperatures (>520–550 °C), a (3×3) reconstruction is observed which is the signature of a nitrided GaAs surface [180]. Low N-content alloys ([N] < 0.5%) only can be grown under these conditions. Higher N-contents inevitably result in a catastrophic evolution of the growth as evidenced by a very spotty RHEED pattern. The optimum growth-temperature for Ga(In)NAs alloys thus results from a compromise between the best temperature for Ga(In)As growth (>500 °C) and the best temperature for N-containing alloys (<500 °C).

For given N_2-plasma conditions two parameters, namely the growth rate and the growth temperature, primarily affect the N-incorporation. Indeed, increasing the temperature from 420 to 580 °C reduces the N-content by about one order of magnitude [178]. On the other hand, the N-incorporation is essentially inversely proportional to the growth rate [173, 175]. In addition, Fig. 15.20

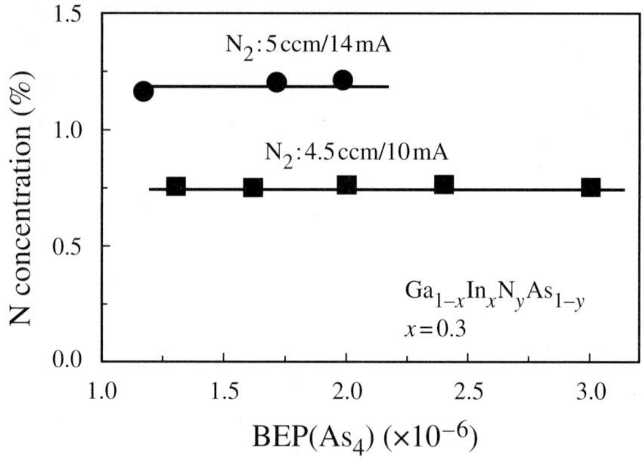

Fig. 15.20 Influence of the As_4 flux on the N composition of GINA layers grown by SS-MBE at fixed N supplies. (Figure reproduced from ref. [173].)

shows that the N-content in GINA layers does not depend on the As-flux during growth [173]. Finally, it has been shown that, in marked contrast to MOVPE growth, the presence of In does not affect the N-incorporation during SS-MBE growth [85, 127, 173]. These results indicate that the N-sticking coefficient depends only on the plasma-cell operating conditions and on the substrate temperature during MBE growth of Ga(In)NAs alloys. Note, however, that the presence of Sb significantly increases the N-incorporation, which indicates a lower-than-unity sticking coefficient in Ga(In)NAs growth [85].

15.3.3 Material properties

All studies up to now show that the PL efficiency of as-grown GaAsN and GINA epilayers or QWs decreases upon an increase of the N-content. It is noticeable that this drop appears to be much more severe for samples grown by MOVPE and related techniques than for MBE-grown samples [127]. In addition, a high concentration of deep levels distributed within the bandgap [181] and persistent photoconductivity [182] have been detected in nominally undoped GINA epitaxial layers grown by MOVPE which exhibit a residual p-type doping in the $10^{17}\,cm^{-3}$ range. Carbon arising from DMHy has often been suspected as a possible source for this high background doping [183]. However, a similar doping is obtained even when using hydrazine which does not contain any C atom [146]. The origin of this unintentional doping remains to be determined since, in contrast, we observe that GINA epitaxial layers grown by SS-MBE generally exhibit a n-type background in the 10^{14} cm^{-3} range. The reduced electron-mobility however, in the 200–300 $cm^2\,V^{-1}\,s^{-1}$, reveals also the presence of defects, as do other electrical characterizations [184, 185]. Finally, oxygen-related deep levels have also been detected in GINA quaternary alloys grown by MOVPE [186]. Unfortunately, the transport properties of Ga(In)AsN alloys have been scarcely investigated up to now which hinders definite conclusions to be drawn.

The reduced PL efficiency upon N-incorporation reveals that non-radiative recombination centers are generated in the material concomitantly with N-uptake. The origin of this is not established. Possible reasons are the fact that despite low N-contents growth temperatures are much lower than for conventional III–V compounds, or the unstable nature of the alloys. In fact, it has been soon recognized that an either *in situ* or *ex situ* post-growth annealing strongly affects the PL properties of Ga(In)AsN-based materials and heterostructures [187–191]. As an illustration we show in Fig. 15.21 the influence of *ex situ* annealing (30 min at 650 °C) on the PL spectra of GaAsN/GaAs (a) and GINA/GaAs (b) QWs grown by SS-MBE. The N-content in both samples is around 3%. In both cases, the PL efficiency is increased by about one order of magnitude and the line-width is slightly reduced upon annealing. However, a major difference between the two materials sytems is evidenced in Fig. 15.21: the energy position of the PL line remains basically unchanged in the case of the GaAsN QW while it is blue-shifted by 90 meV for the GINA QW. This indicates that annealing of GaAsN apparently cures GaAs-related defects induced by the low growth-temperature. On the other hand, PL

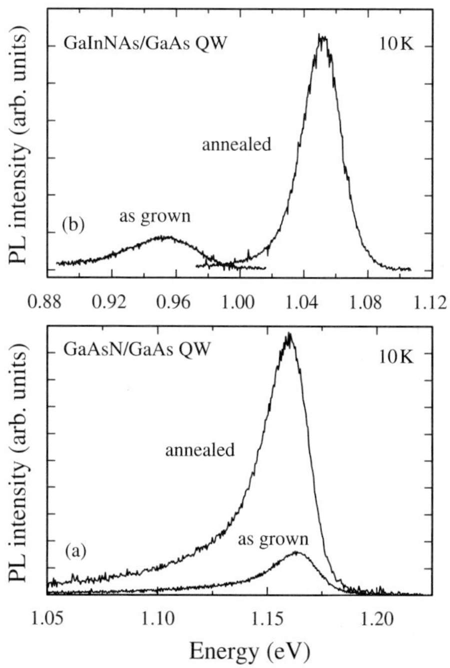

Fig. 15.21 PL spectra taken at 10 K from as-grown and annealed Ga(In)AsN/GaAs QWs grown by SS-MBE. (a) 8-nm wide $GaAs_{0.97}N_{0.03}$ QW, (b) 5-nm wide $Ga_{0.67}In_{0.33}N_{0.03}As_{0.97}$ QW. Annealing was done for 30 min at 650 °C under a flowing-N_2 ambient with GaAs proximity capping.

studies [128, 133, 134] have shown that as-grown GINA exhibits a strong carrier localization which arises from the presence of both In and N in the alloy [134], while theoretical calculations support the formation of In-N-rich clusters [139]. The large blue-shift of the PL line (Fig. 15.21) [187–191] together with the strong decrease of the localization observed upon annealing [135, 136] reveal that annealing not only reduces the non-radiative-defect density but also profoundly affects the structural properties of GINA on a nanoscopic scale. In this respect, cathodoluminescence shows that the lateral uniformity of the PL emission is strongly enhanced for annealed samples [192]. In addition, annealing affects the distribution and position of deep levels in MOVPE-grown GINA [181]. Finally, recent results report that not all N-atoms are incorporated on substitutional group-V sites in MBE-grown GaAsN alloys with N-content above ~1.5% [193, 194]. XPS data displayed on Fig. 15.22 evidence two N-bonding configurations in as grown samples while only the Ga–N bond is detected in annealed samples [193]. A N-induced complex, most probably involving interstitial N, is thus cured upon annealing [193].

15.3.4 Perspectives

The results summarized above reveal that SS-MBE is particularly well appropriate for the growth of Ga(In)NAs. Still, the strong impact of annealing on the material properties clearly indicate that the epitaxial growth of these alloys is not yet fully mastered. Some aspects such as, to cite but a few, the effect of plasma species on the crystal quality during MBE growth, N segregation,

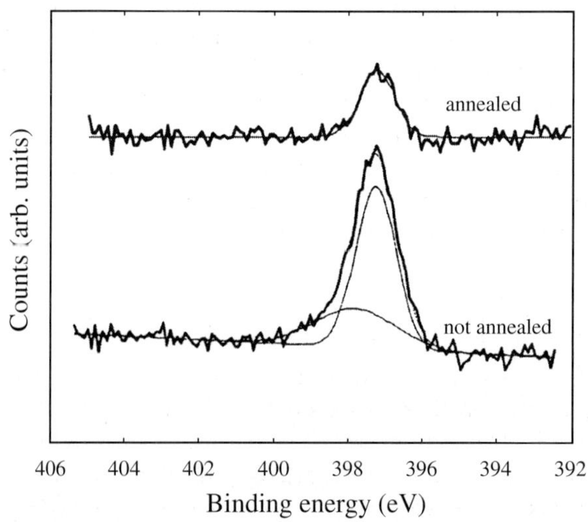

Fig. 15.22 XPS signals from a N(1s) peak from annealed and as-grown $GaAs_{0.96}N_{0.04}$ films. Annealing was done for 30 s at 775 °C under a flowing-N_2 ambient with GaAs proximity capping. (Figure reproduced from ref. [193].)

phase separation, ordering, even remain to be explored. For all growth techniques there appear to be a need to further work on N-sources. Also, transport properties of these alloys should be investigated in more details in connection with the epitaxial-growth conditions. Finally, most work up to now has focused on low N-content (~1%) alloys. The growth of high-quality, high-N-content alloys for fundamental as well as applied purposes remains a real challenge.

15.4 Applications of III–V–N alloys and heterostructures

Among the original driving-force behind the research effort on III–V–N alloys was the perspective to develop optoelectronics devices lattice-matched to Si substrates [77]. Indeed, the Si lattice-parameter is intermediate between those of III–N and conventional III–V compounds. However, beside for GaPN, lattice matching to Si is achieved for high N contents which renders this perspective rather elusive. Still, the fast decrease of the bandgap upon N-incorporation opens the way to very promising applications. Although $GaN_{0.011}P_{0.989}$/GaP red light-emitting diodes have recently been fabricated with interesting characteristics [50], most work is currently being devoted to GINA-based applications.

Fig. 15.23 shows the usual bandgap *vs.* lattice parameter plot which includes GaN. A general trend with conventional III–V compounds is that the smaller the lattice parameter, the wider the bandgap. III–V–N compounds however depart from this rule because they exhibit a giant bowing of their bandgap. Consequently, incorporating In and N into GaAs results in a parallel reduction of the bandgap, but on opposite variations of the lattice parameter. The GINA quaternary alloy can thus be obtained lattice-matched onto GaAs (for [In]/[N] ~ 3) with a lower bandgap than GaAs. Alternatively, strained GINA layers can accommodate more strain than their GaInAs counterpart, and thus have a lower bandgap. The GINA quaternary alloy thus forms the foundation of a new family of GaAs-based narrow-bandgap semiconductor compounds. Most of the work worldwide is directed toward the development of high-efficiency solar-cells and 1.3 μm lasers. We now review the main results.

15.4.1 High efficiency multispectral solar-cells

$Ga_{0.5}In_{0.5}P$/GaAs tandem solar-cell lattice-matched to GaAs and Ge are currently in production because of their record efficiency [195]. Their current market is space photovoltaics, but they are also leading candidate for the high-concentration terrestrial-photovoltaic market which is potentially much larger. Still, to compensate for their high costs, it is highly desirable to further increase the efficiency of such cells. Device simulations (Table 15.1) show that a dramatic improvement is expected for three-junctions or four-junctions solar-cells where the third junction is a 1 eV-semiconductor and the fourth junction

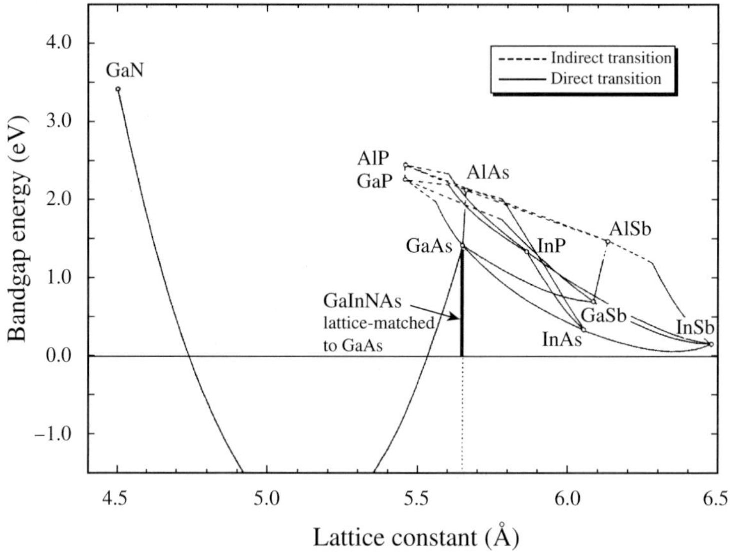

Fig. 15.23 Relationship between the lattice-constant and bandgap energy of III–V compound semiconductors. (Figure reproduced from ref. [202].)].)

Table 15.1 Expected solar cell efficiencies at 300 K for 2, 3, and 4 junction devices [196]. The AM numbers apply to use in space and the AM1.5D 500 suns numbers apply to terrestrial concentrator systems.

Device	AM0 1 sun	AM1.5D 500 suns
2 junctions	31%	36%
3 junctions (Fig. 15.24)	38%	47%
4 junction (Fig. 15.24)	41%	52%

is Ge (Fig. 15.24) [196]. However, the lack of a suitable 1 eV semiconductor lattice-matched to GaAs (and Ge) has up to now hindered the development of these cells. The GINA quaternary alloy has thus rapidly been considered as an appealing candidate for the third-junction material [197–201].

Bulk, lattice-matched GINA epitaxial layers [197, 201] as well as 1-eV solar cells [198–200], all prepared by MOVPE, have been investigated in this context. It appears that the overall material quality of as-grown samples is too low for such application. After annealing, the internal quantum efficiency increases from 20% to 70% which remains low. In addition, the minority carrier diffusion is very short, limited by the carrier localization in the alloy.

The common belief is that due to its singular band-structure and/or crystalline structure the GINA alloy is not suited for solar-cell applications. Still,

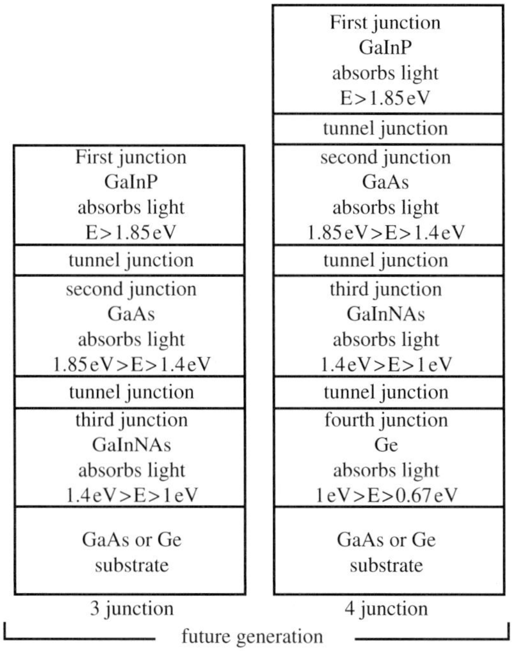

Fig. 15.24 Structures of three and four junction future generation solar-cells. (Figure reproduced from ref. [198].)

we feel that at this stage it is not easy to discriminate between extrinsic effects and intrinsic properties. Some additional work remains to be done before one can definitely conclude.

15.4.2 Long-wavelength lasers

The original idea of Kondow *et al.* when they draw the attention of the community toward the GINA quaternary alloy, was its potential for the development of high-performance GaAs-based laser diodes (LDs) operating at the 1.3 μm telecom wavelength [202, 203]. In fact, current LDs are based on GaInAsP/InP semiconductor heterostructures. They suffer particularly from poor temperature-characteristics at least partly associated to a small conduction-band discontinuity. A scheme of the band lineup for GaAsN/GaAs and GaInAs/GaAs is depicted in Fig. 15.25 [202, 203]. The incorporation of N in GaAs lowers both the conduction and valence bands, with a faster move of the conduction band. In contrast, In in GaAs lowers the conduction band and rises the valence band. Consequently, incorporating In and N into GaAs will induce a large fall of the conduction band while the valence band will be less affected. One thus expects very high conduction-band discontinuities [202, 203]. Now, Fig. 15.26 displays the evolution of the LD characteristic temperature T_0 calculated as a function of the conduction-band discontinuity [202,

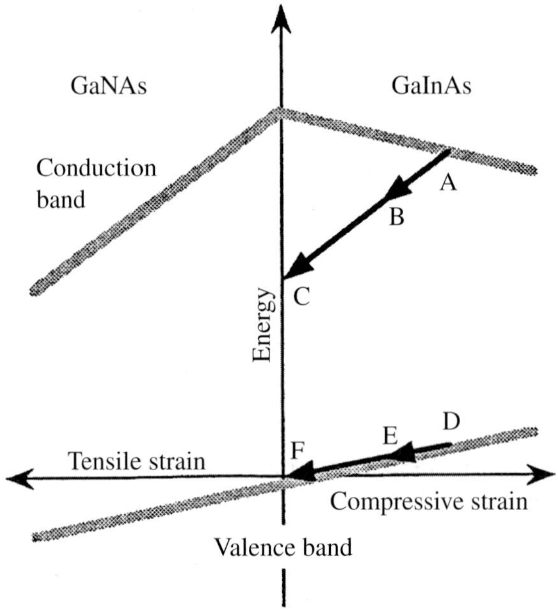

Fig. 15.25 Schematic diagram of the band line-up for GaInAs/GaAs and GaAsN/GaAs heterostructures. (Figure reproduced from ref. [202].)

203]. It shows that T_0 can be improved by using AlGaInAs- instead of GaInAsP-based LDs. However, it reaches its intrinsic value for conduction-band discontinuities as large as 300 meV, which can indeed be achieved with GINA/GaAs heterostructures. GaInNAs-based LDs should thus exhibit record temperature-characteristics [202, 203].

From another point of view, the fabrication of monolithic vertical-cavity surface-emitting lasers (VCSELs) with the InP-based system is extremely difficult. Indeed the low refractive-index contrast between GaInAsP and InP makes it very problematic to obtain high-reflectivity Bragg mirrors. As a consequence, long-wavelength VCSELs fabricated up to now rely on complicated technologies such as wafer fusion, two-step epitaxy, *ex situ* deposition of dielectric mirrors or epitaxy of metamorphic mirrors. In contrast, GINA offers the unprecedented opportunity to fabricate long-wavelength monolithic-VCSELs in a single epitaxy step using the well mastered GaAs/Al(Ga)As Bragg-mirror technology. The real breakthrough in terms of industrial benefit is surely to be expected here.

These perspectives have fostered numerous works on edge-emitting LDs as well as VCSELs. The decrease of the PL efficiency when increasing the N-content (cf. Section 15.3.3) has led to the following strategy to reach the 1.3 μm wavelength range: the In-content is increased to its maximum while the N-amount is maintained at the minimum necessary to prevent strain relaxation. Consequently, most lasers up to now are based on GINA QWs with an In content in the 0.30–0.40 range, a N-content below 1% and a thickness

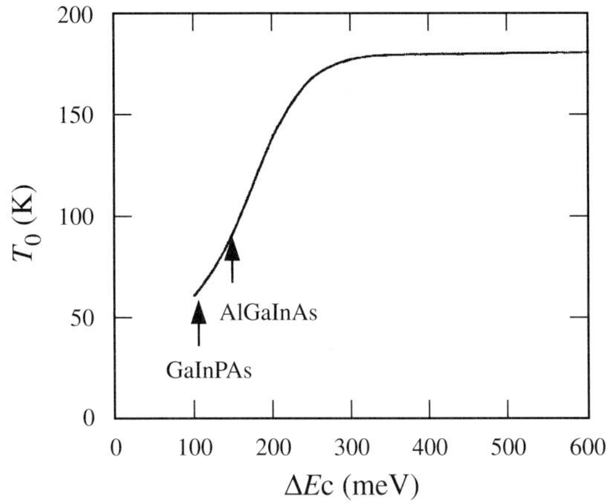

Fig. 15.26 Calculated relationship between the characteristic temperature T_0 and the conduction-band discontinuity ΔEc. (Figure reproduced from ref. [202].)

around 6–8 nm [204]. A problem however in this strategy is the difficulty to accurately control the N-amount which leads to many lasers emitting around 1.2–1.25 μm rather than precisely 1.3 μm [204].

Many methods for growing GINA LDs have been used, including GS-MBE [205], SS-MBE [206–209], CBE [210] and MOVPE [211–214]. Fig. 15.27 shows the threshold current density reported at various wavelengths. It appears clearly that SS-MBE allows fabrication of LDs emitting at the longest wavelength with the lowest threshold. In addition, output powers as high as 8 W with a slope efficiency of ~0.5 W/A under continuous wave operation at 1.3 μm have been demonstrated [209]. The characteristic temperature T_0 of these LDs, however, remains in the 80–100 K range, lower than anticipated [215]. The reason for this is not yet established. In any case, GINA-based LDs operating at 1.3 μm already exhibit performances which compare favorably with their InP-based counterparts.

The results are even more impressive as far as VCSELs are concerned. Again, GS-MBE [216], SS-MBE [215, 217–220], MOVPE [221], and CBE [222] a growth techniques have been used. Again, record performances have been obtained with SS-MBE devices. VCSELs emitting at 1.30 μm at 300 K exhibit threshold currents as low as 4 mA (7 kA/cm^2) and output powers as high as 1 mW under continuous wave operation at 11 mA drive-current [215]. Lasing up to 85 °C with output powers in the mW-range has been obtained under pulsed operation [215]. Finally, the compatibility of GaInNAs-based VCSELs with SONET links has also been demonstrated [220].

15.4.3 Other applications

Preliminary photodetectors have also been fabricated from GaAsN [223, 224] and GINA [225] alloys for detection around 1.0 μm [223, 224] or at 1.3 μm

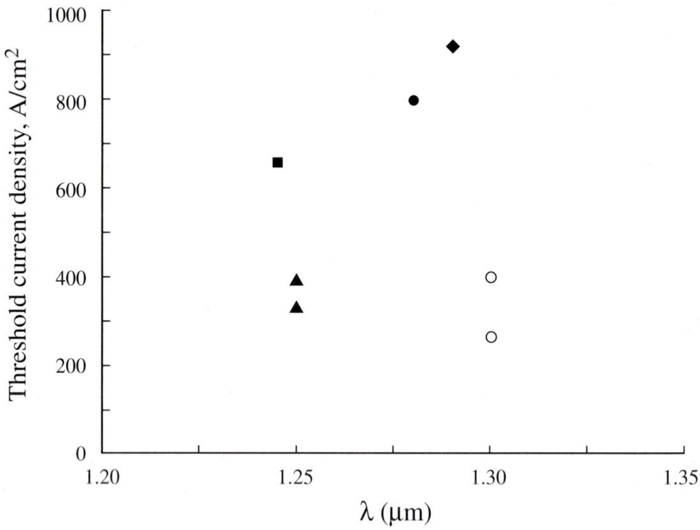

Fig. 15.27 Comparison of threshold current densities of GINA QW LDs grown by MOVPE and SS-MBE. ○ SS-MBE, ref. [209], ● MOVPE, ref. [212], ■ MOVPE, ref. [211], ◆ MOVPE, ref. [213], ▲ MOVPE, ref. [214]. (Figure reproduced from ref. [214].)

[225]. Similar to VCSELs, the possibility to grow monolithic Bragg-mirrors is a very important advantage of this materials system since it allows preparation of resonant-cavity photodetectors [224, 225]. The resonant-cavity enhances the quantum efficiency up to 75% at 1.3 μm [225]. Work is needed to further develop high-speed, high-efficiency resonant-cavity photodetectors operating near 1.3 μm which are highly desirable for wavelength division multiplexing applications.

Finally, GINA alloys are also currently being investigated for the fabrication of heterojunction bipolar transistors (HBTs). The idea is to use a quasi-lattice-matched GINA alloy as the base of the HBT. The narrower bandgap of the base would then allow lower turn-on voltage and thus reduced power dissipation. Only a few works reports on studies of GINA-based heterostructures for HBT applications [226–229]. Currently, the performances appear to be limited by the material quality as well as the band-discontinuity configurations. Again, work remains to be done, possibly by investigating other combinations of materials, before definite conclusions can be drawn.

15.4.4 Perspectives

Most effort on the applications of III–V–N compounds has been directed toward the development of LDs and VCSELs operating at 1.3 μm. As a result, impressive performances have been achieved in a very short period of time. An enticing challenge is to extend the emission wavelength of GINA-based heterostructures toward the 1.55 μm range. Recent results on LDs [230] or QW heterostructures [127] suggests that this might not be out of reach, although much effort should be devoted to this aim before practical devices can be anticipated.

Finally, although most work is being focused on GINA-based heterostructures grown onto GaAs substrates, incorporating N into $Ga_{0.47}In_{0.53}As$, which is lattice-matched onto InP with a 0.74 eV band-gap, or into Sb-based III–V compounds would allow extension of the working wavelength in the mid-infrared range.

15.5 Challenges

All results presented above show that the III–V–N semiconducting alloys are particularly enticing from both points of view of condensed-matter physics as well as telecom–related applications. However, these materials systems remain very challenging. In fact, in spite of the rapid pace of progress in the field of 1.3-μm lasers, the research in this field is still in its infancy.

Most work up to now has focused on the (indeed intriguing) electronic properties of these semiconductors. Other features yet deserve to be explored. To name but a few, the peculiar epitaxial-growth procedures and/or techniques (e.g. plasma cells) which have to be used to obtain these alloys suggest that their microstructure should be investigated in more detail. In the same line, the present N-sources do not appear to be optimized, neither for MOVPE, nor for MBE growth. In particular, the design of plasma cells needs to be improved in order to extract the only species which are useful to MBE growth while shielding the plasma ions.

In addition, the correlation between the microstructure and the electronic properties, and their evolution after post-growth annealing, remains to be established. The study of transport properties deserves also more attention than generally granted. Further theoretical work is also needed before one can rationalize all intriguing properties of these alloys. Close collaboration between experimentalists and theoreticians will then surely allow a definite picture of these semiconducting materials to be retrieved.

Finally, the body of the research work has primarily been concerned up to now with III–As–N alloys because of the immediate applications anticipated. Still, other III–V–N compounds surely deserve fundamental investigations, and can lead to important applications.

Thus we feel that much work must be done before these materials systems can definitely be understood, and, consequently, their properties in dedicated applications can be fully exploited.

Acknowledgements

As far as our own work is concerned we gratefully acknowledge our colleagues appearing in the references. Work at CRHEA is supported by the

Ministère de la Recherche, Réseau Micro et Nano Technologies (projet REGINAL) and by the FET arm of the IST program of the European Commission (project IST-2000-26478 GiNA 1.T). We gratefully thank Thales-RT for the donation of a MBE system. We acknowledge J. Massies for permanent support.

References

1. Thomas, D. G., Hopfield, J. J., and Frosch, C. J. (1965). *Phys. Rev. Lett.* **15**, 857.
2. Thomas, D. G., and Hopfield, J. J. (1966). *Phys. Rev*. **150**, 680.
3. Hopfield, J. J., Thomas, D. G., and Lynch, R. T. (1966). *Phys. Rev. Lett.* **17**, 312.
4. Hopfield, J. J., Dean, P. J., and Thomas, D. G. (1967). **158**, 748.
5. Stringfellow, G. B. (1972). *J. Electrochem. Soc.* **119**, 1780.
6. Wei, S.-H., and Zunger, A. (1996). *Phys. Rev. Lett.* **76**, 664.
7. Ho, I.-H., and Stringfellow, G. B. (1997). *J. Cryst. Growth* **178**, 1.
8. Weyers, M., Sato, M., and Ando, H. (1992). *Jpn. J. Appl. Phys*. **31**, L853.
9. Baillargeon, J. N., Cheng, K. Y., Hofler, G. E., Pearah, P. J., and Hsieh, K. C. (1992). *Appl. Phys. Lett.* **60**, 2540.
10. Sakai, S., Ueta, Y., and Terauchi, Y. (1993). *Jpn. J. Appl. Phys*. **32**, 4413.
11. Neugebauer, J. and Van de Walle, C. G. (1995). *Phys. Rev. B* **51**, 10568.
12. Cohen, E. and Sturge, M. D. (1977). *Phys. Rev. B* **15**, 1039.
13. Scifres, D. R., Holonyak Jr, N., Duke, C. B., Kleiman, G. G., Kunz, A. B., Craford, M. G. *et al.* (1971). *Phys. Rev. Lett.* **27**, 191.
14. Scifres, D. R., Mackey, H. M., Holonyak Jr, N., Dupuis, R. D., Zack, G. W., Duke, C. B. *et al.* (1972). *Phys. Rev. B* **5**, 2206.
15. Mariette, H. (1987). *Physica B* **146**.
16. Mathieu, H., Bayo, L., Camassel, J., and Merle, P. (1980). *Phys. Rev. B* **22**, 4834.
17. Gil, B., Camassel, J., Merle, P., and Mathieu, H. (1982). *Phys. Rev. B* **25**, 3987.
18. Allen, J. W. (1971). *J. Phys.* C. **1**, 1136.
19. Thuselt, F., Kreher, K., and Wünsche, H. J. (1980). *Solid State Communications*, **36**, 563.
20. Benoît à la Guillaume, C. (1983). *J. Phys. Lett.* **44**, L.53.
21. Ming Fu Li, De Qiang Mao, and Shang Yuan Ren (1985). *Phys. Rev. B* **32**, 6907.
22. Faulkner, R. A. (1968). *Phys. Rev*. **175**, 991.
23. Brand, S., and Jaros, M. (1979). *J. Phys.* C. **12**, 2789.
24. Gil, B., Albert, J. P., Camassel, J., Mathieu, H., and Benoit à la Guillaume, C. (1986). *Phys. Rev. B* **33**, 2701.
25. Kent, P. and Zunger, A. (2001) *Phys. Rev. B* **64**, 115208.
26. Gil, B., Camassel, J., Albert, J. P., and Mathieu, H. (1986). *Phys. Rev. B* **33**, 2690.
27. Gil, B. (1987). *Physica* **146 B**, 84.
28. Gil, B. and Mariette, H. (1987). *Phys. Rev. B* **35**, 7999.
29. Grzegory, I., Jun, J., and Karpinski, J. (1986). *Physica* **139 & 140B**, 347.
30. Gil, B., Jun, J., Grzegory, I., and Mariette, H. (1989). *Proceedings of the 11 AIRAPT International Conference on the Physics of High Pressures- Kiev 1987*, vol **1**, p. 433, Naukova Dumka.
31. Leroux-Hugon, P., and Mariette, H. (1984). *Phys. Rev. B* **30**, 1622.
32. Mariette, H. and Gil, B. (1989). *Semiconductors in Science and Technology* **4**, 298.
33. Gil, B., Baj, M., Camassel, J., Mathieu, H., Benoit à la Guillaume, C., Mestres, N., and Pascual, J. (1984). *Phys. Rev. B* **29**, 3398.
34. Zhao, X., Li, G., Han, H., Wang, Z., Tang, R., and Hu, J. (1984). Chinese *Phys. Lett.* **1**, 1.

35. Eremets, M. I., Krasnovskij, D. A., Struzhkin, V. V., and Shirokov, A. M. (1989). *Semicond. Sci. Technol.* **4**, 257.
36. Leroux, M. and Gil, B. (1995). *Annales de Physique*, colloque **C2**, supplément au **n°3**, **vol. 20**, 109.
37. Miyoshi, S., Yaguchi, H., Onabe, K., Ito, R., and Siraki, Y. (1993). *Appl. Phys. Lett.* **63**, 3506.
38. Miyoshi, S., Yaguchi, H., Onabe, K., Siraki, Y., and Ito, R. (1994). *Inst. Phys. Conf. Series* **136**, 637.
39. Miyoshi, S., Yaguchi, H., Onabe, K., Ito, R., and Siraki, Y. (1994). *J. Cryst. Growth* **145**, 87.
40. Yaguchi, H., Miyoshi, S., Arimoto, H., Saito, S., Akiyama, H., Onabe, K. *et al.* (1996). *Inst. Phys. Conf. Series* **145**, 307.
41. Yaguchi, H., Miyoshi, S., Biwa, G., Kibune, M., Onabe, K., Shiraki, Y., and Ito, R. (1997). *J. Cryst. Growth* **170**, 353.
42. Yaguchi, H., Miyoshi, S., Arimoto, H., Saito, S., Akiyama, H., Onabe, K. *et al.* (1997). *Solid State Electronics* **41**, 231.
43. Yaguchi, H., Biwa, G., Miyoshi, S., Aoki, D., Arimoto, K., Onabe, K. *et al.* (1998). *J. Cryst. Growth* **189/190**, 496.
44. Biwa, G., Yaguchi, H., Onabe, K., and Shiraki, Y. (1998). *J. Cryst. Growth* **189/190**, 485.
45. Biwa, G., Yaguchi, H., Onabe, K., and Shiraki, Y. (1998). *J. Cryst. Growth* **195**, 574.
46. Xin, H. P., Tu, C. W., Zhang, Y., and Mascarenhas, A. (2000). *Appl. Phys. Lett.* **76**, 1267.
47. Shan, W., Walukiewicz, W., Yu, K. M., Wu, J., Ager III, J. W., Haller, E. E. *et al.* (2000). *Appl. Phys. Lett.* **76**, 3251.
48. Zhang, Y., Fluegel, B., Mascarenhas, A., Xin, H. P., and Tu, C. W. (2000). *Phys. Rev. B* **62**, 4493.
49. Xin, H. P. and Tu, C. W. (2000). *Appl. Phys. Lett.* **77**, 2180.
50. Xin, H. P., Welty, R. J., and Tu, C. W. (2000). *Appl. Phys. Lett.* **77**, 1946.
51. Bellaiche, L., Wei, S. H., and Zunger, A. (1997). *Appl. Phys. Lett.* **70**, 3558.
52. Bellaiche, L., Wei, S. H., and Zunger, A. (1997). *B* **56**, 10233.
53. Kent, P. R. C. and Zunger, A. (2001). *Phys. Rev. Lett.* **86**, 2613.
54. Nelson, R. J. (1982). *Excitons* (eds E. I. Rashba and M. D Sturge), Chapter 8, p. 319. North Holland, Amsterdam.
55. Wolford, D. J., Bradley, J. A., Fry, K., and Thompson, J. (1987). *Proceedings of the 17th International Conference on the Physics of Semiconductors, Stockholm 1986*, (eds J. D. Chadi and W. A. Harrison), p. 627. Springer, New York.
56. Leroux, M., Neu, G., Contour, J. P., Massies, J., and Vérié, C. (1986). *J. Appl. Phys.* **59**, 2996.
57. Schwabe, R., Seifert, W., Bugge, F., Bindenmann, R., Agekyan, V. F., and Pogarev, S. V. (1985). *Solid State Communications* **55**, 167.
58. Xiao Liu, M. E., Pistol, L., Samuelson, S. Schwetlick, and Seifert, W. (1990). *Appl. Phys. Lett.* **56**, 1451.
59. Makimoto, T., Saito, H., and Kobayashi, N. (1997). *Jpn. J. Appl. Phys.* 1 **36**, 1694.
60. Zhang, Y. and Mascarenhas, A. (2000). *Phys. Rev. B* **61**, 15562.
61. Francoeur, S., Nikishin, S. A., Jin, C., Qiu, Y., and Temkin, H. (1999). *Appl. Phys. Lett.* **75**, 1538.
62. Makimoto, T., Saito, H., Nishida, T., and Kobayashi, N. (1997). *Appl. Phys. Lett.* **70**, 2984.
63. Klar, P. J., Grüning, H., Heimbrodt, W., Koch, J., Höhndorf, F., Stolz, W. *et al.* (2000). *Appl. Phys. Lett.* **76**, 3439.
64. Zunger, A., (1999). *Phys. Stat. Soli.* (b) **216**, 117.
65. Rubio, A. and Cohen, M. L. (1995). *Phys. Rev. B* **51**, 4343.
66. Neugebauer, J. and Van de Walle, C. G. (1995). *Phys. Rev. B* **51**, 10568.

67. Su-Hua Wei, and Zunger, A. (1996). *Phys. Rev. Lett.* **76**, 664.
68. Bellaiche, L., Su-Huai Wei, and Zunger, A. (1996). *Phys. Rev. B* **54**, 17568.
69. Jones, E. D., Modine, N. A., Allermann, A. A., Kurtz, S. R., Wright, A. F., Tozer, S. T., and Wei, X. (1999). *Phys. Rev. B* **60**, 4430.
70. Bellaiche, L. (1999). *Appl. Phys. Lett.* **75**, 2578.
71. Mattila, T., Su-Huai Wei, and Zunger, A. (1999). *Phys. Rev. B* **60**, R11245.
72. Lindsay, A. and O'Reilly, E. P. (1999). *Solid State Commun.* **112**, 443.
73. Gil, B. (2000). *Solid State Commun.* **114**, 623.
74. Jones, E. D., Allerman, A. A., Kurz, S. R., Modine, N. A., Bajaj, K. K., Tozer, S. W., and Wei, X. (2000). *Phys. Rev. B* **62**, 7144.
75. Al-Yacoub, A. and Bellaiche, L. (2000). *Phys. Rev. B* **62**, 10847.
76. Bellaiche, L., Modine, N. A., and Jones, E. D. (2000). *Phys. Rev. B* **62**, 15311.
77. Kondow, M., Uomi, K., Hosomi, K., and Mozume, T. (1994). *Jpn. J. Appl. Phys.* **33**, L1056.
78. Salzmann and Tempkin, R. (1997). *Mater. Sci. Eng. B* **50**, 148.
79. Pozina, G., Ivanov, I., Monemar, B., Thordson, J. V., and Andersson, T. G. (1998). *J. Appl. Phys.* **84**, 3830.
80. Shan, W., Walukiewicz, W., Ager III, J. W., Haller, E. E., Geisz, J. F., Friedman, D. J. *et al.* (1999). *Phys. Rev. Lett.* **82**, 1221.
81. Perkins, J. D., Mascarenhas, A., Yong Zhang, Geisz, J. F., Friedman, D. J., Olson, J. M., and Kurtz, S. R. (1999). *Phys. Rev. Lett.* **82**, 3312.
82. Grüning, H., Chen, L., Hartmann, T., Klar, P. J., Heimbrodt, W., Höhnsdorf, F. *et al.* (1999). *Phys. Stat. Soli.* (b) **215**, 39.
83. Skierbiszewski, C., Perlin, P., Wisnieski, P., Suski, T., Walukiewicz, W., Shan, W., Ager III, J. W., Haller, E. E., Geisz, J. F., Friedman, D. J., Olson, J. M., and Kurtz, S. R. (1999). *Phys. Stat. Soli.* (b) **216**, 135.
84. Skierbiszewski, C., Perlin, P., Wisnieski, P., Knap, W., Suski, T., Walukiewicz, W. *et al.* (2000). *Appl. Phys. Lett.* **76**, 2409.
85. Harmand, J. C., Ungaro, G., Largeau, L., and Le Roux, G. (2000). *Appl. Phys. Lett.* **77**, 2482.
86. Hyeonsik, M., Cheong, Yong Zhang, Mascarenhas, A., and Geisz, J. F. (2000). *Phys. Rev. B* **61**, 13687.
87. Shan, W., Walukiewicz, W., Yu, K. M., Ager III, J. W., Haller, E. E. *et al.* (2000). *Phys. Rev. B.* **62**, 4211.
88. Leibirer, G., Gottschalch, V., Rheinlander, B., Sik, J., and Schubert, M. (2000). *Appl. Phys. Lett.* **77**, 1650.
89. Yu, K. M., Walukiewicz, W., Yu, K. M., Ager III, J. W., Haller, E. E., Geisz, J. F., and Ridway, M. C. (2000). *Appl. Phys. Lett.* **77**, 2858.
90. Hung, W. K., Chern, M. Y., Chen, Y. F., Yang, Z. L., and Huang, Y. S. (2000). *Phys. Rev. B* **62**, 13028.
91. Alt, H. C., Egorov, A. Yu, Riechert, H., Wiedemann, B., Meyer, J. D., Michelmann, R. W., and Bethge, K. (2000). *Appl. Phys. Lett.* **77**, 2858.
92. Thinh, N. Q., Buyanova, I. A., Hai, P. N., Chen, W. M., Xin, H. P., and Tu, C. W. (2001). *Phys. Rev. B* **63**, 033203.
93. Malikova, L., Pollak, F. H., and Bhat, R. (1998). *J. Electron. Mater.* **27**, 484.
94. Sik, J., Schubert, M., Leibirer, G., Gottschalch, V., Kirpal, G., and Humlicek, J. (1999). *Appl. Phys. Lett.* **76**, 2859.
95. Hung, W. K., Chern, M. Y., Chen, Y. F., Fan, J. C., Lin, T. Y., and Chen, Y. F. (1999). *Appl. Phys. Lett.* **74**, 3951.
96. Walukiewicz, W., Shan, W., Yu, K. M., Ager III, J. W., Haller, E. E., Miotkowski, I., Seong, M. J., Alawadhi, H., and Ramdas, A. K. (2000). *Phys. Rev. Lett.* **85**, 1552.
97. Yong Zhang, Mascarenhas, A., Xin, H. P., and Tu, C. W. (2000). *Phys. Rev. B* **61**, 4433.

98. Yu, K. M., Walukiewicz, W., Shan, W., Ager III, J. W., Wu, J. W., Haller, E. E. *et al.* (2000). *Phys. Rev. B* **61**, R13337.
99. Wang, L.-W. (2001). *Appl. Phys. Lett.* **78**, 1565.
100. Pankove, J. I. and Hutchby, J. A. (1976). *J. Appl. Phys.* **47**, 5387.
101. Sakai S, S., Ueta, Y., and Terauchi, Y. (1993). Japan *J. Appl. Phys*. **32**, 4413.
102. Van de Walle, C. G. and Neugebauer, J. (2000). *Appl. Phys. Lett.* **76**, 1009.
103. Mattila, T. T. and Zunger, A. (1998). *Phys. Rev. B* **58**, 1367.
104. Mattila T, T. and Zunger, A. (1999). *Phys. Rev. B* **59**, 9943.
105. Shimaoka, G. and Udagawa T, T. (1999). *Appl. Surf. Sci.* **142**, 237.
106. Li X, X., Kim, S., Reuter, E. E., Bishop, S. G., and Coleman, J. J. (1998). *Appl. Phys. Lett.* **72**, 1990.
107. Tsatsul'nikov, A. F., Ya Ber, B., Kartashova, A. P., Yu Kudryavtsev, A., Ledentsov, N. N., Lundin, V. V. *et al.* (1999). *Semiconductors* **33**, 728.
108. Jin, S. R., Ramsteiner, M., Grahn, H. T., Ploog, K. H., Zhu, Z. Q., Shen, D. X., Li, A. Z., Metev, P., and Guido, L. J. (2000). *J. Cryst. Growth* **212**, 56.
109. Guido, L. J., Mitev, P., Gherasimova, M., and Gaffey, B. (1998). *Appl. Phys. Lett.* **72**, 2005.
110. Okumura, H., Hamaguchi, H., Feuillet, G., Ishida, Y., and Yoshida, S. (1998). *Appl. Phys. Lett.* **72**, 3056.
111. Iwata, K., Asahi, H., Asami, K., Kuroiwa, R., and Gonda, S. (1998). *Jpn. J. Appl. Phys*. **37**, 1436.
112. Cheng, T. S., Jenkins, L. C., Hooper, S. E., Foxon, C. T., Orton, J. W., and Lacklison, D. E. (1995). *Appl. Phys. Lett.* **66**, 1509.
113. Feuillet, G., Hamaguchi, H., Okumura, H., and Yoshida, S. (1999). *Mater. Sci. Eng. B* **59**, 80.
114. Feuillet, G., Hamaguchi, H., Ohta, K., Hacke, P., Okumura, H., and Yoshida, S. (1997). *Appl. Phys. Lett.* **70**, 1025.
115. Winser, A. J., Novikov, S. V., Davis, C. S., Cheng, T. S., Foxon, C. T., and Harrison, I. (2000). *Appl. Phys. Lett.* **77**, 2506.
116. Huang, H. Y., Lin, W. C., Lee, W. H., Shu, C. K., Liao, K. C., Chen, W. K., Lee, M. C., and Lee, Y. Y. (2000). *Appl. Phys. Lett.* **77**, 2819.
117. Novokov, S. V., Winser, A. J., Harrison, I., Davis, C. S., and Foxon, C. T. (2001). *Semicond. Sci. Technol.* **16**, 103.
118. Gil, B., Morel, A., Taliercio, T., Lefebvre, P., Foxon, C. T., Harrison, I., Winser, A. J., and Novikov, S. V. (2001). *Appl. Phys. Lett.* **79**, 69.
119. Bi, W. G. and Tu, C. W. (1996). *J. Appl. Phys*. **80**, 1934.
120. Yu, K. M., Walukiewicz, W., Wu, J., Beeman, J. W., Ager III, J. W., Haller, E. E., Shan, W., Xin, H. P., and Tu, C. W. (2001). *Appl. Phys. Lett.* **78**, 1077.
121. Ungaro, G., Le Roux, G., Teissier, R., and Harmand, J.-C. (1999). *Electron. Lett.* **35**, 1246.
122. Buyanova, I., Chen, W. M., Pozina, G., Monemar, B., Xin, H. P., and Tu, C. W. (1999). *Phys. Stat. Sol.* (b) **216**, 125.
123. Yong Zhang, Mascarenhas, A., Xin, H. P., and Tu, C. W. (2000). *Phys. Rev. B* **61**, 4479.
124. Hetterich, M., Dawson, M. D., Yu Egorov, A., Bernklau, D., and Riechert, H. (2000). *Appl. Phys. Lett.* **76**, 1030.
125. Hai, P. N., Chen, W. M., Buyanova, I., Xin, H. P., and Tu, C. W. (2000). *Appl. Phys. Lett.* **77**, 1843.
126. Sun, B. Q., Jiang, D. S., Luo, X. D., Xu, Z. Y., Pan, Z., Li, L. H., and Wu, R. H. (2000). *Appl. Phys. Lett.* **76**, 2862.
127. Tournié, E., Pinault, M.-A., Vézian, S., Massies, J., and Tottereau, O. (2000). *Appl. Phys. Lett.* **77**, 2189.
128. Grenouillet, L., Bru-Chevallier, C., Guillot, G., Gilet, P., Duvaut, P., Vannuffel, C., Million, A., and Chevenas-Paule, A. (2000). *Appl. Phys. Lett.* **76**, 2241.

129. Polimeni, A., Capizzi, M., Geddo, M., Fischer, M., Reinhardt, M., and Forchel, A. (2000). *Appl. Phys. Lett.* **77**, 2870.
130. Buyanova, I., Pozina, G., Hai, P. N., Thinh, N. Q., Bergman, J. P., Chen, W. M. (2000). *Appl. Phys. Lett.* **77**, 2325.
131. Thinh, N. Q., Buyanova, I. A., Hai, P. N., Chen, W. M., Xin, H. P., and Tu, C. W. (2001). *Phys. Rev. B* **63**, 033203.
132. Gil, B. and Bigenwald, P. (2001). *Phys. Stat. Sol.* (a) **183**, 111.
133. Kashner, A., Lüttgert, T., Born, H., Hoffmann, A., Yu Egorov, A., and Riechert, H. (2001). *Appl. Phys. Lett.* **78**, 1391.
134. Pinault, M.-A., and Tournié, E. (2001). *Appl. Phys. Lett.* **78**, 1564.
135. Uesugi, K., Suemene, I., Hasegawa, T., Akutagawa, T. (2000). *Appl. Phys. Lett.* **76**, 1285.
136. Bru-Chevallier, C. (private communication).
137. Teo, K. L., Colton, J. S., and Yu, P. Y. (1998). *Appl. Phys. Lett.* **73**, 1697.
138. Schenk, H. P. D., Leroux, M., and de Mierry, P. (2000). *J. Appl. Phys.* **88**, 1525.
139. Kim, K. and Zunger, A. (2001). *Phys. Rev. Lett.* **86**, 2609.
140. Zhang, S. B. and Wei, S.-H. (2001). *Phys. Rev. Lett.* **86**, 1789.
141. Beaumont, B., Gibart, P., and Faurie, J.-P. (1995). *J. Cryst. Growth* **156**, 140.
142. Neumayer, D. A. and Ekerdt, J. G. (1996). *Chem. Mater.* **8**, 9.
143. Ougazzaden, A., Le Bellego, Y., Rao, E. V. K., Leprince, L., and Patriarche, G. (1997). *Appl. Phys. Lett.* **70**, 2861.
144. Sato, S. I., Osawa, Y., and Saitoh, T. (1997). *Jpn. J. Appl. Phys.* **36**, 2671.
145. Bhat, R., Caneau, C., Salamanca-Riba, L., Bi, W., and Tu, C. W. (1998). *J. Cryst. Growth* **195**, 427.
146. Friedman, D. J., Norman, A. G., Geisz, J. F., and Kurtz, S. R. (2000). *J. Cryst. Growth* **208**, 11.
147. For a review see, e.g., Stolz, W., and Whitaker, T. (1999). *Compound Semiconductor Magazine* **5**(7), 29.
148. Höhnsdorf, F., Koch, J., Agert, C., and Stolz, W. (1998). *J. Cryst. Growth* **195**, 391.
149. Pan, Zh., Miyamoto, T., Schlencker, D., Koyama, F., and Iga, K. (1999). *Jpn. J. Appl. Phys.* **38**, 1012.
150. Moto, A., Tanaka, S., Ikoma, N., Tanabe, T., Takagishi, S., Takahashi, M., and Katsuyama, T. (1999). *Jpn. J. Appl. Phys.* **38**, 1015.
151. Ougazzaden, A., Rao, E. V. K., Sermage, B., Leprince, L., and Gauneau, M. (1999). *Jpn. J. Appl. Phys.* **38**, 1019.
152. Toivonen, J., Hakkarainen, T., Sopanen, M., and Lipsanen, H. (2000). *J. Cryst. Growth* **221**, 456.
153. Miyamoto, T., Kageyama, T., Makino, S., Schlencker, D., Koyama, F., and Iga, K. (2000). *J. Cryst. Growth* **209**, 339.
154. Friedman, D. J., Geisz, J. F., Kurtz, S. R., Olson, J. M., and Reedy, R. (1998). *J. Cryst. Growth* **195**, 438.
155. Moison, J. M., Guille, C., Houzay, F., Barthe, F., and Rompay, M. V. (1989). *Phys. Rev. B* **40**, 6149.
156. Kao, Y. C., Broekaert, T. P. E., Liu, H. Y., Tang, S., Ho, I. H., and Stringfellow, G. B. (1996). *Mater. Res. Soc. Symp. Proc.* **423**, 335.
157. Naoi, H., Naoi, Y., and Sakai, S. (1997). *Solid. State Electron.* **41**, 319.
158. Beresford, R., Stevens, K. S., and Schwartzmann, A. F. (1998). *J. Vac. Sci. Technol. B* **16**, 1293.
159. Matsuoka, T. (1992). *J. Cryst. Growth* **124**, 433.
160. Bi, W. G., Tu, C., Mathes, D., and Hull, R. (1997). *Mat. Res. Soc. Symp. Proc.* **449**, 203.
161. Uesugi, K. and Suemune, I. (1997). *Jpn. J. Appl. Phys.* **36**, L1572.

162. Takeuchi, K., Miyamoto, T., Kageyama, T., Koyama, F., and Iga, K. (1998). *Jpn. J. Appl. Phys.* **37**, 1603.
163. Qiu, Y., Jin, C., Francoeur, S., Nikishin, S. A., and Temkin, H. (1998). *Appl. Phys. Lett.* **72**, 1999.
164. Miyamoto, T., Takeuchi, K., Kageyama, T., Koyama, F., and Iga, K. (1999). *J. Cryst. Growth* **197**, 67.
165. Kondow, M., Uomi, K., Niwa, A., Kitatani, T., Watahiki, S., and Yazawa, Y. (1996). *Jpn. J. Appl. Phys.* **35**, 1273.
166. Kondow, M., Uomi, K., Kitatani, T., Watahiki, S., and Yazawa, Y. (1996). *J. Cryst. Growth* **164**, 175.
167. Xin, H. P. and Tu, C. W. (1998). *Appl. Phys. Lett.* **72**, 2442.
168. Gilet, P., Chenevas-Paule, A., Duvaut, P., Grenouillet, L. Holliger, P., Million, A. *et al.* (1999). *Phys. Stat. Sol.* (a) **176**, 279.
169. Francoeur, S., Nikishin, S. A., Jin, C., Qiu, Y., and Temkin, H. (1999). *Appl. Phys. Lett.* **75**, 1538.
170. Orton, J. W., Lacklison, D. E., Baba-Ali, N., Foxon, C. T., Cheng, T. S., Novikov, S. V., Johnston, D. F. C., Hooper, S. E., Jenkins, L. C., Challis, L. J., and Tansley, T. L. (1995). *J. Electron. Mater.* **24**, 263.
171. Thordson, J. V., Zsebök, O., Södervall, U., and Andersson, T. G. (1997). Internet *J. Nitride Semicond. Res.* **2**, 8.
172. Yang, X., Jurkovic, M. J., Heroux, J. B., and Wang, W. I. (1999). *Appl. Phys. Lett.* **75**, 178.
173. Pan, Z., Li, L. H., Zhang, W., Lin, Y. W., and Wu, R. H. (2000). *Appl. Phys. Lett.* **77**, 214.
174. Gotthold, D. W., Govindaraju, S., Mattord, T., Holmes, Jr. A.L., and Streetman, B. G. (2000). *J. Vac. Sci. Technol. A* **18**, 461.
175. Spruytte, S. G., Coldren, C. W., Marshall, A. F., Larson, M. C., and Harris, J. J. (2000). Internet *J. Nitride Semicond. Res.* **5S1**, W.8.4.
176. Hughes, W. C., Rowland, Jr. W. H., Johnson, M. A., Fujita, S., Cook, Jr. J. W., Schetzina, J. F., Ren, J., and Edmond, J. A. (1995). *J. Vac. Sci. Technol. B* **13**, 1571.
177. Egorov, A. Yu., Bernklau, D., Livshits, D., Ustinov, V., Alferov, Zh.I., and Riechert, H. (2000). *Inst. Phys. Conf. Ser.* **166**, 359.
178. Pinault, M.-A. and Tournié, E. (2001). *Appl. Phys. Lett.* **79**, 3404.
179. Tournié, E. and Pinault, M.-A. 11th European Workshop on Molecular-Beam Epitaxy, 4–7 February 2001, Hinterzarten (Germany).
180. Lu, J., Haworth, L., Hill, P., Westwood, D. I., and MacDonald, J. E. (1999). *J. Vac. Sci. Technol. B* **17**, 1659.
181. Kwon, D., Kaplar, R. J., Ringel, S. A., Alleman, A. A., Kurtz, S. R., and Jones, E. D. (1999). *Appl. Phys. Lett.* **74**, 2830.
182. Li, J. Z., Lin, J. Y., Jiang, H. X., Geisz, J. F., and Kurtz, S. R. (1999). *Appl. Phys. Lett.* **75**, 1899.
183. Moto, A., Takahashi, M., and Takagishi, S. (2000). *J. Cryst. Growth* **221**, 485.
184. Krispin, P., Spruytte, S. G., Harris, J. S., and Ploog, K. H. (2000). *J. Appl. Phys.* **88**, 4153.
185. Fleck, A., Robinson, B. J., and Thompson, D. A. (2001). *Appl. Phys. Lett.* **78**, 1694.
186. Balcioglu, A., Ahrenkiel, R. K., and Friedman, D. J. (2000). *Appl. Phys. Lett.* **76**, 2397.
187. Rao, E. V. K., Ougazzaden, A., Le Bellego, Y., and Juhel, M. (1998). *Appl. Phys. Lett.* **72**, 1409.
188. Kageyama, T., Miyamoto, T., Makino, S., Koyama, F., and Iga, K. (1999). *Jpn. J. Appl. Phys.* **38**, L298.
189. Yang, X., Heroux, J. B., Jurkovic, M. J., and Wang, W. I. (1999). *J. Vac. Sci. Technol. B* **17**, 1144.

190. Xin, X. P., Kavanagh, K. L., Zhu, Z. Q., and Tu, C. W. (1999). *J. Vac. Sci. Technol. B* **17**, 1649.
191. Pan, Z., Li, L. H., Zhang, W., Lin, Y. W., Wu, R. H., and Ge, W. (2000). *Appl. Phys. Lett.* **77**, 1280.
192. Kitatani, T., Nakahara, K., Kondow, M., Uomi, K., and Tanak, T. (2000). *J. Cryst. Growth* **209**, 345.
193. Spruytte, S. G., Coldren, C. W., Harris, J. S., Wampler, W., Krispin, P., Ploog, K., and Larson, M. C. (2001). *J. Appl. Phys.* **89**, 4401.
194. Li, W., Pessa, M., and Likonen, J. (2001). *Appl. Phys. Lett.* **78**, 2864.
195. Bertness, K. A., Kurtz, S. R., Friedman, D. J., Kibbler, A. E., Cramer, C., and Olson, J. M. (1994). *Appl. Phys. Lett.* **65**, 989.
196. Kurtz, S. R., Myers, D., and Olson, J. M. (1997). *Proc. 26th IEEE Photovoltaic Specialist Conference*, Anaheim, USA, p. 875.
197. Geisz, J. F., Friedman, D. J., Olson, J. M., Kurtz, S. R., and Keyes, B. M. (1998). *J. Cryst. Growth* **195**, 401.
198. Friedman, D. J., Geisz, J. F., Kurtz, S. R., and Olson, J. M. (1998). *J. Cryst. Growth* **195**, 409.
199. Friedman, D. J., Geisz, J. F., Kurtz, S. R., and Olson, J. M. (1998). *Proc. 2nd World Conf. on Photovoltaic Energy Conversion*, New York, USA.
200. Kurtz, S. R., Allerman, A. A., Jones, E. D., Gee, J. M., Banas, J. J., and Hammons, B. E. (1999). *Appl. Phys. Lett.* **74**, 729.
201. Kurtz, S. R., Allerman, A. A., Seager, C. H., Sieg, R. M., and Jones, E. D. (2000). *Appl. Phys. Lett.* **77**, 400.
202. Kondow, M., Uomi, K., Niwa, A., Kitatani, T., Watahiki, S., and Yazawa, Y. (1996). *Jpn. J. Appl. Phys.* **35**, 1273.
203. Kondow, M., Kitatani, T., Nakatsuka, S., Larson, M. C., Nakahara, K., Yazawa, Y. *et al.* (1997). *IEEE J. Sel. Top. Quantum Electron.* **3**, 719.
204. A detailed review of the published data on laser design and emission characteristics is given in: Potter, R., Mazzucato, S., Balkan, N., Adams, M. J., Chalker, P. R., Joyce, T. B., and Bullough, T. J. (2001). *Superlattices and Microstruct.* **29**, 169.
205. Nakahara, K., Kondow, M., Kitatani, T., Larson, M. C., and Uomi, K. (1998). *IEEE Photon. Technol. Lett.* **10**, 487.
206. Yiang, X., Jurkovic, M. J., Heroux, J. B., and Wang, W. I. (1999). *Electron. Lett.* **35**, 1081.
207. Borchert, B., Egorov, A. Y., Illek, S., and Riechert, H. (2000). *IEEE Photon. Technol. Lett.* **12**, 597.
208. Coldren, C. W., Spruytte, S. G., Harris, J. S., and Larson, M. C. (2000). *J. Vac. Sci. Technol. B* **18**, 1480.
209. Livshits, D. A., Egorov, A. Yu., and Riechert, H. (2000). *Electron. Lett.* **36**, 1381.
210. Kageyama, T., Miyamoto, T., Makino, S., Nishiyama, N., Koyama, E., and Iga, K. (2000). *IEEE Photon. Technol. Lett.* **12**, 10.
211. Sato, S. and Satoh, S. (1999). *IEEE J. Sel. Topics Quantum Electron.* **5**, 707.
212. Höhndorf, F., Koch, J., Leu, S., Stolz, W., Borchert, B., and Druminski, M. (1999). *Electron. Lett.* **35**, 571.
213. Sato, S. (2000). *Jpn. J. Appl. Phys.* **15**, 3403.
214. Kawaguchi, M., Gouardes, E., Schlenker, D., Kondo, T., Miyamoto, T., Koyama, F., and Iga, K. (2000). *Electron. Lett.* **36**, 1776.
215. Riechert, H., Egorov, A. Yu., and Steinle, G. 11th European Workshop on Molecular-Beam Epitaxy, 4–7 February 2001, Hinterzarten, Germany.
216. Larson, M. C., Kondow, M., Kitatani, T., Nakahara, K., Tamura, K., Yazawa, Y., Okai, M., Inoue, H., and Uomi, K. (1998). *SPIE Proc.* **3286**, 30.

217. Coldren, C. W., Larson, M. C., Spruytte, S. G., and Harris, J. S. (2000). *Electron. Lett.* **36**, 951.
218. Choquette, K. D., Klem, J. F., Fisher, A. J., Blum, O., Allerman, A. A., Fritz, I. J. *et al.* (2000). *Electron. Lett.* **36**, 1388.
219. Steinle, G., Riechert, H., and Egorov, A. Yu. (2001). *Electron. Lett.* **37**, 93.
220. Jackson, A. W., Naone, R. L., Dalberth, M. J., Smith, J. M., Malone, K. J., Kisker, D. W. *et al.* (2001). *Electron. Lett.* **37**, 355.
221. Sato, S., Nishiyama, N., Miyamoto, T., Takahashi, T., Jikutani, N., Arai, M. *et al.* (2000). *Electron. Lett.* **36**, 2018.
222. Kageyama, T., Miyamoto, T., Makino, S., Ikenaga, Y., Nishiyama, N., Matsutani, A. *et al.* (2001). *Electron. Lett.* **37**, 225.
223. Kinsey, G. S., Gotthold, D. W., Holmes, Jr. A. L., Streetman, B. G., and Campbell, J. C. (2000). *Appl. Phys. Lett.* **76**, 2824.
224. Kinsey, G. S., Gotthold, D. W., Holmes, Jr. A. L., and Campbell, J. C. (2000). *Appl. Phys. Lett.* **77**, 1543.
225. Héroux, J. B., Yang, X., and Wang, W. I. (1999). *Appl. Phys. Lett.* **75**, 2716.
226. Li, N. Y., Chang, P. C., Baca, A. G., Xie, X. M., Sharps, P. R., and Hou, H. Q. (2000). *Electron. Lett.* **36**, 81.
227. Chang, P. C., Baca, A. G., Li, N. Y., Xie, X. M., Hou, H. Q., and Armour, E. (2000). *Appl. Phys. Lett.* **76**, 2262.
228. Chang, P. C., Baca, A. G., Li, N. Y., Sharps, P. R., Hou, H. Q., Laroche, J. R., and Ren, F. (2000). *Appl. Phys. Lett.* **76**, 2788.
229. Monier, C., Baca, A. G., Chang, P. C., Li, N. Y., Hou, H. Q., Ren, F., and Pearton, S. J. (2001). *Electron. Lett.* **37**, 198.
230. Fischer, M., Reinhardt, M., and Forchel, A. (2000). *Electron. Lett.* **36**, 1208.

Index